LONDON MATHEMATICAL SOCIETY LECT

Managing Editor: Professor J.W.S. Cassels, Departmⁱ matical Statistics, University of Cambridge, 16 Mill Lan⸻

The books in the series listed below are available from booⱪ⸻⸻⸻, ⸻, Cambridge University Press.

34   Representation theory of Lie groups, M.F. ATIYAH et al
36   Homological group theory, C.T.C. WALL (ed)
39   Affine sets and affine groups, D.G. NORTHCOTT
40   Introduction to H_p spaces, P.J. KOOSIS
46   p-adic analysis: a short course on recent work, N. KOBLITZ
49   Finite geometries and designs, P. CAMERON, J.W.P. HIRSCHFIELD & D.R. HUGHES (eds)
50   Commutator calculus and groups of homotopy classes, H.J. BAUES
57   Techniques of geometric topology, R.A. FENN
59   Applicable differential geometry, M. CRAMPIN & F.A.E. PIRANI
62   Economics for mathematicians, J.W.S. CASSELS
66   Several complex variables and complex manifolds II, M.J. FIELD
69   Representation theory, I.M. GELFAND et al
74   Symmetric designs: an algebraic approach, E.S. LANDER
76   Spectral theory of linear differential operators and comparison algebras, H.O. CORDES
77   Isolated singular points on complete intersections, E.J.N. LOOIJENGA
79   Probability, statistics and analysis, J.F.C. KINGMAN & G.E.H. REUTER (eds)
80   Introduction to the representation theory of compact and locally compact groups, A. ROBERT
81   Skew fields, P.K. DRAXL
82   Surveys in combinatorics, E.K. LLOYD (ed)
83   Homogeneous structures on Riemannian manifolds, F. TRICERRI & L. VANHECKE
86   Topological topics, I.M. JAMES (ed)
87   Surveys in set theory, A.R.D. MATHIAS (ed)
88   FPF ring theory, C. FAITH & S. PAGE
89   An F-space sampler, N.J. KALTON, N.T. PECK & J.W. ROBERTS
90   Polytopes and symmetry, S.A. ROBERTSON
91   Classgroups of group rings, M.J. TAYLOR
92   Representation of rings over skew fields, A.H. SCHOFIELD
93   Aspects of topology, I.M. JAMES & E.H. KRONHEIMER (eds)
94   Representations of general linear groups, G.D. JAMES
95   Low-dimensional topology 1982, R.A. FENN (ed)
96   Diophantine equations over function fields, R.C. MASON
97   Varieties of constructive mathematics, D.S. BRIDGES & F. RICHMAN
98   Localization in Noetherian rings, A.V. JATEGAONKAR
99   Methods of differential geometry in algebraic topology, M. KAROUBI & C. LERUSTE
100 Stopping time techniques for analysts and probabilists, L. EGGHE
101 Groups and geometry, ROGER C. LYNDON
103 Surveys in combinatorics 1985, I. ANDERSON (ed)
104 Elliptic structures on 3-manifolds, C.B. THOMAS
105 A local spectral theory for closed operators, I. ERDELYI & WANG SHENGWANG
106 Syzygies, E.G. EVANS & P. GRIFFITH
107 Compactification of Siegel moduli schemes, C-L. CHAI
108 Some topics in graph theory, H.P. YAP
109 Diophantine analysis, J. LOXTON & A. VAN DER POORTEN (eds)
110 An introduction to surreal numbers, H. GONSHOR
111 Analytical and geometric aspects of hyperbolic space, D.B.A.EPSTEIN (ed)
113 Lectures on the asymptotic theory of ideals, D. REES
114 Lectures on Bochner-Riesz means, K.M. DAVIS & Y-C. CHANG
115 An introduction to independence for analysts, H.G. DALES & W.H. WOODIN

116 Representations of algebras, P.J. WEBB (ed)
117 Homotopy theory, E. REES & J.D.S. JONES (eds)
118 Skew linear groups, M. SHIRVANI & B. WEHRFRITZ
119 Triangulated categories in the representation theory of finite-dimensional algebras, D. HAPPEL
121 Proceedings of *Groups - St Andrews 1985*, E. ROBERTSON & C. CAMPBELL (eds)
122 Non-classical continuum mechanics, R.J. KNOPS & A.A. LACEY (eds)
124 Lie groupoids and Lie algebroids in differential geometry, K. MACKENZIE
125 Commutator theory for congruence modular varieties, R. FREESE & R. MCKENZIE
126 Van der Corput's method of exponential sums, S.W. GRAHAM & G. KOLESNIK
127 New directions in dynamical systems, T.J. BEDFORD & J.W. SWIFT (eds)
128 Descriptive set theory and the structure of sets of uniqueness, A.S. KECHRIS & A. LOUVEAU
129 The subgroup structure of the finite classical groups, P.B. KLEIDMAN & M.W.LIEBECK
130 Model theory and modules, M. PREST
131 Algebraic, extremal & metric combinatorics, M-M. DEZA, P. FRANKL & I.G. ROSENBERG (eds)
132 Whitehead groups of finite groups, ROBERT OLIVER
133 Linear algebraic monoids, MOHAN S. PUTCHA
134 Number theory and dynamical systems, M. DODSON & J. VICKERS (eds)
135 Operator algebras and applications, 1, D. EVANS & M. TAKESAKI (eds)
136 Operator algebras and applications, 2, D. EVANS & M. TAKESAKI (eds)
137 Analysis at Urbana, I, E. BERKSON, T. PECK, & J. UHL (eds)
138 Analysis at Urbana, II, E. BERKSON, T. PECK, & J. UHL (eds)
139 Advances in homotopy theory, S. SALAMON, B. STEER & W. SUTHERLAND (eds)
140 Geometric aspects of Banach spaces, E.M. PEINADOR and A. RODES (eds)
141 Surveys in combinatorics 1989, J. SIEMONS (ed)
142 The geometry of jet bundles, D.J. SAUNDERS
143 The ergodic theory of discrete groups, PETER J. NICHOLLS
144 Introduction to uniform spaces, I.M. JAMES
145 Homological questions in local algebra, JAN R. STROOKER
146 Cohen-Macaulay modules over Cohen-Macaulay rings, Y. YOSHINO
147 Continuous and discrete modules, S.H. MOHAMED & B.J. MÜLLER
148 Helices and vector bundles, A.N. RUDAKOV et al
149 Solitons, nonlinear evolution equations and inverse scattering, M.A. ABLOWITZ & P.A.CLARKSON
150 Geometry of low-dimensional manifolds 1, S. DONALDSON & C.B. THOMAS (eds)
151 Geometry of low-dimensional manifolds 2, S. DONALDSON & C.B. THOMAS (eds)
152 Oligomorphic permutation groups, P. CAMERON
153 L-functions and arithmetic, J. COATES & M.J. TAYLOR (eds)
154 Number theory and cryptography, J. LOXTON (ed)
155 Classification theories of polarized varieties, TAKAO FUJITA
156 Twistors in mathematics and physics, T.N. BAILEY & R.J. BASTON (eds)
157 Analytic pro-*p* groups, J.D. DIXON, M.P.F. DU SAUTOY, A. MANN & D. SEGAL
158 Geometry of Banach spaces, P.F.X. MÜLLER & W. SCHACHERMAYER (eds)
159 Groups St Andrews 1989 Volume 1, C.M. CAMPBELL & E.F. ROBERTSON (eds)
160 Groups St Andrews 1989 Volume 2, C.M. CAMPBELL & E.F. ROBERTSON (eds)
161 Lectures on block theory, BURKHARD KÜLSHAMMER
162 Harmonic analysis and representation theory for groups acting on homogeneous trees, A. FIGÀ-TALAMANCA & C. NEBBIA
163 Topics in varieties of group representations, S.M. VOVSI
164 Quasi-symmetric designs, M.S. SHRIKANDE & S.S. SANE
165 Groups, combinatorics & geometry, M. LIEBECK & J. SAXL (eds)
166 Surveys in combinatorics, 1991, A.D. KEEDWELL (ed)
167 Stochastic analysis M.T. BARLOW & N.H. BINGHAM (eds)

GROUPS, COMBINATORICS & GEOMETRY
Durham, 1990

edited by

Martin Liebeck
Imperial College, London
and
Jan Saxl
Gonville and Caius College, Cambridge

CAMBRIDGE
UNIVERSITY PRESS

Published by the Press Syndicate of the University of Cambridge
The Pitt Building, Trumpington Street, Cambridge CB2 1RP
40 West 20th Street, New York, NY 10011, USA
10, Stamford Road, Oakleigh, Melbourne 3166, Australia

Printed in Great Britain at the University Press, Cambridge

Library of Congress cataloguing in publication data available
British Library cataloguing in publication data available

ISBN 0 521 40685 4

Contents

Authors' Addresses ix

Introduction xi

Sporadic groups

M. Aschbacher & Y. Segev
Uniqueness of sporadic groups 1

M. Aschbacher & Y. Segev
The study of J_4 via the theory of uniqueness systems 12

J.H. Conway
Y_{555} and all that 22

J.H. Conway & A.D. Pritchard
Hyperbolic reflections for the Bimonster and $3Fi_{24}$ 24

A.A. Ivanov
A geometric characterization of the Monster 46

S.P. Norton
Constructing the Monster 63

Moonshine

M. Akbas & D. Singerman
The signature of the normalizer of $\Gamma_0(N)$ 77

D. Alexander, C. Cummins, J. McKay & C. Simons
Completely replicable functions 87

R.E. Borcherds
Introduction to the Monster Lie algebra 99

G. Mason
Remarks on Moonshine and orbifolds 108

Local and geometric methods in group theory

H. Cuypers & J.I. Hall
The classification of 3-transposition groups with trivial center 121

W. Lempken, C. Parker & P. Rowley
(S_3, S_6)-Amalgams 139

P. Rowley
Pushing down minimal parabolic systems 144

G. Stroth
Nonspherical spheres 151

B. Stellmacher
On the 2-local structure of finite groups 159

F.G. Timmesfeld
Groups generated by k-root subgroups – a survey 183

Geometries and related groups

P.J. Cameron
Finiteness questions for geometries 205

R.W. Carter
Kac-Moody groups and their automorphisms 218

E.E. Shult
Generalized hexagons as geometric hyperplanes of near hexagons 229

L.H. Soicher
On simplicial complexes related to the Suzuki sequence graphs 240

J. Tits
Twin buildings and groups of Kac-Moody type 249

Finite and algebraic groups of Lie type

A.V. Borovik
*Some remarks on the structure of finite subgroups
of simple algebraic groups* 287

R. Lawther
Some (almost) multiplicity-free coset actions 292

G. Röhrle
Orbits in internal Chevalley modules 311

G.M. Seitz
Subgroups of finite and algebraic groups 316

I.D. Suprunenko & A.E. Zalesski
*Irreducible representations of finite Chevalley groups containing
a matrix with a simple spectrum* 327

D.M. Testerman
Overgroups of unipotent elements in simple algebraic groups 333

Finite permutation groups

P.J. Cameron
Some open problems on permutation groups 340

R.M. Guralnick
The genus of a permutation group 351

R.M. Guralnick & J. Saxl
Primitive permutation characters 364

C.E. Praeger
Closures of finite permutation groups and relation algebras 368

Further aspects of simple groups

R.T. Curtis
Symmetric presentations I: Introduction, with particular reference to the Mathieu groups M_{12} and M_{24} 380

B. Hartley
Finite and locally finite groups containing a small subgroup with small centralizer 397

W.M. Kantor
Some topics in asymptotic group theory 403

I.A. Suleiman & R.A. Wilson
The 3-modular characters of the McLaughlin group McL and its automorphism group McL.2 422

Related topics

J.H. Conway
The orbifold notation for surface groups 438

C. Hering
A remark on two Diophantine equations of Peter Cameron 448

D.F. Holt & S. Rees
Testing for isomorphism between finitely presented groups 459

J.G. Thompson
Discrete groups and Galois theory 476

S. Wilson
Smooth coverings of regular maps 480

Participating Authors' Addresses

M. Aschbacher, Dept. of Mathematics, California Institute of Technology, Pasadena, California 91125, USA

R.E. Borcherds, DPMMS, 16 Mill Lane, Cambridge CB2 1SB, England

A.V. Borovik, Komplexny Otdel S.O. AN SSSR, Prospect Mira 19a, Omsk 644050, USSR

P.J. Cameron, Dept. of Mathematics, Queen Mary and Westfield College, Mile End Road, London E1 4NS, England

R.W. Carter, Mathematics Institute, University of Warwick, Coventry CV4 7AL, England

J.H. Conway, Dept. of Mathematics, Princeton University, Princeton, NJ 08544, USA

R.T. Curtis, Dept. of Mathematics, University of Birmingham, Birmingham B15 2TT, England

H. Cuypers, Math. Sem. University, Olshauserstrasse 40–60, 2300 Kiel, Germany

R.M. Guralnick, Dept. of Mathematics, University of Southern California, Los Angeles, CA 90089, USA

J.I. Hall, Dept. of Mathematics, University of Michigan, East Lansing, Michigan 48223, USA

B. Hartley, Dept. of Mathematics, University of Manchester, Manchester M13 9PL, England

C. Hering, Mathematisches Institut, Auf der Morgenstelle 10, 7400 Tubingen 1, Germany

D.F. Holt, Mathematics Institute, University of Warwick, Coventry CV4 7AL, England

A.A. Ivanov, Institute for System Studies of the Academy of Sciences, 9, Prospect 60 Let Oktyabrya, 117312 Moscow, USSR

W.M. Kantor, Dept. of Mathematics, University of Oregon, Eugene, Oregon 97403, USA

R. Lawther, DPMMS, 16 Mill Lane, Cambridge CB2 1SB, England

G. Mason, Dept. of Mathematics, UCSC, Santa Cruz, California 95064, USA

J. McKay, Dept. of Computer Science, Concordia University, Montréal, Canada H3G 1M8

S.P. Norton, DPMMS, 16 Mill Lane, Cambridge CB2 1SB, England

C. Parker, Dept. of Mathematics, University of Wisconsin-Parkside, Wisconsin 53141–2000, USA

C.E. Praeger, Dept. of Mathematics, University of W. Australia, Nedlands, W. Australia 6009

S. Rees, Dept. of Mathematics, The University, Newcastle-upon-Tyne NE1 7RU, England

G. Röhrle, Dept. of Mathematics, University of Southern California, Los Angeles, CA 90089, USA

P. Rowley, Dept of Mathematics, UMIST, Manchester M60 1QD, England

Y. Segev, Dept. of Mathematics, Beersheva University, Israel

G.M. Seitz, Dept. of Mathematics, University of Oregon, Eugene, Oregon 97403, USA

E.E. Shult, Dept. of Mathematics, Kansas State University, Manhattan, Kansas 66506, USA

D. Singerman, Dept. of Mathematics, University of Southampton, Southampton SO9 5NH, England

L.H. Soicher, Dept. of Mathematics, Queen Mary and Westfield College, Mile End Road, London E1 4NS, England

B. Stellmacher, Math. Sem. University, Olshauserstrasse 40–60, 2300 Kiel, Germany

G. Stroth, Institut für Mathematik II, Freie Universität, Arnimallee 3, 1000 Berlin 33, Germany

I.A. Suleiman, Dept. of Mathematics, University of Birmingham, Birmingham B15 2TT, England

D.M. Testerman, Dept. of Mathematics, Wesleyan University, Middletown, CT 06457, USA

J.G. Thompson, DPMMS, 16 Mill Lane, Cambridge CB2 1SB, England

F.G. Timmesfeld, Mathematisches Institut, University of Giessen, Arndtstrasse 2, 6300 Giessen, Germany

J. Tits, Dept. of Mathematics, Collège de France, 11 Place Marcelin Berthelot, 75231 Paris, France

R.A. Wilson, Dept. of Mathematics, University of Birmingham, Birmingham B15 2TT, England

S. Wilson, Dept. of Mathematics, University of Arizona, USA

A.E. Zalesskii, Institute of Mathematics of the Academy of Sciences, Surganova Ulica, 220604 Minsk, USSR

Other Authors' Addresses

M. Akbas, Dept. of Mathematics, University of Southampton, Southampton SO9 5NH, England

D. Alexander, Dept. of Computer Science, Concordia University, Montréal, Canada H3G 1M8

C. Cummins, Dept. of Computer Science, Concordia University, Montréal, Canada H3G 1M8

W. Lempken, Institute for Experimental Mathematics, University of Essen, Ellenstrasse 29, D–4300 Essen 12, Germany

A.D. Pritchard, DPMMS, 16 Mill Lane, Cambridge CB2 1SB, England

C. Simons, Dept. of Computer Science, Concordia University, Montréal, Canada H3G 1M8

I.D. Suprunenko, Institute of Mathematics of the Academy of Sciences, Surganova Ulica, 220604 Minsk, USSR

Introduction

This book contains the proceedings of the L.M.S. Durham Symposium on Groups and Combinatorics, July 5–15, 1990, supported by the Science and Engineering Research Council of Great Britain.

The classification of finite simple groups was completed in 1980, and most of the conference was concerned with trends in group theory and related areas which have come to the fore since then. We have divided the material into eight sections, which we now outline.

(1) Sporadic groups

During the conference, a spectacular proof of the ten-year-old Conway "Y_{555} conjecture" concerning a Coxeter-type presentation of the Monster sporadic group, was achieved by Norton and Ivanov. In their articles, *Norton* and *Ivanov* give details of their proof; Ivanov obtains a new geometric characterization of the Monster, and this is used by Norton to prove the conjecture. *Conway*'s short paper outlines the background. The article of *Conway* and *Pritchard* gives a proof of the Y_{552} presentation for the Fischer group Fi_{24}. In a different vein, the papers of *Aschbacher* and *Segev* concern proofs of uniqueness results for sporadic groups. In their first article they survey their general framework for providing uniqueness results; this uses a graph-theoretical setting and some ideas with a topological flavour. Their second paper discusses in detail the uniqueness proof for the group J_4.

(2) Moonshine

This section concerns the remarkable relationships between the Monster group and certain modular functions. Much insight into moonshine is gained from studying the Monster Lie algebra, which is described by *Borcherds* in his article. The modular functions associated with elements of the Monster satisfy the so-called "completely replicable" property; in their paper, *Alexander, Cummins, McKay* and *Simons* enumerate completely replicable functions. Their article also includes as an appendix a useful bibliography of articles on moonshine. *Akbas* and *Singerman* study certain Fuchsian groups connected with moonshine and *Mason* discusses the topological concept of an orbifold, and its relationship with moonshine.

(3)   Local and geometric methods in group theory

Four of the six papers in this section deal with the theory of parabolic systems and amalgams; these concern groups generated by a collection of parabolic-type subgroups, given certain hypotheses on the intersections of these subgroups. This flourishing area has led to new methods in local group theory and its geometrical ramifications. *Stellmacher* presents new work in this area on 2-local structure of finite groups. The papers of *Rowley*, and *Lempken, Parker* and *Rowley* survey their results on parabolic systems for which the associated diagram is of type o———o ... o===o. *Stroth* presents some results on weak BN-pairs. The section contains two further papers. *Cuypers* and *Hall* give a detailed discussion of their recent classification of 3-transposition groups. And *Timmesfeld* describes his recent results characterizing groups of Lie type over arbitrary fields in terms of root group generation. The proofs in these two papers have a strong geometrical flavour.

(4)   Geometries and related groups

In his article, *Tits* describes his recent theory of twin buildings, developed with Ronan, which provides a natural setting for Kac-Moody groups. *Carter* presents a result determining the automorphisms of Kac-Moody groups in characteristic zero. The other three papers in this section are concerned with geometries. *Cameron* poses a a series of problems on the general theme of finiteness for geometries. *Shult* proves some results on geometric hyperplanes of near hexagons. And *Soicher* introduces some geometries related to the well known Suzuki sequence associated with the sporadic group Suz.

(5)   Finite and algebraic groups of Lie type

The six articles in this section fall into two areas: subgroup structure, and representation theory, of finite and algebraic groups of Lie type. *Seitz* gives a broad survey of recent results on subgroups of both the finite and algebraic groups. Particularly impressive is the determination of the maximal connected subgroups of simple algebraic groups, due to Seitz and Testerman; applications of this work to the finite groups are also discussed. In her article, *Testerman* announces a result proving the existence of some of these maximal subgroups of type A_1. *Borovik* presents some of his work on subgroups. The other three articles concern representations. *Röhrle* discusses an intriguing relationship between the number of orbits of the Levi factor of a parabolic subgroup P of a simple algebraic group G on its "internal Chevalley modules", and the number of double cosets of P in G. *Lawther* presents results on the permutation actions of a finite group of Lie type on the coset space of a subgroup defined over a subfield, or a twisted subgroup; in particular he determines which of these actions are multiplicity-free. And *Suprunenko* and *Zalesski* investigate irreducible representations of finite groups of Lie type such that the image of some element has all eigenvalues distinct.

(6) Finite permutation groups

This section contains four very different articles, all reflecting recent developments in the area. In the past ten years, the classification has been used to settle many classical problems in finite permutation groups. *Cameron* discusses a few of these and goes on to suggest a number of problems which may not lend themselves to this approach. *Guralnick*'s article contains a collection of recent results on monodromy groups of covers of Riemann surfaces; the methods come mainly from recent advances in permutation group theory. *Praeger* presents some results on relation algebras, which are closely related to 2-closed permutation groups. Lastly, *Guralnik* and *Saxl* settle an old question of Wielandt on permutation characters; this work was inspired by lectures by Thompson and Borovik at the conference.

(7) Further aspects of simple groups

In his paper, *Kantor* discusses a range of asymptotic questions on finite groups, mainly concerning simple groups. These involve lengths of presentations and proportions of generating sets of elements. *Curtis* introduces the idea of a symmetric presentation of a group, and shows that some of the Mathieu groups have such presentations. *Hartley*'s article contains a number of results on fixed point subgroups of groups of automorphisms of simple groups, with applications to locally finite groups. And *Suleiman* and *Wilson* calculate the 3-modular character tables of the sporadic group McL and its automorphism group.

(8) Related topics

This final section contains five articles on a variety of topics not covered by the previous sections. *Thompson* discusses Fuchsian groups and Fricke rings, and their relationship with Galois theory. *Conway* discusses the orbifold notation for surface groups and gives many examples. There are also articles by *Hering* on Diophantine equations, by *Holt* and *Rees* on isomorphism testing in computational group theory, and by *Wilson* on smooth coverings of regular maps.

The conference was most enjoyable and successful, as we hope this volume will show. We would like to record our gratitude to the L.M.S and S.E.R.C for their financial support, and to the staff at Durham University for their generous and skilful assistance with the organisation. Thanks are also due to Patrick Johnston of Imperial College for much time-consuming preparation of the text for many of the articles.

Martin Liebeck
Imperial College
London

Jan Saxl
Gonville and Caius College
Cambridge

Uniqueness of sporadic groups

MICHAEL ASCHBACHER AND YOAV SEGEV

Initial work on the sporadic finite simple groups falls into one or more of the following categories:

> Discovery
> Structure
> Existence
> Uniqueness

More precisely let $\mathcal{H}$ be some group theoretic hypothesis. A group theorist begins to investigate groups G satisfying $\mathcal{H}$ and generates information about the structure of such groups. Typical examples of structural information include the group order, the isomorphism type of normalizers of subgroups of prime order, and perhaps eventually the character table of G. When a sufficiently large body of self-consistent structural information has been generated, the group is said to be *discovered*. This is roughly the point where the group theoretic community first becomes convinced that the group exists.

The group actually *exists* when there is a proof that there is at least one group satisfying hypothesis $\mathcal{H}$, while the group is *unique* when there is a proof that, up to isomorphism, there is at most one group satisfying $\mathcal{H}$. More detailed information about the group structure usually comes later and might include the calculation of the automorphism group and Schur multiplier of G, an enumeration of the maximal subgroups of G, and the generation of the modular character tables for G.

As part of the ongoing effort to produce a complete, unified, and accessible proof of the Classification Theorem, Aschbacher has begun to try to write down in one place a complete and fairly self-contained proof that the 26 sporadic groups exist and are unique. The plan is to generate at the same time the basic structural information about each sporadic group necessary for the Classification. This program dovetails with the Gorenstein-Lyons-Solomon effort to "revise" the proof of the Classification, since GLS give themselves the existence, uniqueness, and basic structure of each sporadic group.

This article concerns itself only with the uniqueness question. The first part consists of an exposition of machinery developed in [2] to deal with the

This work was partially supported by BSF 88–00164. The first author is partially supported by NSF DMS–8721480 and NSA MDA9 0–88–H–2032 uniqueness of some of the sporadic groups. To understand and appreciate the statement of the main results from [2] it is first necessary to introduce some graph theoretic and geometric concepts. This is done in section 2. Then the main theorems from [2] are stated in section 3, where there is also a very brief discussion of the proof of these results. Section 4 is devoted to a discussion of how to use the theory in [2] to prove the uniqueness of some of the larger sporadic groups via their local geometries. The final section contains some speculation about possible ways to establish the uniqueness of each of the sporadic groups and the diff iculties involved in such an undertaking.

A final general remark. It seems to us that any good second generation treatment of the uniqueness of the sporadic groups must do several things. It must be simple, clear, and elegant. It should be independent of machine calculation. Finally it should be as uniform as possible with a minimum of case analysis. Given our present understanding of the sporadic groups as 26 independent entities, some amount of case analysis seems unavoidable, but the machinery in [2] gives hope that some differences can be minimized. However the theory in [2] is in its infancy and much remains to be done before a truly uniform treatment of the uniqueness of the sporadic groups exists.

Section 2. Graphs.

In this section Δ is a graph. Let x be a vertex in Δ, write $\Delta(x)$ for the set of vertices distinct from x and adjacent to x in Δ, $x^{\perp} = \Delta(x) \cup \{x\}$, and $\Delta^n(x)$ for the set of vertices at distance n from x. A *morphism* $d : \Delta : \to \Delta'$ of graphs is a map of vertices such that $d(x^{\perp}) \subseteq d(x)^{\perp}$ for all $x \in \Delta$.

Let $P = P(\Delta)$ be the set of paths in Δ. Thus the members of P are the finite sequences $p = x_0 \cdots x_r$ from Δ with $x_{i+1} \in x_i^{\perp}$ for all i. Write $org(p)$, $end(p)$ for the *origin* x_0 and *end* x_r of p, respectively. Write pq for the concatenation of paths p and q such that $end(p) = org(q)$. Write p^{-1} for the path $x_r \cdots x_0$. The path $p = x_0 \cdots x_r$ is a *circuit* if $x_r = x_0$.

Define an equivalence relation $\sim$ on P to be *P-invariant* if the following four conditions are satisfied:

(PI1) If $p \sim q$ then $org(p) = org(q)$ and $end(p) = end(q)$.
(PI2) $rr^{-1} \sim org(r)$ for all $r \in P$.
(PI3) Whenever $p \sim p'$ and $q \sim q'$ with $end(p) = org(q)$, then also $pq \sim p'q'$.
(PI4) $x \sim xx$ for all $x \in \Delta$.

Define the *kernel* of an equivalence relation $\sim$ on P to be the set $ker(\sim)$ of all circuits s such that $s \sim org(s)$. Define a subset S of P to be *closed* if it is the kernel of some invariant equivalence relation and define the *closure* of a set T of circuits to be the intersection of all closed subsets containing T. There is an intrinsic characterization of closed sets in section 2 of [2] which shows the intersection of closed sets is closed, so the closure of T is well-defined.

Given a set S of circuits of Δ, define a relation $\sim_S$ on P by $p \sim_S q$ if p and q have the same origin and end and $pq^{-1} \in S$. It is easy to check that:

(2.1) *Let $\sim$ be a P-invariant equivalence relation. Then $\sim = \sim_{ker(\sim)}$. In particular a set S of circuits of Δ is closed if and only if $\sim_S$ is a P-invariant equivalence relation on P.*

Define the *basic relation* to be the relation $\sim_{Bas}$ where Bas is the smallest closed subset of P. Write $\equiv$ for $\sim_{Bas}$. Notice $\equiv$ is characterized by the property that if $\sim$ is P-invariant and $a \equiv b$ then $a \sim b$.

Write $[P]$ for the set of equivalence classes $[p]$ of the basic relation $\equiv$. For $x \in \Delta$, write $\Sigma(\Delta, x)$ for the set of paths p with origin x and write $\pi_1(\Delta, x)$ for the set of classes $[p] \in [P]$ with p a circuit and $org(p) = x$. As $\equiv$ is P-invariant, $\pi_1(\Delta, x)$ is a group under the product $[p][q] = [pq]$. Of course $\pi_1(\Delta, x)$ is the fundamental group of the graph and is free (*cf.* Section 5.1 in Serre [12]), but we won't need this fact.

Define Δ to be *r-generated* if the closure of the set of all circuits of length at most r is the set of all circuits. We say Δ is *triangulable* if Δ is 3-generated. Intuitively Δ is triangulable if each circuit is the product of triangles, and a given path can usually be seen to be in the closure of the triangles by drawing suitable pictures like those suggested by 2.3 below. More formally:

(2.2) *Δ is triangulable if and only if for each $x \in \Delta$, $\pi_1(\Delta, x)$ is generated by classes $[rtr^{-1}]$, $r \in P$, t triangle, $org(r) = x$, $end(r) = org(t)$.*

Define a morphism $d : \Gamma \rightarrow \Delta$ of graphs to be a *local bijection* if for all $\alpha \in \Gamma$,

$$d_\alpha = d|_{\alpha^\perp} : \alpha^\perp \rightarrow d(\alpha)^\perp$$

is a bijection. Define d to be a *fibering* if d is a surjective local bijection. The fibering is *connected* if its domain Γ is connected. The fibering is a *covering* if $d_\alpha : \alpha^\perp \rightarrow d(\alpha)^\perp$ is an isomorphism for all $\alpha \in \Gamma$. We say Δ is *simply connected* if Δ is connected and Δ possesses no proper connected coverings.

Caution. In the combinatorial group theoretic literature the term "covering" is sometimes used as we use the term fibering. However we prefer to reserve the word covering for a local isomorphism. For example coverings of topological spaces and Tits' coverings of geometries in [16] are local isomorphisms.

Given a P-invariant equivalence relation $\sim$, define $P/\sim\; = \tilde{P}$ to be the set of equivalence classes of $\sim$ and make $\tilde{P}$ into a graph by decreeing that $\tilde{p}$ is adjacent to $\tilde{q}$ if $p \sim qx$, where $x = end(p) \in \Delta(end(q))$. Notice that if $p \sim qx$ then $q \sim p \cdot end(q)$, so our graph is undirected.

Recall $\Sigma(\Delta, x)$ denotes the set of paths with origin x. Write $\Sigma(\Delta, x)/\sim$ for the set of classes $\tilde{p}$ with $p \in \Sigma(\Delta, x)$ and $\tilde{\pi}_1(\Delta, x)$ for the group of all $\tilde{p}$ with $p \in \Sigma(\Delta, x)$ a circuit.

(2.3) *Assume Δ is connected and let T be the closure of the set of triangles of Δ. Then*

(1) $end : \Sigma(\Delta, x)/\equiv \;\to \Delta$ is a universal connected fibering for Δ.

(2) $end : \Sigma(\Delta, x)/\sim_T \;\to \Delta$ is a universal connected covering for Δ.

(3) Δ is simply connected if and only if Δ is triangulable.

It will be important for us to know when certain graphs are simply connected. Lemma 2.3.3 says Δ is simply connected if and only if it is triangulable, and in the graphs we encounter this turns out to be an effective means for proving simple connectivity.

Remark. Let $K \equiv K(\Delta)$ be the simplicial complex whose vertices are the complete subgraphs of Δ. At this conference Tits asked if Δ is simply connected if and only if t he topological space $|K|$ of K is simply connected. The answer is yes. Namely if $\sim \;= \;\sim_T$ where T is the closure of all triangles, then $\tilde{\pi}_1(\Delta, x)$ is the *edge path group* of K (*cf.* Chapter 3, Section 6 of [14]) so by Theorem 3.6.16 in [14] , $\tilde{\pi}_1(\Delta, x) \cong \pi_1(|K|, x)$, the fundamental group of $|K|$.

We close the section with a few elementary lemmas from [2] on triangulation. In each case S is a closed subset of P.

(2.4) *(1) If pq, pr, and $r^{-1}q$ are circuits with pr, $r^{-1}q \in S$, then $pq \in S$.*

(2) Let $a_i, b_i, c_j \in P$, $1 \le i \le n$, $1 \le j < n$, such that $org(a_i) = x$, $end(b_i) = u$, $end(a_i) = org(b_i) = org(c_i) = end(c_{i-1})$ for $1 \le i \le n$. Assume $a_i c_i a_{i+1}^{-1}$ and $b_i^{-1} c_i b_{i+1}$ are in S for $1 \le i < n$. Then $a_n b_n b_1^{-1} a_1^{-1} \in S$.

Given integers n, m with $n \ge 2$, define $|m|_n = r$, where $0 \le r \le n/2$ and $m \equiv r$ or $-r$ mod n. Then define a circuit $p = x_0 \cdots x_n$ of length n to be a *n-gon* if $d(x_i, x_j) = |i - j|_n$, for all i, j, $0 \le i, j \le n$. Define $gon(S)$ to be the least r for which there exists an *S-nontrivial circuit* (*i.e.* a circuit not in S) of length r.

(2.5) *Let $r = gon(S)$ and p be an S-nontrivial circuit of length r. Then p is an r-gon.*

To show S consists of all circuits, by 2.5 it suffices to show that for all $r \leq 2diam(\Delta) + 1$, each r-gon is in S.

(2.6) *Assume $r = gon(S) > 3$ and for each $x \in \Delta$ and $u \in \Delta^2(x)$, $\Delta(x, u)$ is connected. Then $gon(S) > 4$.*

PROOF: By 2.5 we may assume $p = x_0 \ldots x_4$ is an S-nontrivial square. By hypothesis there is a path $x_1 = y_1, \ldots, y_n = x_3$ in $\Delta(x_0, x_3)$. Now appeal to 2.4.2 with $a_i = x_0 y_i$, $b_i = y_i x_2$, and $c_i = y_i y_{i+1}$.

Section 3. The Main Theorems of [2].

Define a *uniqueness system* to be a 4-tuple $\mathcal{U} = (G, H, \Delta, \Delta_H)$ such that G is an edge transitive group of automorphisms of the undirected graph Δ, $H \leq G$, Δ_H is a graph with vertex set xH and edge set $(x, y)H$ for some $x \in \Delta$ and $y \in \Delta(x) \cap xH$, and:

(U) $G = \langle H, G_x \rangle$, $G_x = \langle G_{x,y}, H_x \rangle$, and $H = \langle H(\{x, y\}), H_x \rangle$.

Say Δ_H is a *base* for $\mathcal{U}$ if the closure of the G-conjugates of circuits in Δ_H is the set of all circuits of Δ.

Define a *similarity* of uniqueness systems $\mathcal{U}, \bar{\mathcal{U}}$ to be a pair of isomorphims $\alpha : G_x \to \bar{G}_{\bar{x}}$ and $\zeta : H \to \bar{H}$ such that $\alpha = \zeta$ on H_x, $H_x \zeta = \bar{H}_{\bar{x}}$, $G_{x,y} \alpha = \bar{G}_{\bar{x}, \bar{y}}$, and $H(\{x, y\})\zeta = \bar{H}(\{\bar{x}, \bar{y}\})$ for some edges $(x, y), (\bar{x}, \bar{y})$ of $\Delta_H, \Delta_{\bar{H}}$, respectively. We say the similarity is *with respect to* $(x, y), (\bar{x}, \bar{y})$ if we wish to emphasize the role of those edges. The similarity is an *equivalence* if there exists $t \in H$ with cycle (x, y) such that $(b^t)\alpha = (b\alpha)^{t\zeta}$ for all $b \in G_{x,y}$.

Define a *morphism* of uniqueness systems $\mathcal{U}, \bar{\mathcal{U}}$ to be a gro up homomorphism $d : G \to \bar{G}$ such that the restrictions $d : H \to \bar{H}$ and $d : G_x \to \bar{G}_{\bar{x}}$ are isomorphisms defining a similarity of $\mathcal{U}$ with $\bar{\mathcal{U}}$. Notice d induces a map $d : \Delta \to \bar{\Delta}$ defined by $(xg)d = \bar{x}(gd)$; it turns out this map is a fibering and induces an isomorphism $d : \Delta_H \to \bar{\Delta}_{\bar{H}}$.

We are now in a position to state the principal results of [2]. The Main Theorem is:

THEOREM 1. *Assume $\mathcal{U}, \bar{\mathcal{U}}$ are equivalent uniqueness systems such that $\Delta_H, \bar{\Delta}_{\bar{H}}$ are bases for $\Delta, \bar{\Delta}$, respectively. Then $\mathcal{U} \cong \bar{\mathcal{U}}$.*

COROLLARY. *Assume $\mathcal{U}$ and $\bar{\mathcal{U}}$ are equivalent uniqueness systems, Δ is triangulable, each triangle of Δ is G-conjugate to a triangle of Δ_H, and $\bar{\mathcal{U}}$ also satisfies these hypotheses. Then $\mathcal{U} \cong \bar{\mathcal{U}}$.*

In order to apply Theorem 1 and its Corollary, we need effective means for verifying the equivalence of uniqueness systems. Several such results are contained in [2]; we record two of them as typical:

THEOREM 2. *Assume $\mathcal{U}$ and $\bar{\mathcal{U}}$ are similar uniqueness systems and for some edge (x, y) of Δ_H, $Aut(G_{x,y}) \cap C(H_{x,y}) = 1$. Then $\mathcal{U}$ is equivalent to $\bar{\mathcal{U}}$.*

THEOREM 3. *Assume $\mathcal{U}, \bar{\mathcal{U}}$ are uniqueness systems satisfying Hypothesis V below with respect to edges $(x, y), (\bar{x}, \bar{y})$ and $\alpha : G_x \to \bar{G}_{\bar{x}}$ and $\zeta : H \to \bar{H}$ are isomorphisms such that $G_{x,y}\alpha = \bar{G}_{\bar{x},\bar{y}}$, $H_x\zeta = \bar{H}_{\bar{x}} = H_x\alpha$, and $H(\{x, y\})\zeta = \bar{H}(\{\bar{x}, \bar{y}\})$). Then $\mathcal{U}$ and $\bar{\mathcal{U}}$ are similar.*

HYPOTHESIS V. *The uniqueness system $\mathcal{U} = (G, H, \Delta, \Delta_H)$ satisfies the following four conditions for some edge (x, y) of Δ_H:*
    (V1) $Aut(H_x) = Aut_{Aut(H)}(H_x)Aut_{Aut(G_x)}(H_x)$.
    (V2) $N_{Aut(G_x)}(H_x) \leq N(G_{x,y}^{H_x})C(H_x)$.
    (V3) $N_{Aut(H)}(H_x) \leq N(H_x H(\{x, y\})H_x)C(H_x)$.
    (V4) $N_{H_x}(H_{x,y}) \leq N_{G_x}(G_{x,y})$.

We close this section with a brief discussion of the proof of Theorem 1.

Let $I = \{1, \ldots, n\}$ be a set of finite order n. Recall an *amalgam of rank* n is a family

$$A = (\alpha_{J,K} : P_J \to P_K : J \subset K \subset I)$$

of group homomorphisms such that for all $J \subset K \subset L$, $\alpha_{J,K}\alpha_{K,L} = \alpha_{J,L}$.

There is an obvious notion of morphism of amalgams. A *completion* $\beta : A \to G$ for A is a family $\beta = (\beta_J : P_J \to G)$ of group homomorphisms such that $G = \langle P_J\beta_J : J \subset I \rangle$ and for all $J \subset K \subset I$ the obvious diagram commutes:

$$
\begin{array}{ccc}
& \alpha_{J,K} & \\
P_J & \longrightarrow & P_K \\
& \beta_J \searrow \quad \nearrow \beta_K & \\
& G &
\end{array}
$$

The completion $\beta : A \to G$ is said to be *faithful* if each β_J is an injection.

The free amalgamated product $G(A)$ of A supplies a universal completion $\iota : A \to G(A)$, and if A possesses a faithful completion then the universal completion is faithful. Of course isomorphic amalgams have isomorphic universal completions.

Let $\mathcal{U} = (G, H, \Delta, \Delta_H)$ be a uniqueness system and (x, y) an edge in Δ_H. To avoid the trivial case we assume $x \neq y$. The *amalgam of $\mathcal{U}$* is the rank

3 amalgam $A(\mathcal{U}) = (\alpha_{J,K} : P_J \to P_K)$ defined by $P_{12} = H$, $P_{23} = G_x$, $P_{13} = G(\{x,y\})$, $P_1 = H(\{x,y\})$, $P_2 = H_x$, $P_3 = G_{x,y}$, and $P_\emptyset = H_{x,y}$, with all maps $\alpha_{J,K}$ inclusions.

Observe the inclusion map $\beta : A(\mathcal{U}) \to G$ is a faithful completion of the amalgam $A(\mathcal{U})$. Let $G(A(\mathcal{U})), \iota$ be the universal completion of $A(\mathcal{U})$. Write $\tilde{G}$ for $G(A(\mathcal{U}))$, $\tilde{H}$ for $H\iota$, $\tilde{G}_{\tilde{x}}$ for $G_x\iota$, etc. Let $\tilde{\Delta}$ be the collinearity graph of the rank 3 coset geometry $\tilde{\Gamma}$ of $\tilde{G}$ on the image of the amalgam under ι. Then $\tilde{G}_{\tilde{x}}$ is indeed the stabilizer of some $\tilde{x} \in \tilde{\Delta}$. Let $\tilde{\Delta}_{\tilde{H}}$ be the collinearity graph of the residue of $\tilde{H}$ in $\tilde{\Gamma}$ and $\tilde{\mathcal{U}} = (\tilde{G}, \tilde{H}, \tilde{\Delta}, \tilde{\Delta}_{\tilde{H}})$. Then

(3.1) $\tilde{\mathcal{U}}$ *is a uniqueness system equivalent to* $\mathcal{U}$, *there exists a morphism* $d : \tilde{\mathcal{U}} \to \mathcal{U}$ *of uniqueness systems, and if* Δ_H *is a base for* Δ *then* $\tilde{\mathcal{U}} \cong \mathcal{U}$.

(3.2)   *If* $\mathcal{U}$ *and* $\bar{\mathcal{U}}$ *are equivalent uniqueness systems then* $A(\mathcal{U}) \cong A(\bar{\mathcal{U}})$.

Notice 3.1 and 3.2 establish Theorem 1. Namely under the hypotheses of Theorem 1, 3.2 says $A(\mathcal{U}) \cong A(\bar{\mathcal{U}})$, so that $\tilde{G}$ is also the universal completion of $A(\bar{\mathcal{U}})$. Then by 3.1, $\mathcal{U} \cong \tilde{\mathcal{U}} \cong \bar{\mathcal{U}}$, as desired.

The proofs of the remaining results are more straightforward but also more technical.

Remark. The proof just sketched shows that under the hypotheses of Theorem 1, G is the free amalgamated product of H, G_x, and $G(\{x,y\})$. This observation supplies a presentation for G.

Section 4. p-local geometries.

In this section we adopt the terminology of Tits in [16] and assume:

(Γ0) G is a flag transitive group of automorphisms of a residually connected rank 3 string geometry Γ and (x, l, π) is a flag in Γ.

Consider the following hypotheses:

(Γ1) Each pair of distinct collinear points x, y is on a unique line $x + y$.
(Γ2) If $x, y \in \Gamma_1(\pi)$ are collinear then $x + y \in \Gamma_2(\pi)$.
(Γ3) Each triangle of Δ is incident with a plane.
(Γ4) $G_{\pi,l}$ is 2-transitive on $\Gamma_1(l)$.
(Γ5) $G_{x,l} = \langle G_{x,y,l}, G_{x,l,\pi} \rangle$ for $x \neq y \in \Gamma_1(l)$.

(4.1) *Assume* (G, Γ, x, l, π) *satisfies hypotheses* (Γi) *for* $i = 0, 4, 5$. *Define* Δ *to be the collinearity graph of* Γ, $H = G_\pi$, *and* Δ_H *the collinearity graph of the residue of* π. *Then*

    (1) $\mathcal{U} = (G, H, \Delta, \Delta_H)$ *is a uniqueness system.*

    (2) *If hypotheses* (Γi), $0 \leq i \leq 5$, *hold then each triangle of* Δ *is* G-*conjugate to a triangle of* Δ_H.

Section 7 in [2] contains lemmas which make it possible to verify that suitable truncations of many of the p-local geometries of the sporadic groups (*cf.* [1]) satisfy hypotheses (Γi), $0 \leq i \leq 5$. Combining this machinery with the results in section 3 and checking the various hypotheses are satisfied, we get the following theorem whose proof is not yet written up in preprint form.

(4.2) *Assume* $\bar{G}$ *is a sporadic group with* p-*local geometry* $\bar{\Gamma}$, *where either*

    (a) $p = 3$ *and* $\bar{G}$ *is* Co_1, Sz, Mc, *or* Ly, *or*

    (b) $p = 2$ *and* $\bar{G}$ *is* M_{24} *or* J_4.

Assume further that (G, Γ, x, l, π) *satisfies Hypothesis* $(\Gamma 0)$ *and* $\alpha : G_x \to \bar{G}_{\bar{x}}$ *and* $\zeta : G_\pi \to \bar{G}_{\bar{\pi}}$ *are isomorphisms with* $G_{x,l}\alpha = \bar{G}_{\bar{x},\bar{l}}$, $G_{\pi,l}\zeta = \bar{G}_{\bar{\pi},\bar{l}}$, *and* $G_{x,\pi}\alpha = \bar{G}_{\bar{x},\bar{\pi}}$. *Then* $G \cong \bar{G}$ *if* Δ *and* $\bar{\Delta}$ *are triangulable.*

The p-local geometries for the sporadic groups are discussed in [1]. Other choices of (G, p) are possible, but the choices in (4.2) are particularly nice for a number of reasons. For example the pairs $(Sz, 2)$ and $(Ly, 5)$ are also possibilities. But in these cases the universal covering of the p-local geometry Γ is an affine building, and hence infinite. As these geometries satisfy $(\Gamma 1)$ and $(\Gamma 3)$, coverings of Γ induce coverings of the collinearity graph Δ, so Δ is not simply connected. This does not necessarily cause big problems since one might show that Δ is n-generated for some $n > 3$, but it is at least an inconvenience. In the case of $(Ly, 5)$, one of the object stabilizers is not local, so extra effort must be expended to prove the existence of this stabilizer.

To complete the treatment of the six sporadic groups listed in 4.2 requires work both at the beginning and end of the problem. To begin one must settle on a group theoretic hypothesis $\mathcal{H}$ with which to characterize G. In each of the six cases of 4.2, the optimal choice for $\mathcal{H}$ is presumably the centralizer of a 2-central involution. Thus except for Mc and Ly, this involves a large extraspecial subgroup. The next step is to prove the existence of the p-local geometry and establish the isomorphisms of 4.2. At this point 4.2 reduces the problem to a check that the collinearity graph Δ is triangulable. Because of the novelty of the approach, this last step is at present the most difficult, even though the graph Δ is the most well-behaved of those associated to G.

However there is no reason to believe that techniques can't be developed to make the check of triangulability of suitable graphs easy.

In the next section we go into some of these matters in more depth.

Section 5. Speculation.

We close with some speculation on how best to establish the uniqueness of each of the sporadic groups. But first some general discussion of the factors which need to be considered.

To begin, given a sporadic group $\bar{G}$, we need to settle on a group theoretic hypothesis $\mathcal{H} = \mathcal{H}(\bar{G})$ with which to characterize $\bar{G}$. Probably $\mathcal{H}$ will always be the general structure of the centralizer of an involution. If $\bar{G}$ possesses an involution $\bar{z}$ such that $O_2(C_{\bar{G}}(\bar{z}))$ is a large extraspecial subgroup then $C_{\bar{G}}(\bar{z})$ is probably the best choice.

The next step is to decide upon a means for establishing the uniqueness of groups G satisfying Hypothesis $\mathcal{H}$. Our predisposition is to use a graph theoretic or geometric approach. Thus in essence we must settle on a large maximal subgroup G_x of G and a self paired orbital of G on G/G_x defining a graph Δ admitting the action of G as a group of automorphisms with G_x the stabilizer of some vertex $x \in \Delta$. If possible, Δ is the collinearity graph of some geometry Γ preserved by G.

There are several things to think about in choosing Δ. First Δ will probably be easiest to work with if it is highly symmetric; that is Δ should be of small diameter and G should be of small rank on Δ. For example in using Theorem 1 this will presumably make it easier to prove Δ is triangulable.

Next it would be best if it is easy to prove the existence of G_x starting from Hypothesis $\mathcal{H}$. If G_x is a local subgroup of G then the existence of G_x is usually easy. But often the nicest subgroups are not local and hence are not easy to construct. For example it is probably easier to work with the 2-local geometry for M_{24} than to try to prove the existence of an M_{23}-subgroup starting from the structure of a 2-central involution. On the other hand the only nice graph associated to G may be on the cosets of a nonlocal subgroup, and hence the construction of this subgroup may be necessary.

Of course eventually we would like to prove the existence of all large maximal subgroups of G, so why not put in this effort at the start? There are a number of reasons to avoid such an approach. In the M_{24} example it is fairly easy to construct *some* M_{24} with an M_{23}-subgroup, so if one can prove uniqueness of M_{24} by any means then the existence of the M_{23}-subgroup follows painlessly. Also it is worth making the treatment of the uniqueness of the sporadic groups as simple and self-contained as possible. Modularization of a very large and complex undertaking is always desirable.

Finally in dealing with 26 sporadic groups, it is an advantage to introduce as much uniformity as possible and to reduce case analysis to a minimum.

Thus for example as the majority of the sporadic groups possess a large extraspecial subgroup, it is probably appropriate to emphasize such subgroups.

Now some ideas about the optimal approach to the uniqueness question for each of the sporadic groups. We have already suggested in section 4 that the groups M_{24}, J_4, Mc, Sz, Ly, and Co_1 are perhaps best viewed as acting on an appropriate p-local geometry. In [3], we completely implemented this approach for J_4 as a test case. Segev discusses that work elsewhere in these Proceedings.

We believe He is best represented on a rank 3 geometry with two $S_6/Z_3/E_{64}$ stabilizers and one $S_3/L_3(4)/E_4$ stabilizer, with the collinearity graph defined on the cosets of the final stabilizer. The notion of "uniqueness system" discussed in section 3 is not quite applicable here, but a very slight generalization can be applied.

The groups F_1, F_2, and F_5 are perhaps best represented on a commuting graph on a class of non-2-central involutions. This approach has been implemented by Griess, Meierfrankenfeld, and Segev in [5] for the Monster, and by Segev in [10] and [11] for F_2 and F_5. The results in [2] described in section 4 grew out of attempts to understand the approach in these papers, and those results can be used to greatly simplify the treatments in [5], [10], and [11]. We have written out such a simplification for the Monster in section 8 of [2] as another test case.

The three Fischer groups are probably best viewed as 3-transposition groups. Given the 3-transposition theory, the proof of the uniqueness of the Fischer groups is elegant and easy. Moreover it is worth developing this theory for a variety of other reasons.

The four small Mathieu groups could be handled by any of a number of means. There is great room for ingenuity here. For example several years ago an undergraduate at Caltech named Laura Anderson (now a graduate student at MIT) produced a simple proof of the uniqueness of M_{12}.

Similarly it is not clear which approach is optimal for the three small Janko groups. Janko proves J_1 is unique as a 7-dimensional linear group over $GF(7)$ in [7]. One could also consider the action on an $L_2(11)$-subgroup. Hall and Wales show J_2 is unique in [6] via constructing a $U_3(3)$-subgroup. Another possibility is to consider the commuting graph on 3-central subgroups of order 3. Considering the small size of J_3, We see no attractive way to prove its uniqueness. Frohardt has established uniqueness in [4] via a trilinear form in characteristic 0 but his proof requires some sweat.

One could approach the uniqueness of Co_2 via either its 2 or 3-local geometry. The latter geometry is nicer but it involves the construction of a $Z_2/U_6(2)$-subgroup. The existing proof is due to F. Smith in [13], which takes this approach.

Similarly Co_3 could be approached via its 2 or 5-local geometry, with the latter involving the construction of a $\mathbf{Z}_2/Mc$-subgroup upon which Co_3 acts 2-transitively.

The groups $O'N$ and HS could be viewed as acting on the commuting graph of a cyclic TI-subgroup of order 4. Alternatively the existing uniqueness proof for HS is due to Parrott and Wong in [9], which constructs the M_{22}-subgroup. One could construct an $L_3(7)$-subgroup of $O'N$.

The existing uniqueness proof for Ru is due to Parrott in [8]; it involves the construction of the $^2F_4(2)$-subgroup. One could also consider the commuting graph on 4-subgroups centralizing Sz/E_4.

We see no good way to prove the uniqueness of F_3 via a graph theoretic approach. The existing proof is due to Thompson [15] . Once a group of type F_3 is shown to possess a faithful complex representation of degree 248, Thompson's argument gives a short elegant proof of the uniqueness of such groups. To prove the existence of such a representation, Thompson generates the character table of F_3. This is quite nontrivial.

REFERENCES

1. M. Aschbacher, *Overgroups of Sylow subgroups in sporadic groups*, Memoirs AMS **343** (1986), 1–235.

2. M. Aschbacher & Y. Segev, *Extending morphisms of groups and graphs*, preprint.

3. M. Aschbacher and Y. Segev, *The uniqueness of groups of type J_4*, preprint.

4. D. Frohardt, *A trilinear form for the third Janko group*, J. Alg. **83** (1983), 349–379.

5. R. Griess, U. Meierfrankenfeld, and Y. Segev, *A uniqueness proof for the Monster*, Ann. Math. **130** (1989), 567–602.

6. M. Hall and D. Wales, *The simple group of order 604,800*, J. Alg. **9** (1968), 417–450.

7. Z. Janko, *A new finite simple group with abelian 2-Sylow subgroups and its characterization*, J. Alg. **3** (1966), 147–186.

8. D. Parrott, *A characterization of the Rudvalis simple group*, Proc. London Math. Soc. **32** (1976), 25–51.

9. D. Parrott and S. Wong, *On the Higman-Sims simple group of order 44,352,000*, Pacific J. Math. **32** (1970), 501–516.

10. Y. Segev, *On the uniqueness of Fischer's Baby Monster*, to appear.

11. Y. Segev, *On the uniqueness of the Harada–Norton group*, to appear.

12. J. Serre, "Trees," Springer–Verlag, Berlin, 1980.

13. F. Smith, *A characterization of the .2 Conway simple group*, J. Alg. **31** (1974), 91–116.

14. E. Spanier, "Algebraic Topology," McGraw-Hill, New York, 1966.

15. J. Thompson, *A simple subgroup of $E_8(C)$*, "Finite Groups," Japan Soc. for Promotion of Science, Tokyo, 1976.

16. J. Tits, *A local approach to buildings*, "The Geometric Vein," Springer–Verlag, New York, 1982, pp. 519–547.

The Study of J_4 via the
Theory of Uniqueness Systems *

Michael Aschbacher and Yoav Segev

This article serves as an exposition of the main result of [3]. In [3] we give the first computer free proof of the uniqueness of groups of type J_4. In addition we supply simplified proofs of some properties of such groups, such as the structure of certain subgroups.

A group of type J_4 is a finite group G possessing an involution z such that $H = C_G(z)$ satisfies $F^*(H) = Q$ is extraspecial of order 2^{13}, H/Q is isomorphic to $\mathbb{Z}_3$ extended by $\mathrm{Aut}(M_{22})$, and $z^G \cap Q \neq \{z\}$. We prove:

Main Theorem *Up to isomorphism there exists at most one group of type* J_4.

Janko was the first to consider groups of type J_4 in [5], where he established various properties of such groups. For example Janko showed that each group G of type J_4 is simple, he determined the order of G, and he described the normalizers of all subgroups of G of prime order. However Janko left open the question of whether there exist groups of type J_4 and whether all groups of type J_4 are isomorphic. Thus Janko is said to have *discovered* J_4 and indeed J_4 was the last of the 26 sporadic simple groups to be discovered.

Around 1980, Conway, Norton, Parker, and Thackray proved the existence and uniqueness of J_4 using extensive machine computation. This

* This work was partially supported by BSF 88-00164. The first author is partially supported by NSF DMS-8721480 and NSA MDA90-88-H-2032.

work is discussed briefly in [6]. Norton *et al.* construct J_4 as a linear group in 112 dimensions over the field of order 2. While the notion of 2-*local geometry* did not exist at that time, this geometry plays an implicit role in [6].

In our proof we completely implement the theory of *uniqueness systems* developed in [2] and discussed in Section 1 of [4]. We apply this theory to the 2-local geometry Γ of a group of type J_4. To do this requires various facts about groups of type J_4 and its collinearity graph Δ. Many of those facts already appear in [5] or [6], but others do not — in particular the fact that Δ is *simply connected* in the language of [2]. The emphasis in this expository article is on showing that fact. For completeness and clarity we supply in [3], our own proof for all necessary facts about groups of type J_4, and thus make no appeal to [5] or [6]. One by-product of this approach is better proofs of some properties of J_4.

1.Two lemmas on graphs.

In this section Δ is an undirected graph. For $x, y \in \Delta$, $X, Y \subseteq \Delta$, write $d_\Delta(x,y) = d(x,y)$ for the distance from x to y in Δ and define

$$d_\Delta(X,Y) = d(X,Y) = \min\{d(x,y) : x \in X, y \in Y\}.$$

Given a symmetric relation R on Δ, write $R(x)$ for the set of y such that $(x,y) \in R$. We also write R for the graph with vertex set Δ and edge set R. Thus d_R denotes the distance function in this graph.

Let $\Delta^n(x) = \{y \in \Delta : d(x,y) = n\}$ and $\Delta^{\leq n}(x) = \{y \in \Delta : d(x,y) \leq n\}$. In addition we adopt the notation and terminology of Section 2 and 3 in [2] (see also Section 2 in [4]); in particular the notions of *triangulable graph* and 4-*generated graph* appear there.

Consider the following hypothesis:

Hypothesis A *The following two conditions hold:*

(A1) $R_0 \subseteq R_1 \subseteq \ldots \subseteq R_N$ *is a chain of symmetric relations on Δ such that* $R_0 = \Delta^{\leq 1}$ *and* $R_N = \Delta^{\leq 2}$.

(A2) *For each* $0 \leq k < N, x \in \Delta, y \in R_{k+1}(x) \cap \Delta^2(x)$ *and* $z \in \Delta^2(x)$,

$$d_{Rm}(R_k(y) \cap \Delta(x), \Delta(x,z)) \leq 1,$$

where $m = min\{k, 1\}$.

Our first result in this section is:

(1.1)　*If hypothesis A holds then Δ is 4 - generated.*

Given a path xuy in Δ write $C\Delta(xuy)$ for the connected component of u in $\Delta(x, y)$. A ternary relation $\hat{R}$ on Δ is *symmetric* if $(x, u, y) \in \hat{R}$ implies $(y, u, x) \in \hat{R}$. Given a symmetric ternary relation $\hat{R}$ write R for the symmetric binary relation $(x, y) \in R$ if and only if $(x, u, y) \in \hat{R}$ for some $u \in \Delta$.

Hypothesis B　*The following conditions hold:*

(B1)　$\hat{R}_0 \subseteq \hat{R}_1 \subseteq \ldots \hat{R}_N$ *is a chain of symmetric ternary relations on Δ such that $\hat{R}_0$ is the set of all triangles of Δ and $\hat{R}_N$ is the set of all paths of Δ of length 2.*

(B2)　*For all $0 \leq k < N, (x, u, y) \in \hat{R}_{k+1} - \hat{R}_0$, and $(x, v, z) \in \hat{R}_N - \hat{R}_0$,*

$$d_{R_m}(D_k(xuy), C\Delta(xvz)) \leq 1$$

where $m = min\{k, 1\}$ and $D_k(xuy)$ consists of $C\Delta(xuy)$ together with all $w \in \Delta(x)$ such that $(y, c, w) \in \hat{R}_k$ for some $c \in C\Delta(xuy)$.

(B3)　*If $y \in R_1(x) \cap \Delta^2(x)$ and $w \in \Delta(y) \cap \Delta^2(x)$ then $d_\Delta(\Delta(x, y), w)) \leq 1$.*

Our second result in this section is:

(1.2)　*If Hypothesis B holds then Δ is triangulable.*

We note that by (4.3.3) in [2], Δ is triangulable if and only if it is simply connected. We briefly discuss the proofs of (1.1) and (1.2). Given the relations $R_i, 0 \leq i \leq N$ of hypothesis (A1) let $type(x, y) = min\{i : (x, y) \in R_i\}$, where $(x, y) \in R_N$ and given a path $p = x_0 \cdots x_n$ in Δ let $type(p) = min\{type(x_i, x_{i+2}) : 0 \leq i \leq n - 2\}$. Let S be the closure of all triangles and squares of Δ (see Section 2 in [2] for a definition). Pick a cycle $c \notin S$ of minimal length and subject to this constraint with $type(c)$ minimal. We obtain a contradiction by showing that c must lie in S using hypothesis (A2). The proof of (1.2) is straightforward as well and has a similar flavour.

2. The geometry of M_{24}

In this section G is the Mathieu group M_{24} and Γ is the 2 - local geometry of G. Thus Γ is a geometry on $I = \{1, 2, 3\}$. The objects of type i are the *octads, trios and sextets* permuted by G, for $i = 1, 2, 3$, respectively. To correspond with terminology used in later sections we also refer to trios as *lines* and sextets as *planes*. Given an object $\gamma \in \Gamma$ write $\Gamma(\gamma)$ for the residue geometry of γ in Γ and $Q(\gamma)$ for the kernel of the action of G_γ on $\Gamma(\gamma)$.

Let $\hat{V}$ be the Golay code module for G. Thus $\hat{V}$ is a 12-dimensional module over the field F of order 2 and we view $\hat{V}$ as a subspace of the power set of the set X of 24 points permuted by G with addition equal to symmetric difference. There is a triple of forms P, C, f defined on $\hat{V}$ via

$$P(x) = |x|/4 \bmod 2, \; C(x,y) = |x \cap y|/2 \bmod 2, \; f(x,y,z) = |x \cap y \cap z| \bmod 2.$$

Moreover f is a symmetric trilinear form with $f(x,x,y) = 0$ for all $x, y \in \hat{V}$. (cf. Section 1 in [1]). The radical of $\hat{V}$ is X, so we have the induced forms on $V = \hat{V}/ < X >$ which we also write as P, C, f. Thus V is the 11 - *dimensional Golay code module* or *dual Todd module*. We identify each trio and sextet with the set of octads incident with the object. Subject to this convention the members of Γ are subspaces of V.

We obtain information on Γ using the forms P, C and f. For $x \in V$, f_x is the bilinear form $f_x(y,z) = f(x,y,z)$ and $R(x) = \text{Rad}(f_x)$. Given a trio ℓ let $\zeta\ell) =< R(x) : x \in \ell >$. Among several results which appear in Section 3 of [3] we show:

(2.1) *The stabilizer G_ℓ of ℓ has 5 orbits $\delta_i(\ell)$, $0 \le i \le 4$ on Γ_2 as follows:*

1) $\delta_0(\ell) = \{\ell\}$ and $G_\ell \cong Y(L_3(2) \times S_3)/E_{2^6}$.
2) $\delta_1(\ell) = \{k : k \cap \ell \neq 0\}$ is of order 42.
3) $\delta_2(\ell) = \{k : k + \ell \in \Gamma_3\}$ is of order 56.
4) $\delta_3(\ell) = \{k : \dim(k \cap \zeta\ell)) = 1\}$ of order 1008.
5) $\delta_4(\ell) = \{k : k \cap \zeta\ell) = 0\}$ is of order 2688.

Remark 1

Let U be a plane in Γ and let $\Lambda(U)$ be the dual of $\Gamma(U)$. Thus the points of Λ are the lines of $\Gamma(U)$ namely $\Gamma_2(U)$ and the lines of $\Lambda(U)$ are the points of $\Gamma(U)$ namely $\Gamma_1(U)$. The collinearity graph on $\Lambda_1(U)$ is defined by $\ell, m \in \Lambda_1(U)$ are adjacent if $\ell \cap m \in \Lambda_2(U)$ or alternatively if $m \in \delta_1(\ell)$. We note that $\Lambda(U)$ carry the structure of a 4-*dimensional symplectic geometry over F*. Thus for $\ell \in \Lambda_1(U)$ we have the notions of a *disc* $\text{Disc}(\ell) = \ell^\perp - \{\ell\}$ consisting of six members of $\Lambda_1(U)$ and an *oval* consisting of five pairwise nonadjacent members of $\Lambda_1(U)$. Given a disc $D = \text{Disc}(\ell)$ we say that ℓ is the center of D. We will sometimes abbreviate and say that D is a disc or an oval of lines in U.

For $\ell \in \Gamma_2$ write $\text{Co}(\ell)$ for the set of lines coplanar with ℓ and let Co be the graph on Γ_2 in which lines are adjacent if and only if they are coplanar. Among various results of Section 3 in [3] we show:

(2.2)   *Let U, π be planes and D a disc or oval of lines in U. Let $\ell \in \Gamma_2$.*
Then

(1) $(Co(\ell) \cup \delta_3(\ell)) \cap D \neq \emptyset$.
(2)   *If $\delta_1(\ell) \cap \Gamma_2(U) \neq \emptyset$, then $\delta_1(\ell) \cap D \neq \emptyset$.*
(3) $d_{Co}(D, \Gamma_2(\pi)) \leq 1$.

(2.3)   *Let ℓ, m be lines with $m \in \delta_3(\ell)$. Then*

(1)   *There exists a unique plane $\pi(\ell, m) \in \Gamma_3(\ell)$ such that $\delta_1(m) \cap \pi(\ell, m) \neq$*
     $\emptyset$.
(2) $Co(\ell, m) \subseteq \Gamma_2(\pi(\ell, m)) \cup \Gamma_2(\pi(m, \ell))$.
(3)   *Let $< y >= m \cap \pi(\ell, m)$ and $< x >= l \cap \pi(m, \ell)$. Then*
$$Co(\ell, m) \cap \Gamma_2(\pi(\ell, m)) = \Gamma_2(\pi(\ell, m), y) \subseteq \delta_1(m)$$
     with $x + y \in \delta_1(\ell)$ and the remaining two lines in $\Gamma_2(\pi(\ell, m)), y)$ con-
     tained in $\delta_2(\ell)$.

(2.4)   *Let ℓ, m be lines with $m \in \delta_4(\ell)$. Then $\delta_2(\ell) \cap \delta_2(m) \neq \emptyset$.*

We now turn to the geometry of groups of type J_4.

3. The geometry of a group of type J_4.

From now on G is a group of type J_4. Recall $H = C_G(z)$ and M denotes
the subgroup G_1 below. In Section 4 of [3] we obtain several properties of
G. We show:

(3.1)   *G contains 3 subgroups G_1, G_2 and G_3 such that:*

(1)   *G_1 is a split extension of E by L, where $E \cong E_{2^{11}}$, $L \cong M_{24}$ and E is*
     the Todd module for L.
(2)   *$G_2 = KS$ with $K \cap S = R = O_2(G_2)$ is special of order 2^{15}, $Z(R) \cong$*
     E_8, $K/R \cong L_3(2)$ and $S/R \cong S_5$.
(3)   *$G_3 = H$.*

   Set $G_{i,j} = G_i \cap G_j$. Then
(4)   *$O_2(G_{1,2}) = ER$ has order 2^{17}, $K \leq G_{1,2}$ and $G_{1,2}/O_2(G_{1,2}) \cong L_3(2) \times$*
     S_3.
(5)   *$O_{2,3}(H) \leq G_{1,3}$ and $G_{1,3}/O_{2,3}(G_{1,3}) \cong S_6/E_{16}$.*
(6)   *$O_{2,3}(H) \leq G_{1,2,3}, |G_{1,3} : G_{1,2,3}| = 15$ and $|G_{2,3} : G_{1,2,3}| = 5$.*

   Let Γ be the coset geometry on $\{G_1, G_2, G_3\}$. Thus $\Gamma_i = G/G_i$ is the set of objects of Γ of type i. We call the members of Γ_i *points, lines and planes*, for $i = 1, 2, 3$, respectively. We note that Γ is the geometry first introduced by Ronan and Smith in [7]. We now state below some basic properties of Γ and Δ which we obtain in Section 5 of [3]. Write (x, ℓ, π) for the flag (G_1, G_2, G_3).

(3.2) Γ *satisfies hypotheses $(\Gamma_0) - (\Gamma_5)$ in Section 7 of [2] (see also Section 4 in [4]).*

The reader is advised to compare the next lemma with (2.1).

(3.3) *Let $y \in \Gamma_1(\ell) - \{x\}$. Then $\Gamma(x)$ is the geometry of lines and planes in the M_{24} geometry and*

(1) $G_{x,\ell}$ *has 5 orbits $\delta_i(x, \ell)$, $0 \le i \le 4$, on $\Gamma_2(x)$.*
(2) $\delta_0(x, \ell) = \{\ell\}$.
(3) $\delta_1(x, \ell)| = 42$ *and for $k \in \delta_1(x, \ell)$, $|\Gamma_1(k) \cap \Delta(y)| = 3$.*
(4) $|\delta_2(x, \ell)| = 56$ *and $\Gamma_1(k) \subseteq \Delta(y)$ for $k \in \delta_2(x, \ell)$.*
(5)   *For $i = 3$, 4 $|\delta_i(x, \ell)| = 1008$ and 2688 respectively and for $k \in \delta_i(x, \ell)$, k and ℓ are not coplanar.*
(6)   *For $i > 1$ and $k \in \delta_i(x, \ell)$, $Q(x)_y$ is transitive on $\Gamma_1(k) - \{x\}$.*

(3.4) *Each triangle in Δ is incident with a plane.*

(3.5) *Let $abca$ be a triangle in Δ. Then*

(1) $a + c \in \delta_i(a, a + b)$ *for some $i \le 2$.*
(2) $N_G(\{a, b, c\})$ *induces S_3 on $\{a, b, c\}$.*
(3)   *If $a + c \in \delta_i(a, a + b)$ then $b + c \in \delta_i(b, a + b)$.*

For the next lemma write $\Delta_1^2(x)$ for the set of $u \in \Delta^2(x)$ such that $\Gamma_3(x, u)$ is nonempty.

(3.6) *Let $y \in \Delta_1^2(x) \cap \Gamma_1(\pi)$. Then*

(1) π *is the unique plane in $\Gamma_3(x, y)$.*
(2) $\Delta_1^2(x)$ *is an orbit of G_x.*
(3) $\Delta(x, y) \subseteq \Gamma_1(\pi)$.

We now sketch a proof for the simple connectedness of Δ.

4. The J_4 graph is simply connected.

Continue the hypotheses and notation of Section 3. By (3.3), $\Gamma(x)$ is the geometry of lines and planes of the M_{24}- geometry discussed in Section 2. In particular the notions of an oval or a disc of lines are applicable here (see Remark 1 in Section 2). Furthermore we have the graph $Co(x)$ on $\Gamma_2(x)$ as defined in Section 2. Given a set of points $S \subseteq \Delta(x)$ define $L(x, S) = \{x + u : u \in S\}$ and given a line $\ell \in \Gamma_2$ let

$$L(x, \ell) = \{x + u : u \in \Delta(x) \cap \Gamma_1(\ell)\}.$$

We will use Lemma 1.2 to prove that Δ is triangulable. Thus we need to produce a chain $(\hat{R}_i : 0 \le i \le N)$ of symmetric ternary relations on Δ satisfying hypothesis B.

We adopt the notational conventions of Section 1. Of course $\hat{R}_0$ is the set of triangles of Δ. Define $\hat{R}_1$ to consist of all triples (x, u, y) such that $u \in \Delta(x, y)$ and $u + x$, $u + y$ are coplanar. By (3.3), $\hat{R}_1$ consists of triples (x, u, y) such that $u + y \in \bigcup_{i=0}^{2} \delta_i(u, u + x)$ and $R_1(x) = x^{\perp} \cup \Delta_1^2(x)$. Let $\hat{R}_2$ consist of $\hat{R}_1$ together with triples (x, u, y) such that $u \in \Delta(x, y)$ and $u + y \in \delta_3(u, u + x)$. Finally let $\hat{R}_3$ consist of all paths xuy in Δ. First it is useful to observe that:

(4.1) If $\pi \in \Gamma_3$ and $(x, \ell) \in \Gamma_1(\pi) \times \Gamma_2(\pi)$ with $\Gamma_1(\ell) \subseteq \Delta(x)$, then $L(x, \ell)$ is an oval in $\Gamma_2(x, \pi)$.

(4.2) Let $(x, u, y) \in \hat{R}_2 - \hat{R}_1$. Then there exists $\pi \in \Gamma_3(x, u)$ such that:

(1) $D = \Delta(x, y) \cap \Gamma_1(\pi)$ is a clique of order 6, $L(x, D)$ is a disc in $\Gamma_2(x, \pi)$ and for each $d \in D$ there exists $d_1 \in D$ such that $d + d_1 \in \delta_1(d, d + y)$. Furthermore the center of $L(x, D)$ is in $L(x, d + d_1)$.

(2) $D_1(xuy) \cap \Gamma_1(m) \ne \emptyset$ for all $m \in \Gamma_1(x, \pi)$.

Proof. We sketch a proof. The fact that D is a clique follows from (3.6.3). Then (3.3.3), (3.3.4), the fact that $\Gamma(x, \pi)$ carries a structure of a 4-dimensional symplectic geometry (see Remark 1 in Section 2) and a result on such geometries, which we do not include here, establishes 1. For (2) note that if $m \in L(x, D)$, then the conclusion of (2) is obvious. If m is the center of $L(x, D)$, the conclusion of (2) follows from (1). Otherwise, as $L(x, D)$ is a disc, there exists a point $d \in D$ such that $m \in \delta_2(x, x + d)$. We then focus our attention on d, namely we apply (2.2.2) to $\Gamma_2(d)$. Pick $d_1 \in D$ such that $d + d_1 \in \delta_1(d, d + y)$; then by (2.2.2) and (4.1), there exists a line $k \in L(d, m)$ such that $k \in \delta_1(d, d + y)$. Thus, by definition $\Gamma_1(k) \cap \Gamma_1(m) \subseteq D_1(xuy)$.

(4.3) Let $(x, u, y) \in \hat{R}_3 - \hat{R}_2$. Then there exists $\pi \in \Gamma_3(x, u)$ such that:

(1) $D = \Delta(x, y) \cap \Gamma_1(\pi)$ is a clique and $L(x, D)$ contains an oval in $\Gamma_2(x, \pi)$.

(2) $D_2(xuy) \cap \Gamma_1(m) \neq \emptyset$ for all $m \in \Gamma_1(x, \pi)$.

Proof. We sketch a proof. The fact that D is a clique follows from (3.6.3). Then (2.4) and (4.1) implies (1). For (2) note that if $m \in L(x, D)$ then the conclusion of (2) is obvious. Otherwise by (1) there exists $d \in D$ such that $m \in \delta_2(x, x + d)$.

We then focus our attention on d, namely we apply (2.2.1) to $\Gamma_2(d)$. Since, by (4.1), $L(d, m)$ is an oval in $\Gamma_2(d, \pi)$, (2.2.1) implies that there exists $v \in \Gamma_1(m)$ such that $d + y \notin \delta_4(d, d + v)$. Thus, by definition $v \in D_2(xuy)$.

Now let $0 \leq k < 3$ and let $(x, u, y) \in \hat{R}_{k+1} - \hat{R}_0$ and $(x, v, z) \in \hat{R}_3 - \hat{R}_0$. Then (4.2.1) and (4.3.1) show that $L(x, C\Delta(xvz))$ contains an oval or a disc. Furthermore, if $\pi \in \Gamma_3(x, u)$ is as in (4.2.2) or (4.3.2), then by (2.2.3), $d_{Co(x)}(\Gamma_1(x, \pi), C\Delta(xvz)) \leq 1$. This together with (4.2.2) and (4.3.2) show that:

(4.4) Δ satisfies Hypothesis (B2).

We note now that $y \in R_1(x) \cap \Delta^2(x)$ if and only if $y \in \Delta_1^2(x)$. Then by (3.6.1), there exists a unique plane $\pi \in \Gamma_3(x, y)$. By considering the action of $Z(G_\pi)$ on $\Delta(y)$ we show that Δ satisfies hypothesis (B3) as well. Thus we have sketched a proof of

Theorem 4.5 Δ is simply connected.

5. Proof of the Main Theorem.

We continue the notation and hypotheses of Section 3. Recall the definition of H and M at the begining of Section 3. In addition let $I = H_x$, $\Delta_H = \Gamma_1(\pi)$, $x \neq y \in \Gamma_1(\ell)$ and $\mathcal{U} = (G, H, \Delta, \Delta_H)$.

We wish to prove that, up to isomorphism there exists a unique group of type J_4. We will do so by appealing to the Corollary to Theorem 1 in [2]. In particular the reader is referred to [2] for the definition of various concepts like *uniqueness system* etc.

(5.1) $\mathcal{U}$ is a uniqueness system.

Proof. This follows from (3.1) and Theorem 7.3 in [2].

(5.2)

(1) $|\text{Aut}(H) : \text{Inn}(H)| = 2$ *and* $\text{Aut}(H) = \text{Inn}(H) < \alpha >$ *where* $O^2(H) = C_H(\alpha)$.

(2) $|\text{Aut}(M) : M| \leq 2$ *and* $\text{Aut}(M) = M < \beta >$ *where* $O_2(M) < \beta >= O_2(\text{Aut}(M)) \cong E_{2^{12}}$.

(3) *Hypothesis V of [3] is satisfied.*

Proof. First we mention that Hypothesis V appears in Section 3 of [4]. We sketch a proof. Let α be the automorphism of H with $C_H(\alpha) = O^2(H)$ and $h\alpha = hz$ for $h \in H - O^2(H)$. As $[H, \alpha] =< z >\leq I$, α acts on I and we denote by α_I the restriction of α to I.

Set $E = O_2(M)$. By (3.1.1), there is a complement ℓ to E in M and E is the Todd module for L. Thus there is a 12-dimensional indecomposable E_0 for ℓ with $E = [E_0, L]$. Pick $\beta \in E_0 - E$. Here $[\beta, M] = E \leq I$, so β acts on M and we denote by β_I the restriction of β to I. Note that $[\beta, I] \not\leq Q$. We show that

$$\text{Aut}(I) = \text{Inn}(I) < \alpha_I >< \beta_I >,$$

* $$\text{Aut}(H) = \text{Inn}(H)N_{\text{Aut}(H)}(I),$$

$$\text{Aut}(M) = MN_{\text{Aut}(M)}(I).$$

Notice (*) implies (1) and (2), since I contains a Sylow 2-subgroup of H and M, so by 1.1 in [3], $C_{\text{Aut}(H)}(I) \leq \text{Inn}(H)$ and $C_{\text{Aut}(M)}(I) \leq M$. Of course (*) also shows $\text{Aut}(I) = \text{Aut}_{\text{Aut}(H)}(I)\text{Aut}_{\text{Aut}(M)}(I)$, which is (V1). As I is maximal in H, $I = N_H(I)$. Therefore (*) implies (V3) of Hypothesis V as $[\alpha, H] \leq I$.

Next $M_y = M_\ell^\infty N_{M_\ell}(P)$ for $P \in \text{Syl}_3(C_{M_\ell}(M_\ell^\infty/O_2(M_\ell^\infty)))$, so we may choose β to act on M_y, and hence (*) implies (V2). Finally I acts on the 15 lines in $\Gamma_2(x, \pi)$ with ℓ the unique fixed point of I_y, so $N_I(I_y) \leq I_\ell$. Similarly x, y are the fixed points of I_y on $\Gamma_1(\ell)$, so $I_y = N_I(I_y)$, establishing (V4) and completing the verification of Hypothesis V.

We now begin the proof of the Main Theorem. Assume $\overline{G}$ is a second group of type J_4 with subgroups $\overline{H}, \overline{M}$, etc. By (5.1) applied to $\overline{G}$, we have the uniqueness system $\overline{\mathcal{U}} = (\overline{G}, \overline{H}, \overline{\Delta}, \overline{\Delta_{\overline{H}}})$. Now we show,

(5.3)

(1) *There exists an isomorphism* $\zeta : H \to \overline{H}$ *with* $I\zeta = \overline{I}$, *and*

$$H(\{x,y\})\zeta = H(\{\overline{x},\overline{y}\}.)$$

(2) *There exists an isomorphism* $\alpha : M \to \overline{M}$, *with* $I\alpha = \overline{I}$, *and* $M_y\alpha = \overline{M_{\overline{y}}}$.

Now (5.2.3) and (5.3) establish the hypotheses of Theorem 4 of [2], so that result shows $\mathcal{U}$ is similar to $\overline{\mathcal{U}}$. Then by (7.9.3) in [2], $\mathcal{U}$ is equivalent to $\overline{\mathcal{U}}$. This fact together with Theorem 4.5 and (3.4) establishes the hypotheses of the Corollary to Theorem 1 of [2], and that result says $G \cong \overline{G}$, completing the proof of the main Theorem.

References

1 M. Aschbacher, *The geometry of trilinear forms*, in "Finite Geometries, Buildings, and Related Topics", Oxford University Press, Oxford, 1990, pp. 75 - 84.

2 M. Aschbacher and Y. Segev, *Extending morphisms of groups and graphs*, preprint.

3 M. Aschbacher and Y.Segev, *The uniqueness of groups of type* J_4, preprint.

4 M. Aschbacher and Y. Segev, *Uniqueness of Sporadic Groups*, These proceedings.

5 Z. Janko, *A new finite simple group of order* 86,775,571,046,077,562,880 *which possesses* M_{24} *and the full cover of* M_{22} *as subgroups*, J. Alg. **42** (1976), 564-596.

6 S. Norton, *The construction of* J_4, Proc. Sym. Pure Math. **37** (1980), 271 - 278.

7 M. A. Ronan and S. D. Smith, *2-local geometries for some sporadic groups*, Proc. Sym. Pure Math **37** (1980), 283 - 289.

Y_{555} and All That

J H Conway

About 10 years ago, I conjectured that the Coxeter relations of the Y_{555} diagram

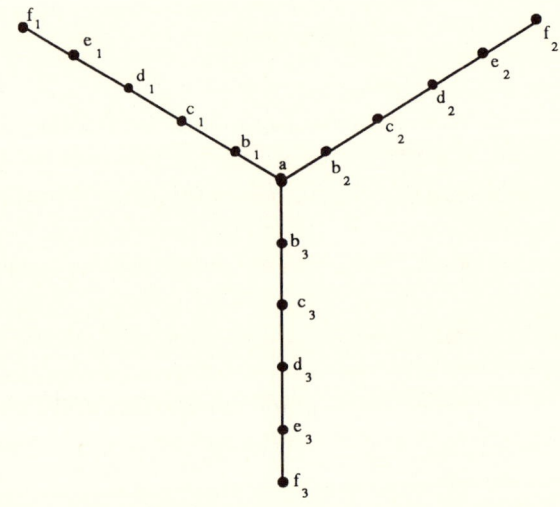

together with the single additional relation

$$(ab_1c_1ab_2c_2ab_3c_3)^{10} = 1,$$

constitute a presentation for the "Bimonster", the wreathed square of the Monster group. The fact that the Bimonster is generated by elements satisfying the above Coxeter relations had already been established by Norton.

Two papers by Ivanov and Norton in this volume present the last stages of the proof of this conjecture. It gives me particular pleasure to congratulate Simon Norton on finishing off the proof, since at no time before we learned (at the Durham meeting) of Ivanov's spectacular result could Simon bring himself to believe that the conjecture might possibly be true!

Roughly, the situation is this. A great deal of work (Conway, Norton, Soicher, Pritchard, Linton) had been done showing the structures of various large subgroups were correctly given by the conjecture, and this culminated with Norton's proof that "the Conway subgroup" of the Bimonster was correct. Then (without reference to this conjecture) Ivanov proved *his* theorem, which asserts in essence that the Monster is presented by a certain very complicated system of generators and relations. At the meeting, Norton deduced these from the Y_{555} system (with the additional relator).

My own lectures at the Durham meeting were rather upset by these developments, since I had intended to talk about the Y_{555} problem, and outline some ideas for its solution. Instead, I briefly described the first few steps in deducing things from the Y_{555} diagram (but from a slightly different axiomatic basis), and then switched to a completely different topic, my "orbifold notation" for surface groups, which is described in another paper in this volume.

In fact my first lecture was essentially taken from the first few pages of a paper by Conway and Pritchard which was written about 6 years ago, but (by my delinquency) has not yet been published. Since this paper provides particularly simple proofs of may consequences of the Y_{555} relations, including some used in Norton's work, it seems appropriate to publish it here. No changes have been made.

Hyperbolic reflections for the Bimonster and $3Fi_{24}$

J. H. Conway and A. D. Pritchard

Contents

0 Introduction   1

1 Establishing the hyperbolic representation   2

2 The monster roots   4

3 The further relations, and some alias groups   8

4 The 26 node theorem   11

5 The identification of Y_{551} 13

6 The reflections of $3Fi_{24}$ 14

7 The element ω 19

8 Presentations for $3Fi_{24}$ 20

0 Introduction

In [1], the "bimonster" (the wreathed square of the Fischer-Griess monster group) was studied in terms of its representation as a quotient of a certain infinite Coxeter group. Here we shall use the representation of this Coxeter group as a hyperbolic reflection group to investigate both the bimonster and its subgroup $3Fi_{24}$.

Throughout the paper, we shall use the notation of [2] for group structures.

In section 1, we give a simple axiomatic definition of a group G, and deduce that G is generated by 16 involutions that satisfy the Coxeter relations of Figure 1. This allows us to represent them in Section 2 by reflections in certain vectors of a hyperbolic space (that is, a space with a Lorentzian metric).

This notation makes it easy to perform calculations with these elements. In Section 2, we shall find some relations that must hold in G, but are not consequences of the Coxeter relations, and will use these to establish many identities in G, which we express in terms of *alias groups*.

Our section 4 contains a short proof of the *26 node theorem* of [1].

The remainder of the paper is devoted to the subgroup Y_{552} of G, which we shall show has the structure $3Fi_{24}$. By way of introduction, Section 5 is used to show that the smaller group Y_{551} has structure $GO_{10}^-(2)$, by completely enumerating its *root vectors*. In Section 6, we describe the root vectors for $3Fi_{24}$, and compute the corresponding alias groups.

A certain element ω is defined in Section 7, and shown to generate a normal subgroup of order 3 in Y_{552}. Modulo this, Section 6 contains verifications that the set of 920808 roots is closed under transformation by the generators of Y_{552}, and this must therefore be a finite group. Section 8 deduces some presentations for $3Fi_{24}$.

The methods of the the paper are almost entirely elementary. Nearly all our results are established by easy verifications that certain vectors transform as they do under particular reflections in others.

1   Establishing the hyperbolic representation

We shall use the fact that the bimonster satisfies the following

Axiom G is a minimal group other than S_{17} that possesses an S_5-subgroup S whose centralizer is a subgroup S_{12} in which a 7 point stabilizer is conjugate to S.

(By *minimal*, we mean merely that no proper subgroup of G has the same property.)

From now on, we shall consider such a group G.

Theorem 1 G is generated by 16 involutions satisfying the Coxeter relations of Figure 1.

Proof. We can suppose that the nodes c_1, d_1, e_1, f_1 in Figure 2 generate the S_5-subgroup S of the Axiom, while

$$f_2, e_2, d_2, c_2, b_2, a, b_3, c_3, d_3, e_3, f_3$$

generate the centralizing S_{12}. Then c_2, d_2, e_2, f_2 will generate a subgroup S_5 conjugate to S, whose centralizer must therefore be another subgroup S_{12}. But c_1, d_1, e_1, f_1 and $a, b_3, c_3, d_3, e_3, f_3$ generate a subgroup $S_5 \times S_7$ of this S_{12}.

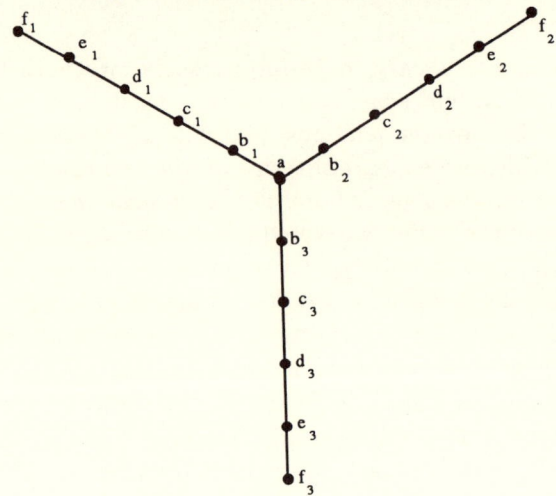

Figure 1: The generators.

But S_{12} has just one conjugacy class of subgroups $S_5 \times S_7$, since the factor S_5 is contained in the centralizer of a Sylow 7-subgroup, and the factor S_7 is determined as the centralizer of the factor S_5.

So we can take

$$f_1 = (0 \quad 1), \ e_1 = (1 \quad 2), \ d_1 = (2 \quad 3), \ c_1 = (3 \quad 4) \text{ and}$$

$$a = (5 \quad 6), \ b_3 = (6 \quad 7), \ c_3 = (7 \quad 8),$$

$$d_3 = (8 \quad 9), \ e_3 = (9 \quad 10), \ f_3 = (10 \quad 11)$$

in this S_{12}. We now adjoin to the diagram a node b_1 representing the transposition $(4 \quad 5)$ of this S_{12}.

We then know that b_3 commutes with all nodes of Figure 2 except a, c_1, and possibly b_2, and that its products with a and c_1 have order 3 (see Figure 3).

Now the nodes

$$f_1, \ e_1, \ d_1, \ c_1, \ b_1, \ a, \ b_2, \ c_2, \ d_2, \ e_2, \ f_2$$

all lie in the S_{12} that centralizes $\langle c_3, d_3, e_3, f_3 \rangle$. Reasoning as before, we see that

$$\langle c_1, d_1, e_1, f_1 \rangle = S_5, \quad \langle a, b_2, c_2, d_2, e_2, f_2 \rangle = S_7$$

and

$$\langle c_2, d_2, e_2, f_2 \rangle = S_5, \quad \langle a, b_1, c_1, d_1, e_1, f_1 \rangle = S_7.$$

In particular, b_1 and b_2 are transpositions of this S_{12}, so that their product must have order 2 or 3. If this order is 2, we have the desired Coxeter group

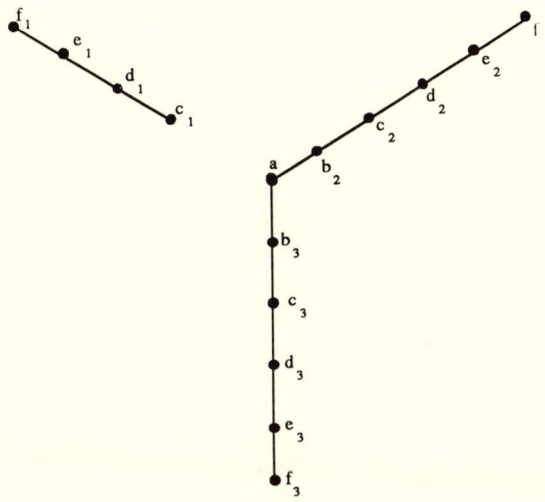

Figure 2

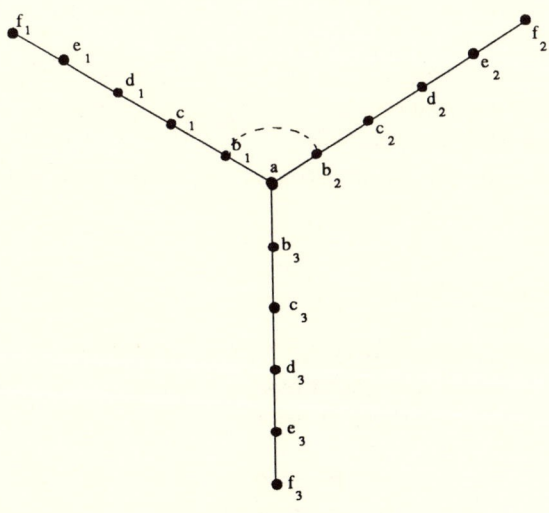

Figure 3

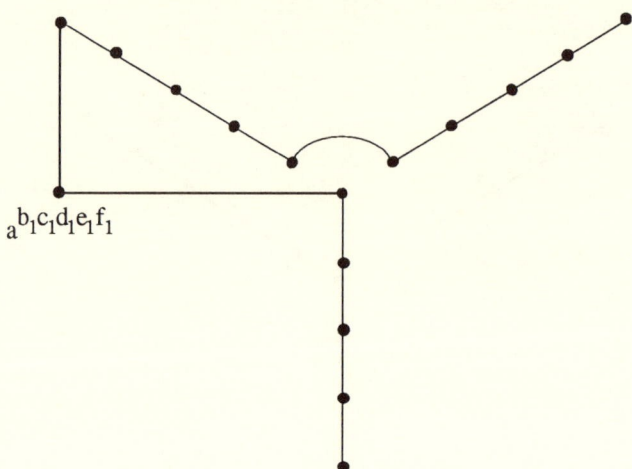

Figure 4: Generators for S_{17}.

of Figure 1; we shall show that if it is 3, then the nodes of Figure 3 generate a group S_{17}.

Supposing that the order is 3, we see that in the S_{12} that centralizes $\langle c_3, d_3, e_3, f_3 \rangle$ we can take

$$f_1 = (A \quad B), e_1 = (B \quad C), d_1 = (C \quad D), c_1 = (D \quad E), b_1 = (E \quad F),$$

$$b_2 = (F \quad G), c_2 = (G \quad H), d_2 = (H \quad I), e_2 = (I \quad J), f_2 = (J \quad K),$$

(since these generate an S_{11}), and then we must have $a = (F \quad X)$, say.

Now we replace a by its conjugate $a^{b_1 c_1 d_1 e_1 f_1}$ as a generator. In the S_{12} just described, this is represented by $(A \quad X)$, while in the S_{12} centralizing $\langle c_2, d_2, e_2, f_2 \rangle$ it is represented by $(0 \quad 5)$ (in our previous numbering). We can now see that all the Coxeter relations of Figure 4 hold; but these define a group S_{17}. The minimality assertion in the Axiom now allows us to deduce Theorem 1.

2   The monster roots

In Figures 5 and 6 we have redrawn Figure 1, and attached certain vectors to its nodes, in two different coordinate systems.

In System 1 (Figure 5), these vectors have the coordinates

$$\begin{array}{llllll} a & b & c & d & e & f \\ g & h & i & j & k & l & t \\ m & n & o & p & q & r \end{array} \tag{1}$$

in a space of 19 coordinates equipped with the quadratic form

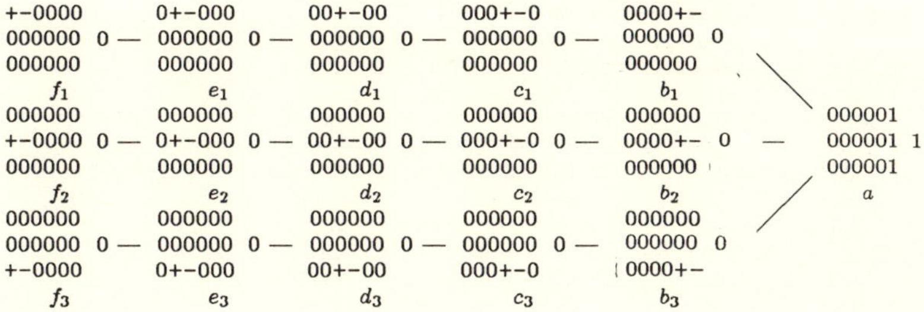

```
+-0000        0+-000        00+-00        000+-0        0000+-
000000  0 —   000000  0 —   000000  0 —   000000  0 —   000000  0              ＼
000000        000000        000000        000000        000000                        000001
  f₁            e₁            d₁            c₁            b₁                   ＼     000001  1
000000        000000        000000        000000        000000                       000001
+-0000  0 —   0+-000  0 —   00+-00  0 —   000+-0  0 —   0000+-  0    —                  a
000000        000000        000000        000000        000000                ／
  f₂            e₂            d₂            c₂            b₂
000000        000000        000000        000000        000000
000000  0 —   000000  0 —   000000  0 —   000000  0 —   000000  0      ／
+-0000        0+-000        00+-00        000+-0        ₍0000+-
  f₃            e₃            d₃            c₃            b₃
```

Figure 5: The fundamental Monster roots in System 1.
$(+ = 1, \; - = -1)$

$$a^2 + b^2 + c^2 + \ldots + p^2 + q^2 + r^2 - t^2. \tag{2}$$

All the vectors also satisfy the relations

$$
\begin{aligned}
a + b + c + d + e + f &= t \\
g + h + i + j + k + l &= t \\
m + n + o + p + q + r &= t
\end{aligned}
\tag{3}
$$

which show that they lie in a subspace of dimension 16. These relations make the coordinate t redundant, and we shall sometimes omit it.

Figure 6 shows the same vectors in an alternative coordinate system (System 2). Here we use just 18 coordinates, in the array

$$
\begin{array}{cccccc}
a & b & c & d & e & f \\
g & h & i & j & k & l & s \\
m & n & o & p & q & *
\end{array}
\tag{4}
$$

with the quadratic form

$$a^2 + b^2 + \ldots + p^2 + q^2 - 5s^2 \tag{5}$$

and relations (which allow us to omit the coordinate s):

$$
\begin{aligned}
a + b + c + d + e + f + g + h + i + j + k + l &= 6s, \\
m + n + o + p + q &= s
\end{aligned}
\tag{6}
$$

The two coordinate systems are linearly related as follows. We have

$$
\begin{array}{cccccc}
a & b & c & d & e & f \\
g & h & i & j & k & l & t \\
m & n & o & p & q & r
\end{array}
=
\begin{array}{cccccc}
a & b & c & d & e & f \\
g' & h' & i' & j' & k' & l' & s \\
m & n & o & p & q & *
\end{array}
$$

```
  +-0000        0+-000        00+-00        000+-0        0000+-
  000000   —    000000   —    000000   —    000000   —    000000
  00000*        00000*        00000*        00000*        00000*          \
    f1            e1            d1            c1            b1
  000000        000000        000000        000000        000000                  00000+
  -+0000   —    0-+000   —    00-+00   —    000-+0   —    0000-+    —             00000-
  00000*        00000*        00000*        00000*        00000*                  00000*
    f2            e2            d2            c2            b2              a
  000000        000000        000000        000000        000000          /
  000000   —    000000   —    000000   —    000000   —    111111
  +-000*        0+-00*        00+-0*        000+-*        00001*
    f3            e3            d3            c3            b3
```

Figure 6: The fundamental roots in System 2.

where
$$g + g' = h + h' = i + i' = j + j' = k + k' = l + l' = s,$$

$$m + n + o + p + q = s = t - r \ .$$

It is easy to check that this does indeed convert the form (2) to the form (5).

The general theory of reflection groups [3] enables us to assert that the group defined by the Coxeter relations of Figure 1 is isomorphic to that generated by the 16 vectors of either Figure 5 or Figure 6, so that G must be a quotient of this group.

We remind the reader that the formula for the reflection in a vector r takes

$$x \quad \text{to} \quad x - \frac{2(x,r)}{(r,r)}r \ .$$

We shall use the term **Monster root** for the vectors obtained from those in Figures 5 and 6 (which are the **fundamental Monster roots**) by repeated reflections in one another. Since every Monster root r has $(r,r) = 2$, the formula for the reflection in one of them simplifies:

$$x \longrightarrow x - (x,r)r \ .$$

The general theory gives us a test for the root vectors in the group generated by the reflections in certain *fundamental* root vectors $r_1, r_2, \ldots, r_n$. In our case, we have $(r_i, r_i) = 2$ and otherwise each (r_i, r_j) is 0 or -1. Supposing this, we select a *Weyl vector* w with $(w, r_i) < 0$ for each r_i. Then the test is

0. if $(r, w) > 0$, replace r by $-r$

1. if $r \in \{\pm r_1, \pm r_2, \ldots, \pm r_n\}$ then r *is* a root vector

2. otherwise replace r by its reflection in any r_i for which (r, r_i) and (r, w) have opposite signs, and repeat.

(If at any stage all the (r, r_i) have the *same* sign as (r, w), then r is *not* a root vector.)

This is easily proved. The reflecting hyperplanes of a discrete reflection group in hyperbolic space divide that space into copies of what is called the fundamental region, which may be taken to be the intersection of certain half-spaces, say $(r_i, x) < 0$. The Weyl vector is any point in this region. The conditions on the inner products of the r_i now ensure that $| (w, r) |$ decreases.

Now the fundamental Monster roots other than a generate a subgroup $S_6 \times S_6 \times S_6$ that appears as coordinate permutations in System 1, while those other than b_3 generate a group $S_{12} \times S_5$ of coordinate permutations in System 2. Using these observations, we can speed up the test for Monster roots.

It can be deduced from this algorithm that in System 1, the Monster roots with $t = 0$ or $t = 1$ are precisely those of the forms

$$0^4 +- \mid 0^6 \mid 0^6 \quad \text{or} \quad 0^5 1 \mid 0^5 1 \mid 0^5 1$$

where the sets separated by $\mid$ are the coordinates, in some order, of the three rows, in some order. To test a vector with $t > 1$, replace it by its reflection in any vector having $t = 1$ with which it has positive inner product, and repeat.

In System 2, the Monster roots with $s = 0$ or $s = 1$ are precisely those of the forms

$$0^{10} +- \mid 0^5 \quad \text{or} \quad 0^{12} \mid 0^3 +-, \quad \text{or} \quad 0^6 1^6 \mid 0^4 1,$$

where $\mid$ separates the coordinates in the first two rows from those of the third. To test a vector with $s > 1$, replace it by its reflection in any root vector having $s = 1$ with which it has positive inner product, and repeat.

(The roots vectors with s or t negative are just the negatives of those for which it is positive.)

For example, in System 1 we verify that

$$
\begin{array}{l}
000024 \\
000222\ 6 \\
111111
\end{array}
$$

is a Monster root by the following sequence of reflections:

00002<u>4</u>	00002<u>3</u>	00002<u>2</u>	00001<u>2</u>	0000<u>11</u>	000001
00022<u>2</u> <u>6</u>	00012<u>2</u> <u>5</u>	00011<u>2</u> <u>4</u>	0001<u>11</u> <u>3</u>	0000<u>11</u> 2	000001 1
<u>1</u>1111<u>1</u>	0<u>1</u>1111	00<u>1</u>111	000<u>1</u>11	0000<u>1</u>1	000001

In each case, we have underlined the coordinates that belong to the support of the reflecting vector to be used next. The process terminates because the coordinates decrease, while remaining integral.

3   The further relations, and some alias groups

Here we shall deduce from our Axiom that G satisfies certain relations that do not follow from the Coxeter relations, and we will explore some of their consequences.

Theorem 2 *The reflections in the vectors*

$$v = \begin{matrix} 000024 \\ 000222 \\ 111111 \end{matrix} 6 \ and \ b_1 = \begin{matrix} 0000+- \\ 000000 \\ 000000 \end{matrix} 0$$

(in System 1) represent the same element of G.

Proof. The element represented by v certainly centralizes the S_5 generated by f_3, e_3, d_3, c_3 in Figure 5, and so must belong to the S_{12} generated by $f_1, e_1, d_1, c_1, b_1, a, b_2, c_2, d_2, e_2, f_2$. But it centralizes the subgroups $\langle f_1, e_1, d_1 \rangle$, $\langle b_2, c_2 \rangle$, $\langle e_2, f_2 \rangle$ of types S_4, S_3, S_3, and the only involution with this property in S_{12} is b_1. □

In future, we shall use the symbol $\doteq$ for this equivalence relation between vectors.

Theorem 3 *We have*
$$\begin{matrix} 000012 \\ 000111 \\ 000111 \end{matrix} 3 \doteq \begin{matrix} 000021 \\ 000111 \\ 111000 \end{matrix} 3 \doteq \begin{matrix} 000012 \\ 111000 \\ 111000 \end{matrix} 3.$$

Proof. We obtain the first of these by transforming the relation of Theorem 2 as follows:

$$\begin{matrix} 000024 \\ 000222 \\ 111111 \end{matrix} 6 \rightarrow \begin{matrix} 000023 \\ 000122 \\ 011111 \end{matrix} 5 \rightarrow \begin{matrix} 000022 \\ 000112 \\ 001111 \end{matrix} 4 \rightarrow \begin{matrix} 000012 \\ 000111 \\ 000111 \end{matrix} 3$$

$$\begin{matrix} 0000+- \\ 000000 \\ 000000 \end{matrix} 0 \rightarrow \begin{matrix} 000010 \\ 000100 \\ 100000 \end{matrix} 1 \rightarrow \begin{matrix} 000011 \\ 000110 \\ 110000 \end{matrix} 2 \rightarrow \begin{matrix} 000021 \\ 000111 \\ 111000 \end{matrix} 3 .$$

The second is similar. □

Theorem 4 *We have*
$$\begin{matrix} 000122 \\ 000122 \\ 000122 \end{matrix} 5 \doteq \begin{matrix} 000211 \\ 000211 \\ 000211 \end{matrix} 4 .$$

Proof. We have

$$\begin{matrix} 000122 \\ 000122 \\ 000122 \end{matrix} 5 = a^b, \ \begin{matrix} 000211 \\ 000211 \\ 000211 \end{matrix} 4 = b^a, \ \text{where} \ a = \begin{matrix} 000100 \\ 000100 \\ 000100 \end{matrix} 1, \ b = \begin{matrix} 000011 \\ 000011 \\ 000011 \end{matrix} 2 .$$

Now on transforming by $\begin{smallmatrix} 000010 \\ 000100 \ 1 \\ 000100 \end{smallmatrix}$, a and b become (using Theorem 3)

$$000+-0 \qquad\qquad 000021 \qquad 000012$$
$$000000 \ \ 0 = A \text{ and } 000111 \ 3 \doteq 000111 \ 3 = B,$$
$$000000 \qquad\qquad 000111 \qquad 111000$$

and we have $A^B = B^A = A + B$ since $(A, B) = -1$. $\qquad\qquad\square$

Many of our identities assert that various vectors are unaffected (up to the relation $\doteq$) by certain alterations of their coordinates. For example we shall show that we have (omitting the redundant coordinate s)

$$000111 \qquad 111000$$
$$000111 \ \doteq \ 111000$$
$$00001* \qquad 00001*$$

(in System 2) which in view of our $S_{12} \times S_5$ of coordinate permutations implies that the reflection in any vector of shape $0^6 1^6 \mid 0^4 1$ represents the same element of G as that in the corresponding vector $1^6 0^6 \mid 0^4 1$ obtained by interchanging digits "0" and "1" before the $\mid$, but fixing those after it.

We shall say that $1^6 0^6 \mid 0^4 1$ is an **alias** for $0^6 1^6 \mid 0^4 1$, and that the **digit permutation** $(0 \quad 1) \mid \sim$ belongs to the **alias group** for this vector. The digit-permutation $\pi \mid \pi'$ is the transformation obtained by applying π to digits before the $\mid$, and π' to those after it, and "$\sim$" indicates the identity permutation. (Note that digit-permutations apply to the *values* of the coordinates, rather than to their *positions*.)

Theorem 5 *The alias groups for vectors of the forms*

$$0^6 1^6 \mid 0^4 1 \qquad 0^4 1^4 2^4 \mid 0^3 1^2 \qquad 0^3 1^3 2^3 3^3 \mid 0^3 1 2$$

contain the respective digit-permutations

$$(0 \quad 1) \mid \sim \qquad \begin{matrix} (0 \quad 1) \mid \sim \\ (1 \quad 2) \mid \sim \end{matrix} \qquad \begin{matrix} (0 \quad 1) \mid (1 \quad 2) \\ (1 \quad 2) \mid (1 \quad 2) \\ (2 \quad 3) \mid (1 \quad 2) \end{matrix} \ .$$

Proof. Table 1 gives the required calculations, which consist largely of conversions between Systems 1 and 2. We have allowed ourselves to use Theorems 2–4 and earlier digit permutations without explicit mention. $\qquad\square$

$$\begin{matrix} 000111 \\ 000111 \\ 00001* \end{matrix} = \begin{matrix} 000111 \\ 111000 \\ 000012 \end{matrix} \; 3 \; \doteq \; \begin{matrix} 111000 \\ 000111 \\ 000012 \end{matrix} \; 3 \; = \; \begin{matrix} 111000 \\ 111000 \\ 00001* \end{matrix} \quad \text{gives} \; (0 \;\; 1)\,|\!\sim \quad \text{for} \; 0^6 1^6 \mid 0^4 1$$

$$\begin{matrix} 000111 \\ 222210 \\ 00011* \end{matrix} = \begin{matrix} 000111 \\ 000012 \\ 000111 \end{matrix} \; 3 \; \doteq \; \begin{matrix} 111000 \\ 000021 \\ 000111 \end{matrix} \; 3 \; = \; \begin{matrix} 111000 \\ 222201 \\ 00011* \end{matrix} \quad (0 \;\; 1)\,|\!\sim$$

$$\begin{matrix} 000012 \\ 222111 \\ 00011* \end{matrix} = \begin{matrix} 000012 \\ 000111 \\ 000111 \end{matrix} \; 3 \; \doteq \; \begin{matrix} 000021 \\ 111000 \\ 000111 \end{matrix} \; 3 \; = \; \begin{matrix} 000021 \\ 111222 \\ 00011* \end{matrix} \quad (1 \;\; 2)\,|\!\sim$$

$$\text{for} \; 0^4 1^4 2^4 \mid 0^3 1^2$$

$$\begin{matrix} 000111 \\ 333222 \\ 00012* \end{matrix} = \begin{matrix} 000111 \\ 000111 \\ 000120 \end{matrix} \; 3 \; \doteq \; \begin{matrix} 111000 \\ 000111 \\ 000210 \end{matrix} \; 3 \; = \; \begin{matrix} 111000 \\ 333222 \\ 00021* \end{matrix} \quad (0 \;\; 1)\,|\,(1 \;\; 2)$$

$$\begin{matrix} 000112 \\ 333221 \\ 00012* \end{matrix} = \begin{matrix} 000112 \\ 000112 \\ 000121 \end{matrix} \; 4 \; \doteq \; \begin{matrix} 000221 \\ 000221 \\ 000212 \end{matrix} \; 5 \; = \; \begin{matrix} 000221 \\ 333112 \\ 00021* \end{matrix} \quad (1 \;\; 2)\,|\,(1 \;\; 2)$$

$$\begin{matrix} 000111 \\ 333222 \\ 00012* \end{matrix} = \begin{matrix} 000111 \\ 000111 \\ 000120 \end{matrix} \; 3 \; \doteq \; \begin{matrix} 000111 \\ 111000 \\ 000210 \end{matrix} \; 3 \; = \; \begin{matrix} 000111 \\ 222333 \\ 00021* \end{matrix} \quad (2 \;\; 3)\,|\,(1 \;\; 2)$$

$$\text{for} \; 0^3 1^3 2^3 3^3 \mid 0^3 1 2$$

Table 1. Calculations for some alias groups.

4   The 26 node theorem

In [1] it was shown that the 16 generators of Figure 1 are included in a set of 26 involutions that satisfy Coxeter relations corresponding to the incidence graph of a projective plane of order 3. We provide a simple proof here.

Theorem 6 *The set of Monster roots displayed in Tables 2 and 3 is invariant under a subgroup of G which induces upon them all the symmetries of the incidence graph of the projective plane of order 3.*

Proof. Table 2 shows these vectors in System 1, and reveals that they are invariant under the group S_3 that bodily permutes the three rows of coordinates. (Recall that we may omit the last coordinate.) Table 3 translates them into System 2, where they can be seen to be invariant under the element π_{12} of G of order 12 that performs the coordinate-permutation

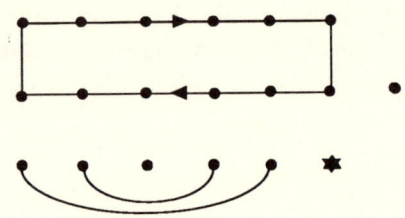

which acts as

$$(f_1 e_1 d_1 c_1 b_1 a b_2 c_2 d_2 e_2 f_2 a_3)(b_3 a_2 z_1 g_3 z_2 a_1)(z_3 g_1 f g_2)(c_3 f_3)(d_3 e_3)$$

(the square of this is the permutation π of [1]).

Since the above S_3 acts by permuting the subscripts on our names for the 26 nodes, there is an element π_{ij} of G that acts as

$$(f_i e_i d_i c_i b_i a b_j c_j d_j e_j f_j a_k)(b_k a_j z_i g_k z_j a_i)(z_k g_i f g_j)(c_k f_k)(d_k e_k)$$

whenever $\{i, j, k\} = \{1, 2, 3\}$. It is easy to check that these permutations generate a group with the desired action on the 26 nodes. □

We remark that it can in fact be shown that the π_{ij} generate a subgroup of G isomorphic to $P\Gamma L_3(3)$ (in other words, the action has trivial kernel). This was not shown in [1].

The form of the 26 node theorem in [1] is rather different. However, all the relations used in either [1] or the present paper are derivable from relations inside subgroups which are known in each paper to have the structure $GO_{10}^-(2)$, and the equivalence of the two forms follows from this.

g_i	f_i	e_i	d_i	c_i	b_i	a_i	z_i	i
200001	+-0000	0+-000	00+-00	000+-0	0000+-	000001	110000	
000111	000000	000000	000000	000000	000000	100000	000011	1
000111	000000	000000	000000	000000	000000	100000	000011	
000111	000000	000000	000000	000000	000000	100000	000011	
200001	+-0000	0+-000	00+-00	000+-0	0000+-	000001	110000	2
000111	000000	000000	000000	000000	000000	100000	000011	
000111	000000	000000	000000	000000	000000	100000	000011	
111000	000000	000000	000000	000000	000000	100000	000011	3
100002	+-0000	0+-000	00+-00	000+-0	0000+-	000001	110000	
	110000					000001		
	110000					000001		
	110000					000001		
	f					a		

Table 2. The 26 nodes, in System 1 coordinates.

g_i	f_i	e_i	d_i	c_i	b_i	a_i	z_i	i
200001	+-0000	0+-000	00+-00	000+-0	0000+-	000001	110000	
222111	000000	000000	000000	000000	000000	011111	111100	1
00011*	00000*	00000*	00000*	00000*	00000*	10000*	00001*	
000111	000000	000000	000000	000000	000000	100000	000011	
022221	-+0000	0-+000	00-+00	000-+0	0000-+	111110	001111	2
00011*	00000*	00000*	00000*	00000*	00000*	10000*	00001*	
000111	000000	000000	000000	000000	000000	+00000	000011	
000111	000000	000000	000000	000000	000000	-00000	222211	3
10000*	+-000*	0+-00*	00+-0*	000+-*	00001*	00000*	11000*	
	110000					00000+		
	112222					00000-		
	11000*					00000*		
	f					a		

Table 3. The 26 nodes, in System 2 coordinates.

5   The identification of Y_{551}

We define Y_{pqr} to be the subgroup of G generated by the node a together with

the first p of b_1, c_1, d_1, e_1, f_1
the first q of b_2, c_2, d_2, e_2, f_2
the first r of b_3, c_3, d_3, e_3, f_3.

The main aim of this paper is to identify Y_{552} with the group $3Fi_{24}$, by constructing it explicitly as a group of permutations of 920808 vectors, modulo the relation $\doteq$. In this section we shall identify Y_{551}.

We shall abbreviate $abcdefghijkl \mid 000pq$ to $abcdefghijkl \mid pq$.

As in [1], we shall use the term **root element** of a group for a member of the conjugacy class that contains the generators by which that group was defined.

Theorem 7 *The group Y_{551} has 528 root elements, namely the images of $0^{10}+- \mid 00$ and $0^6 1^6 \mid 01$ under the S_{12} that permutes the first 12 coordinates. It is isomorphic to $GO_{10}^-(2)$.*

Proof. We note that since $0^{10}+- \mid 00 \doteq 0^{10}-+ \mid 00$, it has at most $\binom{12}{2} = 66$ images under this S_{12} (up to $\doteq$), while $0^6 1^6 \mid 01 \doteq 1^6 0^6 \mid 01$ has at most $\frac{1}{2}\binom{12}{6} = 462$, so the indicated set has indeed at most 528 elements. It has been obtained by closing the set of generators Y_{551} under S_{12}, and so to prove that it is the set of roots it will suffice to show that it is closed under reflection in the particular vector $v = 000000111111 \mid 01$. Now reflection in v

fixes itself,
interchanges $000001011111 \mid 01$ with $00000+-00000 \mid 00$,
fixes $000011001111 \mid 01$ and $0000000000+- \mid 00$, and
reflects $000111000111 \mid 01$ into $000111111222 \mid 02$.

However, the last printed vector is an alias for $000111111000 \mid 01$, which we deduce from Theorem 3 as follows

$$\begin{array}{cccc} 000111 & 000111 & 000111 & 000111 \\ 222111 = 000111 & \doteq & 111000 = 000111 \\ 00002* & 000021 & 000012 & 00001* \end{array}.$$

(In view of the relation $0^6 1^6 \mid 01 \doteq 1^6 0^6 \mid 01$, the vectors we have considered are sufficient.)

The transformation rules we have established for these vectors $abcdefghijkl \mid pq$ are quite simple. Ignoring p and q, and reading $a, b, c, d, e, f, g, h, i, j, k, l$ modulo 2, the reflection rule becomes

$$r : v \longrightarrow v + (v, r)r$$

the root vectors being those whose weights are congruent to 2, modulo 4, considered modulo the all 1s vector 111111111111. The **weight** of a vector is the number of non-zero entries $a, b, \ldots k, l$ (modulo 2). It is known that these operations generate the group $GO_{10}^-(2)$, which contains the simple group $O_{10}^-(2)$ to index 2.

The group consisting of the even products of our reflections (modulo $\doteq$) is therefore a central extension of $O_{10}^-(2)$, and is easily seen to be perfect. Since the multiplier of $O_{10}^-(2)$ is trivial, this group must be $O_{10}^-(2)$ itself, and Y_{551} is therefore $GO_{10}^-(2)$. □

6   The reflections of $3Fi_{24}$

In the remainder of this paper, we shall show that the Monster roots of form $abcdefghijkl \mid pq$, meaning $abcdefghijkl \mid 000pq$, generate a subgroup of G having the structure $3Fi_{24}$.

Theorem 8 *The group Y_{552} has 920808 root elements, namely the vectors $\omega^i v$, where ω is the element of order 3 studied in Section 7, and v is an image of one of the vectors in Table 4 under the group $S_{12} \times S_2$ that separately permutes the 12 coordinates before the $\mid$ and the 2 after it. The group Y_{552} has $\langle \omega \rangle$ as a normal subgroup of order 3, and $Y_{552}/\langle \omega \rangle$ is isomorphic to Fischer's 3-transposition group Fi_{24}.*

Proof. Some of the alias groups for the vectors in Table 4 have already been established in Theorem 5. Table 5 contains calculations which establish the others, and also that for the vector

$$0 \quad 1 \quad 2 \quad 12 \quad 4 \quad 14 \quad 6 \quad 16 \quad 8 \quad 18 \quad 19 \quad 20 \mid 10 \quad 10$$

which will be needed in Section 7. The generators for the groups found in Table 5 are not the canonical ones of Table 4 – we regard calculations within these small permutation groups as trivial, and leave them as exercises for the reader.

The argument is like that of Theorem 7, and the reader will find it easier to follow if we work modulo $\langle \omega \rangle$ for the moment, leaving the justification of this procedure to Section 7. Modulo $\langle \omega \rangle$, there are (at most)

$$66 + 1 + 924 + 5775 + 30800 + 124740 + 5040 + 20790 + 118800 = 306936$$

images under $S_{12} \times S_2$ of the vectors in Table 4. The only generator of Y_{552} not in $S_{12} \times S_2$ is $b_3 = 000000111111 \mid 01$, and Table 6 shows that the set is closed under transformation by this vector. Most of the entries in Table 6 are direct reflections. Table 7 contains calculations which handle those of the exceptions that don't involve ω. All calculations involving ω are dealt with in Section 7.

vector v	number	$S(v)$	other identities
00000000001$\bar{1}$ | 00	66	C_2	none
000000000000 | 1$\bar{1}$	1	M_1	$-v \doteq v$
000000111111 | 10	924	M_2	none
000011112222 | 11	5775	M_3	none
000111222333 | 12	30800	M_4	000222111333 | 21 $\doteq v$
001122334455 | 14	124740	M_6	none
0123456789Xx | 1X	5040	M_{12}	0123456789Xx | X1 $= \omega v$
001224466788 | 44	20790	A_6	none
001234567899 | 45	118800	Q_{10}	002135468799 | 54 $\doteq v$ 901234567890 | 45 $\doteq \omega v$

Table 4. The nine types of root vectors for Y_{552}.

($\bar{1}$ means -1, X means "ten", x means "eleven")

Notes. The nine types are the orbits under $S_{12} \times S_2$.

   $S(v)$ is the part of the alias group that fixes the last 2 coordinates

   C_2 is generated by $(1 \ \ \bar{1})$

   M_n is generated by $\begin{cases} R_n: t \to n - 1 - t \\ S_n: t \to \min{(2t, 2n - 1 - 2t)} \end{cases}$

   Q_{10} is generated by $(1234)(5678)$ and $(543)(678)(90)$

   A_6 consists of all even permutations of 0,2,4,6,8,odd,

where "odd" refers collectively to 1 and 7, which may be interchanged freely. M_{12} is Mathieu's group, and we have

$$M_1 = S_1, \ M_2 = S_2, \ M_3 = S_3, \ M_4 = A_4, \ M_6 = PGL_2(5),$$

where M_6 acts on 0,1,2,3,4,5 as $PGL_2(5)$ does on $0,\infty,1,2,3,4$. Also, $Q_{10} = S_2 \times PSL_2(7)$, where the factor S_2 interchanges 0 and 9, and Q_{10} permutes 1,2,3,4,5,6,7,8 as $PSL_2(7)$ permutes $0,1,3,4,5,2,6,\infty$.

   The *number* column contains the number of images under $S_{12} \times S_2$, counted modulo $\langle \omega \rangle$, and can be computed from the two subsequent columns.

$\begin{pmatrix} 0 & 1 \\ 1 & 2 \end{pmatrix}\begin{pmatrix} 3 & 4 \\ 4 & 5 \end{pmatrix}\bigg|-$

$\left(\begin{array}{l} 001122\overline{334455} \\ 001120312233 \\ 110031\underline{203322} \\ 110033225544 \end{array}\ \middle|\ \begin{array}{l} 1\overline{4} \\ 12 \\ 12 \\ 14 \end{array}\right)$

$(0\ \ 1)(2\ \ 3)(4\ \ 5)\,|-$

$\begin{pmatrix} 6 & 8 \\ 0 & 2 \end{pmatrix}\begin{pmatrix} 4 & \text{odd} \\ 4 & \text{odd} \end{pmatrix}\bigg|-$

$\left(\begin{array}{l} 100\underline{22}4\overline{466887} \\ 100202\underline{242443} \\ 100202131332 \\ 200101\underline{232331} \\ 200101121220 \\ 011\underline{212202}001 \\ 012323312001 \\ 021\underline{313321}002 \\ \underline{022414431}002 \\ 244668871002 \end{array}\ \middle|\ \begin{array}{l} 4\overline{4} \\ 22 \\ 21 \\ 12 \\ 11 \\ 11 \\ 12 \\ 21 \\ 22 \\ 44 \end{array}\right)$

$(0\ \ 4\ \ 8)(2\ \ 6\ \ \text{odd})\,|-$

$\begin{pmatrix} 0 & 1 \\ 1 & 2 \end{pmatrix}\begin{pmatrix} 2 & 3 \\ 4 & 3 \end{pmatrix}\begin{pmatrix} 4 & 5 \\ 5 \end{pmatrix}\begin{pmatrix} 6 & 7 \\ 7 & 8 \end{pmatrix}\begin{pmatrix} 8 & 9 \\ X & 9 \end{pmatrix}\begin{pmatrix} X & x \\ x \end{pmatrix}\bigg|-$

$\left(\begin{array}{l} 012345\overline{6789Xx} \\ 012340623456 \\ 012310320123 \\ 120321\underline{301203} \\ 120341\underline{503425} \\ 120347569X8x \end{array}\ \middle|\ \begin{array}{l} 1\overline{X} \\ 15 \\ 12 \\ 12 \\ 14 \\ 1X \end{array}\right)$

$(0\ \ 1\ \ 2)(7\ \ 6\ \ 5)(8\ \ 9\ \ X)\,|-$

$\begin{pmatrix} 0 & 9 \\ 1 & 2 \end{pmatrix}\begin{pmatrix} 1 & 8 \\ 3 \end{pmatrix}\begin{pmatrix} 2 & 7 \\ 4 \end{pmatrix}\begin{pmatrix} 3 & 6 \\ 5 \end{pmatrix}\begin{pmatrix} 4 & 5 \\ 6 \end{pmatrix}\begin{pmatrix} 7 & 8 \end{pmatrix}\bigg|-$

$\left(\begin{array}{l} 001234\overline{567899} \\ 001231\underline{534566} \\ 001231201233 \\ 002132\underline{102133} \\ 002132\underline{435466} \\ 002135468799 \end{array}\ \middle|\ \begin{array}{l} 4\underline{5} \\ 42 \\ 12 \\ 21 \\ 51 \\ 54 \end{array}\right)$

$(1\ \ 2)(4\ \ 5)(7\ \ 8)\,|\,(4\ \ 5)$

$\begin{pmatrix} 0 & 1 \\ 0 & 4 \end{pmatrix}\begin{pmatrix} 2 & 6 \\ 2 & 6 \end{pmatrix}\begin{pmatrix} 4 & 8 \\ 8 \end{pmatrix}\begin{pmatrix} 19 & 20 \\ 20 & 16 \end{pmatrix}\begin{pmatrix} & \\ 18 & 14 & 12 \end{pmatrix}\bigg|-$

0	1	2	$\overline{12}$	4	$\overline{14}$	6	$\overline{16}$	8	$\overline{18}$	$\overline{19}$	$\overline{20}$	10	$\overline{10}$
0	1	2	6	4	8	0	4	2	6	7	8	4	4
4	2	1	$\underline{6}$	0	$\underline{8}$	$\underline{4}$	$\underline{0}$	$\underline{7}$	$\underline{6}$	2	$\underline{8}$	4	4
4	2	1	18	0	20	16	6	19	12	8	14	10	10

$(0\ \ 4)(1\ \ 2)(6\ \ 16)(8\ \ 19)(12\ \ 18)(14\ \ 20)\,|-$

Table 5. The harder alias groups.

Notes: The leading vector in any of the five portions may be obtained from a simpler vector by reflection in the root vector of shape $0^6 1^6\ |\ 01$ or $0^4 1^4 2^4\ |\ 11$ indicated by the overlinings. (The number of lines in any place indicates the appropriate coordinate of the reflecting vector). Two members of its alias group arise from the symmetries of this simpler vector that survive this reflection, and are given before the leading vector.

The last vector in any portion is proved as before to be an alias of the leading vector by the cancelling sequence of reflections indicated by the interlinings. Parentheses link aliases that are already known.

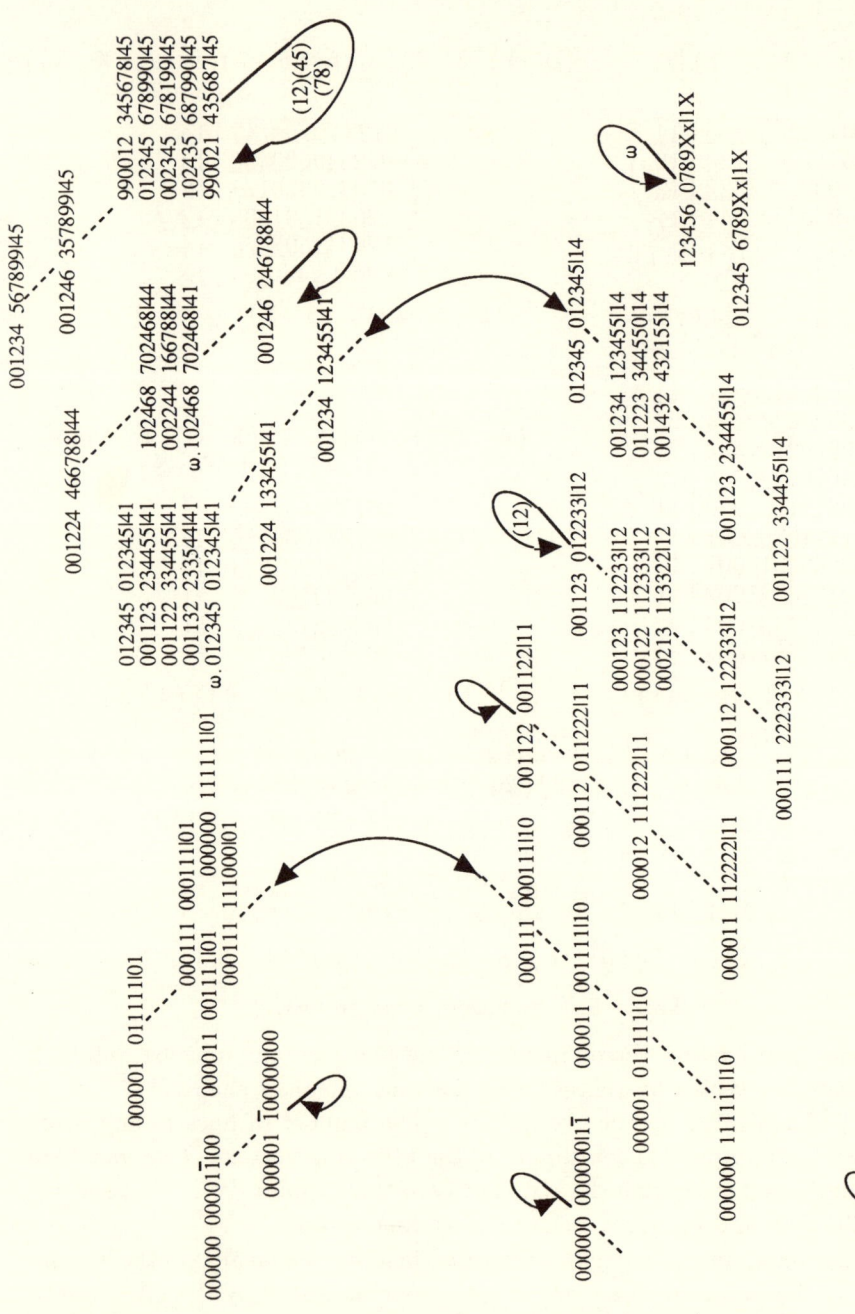

Table 6. Action of 000000 111111 | 01 and 000000 000000 | 1$\bar{1}$ on the transpositions of $3Fi_{24}$.

An arrow such as $\overset{x}{\curvearrowright}$ means $\overset{s}{v} = \overset{x}{v}$.

$\bar{1}$ means -1, X means "ten", x means "eleven"

Notes. Here 000000 111111 | 01 acts by reversing the order within each column in the upper and lower parts of the diagram, while the arrows sketch the action of 000000 000000 | 1$\bar{1}$ = s. Each case is an S_{12} image of one of the cases displayed.

It is easy to see that the product of any of these 306936 elements with 000000000000 | 1$\bar{1}$ has order at most 3 (still working modulo $\langle\omega\rangle$). It follows that $Y_{552}/\langle\omega\rangle$ is a 3-transposition group, and Fi_{24} is the only possible candidate on Fischer's list. (We can use the fact that Y_{552} has a perfect subgroup of index 2.).   □

```
    0 0 0 1 2 3 1 1 2 2 3 3 | 1 2
    0 0 0 1 2 3 2 2 3 3 4 4 | 1 3
  ( 0 0 0 1 2 1 2 0 1 1 2 2 | 1 1 )
  ( 0 0 0 2 1 2 1 0 2 2 1 1 | 1 1 )
    0 0 0 2 1 3 1 1 3 3 2 2 | 1 2

    0 0 1 2 3 4 1 2 3 4 5 5 | 1 4
    0 0 1 2 3 4 2 3 4 5 6 6 | 1 5
  ( 0 0 1 2 3 1 2 0 1 2 3 3 | 1 2 )
  ( 0 0 3 1 2 3 1 0 3 1 2 2 | 1 2 )
  ( 0 0 3 1 2 5 1 2 5 3 4 4 | 1 4 )
  ( 0 0 1 4 3 2 4 3 2 1 5 5 | 1 4 )

    0 0 1 1 2 3 2 3 4 4 5 5 | 4 1
    0 0 1 1 2 3 3 4 5 5 6 6 | 4 2
  ( 0 0 1 1 2 3 0 1 2 2 3 3 | 1 2 )
  ( 0 0 1 1 3 2 0 1 3 3 2 2 | 2 1 )
    0 0 1 1 3 2 2 3 5 5 4 4 | 4 1

    0 1 2 3 4 5 6 7 8 9 9 0 | 4 5
    0 1 2 3 4 5 7 8 9 X X 1 | 4 6
  ( 0 1 2 3 4 0 2 3 4 5 5 1 | 4 1 )
  ( 1 0 2 4 3 1 2 4 3 5 5 0 | 4 1 )
    1 0 2 4 3 5 6 8 7 9 9 0 | 4 5

  ( 9 9 0 0 1 2 3 4 5 6 7 8 | 4 5 )
  ( 5 5 0 0 1 2 3 4 1 2 3 4 | 4 1 )
  ( 0 0 1 1 2 3 5 4 2 3 5 4 | 4 1 )
    0 0 1 1 2 3 6 5 3 4 6 5 | 4 2
    0 0 1 1 2 3 3 2 0 1 3 2 | 1 2
    0 0 1 1 3 2 2 3 0 1 2 3 | 2 1
  ( 0 0 1 1 3 2 4 5 2 3 4 5 | 4 1 )
  ( 5 5 0 0 2 1 4 3 1 2 4 3 | 4 1 )
    9 9 0 0 2 1 4 3 5 6 8 7 | 4 5
```

Table 7. The hard cases of Table 6 that do not involve ω.
Notes: [The conventions are as in Table 5. The product of the indicated transformations is the desired reflection in 000000111111 | 01.]

7 The element ω

We define $\omega = yx$, so that $\omega^v = zx$, where

$$
\begin{array}{lcccccccccccccccc}
v = & 0 & 0 & 0 & 1 & 0 & 1 & 0 & 1 & 0 & 1 & 1 & 1 & | & 0 & 1 \\
x = & 0 & 1 & 2 & 3 & 4 & 5 & 6 & 7 & 8 & 9 & 10 & 11 & | & 1 & 10 \\
y = & 0 & 1 & 2 & 3 & 4 & 5 & 6 & 7 & 8 & 9 & 10 & 11 & | & 10 & 1 \\
z = & 0 & 1 & 2 & 12 & 4 & 14 & 6 & 16 & 8 & 18 & 19 & 20 & | & 10 & 10 & = y^v.
\end{array}
$$

Then the alias group of z, which was found in Table 5, is a group M_{12} which is identified with the one fixing x when we subtract 9 from the coordinates 12, 14, 16, 18, 19, 20 of z. It follows that there is a group $M \cong M_{12}$ of permutations of the 12 coordinate places before the | that replace x, y, and z by aliases of themselves, and so fix the corresponding elements of G. So M fixes both ω and ω^v. (Note that the elements of M are permutations of the coordinate positions rather than of their values, which are what the alias groups permute.)

We now know that ω and ω^v are centralized by M and M^v. The calculations

$$
\begin{array}{ll}
0123456789\text{X}\text{x} \mid 1\text{X} & 000000\underline{000000} \mid +\underline{-} \\
012\underline{345}012\underline{345} \mid 14 & 00\underline{0000}111\underline{111} \mid 1\underline{0} \\
01\underline{2012012012} \mid 1\underline{1} & 00\underline{0}222111\underline{333} \mid 12 \\
0\underline{11001011001} \mid 1\underline{0} & 00\underline{2244}11\underline{3355} \mid 14 \\
000000000000 \mid +- & 06824\text{X}17935\text{x} \mid 1\text{X} \\
\quad\longleftarrow \pi \longrightarrow & \quad\longleftarrow \pi \longrightarrow \\
000000000000 \mid +- & 0123456789\text{X}\text{x} \mid 1\text{X}
\end{array}
$$

show that x and $s = 0^{12} \mid +-$ are interchanged by p, the product of the reflections in the four vectors

$$
\begin{array}{l}
000000111111 \mid 01 \\
000111000111 \mid 01 \\
001011001011 \mid 01 \\
011001011001 \mid 01
\end{array}
$$

with the permutation π of the coordinate places that restores $06824\text{X}17935\text{x} \mid 1\text{X}$ to its rightful order. Since $y = x^s$, we have $\omega = (sx)^2$, so that ω is inverted by p. The further calculation

$$
\begin{array}{l}
000101\underline{010111} \mid 0\underline{1} \\
000\underline{101}010\underline{111} \mid 01 \\
000\underline{0}-00+\underline{0000} \mid 00 \\
0\underline{01001011011} \mid 0\underline{1} \\
0-00000000+0 \mid 00 \\
\quad\longleftarrow \pi \longrightarrow \\
00000+-00000 \mid 00
\end{array}
$$

shows that p transforms v to the coordinate permutation t that interchanges the middle two of the twelve coordinates before the |. Now inside $Y_{551} = GO_{10}^-(2)$, the stabiliser of s is the obvious group $S_{12} = S$ of coordinate permutations, and so that of x is S^p, a conjugate S_{12}, and we have $S \cap S^p = M$. It follows that $M^p = M$ and $(M^v)^p = M^t$, and that both these groups centralize ω. But since M and M^t are (distinct) maximal subgroups of the subgroup $A = A_{12}$ of S, we see that $A = \langle M, M^t \rangle$ also centralizes ω. Again, A and A^p are distinct maximal subgroups of the simple subgroup $O_{10}^-(2)$ of index 2 in Y_{551}, and so, since p inverts ω, the group $O_{10}^-(2) = \langle A, A^p \rangle$ fixes ω. Since p is the product of an *odd* number of reflections, we see that $Y_{551} = \langle O_{10}^-(2), p \rangle$ normalises $\langle \omega \rangle$, and then so does $Y_{552} = \langle Y_{551}, s \rangle$.

We now know that every refelction of Y_{552} inverts ω. Transforming the relation $\omega = yx$ successively by v and s, we obtain $\omega^{-1} \doteq zx$ and $\omega = zy$, so that $\omega^2 = zy.yx = zx = \omega^{-1}$, proving that ω has order 3.

The entries of Tables 4 and 6 that involve ω can be written as assertions about commutators, as follows:-

$$[0123456789Xx|1X, \ 000000000000|+-] = \omega$$
$$[\,012345012345|41\,, \ 000000111111|01\,] = \omega$$
$$[\,102468702468|44\,, \ 000000111111|01\,] = \omega$$
$$[\,001234567899|45\,, \ +0000000000-|00\,] = \omega.$$

The first is immediately equivalent to our definition of ω, and the others are verified below. In each case, the top line is the desired commutator, which we reflect twice to get a commutator whose value is already known to be ω.

$$[012345\underline{012345}|41, \ 000000\underline{111111}|\underline{01}]$$
$$[0123456789Xx|\underline{X1}, \ 000000000000|\underline{-+}]$$
$$[0123456789Xx|1X, \ 000000000000|+-] = \omega,$$

$$[102\underline{468}702\underline{468}|44, \ 0000\underline{00}111111|01]$$
$$[1024\underline{35}402\underline{135}|41, \ 000000\underline{111111}|\underline{01}]$$
$$[102435X6879x|X1, \ 000000000000|-+] = \omega,$$

$$[001234\underline{567899}|4\underline{5}, \ +00000\underline{00000}-|0\underline{0}]$$
$$[\underline{0}01234\underline{123455}|41, \ \underline{1}00000\underline{111110}|\underline{01}]$$
$$[601234789015|01, \ 000000000000|-+] = \omega.$$

8   Presentations for $3Fi_{24}$

One of the conjectured presentations of [2] is that the single relation

$$(ab_1 c_1 ab_2 c_2 ab_3 c_3)^{10} = 1$$

completes the Coxeter relations of Y_{442} to a presentation of $3Fi_{24}$. This was first verified by Z. Djokovic, who used a machine enumeration of 920808 cosets that required 16 hours of computing time. In this section, we shall give an elementary proof of another presentation of $3Fi_{24}$, and describe its relation to Djokovic's result.

We first discuss the geometrical significance of the relation we found in Section 3, namely $L = R$, where

$$L = \begin{matrix} 000111 \\ 000111 \\ 000012 \end{matrix} \ , \ R = \begin{matrix} 111000 \\ 000111 \\ 000021 \end{matrix} \ , \text{ so that } L + R = \begin{matrix} 111111 \\ 000222 \\ 000033 \end{matrix} \ = N, \text{ say.}$$

This is equivalent to any equation of the form $u \doteq v$, where u and v are any two Monster roots whose sum or difference is N. Geometrically, the fundamental root vectors of Y_{521} define a semidefinite space in which the unique null vector (up to scalar factors) is N. The corresponding abstract Coxeter group is an infinite group of structure $\mathbf{Z}^8 : W$, where W, the Weyl group of E_8, is the finite group defined by the Coxeter relations of Y_{421}.

Our relation expresses the fact that the translation subgroup $\mathbf{Z}^8$ is in the kernel of the homomorphism onto $3Fi_{24}$. (In fact the central involution of W is also in the kernel, as can be seen inside $Y_{551} = GO_{10}^-(2)$, but we do not use this.)

Theorem 9 *The Coxeter relations of Y_{552}, together with the four relations of the above type for the subgroups $Y_{521}, Y_{512}, Y_{251}, Y_{152}$, form a presentation for $3Fi_{24}$.*

Proof. One need only check that our identification of Y_{552} with $3Fi_{24}$ has used just these relations.   □

The equivalence with Djokovic's result requires a few (small) coset enumerations, and is essentially present in [1]. On the one hand, it follows (without coset enumeration) from the above relations that the subgroup Y_{522} of Y_{552} is $O_8^+(2) : 2$, in which one can verify the relation $(ab_1c_1ab_2c_2ab_3c_3)^{10} = 1$. On the other hand, the early results of [1] show that in Y_{442} with this latter relation one can define f_1 and f_2 so that the relations of Theorem 9 hold. An extra relation that reduces the group to Fi_{24} is $(ab_1c_1d_1e_1f_1b_2c_2d_2e_2b_3c_3)^{33} = 1$.

Acknowledgement.

The vectors in our Table 6 have much the same structure as the combinatorial symbols in a table first produced by Simon Norton, and used by him to calculate with the transpositions of F_{24}. Norton also gave an outline proof of the sufficiency of the additional relation $(ab_1c_1ab_2c_2ab_3c_3)^{10} = 1$, differing somewhat from ours. We thank him for many interesting conversations.

References

[1] J.H. Conway, S.P. Norton, and L.H. Soicher, *The Bimonster, the group Y_{555}, and the projective plane of order 3*. Proceedings of the Chicago conference on Group theory and computation, December 1985, in Proceedings of Symposia in Pure Mathematics.

[2] J.H. Conway, R.T. Curtis, S.P. Norton, R.A. Parker, and R.A. Wilson, *An Atlas of Finite Groups*, Oxford University Press, 1985.

Note added 1991: the following book is a useful reference for reflection groups.

[3] J.E. Humphreys, *Reflection Groups and Coxeter Groups,* Cambridge University Press 1990.

A Geometric Characterization of the Monster

A.A. Ivanov

We present a characterization of the Monster sporadic simple group in terms of its 2-local parabolic geometry.

1. Introduction

We consider the largest sporadic simple group which is called the Fischer-Griess Monster or the Friendly Giant and is denoted by F_1, M or FG. Here we follow the monster terminology and the notation F_1.

Let $\mathcal{G}(F_1)$ be the minimal 2-local parabolic geometry of F_1 constructed in [RSt]. Then $\mathcal{G}(F_1)$ has rank 5 and belongs to a string diagram all whose nonempty edges except one are projective planes of order 2 and one terminal edge is the triple cover of the generalized quadrangle of order $(2,2)$, related to the nonsplit extension $3 \cdot S_6$. The geometries having diagrams of this shape are called T-geometries.

The geometry $\mathcal{G}(F_1)$ can be described as follows. The group $H \cong F_1$ contains an elementary abelian subgroup E of order 2^5 such that $N_H(E)/C_H(E) \cong L_5(2)$. Let $E_1 < E_2 < \ldots < E_5 = E$ be a chain of subgroups of E, where $|E_i| = 2^i, 1 \le i \le 5$. Then the elements of type i in $\mathcal{G}(F_1)$ are the subgroups of H which are conjugate to E_i; two elements are incident if one of the subgroups contains the other.

Let $\{\alpha_1, \alpha_2, \ldots, \alpha_5\}$ be a maximal flag in $\mathcal{G}(F_1)$ and H_i be the stabilizer of α_i in H. Then H_i are called the *maximal parabolic subgroups* associated with the action of H on $\mathcal{G}(F_1)$. Without loss of generality we can assume that $\alpha_i = E_i$ (clearly $H_i = N_H(E_i)$ in this case), $1 \le i \le 5$. Below we present a diagram of stabilizers where under the node of type i the structure of H_i is indicated.

$$
\begin{array}{ccccc}
\overset{1}{\circ}\!\!-\!\!\!-\!\!\!-\!\!\!-\!\!\!-\!\!\!-\!\!\overset{2}{\circ}\!\!-\!\!\!-\!\!\!-\!\!\!-\!\!\!-\!\!\!-\!\!\overset{3}{\circ}\!\!-\!\!\!-\!\!\!-\!\!\!-\!\!\!-\!\!\!-\!\!\overset{4}{\circ}\!\!=\!\!\!=\!\!\!=\!\!\!=\!\!\overset{5}{\circ} \\
\end{array}
$$

Co_1	$S_3 \times M_{24}$	$L_3(2) \times 3 \cdot S_6$	$L_4(2) \times S_3$	$L_5(2)$
2^{1+24}	$2^{2+11+22}$	$2^{3+6+12+18}$	$2^{4+1+2+8+8+12+4}$	$2^{5+1+5+10+10+5}$

The original motivation of the present work was the classification problem for the flag-transitive T-geometries and the result was that $\mathcal{G}(F_1)$ is simply connected i.e. that it does not admit a proper covering. This result was announced at the Durham symposium (see also [Ivn4]). S.Norton [Nor2] in an application of this result has proved that Y_{555} is a presentation for the Bimonster, i.e. for the wreath product $(F_1 wr Z_2)$. For this reason we formulate our main result here just in the form it is used in [Nor2].

Theorem A. *Let H be a group satisfying the following properties:*

(a) It is generated by subgroups H_1, H_2 and H_3 of shapes $2^{1+24}.Co_1$, $2^{2+11+22}.(S_3 \times M_{24})$ and $2^{3+6+12+18}.(L_3(2) \times 3 \cdot S_6)$ respectively. In H_2 and H_3, on the elementary abelian normal subgroups of orders 2^2 and 2^3 their full automorphism groups are induced.

(b) $H_1 \cap H_2$ has index 3 in H_2.

(c) $H_1 \cap H_3$ and $H_2 \cap H_3$ both have index 7 in H_3 and they correspond to an incident point-line pair of a projective plane of order 2 acted on by the composition factor $L_3(2)$ of H_3.

Then $H \cong F_1$. □

Let us take in $\mathcal{G}(F_1)$ a triple of pairwise incident elements of types 1, 2 and 3. For $H = F_1$ let H_1, H_2, H_3 be the stabilizers in H of the chosen elements. Then one can see just from the above diagram of stabilizers that the conditions in Theorem A are satisfied. So the simple connectedness of $\mathcal{G}(F_1)$ is a direct consequence of Theorem A. In fact this theorem implies a more deep conclusion about the structure of flag-transitive T-geometries.

Let $\mathcal{H}$ be a T-geometry of rank 5 and H act flag-transitively on $\mathcal{H}$. Let $\{\alpha_1, \alpha_2, \ldots, \alpha_5\}$ be a maximal flag in $\mathcal{H}$ and H_i be the stabilizer of α_i in $H, 1 \leq i \leq 5$. Let K be the action induced by $H_1 \cap H_2$ on the residue of $\{\alpha_1, \alpha_2\}$ and let S be the Sylow 2-subgroup of K. Notice that $K \cong M_{24}$ in the case $\mathcal{H} \cong \mathcal{G}(F_1)$. Suppose that $|S| = 2^{10}$. Then it follows from results in [Row1], [Row2], [Row3], [Hei], [Ivn1] and some unpublished results by S.V.Shpectorov and the author that the conditions of Theorem A are satisfied. So we have the following

Theorem B. *Let $\mathcal{H}$ be a flag-transitive T-geometry of rank 5. Suppose that the Sylow 2-subgroup of the action induced on the residue of a flag of type $\{1, 2\}$ by its stabilizer is of order 2^{10}. Then $\mathcal{H}$ is isomorphic to $\mathcal{G}(F_1)$.* □

It follows from results in [IS1] that $\mathcal{G}(F_1)$ can not arise as a residue in T-geometries of rank 6.

In fact there is just one additional possibility for the order of S, namely 2^9. This possibility is realized in an infinite series of flag-transitive T-geometries constructed in [IS2]. The automorphism group of the rank n member of this series is a nonsplit extension $3^{\alpha(n)} \cdot Sp_{2n}(2)$ where $\alpha(n) = (2^n - 1)(2^n - 2)/6$ is the number of lines in an n-dimensional vector space over $GF(2)$.

The subgroups H_1, H_2 and H_3 in Theorem A are also parabolics for the maximal 2-local parabolic geometry of F_1 constructed in [RSt]. So the simple connectedness of this geometry also follows from Theorem A. The subgroups H_1 and H_2 played a very important role both in construction of F_1[Gri1],

[Con2], [Tit2] and in its uniqueness proof [Tho]. In addition H_1 is the centralizer of an involution in F_1. Starting with the fact that H contains an involution a with $C_H(a) \cong 2^{1+24}.Co_1$ and with certain information about fusing in H of involutions from $O_2(C_H(a))$ it is more or less straightforward to show that H contains subgroups H_1, H_2 and H_3 satisfying Theorem A. The information about fusing of involutions can be easily deduced from existence of another involution b in H with $C_H(b) \cong 2 \cdot F_2$. So Theorem A has a close relation to the characterization of F_1 in terms of centralizers of involutions (cf. [Smi] and [GMS]). On the other hand we do not assume that H_1 is the full centralizer of an involution in H and make no restrictions concerning finiteness or simplicity of H.

The paper is organized as follows. In Section 2 we collect a number of known facts on the Leech lattice and related groups as well as on the Monster F_1 and the Baby Monster F_2. In Section 3 we establish certain properties of an amalgam $\{H_1, H_2, H_3\}$ satisfying the hypothesis of Theorem A. One of the main results there is reconstruction of H_1 up to isomorphism. In Section 4 we define a subamalgam in the considered amalgam by the following rule. We show that H_i contains a conjugacy class C_i of involutions and $C_1 \supset C_2 \supset C_3$. Let us pick an arbitrary element $\sigma \in C_3$ and consider the subamalgam of the subgroups $L_i = C_{H_i}(\sigma), i = 1, 2, 3$. The results in Section 3 enable us to deduce certain properties of the subamalgam. By Proposition 2.12 these properties imply that the subgroup $L = \langle L_1, L_2, L_3 \rangle$ in H is the nonsplit extension $2 \cdot F_2$. On this step we also use the description of the Schur multiplier of F_2 due to R.Griess. In Section 5 we consider the action of H on the set Γ of cosets of $L \cong 2 \cdot F_2$ in H. We show that this action preserves a graph $\Gamma(H)$ in which the setwise stabilizer of an edge is isomorphic to $2^{2 \cdot 2} E_6(2).2$. We define a class of triangles in $\Gamma(H)$. The stabilizer X of a triangle from this class is isomorphic to $2^{2 \cdot 2} E_6(2).S_3$. We consider in H the amalgam $\mathcal{A} = \{H_1, L, X\}$ and show that the conditions on the intersections specify this amalgam up to isomorphism. So this amalgam is isomorphic to an analogous amalgam in the Monster group F_1. This implies that $\Gamma(H)$ and $\Gamma(F_1)$ have a common cover $\tilde{\Gamma}$. Moreover with respect to the covering $\tilde{\Gamma} \to \Gamma(F_1)$ all triangles are contractible. Now to complete the proof it is sufficient to use the fact that the fundamental group of the Monster graph is generated by its triangles. The latter result (Proposition 2.17) was proved by the author when preparing the first version of this paper. The proof is similar to that given in [Ivn2] and dealing with a graph related to Janko's group J_4. An independent triangulability proof for the Monster graph was given in [AS]. So the proof is excluded from the present version of the paper.

Proposition 2.12 plays a crucial role in our proof. This proposition is the main result of [Ivn3]. The structure of [Ivn3] is similar to that of the present paper. Namely, in the amalgam $\{K_1, K_2, K_3\}$ in question we consider a subamalgam consisting of the centralizers in K_i of a certain involution from their common intersection. The subgroup in K generated by this subamalgam is identified with $2 \cdot {}^2 E_6(2).2$, using Tits' local characterization of buildings [Tit1]. After that we consider a graph on the cosets of the reconstructed subgroup and show that this graph has a common cover with a graph $\Gamma(F_2)$ on $\{3, 4\}$-transpositions of the Baby Monster F_2. Two transpositions are adjacent in $\Gamma(F_2)$ if their product is a central involution. On the final step we prove that the triangles in $\Gamma(F_2)$ generate the fundamental group of the graph.

2. Preliminary results

First of all we recall some known properties of the Leech lattice and related groups (cf. [Con1], [ATLAS], [Wil]).

The Mathieu group M_{24} has exactly two irreducible 11-dimensional $GF(2)$-modules: a factor module of the Golay code and a submodule in the Golay cocode. To simplify the terminology we will call these irreducible modules Golay code and Golay cocode, respectively. The orbit lengths on nonzero vectors are 759, 1288 in the code and 276, 1771 in the cocode.

Let A be a subgroup of order 3 in M_{24} whose normalizer is isomorphic to $3 \cdot S_6$.

Lemma 2.1. *Let V be either the Golay code or the Golay cocode. Then $C_V(A)$ is a 5-dimensional indecomposable module for S_6.* $\square$

Let Λ be the Leech lattice, $\Lambda_n = \{\lambda | \lambda \in \Lambda, (\lambda, \lambda) = 16n\}$ and $\overline{\Lambda} = \Lambda/2\Lambda$. Then $\overline{\Lambda}$ carries the structure of a 24-dimensional vector space over $GF(2)$ and $\overline{\Lambda} = \overline{\Lambda}_0 \cup \overline{\Lambda}_2 \cup \overline{\Lambda}_3 \cup \overline{\Lambda}_4$ (recall that $\Lambda_1 = \emptyset$). Moreover, $\overline{\Lambda}$ is an irreducible selfdual module for the Conway group Co_1. The group Co_1 acting on $\overline{\Lambda}$ preserves a unique nontrivial quadratic form f where $f(\overline{\lambda}) = 0$ if and only if $\overline{\lambda} \in \overline{\Lambda}_i$ and i is even.

Lemma 2.2. $\overline{\Lambda}_i, i = 0, 2, 3, 4$ *are the orbits of Co_1 on $\overline{\Lambda}$. The corresponding stabilizers are isomorphic to Co_1, Co_2, Co_3, $2^{11}.M_{24}$ respectively.* $\square$

Lemma 2.3. *Let M be a subgroup in Co_1 of the shape $2^{11}.M_{24}$. Then M stabilizes an element from $\overline{\Lambda}_4$ and $O_2(M)$ is the Golay code.* $\square$

If $\overline{\lambda} \in \overline{\Lambda}_i$ for $i = 2, 3$ then $\overline{\lambda}$ is the image of exactly two vectors from Λ_i which differ from each other by multiplication on -1. Each vector from $\overline{\Lambda}_4$ is the image of exactly 48 vectors from Λ_4 which form a coordinate frame.

Let Λ_4^8 be the set of vectors in Λ_4 with one nonzero coordinate equal to ± 8 and Λ_4^4 be the set of vectors in Λ_4 with supports of size 4 and with nonzero coordinates equal to ± 4. Then $\overline{\Lambda}_4^8$ consists of a unique element which will be denoted by $\overline{\lambda}^\infty$. If $\overline{\lambda} \in \overline{\Lambda}_4^4$ then all preimages of $\overline{\lambda}$ in Λ_4 are contained in Λ_4^4 and they can be characterized as follows: the supports are tetrads of some sextet and the number of minuses is of a fixed parity. This means that there is an equivalence relation on $\overline{\Lambda}_4^4$ with classes of size 2 and these classes are indexed by the sextets. If $\overline{\mu}$, $\overline{\nu}$ are equivalent vertices from $\overline{\Lambda}_4^4$ then $\{0, \overline{\lambda}^\infty, \overline{\mu}, \overline{\nu}\}$ is a subspace.

Let M be the stabilizer of $\overline{\lambda}^\infty$ in Co_1, $M \cong 2^{11}.M_{24}$.

Lemma 2.4. *M has exactly 3 orbits on $\overline{\Lambda}_2$ denoted by $\overline{\Lambda}_2^4$, $\overline{\Lambda}_2^3$ and $\overline{\Lambda}_2^2$. These orbits contain images of vectors of the shapes $(4^2, 0^{22})$, $(-3, 1^{23})$ and $(2^8, 0^{16})$. The corresponding stabilizers are isomorphic to $2^{10}.\mathrm{Aut}(M_{22})$, M_{23} and $2^{1+8}.A_8$, respectively.* □

Lemma 2.5. *M has exactly 5 orbits on $\overline{\Lambda}_4 - \{\overline{\lambda}^\infty\}$. The corresponding stabilizers are isomorphic to $2^{4+12}.3 \cdot S_6$, $M_{12}.2$, $[2^{11}].L_3(2)$, $L_3(4).S_3$ and $2^{1+8}.A_8$. Here the first of the stabilizers corresponds to $\overline{\Lambda}_4^4$.* □

The next two lemmas are consequences of Lemmas 2.3 and 2.5.

Lemma 2.6. *Let $\{\overline{\lambda}^\infty, \overline{\mu}, \overline{\nu}\}$ be a triple of vectors from $\overline{\Lambda}_4$ and suppose that its setwise stabilizer in Co_1 is of the shape $[2^{16}].(3 \cdot S_6 \times S_3)$. Then $\overline{\mu}$ and $\overline{\nu}$ are equivalent vertices from $\overline{\Lambda}_4^4$.* □

The images under Co_1 of the triple from Lemma 2.6 will be called *special triples*.

Lemma 2.7. *Let $J \cong 2^{4+12}.(3 \cdot S_6 \times S_3)$ be the stabilizer of a special triple and D be a Sylow 3-subgroup in $O_{2,3}(J)$. Then the centralizer of D in $O_2(J)$ is a 4-dimensional irreducible module for S_6.* □

The special triples are closely related to both minimal and maximal 2-local parabolic geometries of Co_1 (cf. [RSm], [RSt]). Namely, in those geometries the elements of $\overline{\Lambda}_4$ are points while the special triples are lines. The following proposition is a consequence of the description of the natural representations of the 2-local geometries of Co_1. This description was obtained independently by S. Smith (unpublished) and by the author (see Section 4 in [Ivn1]).

Proposition 2.8. *Let W be a $GF(2)$-module for Co_1 which is generated by a set of 1-dimensional subspaces indexed by the elements of $\overline{\Lambda}_4$ and suppose that the subspaces corresponding to a special triple generate a 2-dimensional subspace. Then W is isomorphic to $\overline{\Lambda}$.* □

The following result follows from a description of the orbits of Co_2 on $\overline{\Lambda}$.

Lemma 2.9. *Let $\overline{\lambda} \in \overline{\Lambda}_2$. Then the subspace $\langle \overline{\lambda} \rangle^\perp$ of $\overline{\Lambda}$ dual to $\langle \overline{\lambda} \rangle$ is an indecomposable module for Co_2.* □

Lemma 2.10. *Co_1 contains exactly three conjugacy classes of involutions. The centralizers in Co_1 are isomorphic to $2^{1+8}.D_4(2)$, $2^{11}.\mathrm{Aut}(M_{12})$, $(2^2 \times G_2(4)).2$ and the centralizers in $\overline{\Lambda}$ are of orders $2^{16}, 2^{12}, 2^{12}$, respectively.* □

Lemma 2.11. Co_1 acting on $\overline{\Lambda}_2$ has rank 4 with subdegrees 1, 4600, 46575, 47104 and with stabilizers isomorphic to $Co_2, U_6(2), 2^{10}.\mathrm{Aut}(M_{22})$, McL, respectively. The 2-orbits of Co_1 on $\overline{\Lambda}_2$ are characterized by the inner products between the corresponding preimages in Λ_2. These inner products are $\pm 32, \pm 16, 0, \pm 8$, respectively. □

A principal step in our proof is the construction in H of a subgroup L isomorphic to $2 \cdot F_2$. To identify this subgroup we will use the main result of [Ivn3] which is similar to Theorem A in the present paper.

Proposition 2.12. Let K be a group satisfying the following properties:

(a) It is generated by subgroups K_1, K_2 and K_3 of shapes $2^{1+22}.Co_2, 2^2.[2^{30}].(S_3 \times \mathrm{Aut}(M_{22}))$ and $2^3.[2^{32}].(L_3(2) \times S_5)$ respectively. In K_2 and K_3, on the elementary abelian normal subgroups of orders 2^2 and 2^3 their full automorphism groups are induced.

(b) $K_1 \cap K_2$ has index 3 in K_2.

(c) $K_1 \cap K_3$ and $K_2 \cap K_3$ both have index 7 in K_3 and they correspond to an incident point-line pair of a projective plane of order 2 acted on by the composition factor $L_3(2)$ of K_3.

Then $K \cong F_2$. □

Let $K_1 \cong 2^{1+22}.Co_2$ be the subgroup of $K \cong F_2$ from Proposition 2.12. Let $P = O_2(K_1)$, $\tilde{P} = P/Z(P)$. Then $\tilde{P}$, as a $GF(2)$-module for Co_2 is isomorphic to $\langle \overline{\lambda} \rangle^\perp / \langle \overline{\lambda} \rangle$ where $\overline{\lambda} \in \overline{\Lambda}_2$ is the vector stabilized by Co_2. Let ψ be the mapping of $\langle \overline{\lambda} \rangle^\perp$ onto $\tilde{P}$ which commutes with the action of Co_2. The following lemma is an implication of the fusion pattern of involution from $O_2(K_1)$(see for instance [Seg]).

Lemma 2.13. Let $\overline{\mu} \in \overline{\Lambda}_2$ and suppose that $\overline{\mu}$ is contained in the orbit of length 4600 of Co_2 on $\overline{\Lambda}_2$. Let τ be an element from P whose image in $\tilde{P}$ coincides with $\psi(\overline{\mu})$. Then τ is an involution and $C_K(\tau) \cong 2 \cdot {}^2E_6(2).2$. □

It is shown in [Gri2] that the Schur multiplier of F_2 is of order at most 2 and from existence of the Monster it follows that this order is exactly 2. So we have the following.

Lemma 2.14. Up to isomorphism there is a unique nonsplit extension $2 \cdot F_2$. □

Lemma 2.15. Let σ be a preimage in $L \cong 2 \cdot F_2$ of the involution τ from Lemma 2.13. Then σ is an involution and $C_L(\sigma) \cong 2^2 \cdot {}^2E_6(2)$. □

The *Monster graph* $\Gamma(F_1)$ is by definition a graph on the involutions of type $2A$ in F_1(the centralizer of a $2A$-involution of is isomorphic to $2 \cdot F_2$), where two involutions are adjacent if their product is again a $2A$-involution. F_1 contains a unique conjugacy class of $2A$-pure subgroups of order 2^2 and the normalizer of such a subgroup is isomorphic to $2^2 \cdot {}^2E_6(2).S_3$. A $2A$-pure subgroup obviously corresponds to a triangle in $\Gamma(F_1)$. The triangles arising in this way will be called *lines*. It is clear that each edge lies in a unique line. We will use the following property of $\Gamma(F_1)$ (cf. [Nor1], [GMS]).

Lemma 2.16. F_1 *acts transitively both on the set of lines and on the set of triangles distinct from lines.* □

The final proposition in this section was proved independently by M. Aschbacher and Y. Segev (see Section 8 in [AS]) and by the author.

Proposition 2.17. *The fundamental group of the Monster graph $\Gamma(F_1)$ is generated by its triangles.* □

3. On the structure of H_1, H_2 and H_3

Let H be a group and H_i $(i = 1, 2, 3)$ be its subgroups satisfying the hypothesis of Theorem A. In this section we deduce some information about the structure of these subgroups. In particular the subgroup H_1 will be identified up to isomorphism. This information will be used in the next section for construction of a subamalgam of $2 \cdot F_2$-type in H.

Let E_i be the normal subgroup of order 2^i in H_i, $1 \leq i \leq 3$. Let $Q = O_2(H_1)$ be the extraspecial group with the center E_1 and $\tilde{Q} = Q/E_1$ be the rank 24 elementary abelian 2-group. The well known properties of extraspecial groups imply

Lemma 3.1. *Let T be a subgroup of H_1 containing $O_2(H_1)$. Then E_1 is the center of T.* □

Since H_2 contains a unique conjugacy class of subgroups of index 3, $H_1 \cap H_2$ centralizes an involution from E_2. So we have

Lemma 3.2. $E_1 \leq E_2$; *in particular E_1 is not normal in H_2.* □

The following lemma is a direct consequence of the condition (c).

Lemma 3.3. H_1 *and H_2 generate H.* □

Let Δ be a graph on the set of (right) cosets of H_1 in H where two cosets are adjacent if they both have nonempty intersection with a coset of H_2. Let $\alpha \in \Delta$ be the coset containing the identity element. Then H_1 is the stabilizer of α in H and H_2 stabilizes a triangle $t = \{\alpha, \beta, \gamma\}$ and induces S_3 on t. Let T be the set of all images of t under H and $T(\alpha)$ be the set of triangles from T passing through α.

Lemma 3.4. *The action induced by H_1 on $T(\alpha)$ is similar to the primitive action of Co_1 on the cosets of $2^{11}.M_{24}$.* □

By Lemma 2.3 the action of Co_1 on $T(\alpha)$ is similar to its action on $\overline{\Lambda}_4$.

Lemma 3.5. *The subgroup $Q = O_2(H_1)$ induces a nontrivial action on the set $\Delta(\alpha)$ of vertices adjacent to α in Δ.*

Proof. Since H_2 induces S_3 on t, it contains an element h which stabilizes γ and permutes α and β. By Lemma 3.3, $H = \langle H_1, h \rangle$. Suppose that Q does not act on $\Delta(\alpha)$. Then Q is contained in the elementwise stabilizer $H(t)$ of the triangle t. By Lemma 3.1 the center of $H(t)$ is E_1. This means that E_1 is normal in H. Hence E_1 is normal in H_2, a contradiction with Lemma 3.2. $\square$

Thus Q acts nontrivially on $\Delta(\alpha)$. Then it is clear that the orbits of this action are of length 2. Moreover, if $\{\epsilon, \delta\}$ is such an orbit then $\{\alpha, \epsilon, \delta\} \in T(\alpha)$ and each triangle from $T(\alpha)$ can be obtained in this manner.

Now let us turn to the action of H_3 on Δ. Let Ξ be the orbit of H_3 containing α.

By condition (c) $|\Xi| = 7$ and $H_2 \cap H_3$ has on Ξ an orbit of length 3 which contains α. It is clear that the latter orbit coincides with the triangle t which is stabilized by H_2. Hence Ξ contains exactly 7 triangles from T and these triangles are the lines of a projective plane on Ξ. Let $t_1 = t, t_2, t_3$ be the triangles from Ξ which lie in $T(\alpha)$. Then $H_1 \cap H_3$ stabilizes $\{t_1, t_2, t_3\}$ as a whole and by Lemma 2.4 the triple $\{t_1, t_2, t_3\}$ corresponds to a special triple of elements in $\overline{\Lambda}_4$. In particular we have the following

Lemma 3.6. H_3 *is the full stabilizer of* Ξ *in* H. $\square$

The subgroup $H_1 \cap H_3$ contains Q and it induces S_4 on Ξ.

Lemma 3.7. Q *induces on* Ξ *a group of order 4.* $\square$

Now we come to one of the central results of the section.

Proposition 3.8. *As a* $GF(2)$-*module for* Co_1, $\tilde{Q}$ *is isomorphic to* $\overline{\Lambda} = \Lambda/2\Lambda$ *where* Λ *is the Leech lattice.*

Proof. From the description of the orbits of Q on $\Delta(\alpha)$ it follows that the module dual to $\tilde{Q}$ contains a collection of 1-dimensional subspaces indexed by the elements of $\overline{\Lambda}_4$. The subspaces corresponding to a special triple generate a 2-dimensional subspace. Now the claim follows from Proposition 2.8 and the selfduality of $\overline{\Lambda}$. $\square$

So H_1 is 2-constrained. By Proposition (2.6) in [GMS], up to isomorphism there are exactly two 2-constrained groups of the shape $2^{1+24}.Co_1$. Moreover we meet the following situation (cf. Section 4 in [Gri1]). There exists a group X_0 of the shape $2^2.2^{24}.Co_1$ with elementary abelian center of order 4. Let Z_i, $1 \leq i \leq 3$ be the subgroups of order 2 in the center of X_0 and $X_i = X_0/Z_i$. Then up to a reordering, X_1 and X_2 are the nonisomorphic 2-constrained groups of shape $2^{1+24}.Co_1$ and $X_3 \cong 2^{24}.Co_0 \cong 2^{25}.Co_1$. Let J and D be subgroups of Co_1 as in Lemma 2.7. We claim that D fixes in the module $\overline{\Lambda}$ only the zero vector. One of the ways to see this is the following (compare Lemma 3.14). By Lemma 2.7 $N_J(D)$ is of the shape $2^4.(S_3 \times 3 \cdot S_6)$.

Consideration of the centralizers of the elements of order 3 in Co_1 (cf. [AT-LAS], [Wil]) shows that $D = O_3(T)$ where $T \cong 3^2 \cdot U_4(3).2$ is the centralizer of a $3B$-element in Co_1. Now the claim follows from the facts that the smallest nontrivial $GF(2)$-representation of $U_4(3)$ is of dimension 20 [Par] and that obviously $U_4(3)$ is not involved in $L_4(2)$.

Let J_i be the preimage of J in X_i, D_i be a Sylow 3-subgroup in $O_{2,3}(J_i)$ and U_i be the centralizer of D_i in $O_2(J_i)$, $0 \leq i \leq 3$. Then, without loss of generality, we can assume that U_i is the image of U_0 under the natural homomorphism of X_0 onto X_i, $1 \leq i \leq 3$. Since D acts fixed point freely on $\overline{\Lambda}$ (see the previous paragraph), by Lemma 2.7 $\mid U_0 \mid = 2^6$ and $\mid U_i \mid = 2^5$, $1 \leq i \leq 3$. Notice that U_0 contains the center of X_0. By Lemma 2.7 the centralizer of D in $O_2(J)$ is a 4-dimensional irreducible module for S_6. This implies that U_i is abelian for $1 \leq i \leq 3$ and so U_0 is abelian as well. Now, U_3 can be mapped isomorphically into Co_0, moreover into the subgroup $2^{12}.M_{24}$ of Co_0 and one can see in this group that U_3 is an indecomposable $GF(2)$-module for S_6. On the other hand, it is not difficult to see that the 6-dimensional module U_0 should definitely be decomposable. So we conclude that among the modules U_1, U_2 and U_3 one is decomposable and two are indecomposable. We have already seen that U_3 is indecomposable. So we come to a criterion which distinguishes X_1 and X_2.

Proposition 3.9. *Up to isomorphism there are exactly two 2-constrained groups of shape $2^{1+24}.Co_1$. Let X_1 and X_2 be representatives of the isomorphism classes. Let J_i be the preimage in X_i of the stabilizer in Co_1 of a special triple, D_i be the Sylow 3-subgroup in $O_{2,3}(J_i)$ and U_i be the centralizer of D_i in $O_2(J_i)$, $i = 1, 2$. Then in exactly one case U_i is an indecomposable 5-dimensional $GF(2)$-module for S_6.* $\square$

Later we will show that for H_1 the 5-dimensional module in Proposition 3.9 is indecomposable but for this purpose we need some information about H_2 and H_3.

Let $\phi : Q \rightarrow \overline{\Lambda}$ be the surjective mapping which commutes with the action of $Co_1 \cong H_1/Q$. Notice that $\phi^{-1}(\overline{\alpha})$ consists of two involutions if $\overline{\alpha} \in \overline{\Lambda}_2 \cup \overline{\Lambda}_4$ and of two elements of order 4 if $\overline{\alpha} \in \overline{\Lambda}_3$. Let us show that E_2 and E_3 are contained in Q. Really, $C_{H_1}(E_2) \cong 2^{2+11+22}.M_{24}$ and by Lemma 2.10 $E_2 \leq Q$. Since $H_1 \cap H_3$ acts transitively on $E_3 - E_1$ this implies $E_3 \leq Q$. Now by Lemmas 2.2 and 2.6 we have the following

Lemma 3.10. *$\phi(E_2)$ is an element of $\overline{\Lambda}_4$ while $\phi(E_3)$ is a special triple.* $\square$

Without loss of generality we will assume that $\phi(E_2) = \overline{\lambda}^\infty$. In this case $\phi(E_3) - \{\overline{\lambda}^\infty\}$ is a pair of equivalent elements from $\overline{\Lambda}_4^4$. The corresponding sextet will be denoted by S.

Lemma 3.11. *Let $R = E_2 \cup \phi^{-1}(\overline{\Lambda}_4^4) \cup \phi^{-1}(\overline{\Lambda}_2^4)$. Then R is an elementary*

abelian subgroup of order 2^{13} in Q.

Proof. A direct calculation in the Leech lattice shows that $\overline{M} = \{0\} \cup \overline{\Lambda}_4^8 \cup \overline{\Lambda}_4^4 \cup \overline{\Lambda}_2^4$ is a subspace of $\overline{\Lambda}$. So R is closed under multiplication. In addition the quadratic form f vanishes on $\overline{M}$ so R is elementary abelian. $\square$

Notice that E_3 is contained in R.

Lemma 3.12. R *is a normal subgroup of* H_2.

Proof. First it is clear that R is generated by E_2 and $\phi^{-1}(\overline{\Lambda}_2^4)$. Let $N = C_{H_2}(E_2)$. Then $N \cong 2^{2+11+22}.M_{24}$ and N is an index 2 subgroup of $H_1 \cap H_2$. Let $\sigma \in \phi^{-1}(\overline{\Lambda}_2^4)$. By Lemma 2.4 $C_N(\sigma) \cong [2^{33}].\mathrm{Aut}(M_{22})$. Let ϵ be any other involution from H_2 which does not lie in $\phi^{-1}(\overline{\Lambda}_2^4)$. Then by Lemmas 2.4, 2.5 and 2.10 $C_N(\epsilon) \not\cong C_N(\sigma)$. Since N is normal in H_2, the claim follows. $\square$

Let $\tilde{R} = R/E_2$. Then it is easy to see that $\tilde{R}$ is an irreducible $GF(2)$-module for $M_{24} \cong N/O_2(N)$, isomorphic to the Golay cocode. This implies the following

Lemma 3.13. $O_{2,3}(H_2)$ *commutes with* $\tilde{R}$. $\square$

Let us turn to the group H_3. Let D be a subgroup of order 3^2 in H_3 such that $N_{H_3}(D)$ modulo $O_2(H_3)$ is isomorphic to $S_3 \times 3 \cdot S_6$. It follows directly from the shape of H_3 that all such subgroups are conjugate and that D induces a group of order 3 on E_3. In particular the conjugacy class of D in H_3 contains subgroups D_1 and D_2 such that $D_1 \le H_1$ and $D_2 \le H_2$.

Lemma 3.14. *Let* $U(D) = C_{H_3}(D) \cap O_2(H_3)$. *Then* $U(D)$ *is a 5-dimensional indecomposable* $GF(2)$-*module for* S_6.

Proof. We can assume that $D = D_2$, i.e. that $D \le H_2$. Let us present D as a direct product $D = A \times B$ where $A \le O_{2,3}(H_2)$. Now an application of Lemma 3.13 shows that $C_{H_2}(A)/A \cong 2^{11}.M_{24}$ and $O_2(C_{H_2}(A))$ is the Golay cocode. Put $P = C_{H_2}(A) \cap H_3$. By Lemma 3.6 $H_2 \cap H_3 = N_{H_2}(E_3)$ and the subgroups conjugate to E_3 in H_2 are in a bijection with the images in $\tilde{R}$ of the elements from $\phi^{-1}(\overline{\Lambda}_4^4)$. So it is easy to see that $P/A \cong 2^{11}.T$ where $T \cong 2^6.3 \cdot S_6$ is the stabilizer of a sextet. Now B is a Sylow 3-subgroup in $O_{2,3}(P)$. Since B acts fixed point freely on $O_2(T)$ to prove the claim it is sufficient to apply Lemma 2.1. $\square$

Let us consider $U(D_1)$. Clearly $E_1 \le U(D_1)$. The image J of $H_1 \cap H_3$ in H_1/Q is the stabilizer in Co_1 of a special triple and by Lemma 2.6, $J \cong 2^{4+12}.(3 \cdot S_6 \times S_3)$. Hence by Lemma 2.7 the image of $U(D_1)$ in H_1/Q has dimension 4. By Lemma 3.14 $U(D_1)$ is indecomposable. By Proposition 3.9 this characterizes H_1, up to isomorphism.

Proposition 3.15. H_1 *is isomorphic to the unique 2-constrained group of shape* $2^{1+24}.Co_1$ *for which the 5-dimensional* $GF(2)$*-module in Proposition 3.9 is indecomposable.* □

The above arguments characterizing H_1 up to isomorphism are quite similar to arguments in Section 8 of [Shp] where an analogous situation in Janko's group J_4 was considered.

Remark. At this point it is straightforward to construct a T-geometry $\mathcal{G}(H)$ on which H acts flag-transitively with H_1, H_2 and H_3 being maximal parabolic subgroups. Namely let us consider the T-geometry $\mathcal{G}(Co_1)$ of the Conway group in its natural representation in $\overline{\Lambda}$. Let $\{\alpha_2, \alpha_3, \alpha_4, \alpha_5\}$ be a maximal flag such that $\phi^{-1}(\alpha_i) = E_i$ for $i = 2, 3$. Put $E_j = \phi^{-1}(\alpha_j)$ for $j = 4, 5$. Then it follows from the definition of $\mathcal{G}(Co_1)$ (cf. [Ivn1]) that $E_4, E_5 \in \phi^{-1}(\overline{\Lambda}_4^4)$ and by Lemma 3.13 these subgroups are normalized by $O_{2,3}(H_2)$. Let H_j be the subgroup generated by the normalizers of E_j in H_i for $1 \le i \le j - 1$. Then $C_{H_j}(E_j) \le H_1 \cap H_2 \cap H_3$ and H_j induces $L_j(2)$ on E_j. Now the desired T-geometry $\mathcal{G}(H)$ can be defined on the cosets of H_i in $H, 1 \le i \le 5$ with two cosets being incident if they have a nonempty intersection. Notice that in the subsequent exposition we will not make any use of the geometry $\mathcal{G}(H)$.

4. A subamalgam of $2 \cdot F_2$-type

As above let ϕ be the mapping of Q onto $\overline{\Lambda}$ commuting with the action of Co_1. It follows from Lemma 2.4 that the stabilizer of an element $\overline{\alpha} \in \overline{\Lambda}_4$ has exactly one orbit of length $2 \cdot \binom{24}{2} = 552$ on $\overline{\Lambda}_2$. This orbit will be denoted by $\overline{\Lambda}_2(\overline{\alpha})$. Clearly $\overline{\Lambda}_2(\overline{\lambda}^\infty) = \overline{\Lambda}_2^4$.

Let us define three sets of involutions in Q: $C_1 = \phi^{-1}(\overline{\Lambda}_2), C_2 = \phi^{-1}(\overline{\Lambda}_2^4)$ and $C_3 = \phi^{-1}(\overline{\Lambda}_2^4 \cap \overline{\Lambda}_2(\overline{\mu}) \cap \overline{\Lambda}_2(\overline{\nu}))$ where $\{\overline{\lambda}^\infty, \overline{\mu}, \overline{\nu}\} = \phi(E_3)$. It follows by definition that $C_3 \subseteq C_2 \subseteq C_1$.

Lemma 4.1. C_i *is a conjugacy class of* H_i *for i=1,2 and 3.*

Proof. For $i = 1$ it is clear since $\overline{\Lambda}_2$ is an orbit of Co_1 on $\overline{\Lambda}$. For $i = 2$ the claim follows from Lemma 3.12. Let $i = 3$. It is easy to show that $\overline{M} = \overline{\Lambda}_2^4 \cap \overline{\Lambda}_2(\overline{\mu}) \cap \overline{\Lambda}_2(\overline{\nu})$ consists of the images in $\overline{\Lambda}$ of the vectors of the shape $(4^2, 0^{22})$, whose supports are contained in a tetrad of the sextet S corresponding to E_3. In particular $|\overline{M}| = 72$. The image of $H_1 \cap H_3$ in H_1/Q acts transitively on $\overline{M}$. On the other hand $H_1 \cap H_3$ contains a Sylow 2-subgroup of H_1 so the length of each of its orbits on C_1 should be divisible by 2^4. Now by Lemma 3.13 $O_{2,3}(H_2)$ normalizes E_3 and so $O_{2,3}(H_2) \le H_3$. Also by Lemma 3.13 $O_{2,3}(H_2)$ stabilizes C_3 as a whole. Since $H_3 = \langle H_1 \cap H_3, O_{2,3}(H_2) \rangle$ the result follows. □

Let $\sigma \in C_3$, $L_i = C_{H_i}(\sigma)$, $K_i = L_i/\langle\sigma\rangle$, $1 \le i \le 3$, $L = \langle L_1, L_2, L_3\rangle$, $K = L/\langle\sigma\rangle$.

Proposition 4.2. $K \cong F_2$.

Proof. It is sufficient to show that for K and its subgroups K_1, K_2 and K_3 the conditions in Proposition 2.12 are satisfied.

Since $H_1 \cap H_3$ is transitive on C_3, it is easy to see that L_3 covers the factor group of H_3 isomorphic to $L_3(2)$. In addition L_3 contains E_3. Hence $L_1 \cap L_3 = C_{L_3}(E_1)$, $L_2 \cap L_3 = N_{L_3}(E_2)$ and the condition (c) follows. Analogously it is clear that $L_1 \cap L_2 = C_{L_2}(E_1)$ and the condition (b) is also satisfied.

So it is sufficient to determine the structure of the subgroups $K_i, i = 1, 2, 3$. For $i = 1$ the proper structure follows from the fact that the stabilizer in Co_1 of a vector from $\overline{\Lambda}_2$ is isomorphic to Co_2. For $i = 2$ the claim follows from Lemmas 2.4 and 3.13. By the arguments in the above paragraph, K_3 has an $L_3(2)$-factor group acting faithfully on E_3. In addition $K_2 \cap K_3$ has index 231 in K_2. It is known that $\mathrm{Aut}(M_{22})$ has a unique class of index 231 subgroups and these subgroups are isomorphic to $2^5.S_5$. □

Proposition 4.3 $L \cong 2 \cdot F_2$.

Proof. In view of Proposition 4.2 it is sufficient to show that $\langle\sigma\rangle$ is not complemented in L. L_1 contains a Sylow 2-subgroup of L and by Lemma 2.9 $\langle\sigma\rangle$ is not complemented in L_1. So L is a nonsplit extension of F_2 by a center of order 2. By Lemma 2.14 this identifies L up to isomorphism. □

Remark. In geometric terms the main result of this section has the following meaning. The subgroup L is the stabilizer in H of a P-subgeometry in the geometry $\mathcal{G}(H)$. By Proposition 4.2 this subgeometry is identified with the geometry $\mathcal{G}(F_2)$ of the Baby Monster group.

5. The Monster graph

In this section we define a graph $\Gamma(H)$ on which H acts vertex- and edge-transitively with $L \cong 2 \cdot F_2$ being the stabilizer of a vertex and with $2^{2\cdot2}E_6(2).S_3$ being the stabilizer of a suitable triangle. If $H \cong F_1$ then $\Gamma(H)$ is just the Monster graph mentioned in Section 2.

Let Γ be the set of cosets of $L \cong 2 \cdot F_2$ in H and Ψ be the subset of Γ consisting of the cosets having nonempty intersections with H_1. Then H acts naturally on Γ and Ψ is an orbit of H_1 in this action. The action of H_1 on Ψ is similar to its action on the cosets of $L_1 = L \cap H_1 \cong 2^{2+22}.Co_2$ or, equivalently, to its action on $\phi^{-1}(\overline{\Lambda}_2)$ by conjugation. In particular E_1 is in the kernel of the action. Since $\overline{\Lambda}$ is irreducible and Co_2 is maximal in Co_1, it is easy to see the following. For distinct elements $\overline{\alpha}, \overline{\beta} \in \overline{\Lambda}_2$ the action of Q on $\phi^{-1}(\overline{\alpha} \cup \overline{\beta})$ is of order 4. So by Lemma 2.11 we have the following

Lemma 5.1. H_1 *acting on* Ψ *has rank five with subdegrees* 1, 1, 9200, 93150 *and* 94208. □

Let $x, x' \in \Psi, L$ and L' be the stabilizers of x and x' in H respectively, and σ, σ' be the corresponding involutions from $\phi^{-1}(\overline{\Lambda}_2)$. Suppose that the pair $\{x, x'\}$ corresponds to the subdegree 9200. By Lemma 2.11 we can assume that $\phi(\sigma)$ and $\phi(\sigma')$ coincide with the images in $\overline{\Lambda}_2$ of the vectors $(4, 4, 0, 0^{21})$ and $(0, 4, 4, 0^{21})$, respectively. This means that $[\sigma, \sigma'] = 1$, i.e. that $\sigma \in L', \sigma' \in L$. Moreover, without loss of generality we can assume that $\sigma, \sigma' \in C_3$.

Let $M = L \cap L'$. Since σ' lies in the center of L', it is clear that $M \leq C_L(\sigma')$ and by Lemma 2.15 $C_L(\sigma') \cong 2^2 \cdot {}^2E_6(2)$. On the other hand if $M_i = L \cap L' \cap H_i$ then M contains M_i for $i = 1, 2, 3$. By definition $M_i = C_{H_i}(\langle \sigma, \sigma' \rangle)$. By consideration of the action of H_i on C_i, it is straightforward to show that $M_1 \cong 2.[2^{22}].U_6(2)$, $M_2 \cong 2^2.[2^{29}].(S_3 \times L_3(4))$, $M_3 \cong 2^3.[2^{29}].(L_3(2) \times S_5)$. Now it is easy to see that M_1, M_2 and M_3 are maximal parabolic subgroups associated with the action of $C_L(\sigma')$ on the building of the group ${}^2E_6(2)$. So we come to the following

Lemma 5.2. $M = \langle M_1, M_2, M_3 \rangle = C_L(\sigma') \cong 2^2 \cdot {}^2E_6(2)$ and $O_2(M) = \langle \sigma, \sigma' \rangle$. □

Remark. The fact that the subamalgam $\{M_1, M_2, M_3\}$ of the amalgam $\{L_1, L_2, L_3\}$ generates in L the subgroup $2^2 \cdot {}^2E_6(2)$ can be deduced from Tits's local characterization of buildings [Tit1]. Such kind of arguments were used in [Ivn3] for the proof of Proposition 2.12.

Let $\Gamma(H)$ be a graph on Γ whose edges are the images of $\{x, x'\}$ under H. Let $\sigma'' = \sigma\sigma'$. On one hand $\sigma'' \in O_2(M)$. On the other hand $\sigma'' \in O_2(H_1)$, moreover $\sigma'' \in \phi^{-1}(\overline{\Lambda}_2)$. With the above choice of σ and σ' the element $\phi^{-1}(\sigma'')$ coincides with the image in $\overline{\Lambda}_2$ of the vector $(4, 0, 4, 0^{21})$. This means that σ'' corresponds to an element $x'' \in \Psi$. It is easy to see that the stabilizer in H_1 of the triple $\{x, x', x''\}$ induces S_3 on this triple. So this triple corresponds to a triangle in $\Gamma(H)$. This triangle and all its images under H will be called *lines*. The lines can be characterized as follows. Let $t = \{x_1, x_2, x_3\}$ be a triangle in $\Gamma(H)$. Let $H(x_i)$ be the stabilizer of x_i in H and τ_i be the nontrivial element of $O_2(H(x_i))$(recall that $H(x_i) \cong 2 \cdot F_2$), $1 \leq i \leq 3$. Then t is a line if and only if $\tau_i \in O_2(H(x_j) \cap H(x_k))$ for all $1 \leq i, j, k \leq 3$. Now by Lemmas 5.1, 5.2 we have the following.

Proposition 5.3. *Each edge of $\Gamma(H)$ is contained in a unique line. If X is the stabilizer in H of a line as a whole then $X \cong 2^2 \cdot {}^2E_6(2).S_3$. The extension of $X/O_2(X)$ by $O_2(X)$ is nonsplit and X acting by conjugation induces S_3 on $O_2(X)$.* □

In view of the properties of ${}^2E_6(2)$ which can be found for instance in [ATLAS], the conditions in Lemma 5.3 specify X up to isomorphism.

A direct calculation in the Leech lattice shows that the graph of valency

9200 on Ψ invariant under H_1, has more than one triangle on an edge. This implies the following.

Lemma 5.4. *The subgraph of $\Gamma(H)$ induced by Ψ contains lines as well as triangles distinct from lines.* □

Let $x \in \Gamma$ be the vertex whose coset contains the identity, t be a line containing x and lying in Ψ and X be the stabilizer of t in H. Let us consider the subamalgam $\mathcal{A}$ in H consisting of the subgroups H_1, L and X. By Lemmas 2.11, 5.1 and 5.4 we have the following

$$H_1 \cap L \cong 2^{2+22}.Co_2, \ H_1 \cap X \cong [2^{23}].U_6(2).S_3$$

$$L \cap X \cong 2^2 \cdot {}^2E_6(2).2, \ H_1 \cap X \cap L \cong [2^{23}].U_6(2).2$$

We intend to show that $\mathcal{A}$ is determined uniquely up to isomorphism. First let us recall the terminology and basic facts concerning amalgams (cf. [Shp], [IS3]).

An *amalgam* is a collection of groups with common identity element and with multiplications and operations of inverse coinciding on intersections. The groups which constitute an amalgam are called the *members* of the amalgam and elements of the members are called *elements* of the amalgam. The *universal closure* of an amalgam has all its elements as generators and all equalities $abc = 1$ valid in members of the amalgam as relations. An *automorphism* of an amalgam is a permutation of its elements which preserves each member as a whole and which, upon restriction to any one of the members, produces an automorphism.

Let $\mathcal{K}_1$ and $\mathcal{K}_2$ be amalgams with distinguished subamalgams $\mathcal{L}_1$ and $\mathcal{L}_2$ both isomorphic to an amalgam $\mathcal{L}$. Let A_i be the normalizer of $\mathcal{L}_i$ in the automorphism group of $\mathcal{K}_i$ and A_i^* be the image of A_i under the natural homomorphism of A_i into the automorphism group $A^* = \text{Aut}(\mathcal{L})$ of $\mathcal{L}, i = 1, 2$.

Lemma 5.5. (Goldschmidt) *The number of isomorphism types of amalgams which can be obtained by union of $\mathcal{K}_1$ with $\mathcal{K}_2$ via some identification of $\mathcal{L}_1$ and $\mathcal{L}_2$, is equal to the number of double cosets of the subgroups A_1^* and A_2^* in A^*.* □

In [Gol] this lemma was proved in the case when $\mathcal{K}_1, \mathcal{K}_2$ and $\mathcal{L}$ are groups. The above extended version was shown to me by S.V.Shpectorov.

Now we are ready to prove the following.

Proposition 5.6. *The amalgam $\mathcal{A} = \{H_1, L, X\}$ is determined uniquely up to isomorphism.*

Proof. The isomorphism types of H_1, L and X are established in Propositions 3.15, 4.3 and 5.3, respectively. Let us show that the amalgam $\mathcal{B} =$

$\{L, X\}$ is determined up to isomorphism. Let $Y = L \cap X$. Then Y is a subgroup of index 3 in X, and all such subgroups are conjugate in X. On the other hand Y is the centralizer in L of a $\{3, 4\}$-transposition in $L/O_2(L) \cong F_2$. By Proposition 2.5.4 in [GMS] all automorphisms of Y are inner. By Goldschmidt's lemma $\mathcal{B}$ is unique up to isomorphism.

Now let us adjoin H_1 to $\mathcal{B} = \{L, X\}$. The subgroups $L \cap H_1$ and $X \cap H_1$ are the centralizers of an involution $\tau \in Y$ in L and X, respectively. All involutions τ with the given structure of centralizers are conjugate in Y. The subamalgam $\mathcal{C} = \{H_1 \cap L, H_1 \cap X\}$ in H_1 can be specified as follows. Let us take an elementary abelian subgroup A of order 2^2 in $Q = O_2(H_1)$ such that $\phi(A)$ consists of three vectors from $\overline{\Lambda}_2$ and $N_{H_1}(A)$ contains a section $U_6(2)$. All such subgroups are conjugate in H_1 and $N_{H_1}(A)/C_{H_1}(A) \cong S_3$. Then for a suitable choice of A and $\sigma \in A^{\#}$ we have $H_1 \cap L = C_{H_1}(\sigma), H_1 \cap X = N_{H_1}(A)$.

Now to apply Goldschmidt's lemma we should study the automorphism group of $\mathcal{C}$. By Proposition 2.5.4 in [GMS] $\mathrm{Out}(H_1 \cap L) \cong Z_2$ and an outer automorphism acts nontrivially on the center of $H_1 \cap L$. It is easy to see that all these automorphism are realized in the normalizer of $H_1 \cap L$ in H_1 which is the preimage of Co_2 under the natural homomorphism $H_1 \to H_1/Q$. Since $H_1 \cap X \cap L$ has index 3 in $H_1 \cap L$ there is at most two ways to extend a given automorphism of $H_1 \cap L$ for an automorphism of $\mathcal{C}$. On the other hand there is an automorphism of $\mathcal{C}$ which is trivial on $H_1 \cap L$ and nontrivial on $H_1 \cap X$, namely the automorphism induced by an element which is in the center of L but not in the center of X. As a result we see that each automorphism of $\mathcal{C}$ can be realized in the normalizer of $\mathcal{C}$ in H_1 and the result follows. □

Remark. Our proof of the above proposition is similar to arguments in Section 8 of [AS]. Nevertheless the situation considered in that paper is slightly different.

Now let $\tilde{H}$ be the universal closure of the amalgam $\mathcal{A}$. Let $\tilde{\Gamma} = \Gamma(\tilde{H})$ be a graph on the cosets of L in $\tilde{H}$ with two vertices being adjacent if the corresponding cosets both have a nonempty intersection with a coset of X. Then by Proposition 5.6 there is a covering $\xi(H) : \tilde{\Gamma} \to \Gamma(H)$. The subgroup H_1 acting on $\tilde{\Gamma}$ has an orbit $\tilde{\Psi}$. The action of H_1 on $\tilde{\Psi}$ is similar to its action on the cosets of $H_1 \cap L$. It is clear that the restriction of $\xi(H)$ on the subgraph of $\tilde{\Gamma}$ induced by $\tilde{\Psi}$ is an isomorphism onto the subgraph of $\Gamma(H)$ induced by Ψ. By Lemma 5.4 lines and some triangle distinct from lines are contractible with respect to $\xi(H)$. Clearly the contractible triangles form a union of orbits under H. Now let us put $H \cong F_1$. By Lemma 2.16 $\xi(F_1)$ is a covering of $\Gamma(F_1)$ with respect to which all triangles are contractible. By Proposition 2.17 $\xi(F_1)$ is an isomorphism. This completes the proof of Theorem A.

Acknowledgement

This paper is a part of our joint work with Sergei V. Shpectorov on classification of P- and T-geometries. I am very thankful to Sergei for fruitful discussions and encouragement.

References

[AS]   Aschbacher, M., Segev, Y.: Extending morphisms of groups and graphs. Preprint 1990.

[Con1]   Conway, J.H.: Three lectures on exceptional groups. In: Powell, M.B. and Higman, G. (eds.) Finite Simple Groups. Acad. Press, 1971, pp.215-247.

[Con2]   Conway, J.H.: A simple construction for the Fischer - Griess monster group, Invent. Math., **79** (1985), 513-540.

[ATLAS]   Conway, J. et al.: Atlas of Finite Groups. Oxford Univ. Press, 1985.

[Gol]   Goldschmidt, D.: Automorphisms of trivalent graphs, Ann. Math., **111** (1980), 377-406.

[Gri1]   Griess, R.L.: The Friendly Giant, Invent. Math., **69** (1982), 1-102.

[Gri2]   Griess, R.L.: Schur multipliers of the known simple groups III. In: Proc. of the Rutgers Group Theory Year 1983-1984, pp. 69-80, Cambridge Univ. Press, Cambridge 1985.

[GMS]   Griess, R.L., Meierfrankenfeld, U., Segev Y.: A uniqueness proof for the Monster, Ann. Math., **130** (1989), 567-602.

[Hei]   Heiss, S.: On a parabolic system of type M_{24}, J.Algebra, to appear.

[Ivn1]   Ivanov, A.A.: The minimal parabolic geometry of the Conway group Co_1 is simply connected. Preprint 1990.

[Ivn2]   Ivanov, A.A.: A presentation for J_4. Proc. London Math. Soc. (to appear)

[Ivn3]   Ivanov, A.A.: A geometric characterization of Fischer's Baby Monster. Preprint, 1991.

[Ivn4]   Ivanov, A.A.: Geometric presentations of groups with an application to the Monster. Proc. ICM-90, Kyoto August 1990 (to appear)

[IS1]   Ivanov, A.A., Shpectorov, S.V.: Universal representations of P-geometries from F_2-series. J. Algebra (to appear)

[IS2]   Ivanov, A.A., Shpectorov, S.V.: An infinite series of flag-transitive T-geometries. Invent. Math. (submitted)

[Nor1]   Norton, S.P.: The uniqueness of the Fischer - Griess Monster, Contemp. Math. **45** (1985), 271-285

[Nor2]   Norton, S.P.: Constructing the Monster. In these proceedings.

[Par]   Parker R.A.: A collection of modular characters, Preprint, Univ. of Cambridge, 1989.

[RSm]   Ronan, M.A., Smith, S.: 2-local geometries for some sporadic groups. In: Proc. Symp. Pure Math. no. 37 (Santa Cruz, 1979, ed. Cooperstein - Mason). Providence AMS, 1980, pp.283-289.

[RSt]   Ronan, M.A., Stroth, G.: Minimal parabolic geometries for the sporadic groups. Europ. J. Combin. **5** (1984) 59-91.

[Row1]   Rowley, P.: On the minimal parabolic system related to M_{24}, J. London Math. Soc., **40** (1989), 40-56.

[Row2]   Rowley, P.: Minimal parabolic systems with diagram o——o——o══o. Preprint 1989.

[Row3]   Rowley, P.: Parabolic systems over $GF(2)$. Lect. at Oberwolfach Conf. "Groups and Geometries", 1988.

[Seg]   Segev, Y.: On the uniqueness of Fischer's Baby Monster. Preprint, 1990.

[Shp]   Shpectorov, S.V.: On geometries with the diagram P^n. Preprint, 1988 (in Russian).

[Smi]   Smith, S.: Large extraspecial subgroups of widths 4 and 6, J. Algebra **58** (1979), 251-281.

[Tho]   Thompson, J.G. On the uniqueness of the Fischer - Griess Monster, Bull. London Math. Soc., **11** (1979), 340-346.

[Tit1]   Tits, J.: A local approach to buildings. In: The Geometric Vein, pp. 519-547, Springer Verlag, 1981.

[Tit2]   Tits, J.: Le Monstre, Seminaire Bourbaki, expose no 620, 1983/84, Asterisque 121-122 (1985), 105-122.

[Wil]   Wilson, R.A.: The maximal subgroups of Conway's group Co_1, J. Algebra, **85** (1983), 144-165.

Constructing the Monster

S. P. Norton

Abstract

We complete the proof that Y_{555} is a presentation of the Bimonster.

1. Introduction

Recently the author published a paper [10] which showed how progress had been made towards proving that Y_{555} (which we redefine below) is a presentation for the wreath square of the Fischer-Griess Monster (which we call the Bimonster) and outlined a possible method of completing the proof. Since then the proof has indeed been completed, but by a different method: results [5-7] announced by A. Ivanov at the 1990 Durham Conference, proved by showing the simple connectedness of a certain simplicial complex, meant that a slight strengthening of the results of [10] was sufficient to complete the proof. This was achieved during the conference, and it therefore seems appropriate to publish it here in the conference proceedings.

We also take the opportunity to present proofs of two other results needed for which no full published version currently exists.

Summary of [10]

We start by recalling some of the notation, terminology and (without proof) results of [10]. Note that the numbering of the theorems has been changed. References [1-3,10-13] contain many other useful results about subgroups of Y_{555}.

We recall that a *Coxeter group* is generated by involutions corresponding to the nodes of a (Coxeter) diagram. The product of two generators has order 2 or 3 according as the corresponding nodes are unjoined or joined by a single unlabelled edge. (Other product orders are possible and correspond to other types of join.) We use the standard notation A_n, D_n, E_n for Dynkin diagrams of these types, and also for the corresponding Coxeter or Weyl groups, and $\tilde{A}_n$, $\tilde{D}_n$, $\tilde{D}_n$ for the corresponding extended Dynkin diagrams.

We define Y_{pqr} ($p, q, r > 1$) in terms of the diagram with three arms of lengths p, q, r radiating from a central node. In the notation of Figure 1 (which shows the case Y_{555}), its presentation has the Coxeter relations plus an additional relator $(ab_1c_1ab_2c_2ab_3c_3)^{10}$. NB: in some of the references cited the definition of Y_{pqr} is slightly different.

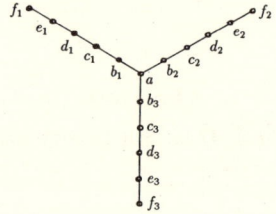

Figure 1

In [2] it is shown on certain assumptions that the group Y_{555} contains 26 involutions, including the generators, satisfying the Coxeter relations of the incidence graph of the projective plane of order 3. Furthermore Y_{555} has all the symmetries of this graph as automorphisms. For the rest of this paper we do not distinguish between nodes of our projective plane and the corresponding group elements, and call the nodes *points* or *lines* according to their status in the plane. We label the nodes using the convention of [2,12,13], based on Figure 2, where subscripts range from 1 to 3 and double and single edges respectively mean that nodes with and without a change of subscript are joined.

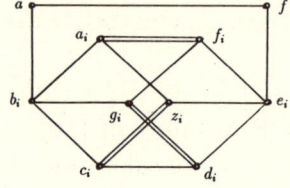

Figure 2

For some purposes it is useful to choose a 17-node subgraph of shape shown in Figure 3 and denote its 12 points and 5 lines as shown; the thirteenth point is called X, and the other 8 lines are unnamed. We call the group generated by this subgraph X_{3333}.

We define a *cog* to be an element of type $(P_1 P_2)^1$, where P_1, P_2 are two points and l the (unique) line containing them. The reason for this name is that by Theorem 2 the cogs determine a COnway Group).

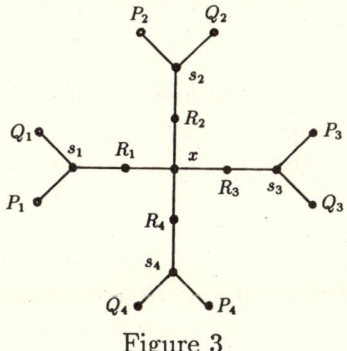

Figure 3

Lemma 1 *The centre of the group generated by a D_4-diagram depends only on the (unique) node that completes this diagram to a $\tilde{D}_4$.*

Notation The central involution of a D_4 completed to a $\tilde{D}_4$ by a node N will be called $N*$. We may call elements of this type point-stars or line-stars according as N is a point or line, but as only point-stars are encountered in this paper we abbreviate this to *stars*.

Lemma 2 *If A and B are two distinct points, then $A*$ and $[A, A*] = [B, B*]$ have order 2. Also $[A, B*] = [A*, B*] = I$. It follows that the points and stars generate an extra-special group 2^{1+26}, whose central involution we call π. Note that for all points $A, [A, A*] = \pi$.*

Theorem 1 $Y'_{555} \cong Y'_{553} \times Y'_{553}$.

Note. In [12] Soicher showed that this result followed from a condition which was subsequently proved in [10].

If g is an even element of Y_{555} (i.e. an element of Y'_{555}), we define its left and right *parts*, which we denote by g_L and g_R, to be the corresponding elements of the two factors isomorphic to Y'_{553}. The left and right parts of a subgroup of Y_{555} are then defined to be the groups consisting of the left or right parts of all its even elements.

In [10] we defined *left* and *right* as in [12]. Later we showed that the product of the thirteen stars is either the left or right part of π. As we made no further use of the original distinction between left and right, we were able to assume without loss of generality that the product of the thirteen stars is the *left* part of π. In this paper we use this assumption as a *definition* of left and right.

Lemma 3 *If P and L are a point and line, then $P^{*L} \in G$, where G is the group generated by the points, stars and cogs.*
Theorem 2 $G \cong 2^{1+26}.(2^{24} : \mathrm{Co}_1)$.

Lemma 4 *The quotient of X_{3333} by the subgroup Π normally generated by the $P_i Q_i$ and π (an element of $< x, R_i >$) has form $O_{10}^+(2) : 2$. Also, the subgroup O normally generated by the $P_i Q_i$ is elementary abelian of order 2^{10}, centralizes Π and is a natural module for X_{3333}/Π.*

Lemma 5 Π *is generated by the π, where s ranges over the products of all combinations of x and the s_i. Also, the commutators of these 32 conjugates of π generate the product of the left and right parts of O. More precisely:*

(1) $\pi_L = \pi^{s_1} \pi^{s_2} \pi^{s_3} \pi^{s_4}.X^*$

(2) π commutes with its conjugates by up to two of the five lines

(3) $[\pi, \pi^{s_1 s_2 s_3}] = X^*$

(4) $[\pi, \pi^{s_1 s_2 x}] = P_3 Q_3 P_4 Q_4$

(5) $[\pi, \pi^{s_1 s_2 s_3 s_4}] = X_R^*$

(6) $[\pi, \pi^{s_1 s_2 s_3 x}]$ is the product of the four $P_i Q_i (P_i Q_i)^{s_i}$'s times the right part of $P_4 Q_4$

(7) $[\pi, \pi^{s_1 s_2 s - 3s - 4x}]$ is the product of the $(P_i Q_i)_L$'s with the $(P_i Q_i)_R^{s_i}$'s.

Theorem 3 $X_{3333} \cong (2^{10+16})^2.O_{10}^+(2) : 2$.

This concludes the list of results quoted from [10].

3. Ivanov's Theorem

We now state the result of Ivanov, proved in [6], and identify the missing links in the proof that Y_{555} is the Bimonster.

Theorem 4 (Ivanov). *Let H be a group satisfying the following properties:*

(a) *It is generated by subgroups H_1, H_2, H_3 of shapes*

$$2^{1+24}.Co_1, \quad 2^{2+11+22}.(S_3 \times M_{24})$$

and

$$2^{3+6+12+18}.(L_3(2) \times 3S_6)$$

respectively. In H_2 and H_3, the elementary abelian normal subgroups of orders 2^2 and 2^3 are fully normalized.

(b) *$H_1 \cap H_2$ has index 3 in H_2.*

(c) *$H_1 \cap H_3$ and $H_2 \cap H_3$ both have index 7 in H_3, corresponding to the points and lines of a projective plane of order 2 acted on by the composition factor $L_3(2)$ of H_3.*

Then $H \cong M$.

We note that H_i, $(1 \le i \le 3)$ is the normalizer in the Monster of an elementary abelian 2-group of order 2^i. Furthermore, these groups are all $2B$-pure (i.e. all their involutions have class $2B$ in the notation of [1]). To apply the theorem, we take $H = (Y_{555})_L$ and $H_1 = G_L$. We then obtain H_2 by taking the intersection of H_1 with one of its conjugates and adjoining an outer automorphism group of type S_3, and H_3 by taking the intersection of the above group (without the outer automorphisms) with a second conjugate of H_1 and adjoining an outer automorphism group of type $L_3(2)$. To complete the proof that H satisfies the conditions of Theorem 4, we need the following:

Theorem 5

(1) *H_2 and H_3 do indeed have the structures referred to.*

(2) *The quotient of $H_1 \cap H_2$ by the intersection of G_L and the conjugate of this group used to define H_2 has order 2.*

(3) *The quotients of $H_1 \cap H_3$ and $H_2 \cap H_3$ by the intersection used to define H_3 both have structure S_4, and correspond to points and lines of the projective plane referred to above.*

(4) *$H =< H_1, H_2 >$. (Since it follows from (3) that H_3 is generated by its intersections with H_1 and H_2, this is equivalent to $H =< H_1, H_2, H_3 >$.)*

We mention here the two other missing links referred to at the end of Section 1:

Lemma 6   *The Bimonster is generated by involutions satisfying the relations of Y_{555}.*

Lemma 7 $^2E_6(2)$ *contains a unique subgroup of type* Fi_{22} *up to conjugation by an outer automorphism.*

These are used in [2] and [7] respectively.

4. A conjugate of G_L.

We define a group to be π-*correct* if every group element commuting with π lies in $< G, G_L >$. We start by proving the following lemma:

Lemma 8 X_{3333} *is π-correct.*

Proof: This is certainly true for O, which is generated by elements of type P_iQ_i, $(P_iQ_i)^{s_i}$, X^* and X^{*x}, all of which lie in G (using Lemma 3). It follows immediately that it is true for $O_L \times O_R$.

Modulo $O_L \times O_R$, Π is elementary abelian of order 2^{32}, and it can be seen from Lemma 5 that $[\Pi, \pi]$ has order 2^{10}. To show that Π is π-correct, it therefore suffices to find 22 independent elements of Π commuting with π and known to lie in $< G, G_L >$. Sixteen of these may be taken to be π and its conjugates by up to two elements of $\{x, s_i\}$; and it turns out, using (1) of Lemma 5, that the right parts of these yield just six more.

We now show that the points, stars and cogs visible in Figure 3 generate, modulo Π, a group of type $[2^{15}]. A_8$. The twelve points in the diagram generate a group of order 2^{12}, of which only $< P_iQ_i >$, of order 2^4, lies in the kernel Π. (This can be seen by exhibiting 10×10 matrices corresponding to the nodes of the diagram.) Adjoining the stars gives rise to an additional 2^9 (containing X^*, the R_i^* and the $P_i^*Q_i^*$), of which 2^3 survive the quotienting out of Π. And by using Table 1 of [10], we may calculate that the cogs generate an additional $2^4. A_8$, which is exactly what is needed.

By examining the 2-local structure of $O^+ + 10(2). 2$, one may deduce that this group of type $[2^{15}]. A_8$ is actually a maximal subgroup (see [1]), so any element of X_{3333}/Π that fixes π must lie in this group. This is what is needed to complete the proof of Lemma 8.

Theorem 6   *Let G' be the group generated by all the points plus the three lines c_i and even products of the three stars f_i. Then G'_L is conjugate to G_L. Moreover, G' is π-correct.*

Proof: It follows from [2] that $Y_{333} = < a, b_i, c_i, d_i >$ contains all the points as well as the four lines mentioned, and Linton has shown [8] that this group is isomorphic to $2^3 \cdot {}^2E_6(2)$. If one draws the diagram consisting of these 17 nodes, one can readily calculate, either by coset enumeration or by working out 27×27 matrices over $GF(4)$ corresponding to the various nodes, that the points, stars and cogs visible in it generate a subgroup U of shape $2^4 \cdot 2^{20} \cdot U_6(2)$, which is in fact the centralizer of π. (Note: this implies that Y_{333} is π-correct - a result we shall use later.) Similar methods will also show that if, instead, we omit the node a, we get a group Q of the same shape which is in fact the centralizer of the node f (or equivalently the element $ff_1f_2f_3$, which we call ρ). It follows that any element of Y_{333} that conjugates ρ to π will conjugate Q to U. Given that both f^* ($= \alpha$, say) and $\pi^{c_1 c_2 c_3}$ ($= \beta$, say) normalize $< \pi, \rho >$ in a way readily determinable from Lemma 2 and (4) of Lemma 5, one may see that $\alpha\beta$ is such an element.

As $F_1^* f_2^*$ commutes with ρ, its conjugate by $\alpha\beta$ – or equivalently β – commutes with π. But $(f_1^* F_2^{'*})^\beta$ is generated by π, the c_i's and $f_1^* f_2^*$ – $\pi\pi^a 3 \cdot b_3^* z_3^*$; and these are all in the (unique) group of type X_{3333} where we identify the nodes of Figure 3 in such a way that X (the only point not in the graph) is z_3 and x is c_3. So by Lemma 8 $(f_1^* f_2^*)_L^{\alpha\beta}$ – and similarly – $(f_1^* f_3^*)_L^{\alpha\beta}$ is in G_L.

Proper subgroups of G_L containing U_L can easily be enumerated, and it can be seen that each of them normalizes a 2-group which is not normalized by every generator of $G'^{\alpha\beta}$; so $\alpha\beta$ must conjugate G_L' to the whole of G_L. This completes the first part of Theorem 6.

All the generators of G' commute with $f_1 f_2 f_3$; since this is odd, any two even elements of G' are equal if and only if their left (or right) parts are equal, and the even part of G' is isomorphic to G_L'. So $G_L' \cong G_L \cong 2^{1+24} \cdot Co_1$, and G' has shape $2^{2+24} \cdot Co_1$; one may then deduce that $C_{G'}(\pi)$ has shape $2^{3+11+22} \cdot M_{24}$. Now the points, stars and cogs of U generate a group of shape $2^{5+9+18} \cdot L_3(4)$; and since $G' \cap G$ also contains elements of type $(f_i^* f_j^*)^{c_i}$, one may calculate that it (and hence G) must contain the whole of $C_{G'}(\pi)$. This is the required result.

5. The remaining steps

We now prove the following:

Theorem 7 *Let us define $H_2 = <G_L \cap G'_L, \alpha_L, \beta_L>$. Then H_2 satisfies conditions (1), (2) and (4) of Theorem 5.*

Proof: We already know from the proof of Theorem 6 that $(G \cap G)_L = G_L \cap G'_L$ has shape $2^{2+11+22}.M_{24}$. The π-correctness of G' implies that this group is the centralizer of π - or equivalently $\pi\rho$ - in G'_L; hence it is normalized by β. Furthermore, as β fixes G' and $\alpha\beta$ takes it to G, $\beta\alpha\beta$ (an involution) must interchange G and G', hence fix $G \cap G'$. So both α_L and β_L normalize $G_L \cap G'_L$. We digress from the proof of Theorem 7 to state:

Lemma 9 *$C_G(\rho)$ has the same left part as $G \cap G'$.*

Proof: The π-correctness of G' means that $C'_G(\pi)$ has the same left part as $G \cap G'$. Conjugation by $\beta\alpha\beta$ gives the required result. By considering the action of α and β on $<\pi, \rho>$, we see that $(\alpha\beta)^3_L$ commutes with π; hence, by the π-correctness of Y_{333}, it lies in G_L, and also in its $(\beta\alpha)_L$-image G'_L, so in their intersection. So since α_L and β_L are involutions, we see that $H_2/(G_L \cap G'_L) \cong S_3$, and it is also clear that this group fully normalizes $Z(G_L \cap G'_L) = <\pi, \rho>_L$. This shows that H_2 has the right structure and completes the proof of (1).

The subgroup $<G_L \cap G'_L, \alpha_L>$ clearly lies in H_1 and, from the above, has index 3 in H_2. Given that β_L is not in H_1, this proves (2).

We now prove the following lemma which will also be of use later.

Lemma 10 *If Γ is a collineation of the projective plane, then G contains an element g such that for each node N, $N^\Gamma = N^g$.*

Proof: In [2] it is shown that if g is an element of the group generated by a dodecagon (i.e. a graph of type $\tilde{A}_{11}$) corresponding to an even cyclic shift, then g permutes the nodes of the projective plane, and this permutation is a collineation. Since $L_3(3)$ is simple, and so generated by any non-trivial conjugacy class, it is sufficient to show that this particular g lies in G.

Now the points and cogs of $\tilde{A}_{11} \cong S_{12}$ generate a (maximal) subgroup $2^6.S_6$. This is the full S_{12}-normalizer of the subgroup generated by the points. So any even cyclic shift, which certainly preserves the set of points, lies in G. This completes the proof of Lemma 10.

We identify any collineation Γ with the corresponding Y_{555}-element g. $<H_1, H_2>$ contains G_L (hence the left part of any collineation); and also its transform by $(\beta\alpha)_L$, an element of H_2. The latter is G'_L which contains a member of each orbit of $L_3(3)$ on the left parts of the products of two nodes. So $< H_1, H_2 >$ contains the left part of the product of any two nodes, hence is the whole of $(Y_{555})_L = H$. This proves (4) and Theorem 7.

We now make several definitions: γ and δ are the elements corresponding to the collineations whose actions on the points are $(b_1 f_2)(b_2 d_2)(b_3 z_2)(d_1 f_1)$ and $(b_2 d_2)(b_3 d_3)(f f_1)(f_2 f_3)$ respectively; $G'' = G'^{\gamma\delta}$, with central involution $\sigma = b_1 d_1 f_1 f_2$; and $H_3 = < H_2 \cap G''_L, \gamma_L, \delta_L >$. Then we have:

Theorem 8 H_3 *satisfies the conditions (1) and (3) of Theorem 5.*

Proof: First we note that G'', as the transform of G' by a collineation, can be given a similar definition: in fact it is generated by thirteen points, the three lines c_2, c_3 and e_3, and even products of the three stars b_{*1}, d_{*1} and f_{*2}. It therefore contains α and $\beta = (\pi b_1^* d_1^* z_2^* z_3^*)^{c_2 c_3}$. So we can write H_3 as $\,_{\mathsf i}G_L \cap G'_L \cap G''_L, \alpha_L, \beta_L, \gamma_L, \delta_L >$ where $G_L \cap G'_L \cap G''_L$ can be seen to have shape $2^{3+6+12+18} . 3S_6$. To complete the proof, we need to study the actions of the last four generators on this group more closely.

Lemma 11 $\alpha_L, \beta_L, \gamma_L$ *and* δ_L *all normalize* $G_L \cap G'_L \cap G''_L$. *Their actions on the centre of this group are as follows:*

α_L *fixes* π_L *and* σ_L, *and multiplies* ρ_L *by* π_L.
β_L *fixes* ρ_L *and* σ_L, *and multiplies* π_L *by* ρ_L.
γ_L *fixes* π_L *and* σ_L, *and multiplies* ρ_L *by* σ_L.
δ_L *fixes* π_L *and* ρ_L, *and multiplies* σ_L *by* ρ_L.

Proof: α and β are known to normalize $G_L \cap G'_L$ and lie in G''_L; and they normalize the centres of these groups in the way stated. It is also straightforward to compute the action of γ and δ, which are collineations permuting the nodes, on the centre. Furthermore, these elements lie in G_L and therefore, by Lemma 9, normalize $G_L \cap G'_L \cap G''_L$, which is the centralizer of $< \rho_L, \sigma_L >$ in G_L.

An immediate corollary is that the centre of $G_L \cap G'_L \cap G''_L$ is fully normalized in H_3.

Lemma 12   *Modulo* $G_L \cap G'_L \cap G''_L$, *the elements* $\alpha_L, \beta_L, \gamma_L$ *and* δ_L *satisfy the Coxeter relations of the diagram shown in Figure 4, where a double bond means that the product has order 4. Furthermore, any three of them generate* *an* S_4.

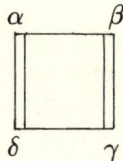

Figure 4

Proof: We already know that the elements are involutions. It is easily checked that all the relations are true in terms of the action group on the centre, i.e. that the relators all centralize $< \pi_L, \rho_L, \sigma_L >$. Because of Lemma 9 and the π-correctness of G' (and therefore all its images by collineations), it is sufficient to show that the relators lie in one of G_L, G'_L and G''_L. $(\alpha_L\beta_L)^3$ is already known to lie in both G_L and G'_L. The other Coxeter relators are easily shown to be trivial, as $\gamma\delta$ is a collineation of order 3, and each of γ or δ either fixes α_L or β_L or transforms it to something that commutes with it, according as the product is shown in Figure 4 as having order 2 or 4.

The relators required to equate the groups generated by three of our elements to S_4 are the cubes of the products of the three elements (in any order). We deal with them as follows: The cube of $\alpha_L\gamma_L\delta_L$ is the product of α_L and its images by the products, in each order, of the collineations γ and δ. Each of these, and hence their product, obviously lies in G_L. The same is true if α is replaced by β and G by G'. The left parts of α, β, and γ each fix that of σ, hence so does the cube of their product. The left parts of α, β and δ each lie in H_2; since the cube of their product fixes the centre of $G_L \cap G'_L$, it will therefore lie in this group. In each case we have the required result, thereby proving Lemma 12.

An easy coset enumeration now shows that the relations of Lemma 12 determine a group $L_3(2)$, and that this is therefore what $H_3/(G_L \cap G'_L \cap G''_L)$ is; and that $H_1 \cap H_3$ (which is obtained by omitting the generator β_L) and $H_2 \cap H_3$ (which is obtained by omitting the generator γ_L) have index 7. It may also be seen that they correspond to dual subgroups. This shows that H_3 satisfies (1) and (3) of Theorem 5, thus completing the proofs of Theorems 8 and 5.

6. Proof of Lemmas 6 and 7.

We start with:

Lemma 13 *There is a unique A_5 of $(2A, 3A, 5A)$-type in M (using the notation of [1]), and its centralizer is a group of type A_{12} in which the 7-point stabilizer is conjugate to the original A_5.*

Note. A proof using an unpublished result of B. Fischer is given in [9].

Proof. From its construction [4] the Monster has a representation of degree 196883, and it was by using this that Fischer, Livingstone and Thorne determined the character table printed in [1]. From this, the $(2A, 3A, 5A)$ structure constant can be used to determine the number of A_5's of this type, and it turns out that this number is consistent with the result that there is a unique class whose centralizer is A_{12}.

To prove this result, we note that any such A_5 must centralize a subgroup of $C_M(5A) \cong 5 \times HN$, where HN is the Harada-Norton group. The only subgroup of this whose order is large enough and divides that of $C_M(3A) \cong 3 . Fi'_{24}$ is A_{12}. Consider an element of order 7 in this A_{12}. Its centralizers in M and in the $A_5 \times A_{12}$ we have just constructed are the Held group and $A_5 \times A_5$ respectively. But the former only contains one subgroup of the latter type, and it also contains an involution interchanging the two A_5's in this. This shows that they are conjugate in M, thus proving Lemma 13.

By considering a diagonal subgroup of $M \times M$ and adjoining relevant involutions, we may see that Lemma 13 also holds if M, A_5 and A_{12} are replaced by the Bimonster, S_5 and S_{12} respectively. The argument of [3] now shows that the $S_5 \times S_{12}$ extends to a subgroup of the Bimonster generated by involutions satisfying the relations of Y_{555}, and this subgroup cannot be anything other than the Bimonster itself. This proves Lemma 6.

To prove Lemma 7, we consider the group $Y'_{222} \cong 3^5 . O_5(3)$. This group has a triple cover with a faithful 27-dimensional representation over $GF(4)$, unique up to complex conjugation; this is a monomial representation obtained by inducing up a linear character of $3^5 . 2^4 . A_5$, a subgroup of index 27. It is also easy to see that this representation supports a 2-space of invariant cubic forms.

From the splitting of this representation over Y'_{221} ($\cong O_5(3) \cong U_4(2)$) as the direct sum of the trivial representation and a uniserial 6+14+6, it may be seen that there are just three involutions in $L_2 7(4)$ commuting with each subgroup Y'_{221}. Of these, just one extends Y'_{222} to Y'_{322} (such an extension exists because $Y_{322} \cong 2 \times O_7(3)$ has its derived group as a direct factor), and this group may be verified to fix just a 1-space of the cubic forms. Similarly we may extend a different arm of the Y_{222} uniquely. If we do both, we get a group $Y'_{322} \cong \mathrm{Fi}_{22}$. It follows that Fi_{22} has a unique 27-dimensional representation over $GF(4)$ preserving at most one cubic form (up to scalar factor), so it can be found inside $^2E_6(2)$ in at most one way up to automorphism. This proves Lemma 7, and hence:

Theorem 9 $Y_{555} \cong M \ wr \ 2$.

7. Is the Monster Fabulous?

The notion of a "fabulous group" was introduced by Conway some time ago to generalize of the concept of Coxeter group. This notion has been defined in more than one way, but we adopt the following here:

Definition   A finite group is *fabulous* if it has a set of generators g_1, $g_2, \ldots, g_n$ such that if S is the subgroup generated by any subset Σ of these generators, and S' is the group presented by elements corresponding to these generators subject to the equivalents of all relations that hold in S and involve only a proper subset of Σ, then the kernel of the natural homomorphism from S' ¿to the (obviously finite) group S is abelian.

The word "fabulous" derives from the occurrence of the words "*F*inite" and "*A B*elian" in the above definition.

If Σ has cardinality 2, then S' will be the free product of two cyclic groups of orders equal to the orders of our two generators, because the presentation of S' contains only relators involving one generator. As the free product of two non-trivial cyclic groups is hyperbolic (and therefore not abelian by finite) unless they both have order 2, it follows that in any non-cyclic fabulous group all the generators are involutions, and that the relators involving two generators specify the orders of the products of pairs of these generators. This implies that any finite Coxeter group is fabulous.

A similar argument with 3-generator subgroups shows that if the product of two generators has order greater than 6, the dihedral group they generate must be a central factor of the full group.

An example of a non-Coxeter 4-generator fabulous group is $L_3(2)$, with the generators shown in Figure 4. The Coxeter relations in each 3-generator subgroup define the group $S' = 2 \times S_4$, and we quotient out the centre to get $S = S_4$.

By collating various results proved by Conway, Linton, Pritchard, Soicher and the author, one finds that in any subdiagram of Y_{555}, relators that are either central elements of finite Coxeter groups or translations of Euclidean Coxeter groups are sufficient to present all corresponding subgroups of Y_{555} (up to a possible finite central extension), except possibly for the cases Y_{433} and Y_{533}, which correspond to the Baby Monster. So, if we can prove* that either Y_{433} or Y_{533} (which are known to be isomorphic) is a finite abelian extension of $2 \times 2B$, where B is the Baby Monster, it would follow that Y_{553} and Y_{555} are fabulous systems of generators for the Monster and Bimonster respectively.

Question Is the Bimonster fabulously generated by the entire projective plane? This would imply, for example, that $\{a, b_1, b_2, b_3, e_1, e_2, f\}$, with certain translations of the subgraphs $\tilde{D}_4$ and $\tilde{D}_5$ as relators, presents an abelian extension of $\tilde{O}_{12}(2)$. This seems unlikely at first sight, but so did Theorem 9 at one time.

8. Concluding remarks.

The methods of this paper and the references cited mean that the definition of the Bimonster in terms of the Y_{555}-presentation (as it is now known to be) actually affords a reasonably efficient way of examining much of the subgroup structure of the Monster. It is therefore not unreasonable to regard this paper, together with its predecessors, as exhibiting a *construction* of the Monster (hence the title of the paper). It is hoped to develop this method further in subsequent papers.

References:

[1] J. H. Conway, R. T. Curtis, S. P. Norton, R. A. Parker and R. A. Wilson, *Atlas of Finite Groups*, Oxford Univ. Press, 1985.

[2] J. H. Conway, S. P. Norton and L. H. Soicher, The Bimonster, the Group Y_{555}, and the Projective Plane of Order 3, *Computers in Algebra* ed. M. C. Tangora, Marcel Dekker, 1988.

[3] J. H. Conway and A. D. Pritchard, Hyperbolic reflections for the Bimonster and $3\mathrm{Fi}_{24}$, *these proceedings*.

[4] R. L. Griess, The Friendly Giant, *Inv. Math.* **69** (1982) pp 1-102.

[5] A. A. Ivanov, Geometric Presentations of Groups with an application to the Monster, *to appear in* Proc. ICM Kyoto 1990.

[6] A. A. Ivanov, A geometric characterization of the Monster, *these proceedings*.

[7] A. A. Ivanov, A geometric characterization of Fischer's Baby Monster, *submitted to J. Alg. Combin.*.

[8] S. Linton, Ph. D. Thesis, Cambridge 1989.

[9] S. P. Norton, The uniqueness of the Fischer-Griess Monster, in *Finite Groups – Coming of age*, ed. J. McKay, 1985 (Contemp. Math. Series, **45**), pp. 271-285.

[10] S. P. Norton, Presenting the Monster? , *Bull. Soc. Math. Belg.* **62** (1990) pp. 595-605.

[11] A. D. Pritchard, Ph. D. Thesis, Cambridge 1989.

[12] L. H. Soicher, From the Monster to the Bimonster, *J. Alg.* **121** (2) 1989, pp 275-280.

[13] L. H. Soicher, More on the group Y_{555} and the projective plane of order 3, *J. Alg.* **136** (1) 1991, pp. 168-174.

The signature of the normalizer of $\Gamma_0(N)$

M. Akbas and D. Singerman

1 Preliminaries

Let $\Gamma_0(N)$ denote the subgroup of the classical modular group consisting of the Möbius transformations

$$z \to \frac{az+b}{cz+d}, \quad a,b,c,d \in \mathbf{Z}, \ ad-bc=1, \ c \equiv 0 \bmod N.$$

Let $\Gamma_B(N)$ be the normalizer of $\Gamma_0(N)$ in $\mathrm{PSL}(2,\mathbf{R})$. Then as $\Gamma_0(N)$ is a Fuchsian group whose fundamental domain has finite area, the same is true of $\Gamma_B(N)$ and so $\Gamma_B(N)$ has a signature

$$(g; m_1, \ldots, m_r; t) \tag{1}$$

where g is the genus of the compactified quotient space, $m_1, \ldots, m_r$ are the periods of the elliptic elements and t is the parabolic class number or *cusp number*.

The signature of $\Gamma_B(N)$ has been determined for square-free N by Maclachlan [7] and Helling [4]. We investigate the signature for arbitrary positive integers N and while we do not solve the whole problem we do enough to calculate the signature at least for $1 \le N \le 74$. In fact for all $N \ge 1$ we calculate the cusp number and the number of periods not equal to 2 . We find all values for N for which $\Gamma_B(N)$ acts transitively on the extended rationals and also all values of N for which $\Gamma_B(N)$ is a triangle group - there are 26 of them.

The normalizer of $\Gamma_0(N)$ was first investigated by Fricke and has appeared in many papers in recent years. In particular it arises in Moonshine [3, 8] and we find that the signatures do have some connections with the Monster. We show that the first three values of N for which the cusp number of $\Gamma_B(N)$ is greater than 1 are $N = 25, 49, 50$ corresponding to the "ghost" classes of the Monster [3].

2 The normalizer of $\Gamma_0(N)$

The group $\mathrm{PSL}(2,\mathbf{R})$ consists of the real Möbius transformations

$$z \rightarrow \frac{\alpha z + \beta}{\gamma z + \delta} \quad \alpha, \beta, \gamma, \delta \in \mathbf{R}, \ \alpha\delta - \beta\gamma > 0. \tag{2}$$

Each such transformation maps the upper half-plane U and also the extended real line $\mathbf{R} \cup \{\infty\}$ to itself. As usual we represent the transformation (2) by the matrix

$$A = \begin{pmatrix} \alpha & \beta \\ \gamma & \delta \end{pmatrix}$$

noting that if $k \in \mathbf{R} - \{0\}$ then A and kA represent the same transformation.

For each non-zero integer M we let $h(M)$ denote the largest divisor of 24 such that $h^2 \mid M$ and we usually write $h = h(N)$. Then from [3] we know that $\Gamma_B(N)$ consists of the transformations of the form

$$T = \begin{pmatrix} ae & b/h \\ cN/h & de \end{pmatrix} \tag{3}$$

where all letters are integers, $e\|N/h^2$ and the determinant of the matrix is $e > 0$. As in [1] we call e the *eterminant* of the transformation (3). ($e \parallel M$ means that $e \mid M$ and $(e, M/e) = 1$. We call e an exact divisor of M.)

It is shown in [1] that the set of exact divisors of M form a group $\mathrm{Ex}(M)$ with respect to the binary operation $*$ defined by $e_1 * e_2 = e_1 e_2/(e_1, e_2)^2$, and $\mathrm{Ex}(M) \cong C_2^r$ where r is the number of prime factors of M. Moreover, the map $E : \Gamma_B(N) \rightarrow \mathrm{Ex}(N/h^2)$ which takes every $T \in \Gamma_B(N)$ to its eterminant is an epimorphism. Thus if $\Gamma_C(N)$ is the set of transformations of $\Gamma_B(N)$ of eterminant 1 then $\Gamma_C(N)\Gamma_B(N)$ and

$$\Gamma_B(N)/\Gamma_C(N) \cong C_2^\rho$$

where ρ is the number of prime factors of N/h^2. By direct calculation

$$\Gamma_0(N/h^2) = H\Gamma_C(N)H^{-1} \tag{4}$$

where $H = \begin{pmatrix} h & 0 \\ 0 & 1 \end{pmatrix}$ and using this relation we obtain

$$\mid \Gamma_B(N) : \Gamma_0(N) \mid = 2^\rho h^2 \tau \tag{5}$$

where $\tau = \tau(N) = (\frac{3}{2})^{\epsilon_1} (\frac{4}{3})^{\epsilon_2}$ and where

$$\epsilon_1 = \begin{cases} 1 \text{ if } 2^2, 2^4, 2^6 \parallel N \\ 0 \text{ otherwise} \end{cases} , \quad \epsilon_2 = \begin{cases} 1 \text{ if } 9 \parallel N \\ 0 \text{ otherwise} \end{cases}$$

(For more details see [1] which also contains a detailed description of the quotient group $\Gamma_B(N)/\Gamma_0(N)$.)

Another important subgroup of $\Gamma_B(N)$ described in [3] and [1] is $\Gamma_W(N)$. For each $e \parallel N$ we define

$$W_e = \begin{pmatrix} ae & b \\ cN & de \end{pmatrix} \text{ of determinant } e.$$

Then $W_e \in \Gamma_B(N)$ of eterminant $e/h(e)^2$ and also $W_e^2 \in \Gamma_0(N)$. The W_e are called Atkin-Lehner involutions, [2] . The quotient group $\Gamma_W(N)/\Gamma_0(N) \cong C_2^r$ where r is the number of prime factors of N so that $\Gamma_W(N)$ has index $2^{\rho-r}h^2\tau$ in $\Gamma_B(N)$. If $h = 1$ then $\Gamma_W(N) = \Gamma_B(N)$. By direct calculation we have

Lemma 1 (a) $H\Gamma_B(N)H^{-1} = \Gamma_W(N/h^2)$

(b) If $h(N/h^2) = 1$ then $H\Gamma_B(N)H^{-1} = \Gamma_B(N/h^2)$.

3   Signatures

A group Λ of signature (1) has a presentation of the form

$$\langle A_1, B_1, .., A_g, B_g, X_1, .., X_r, P_1, .., P_t \mid$$

$$X_1^{m_1} = .. = X_r^{m_r} = X_1 X_2 .. X_r P_1 .. P_t \prod_{i=1}^{g} [A_i, B_i] = 1 \rangle \qquad (6)$$

The transformations A_i, B_i are hyperbolic, X_i elliptic and P_i parabolic. We recall that the transformation (2) is parabolic, if when normalized so that $\alpha\delta - \beta\gamma = 1$, it has trace ± 2. It then fixes a unique point in $\mathbf{R} \cup \{\infty\}$. If Λ has signature (1) and presentation (6) then every maximal parabolic cyclic group is conjugate to $\langle P_j \rangle$ $(j = 1, \ldots t)$. Each such $\langle P_i \rangle$ is just the stabilizer of a point $r \in \mathbf{R} \cup \{\infty\}$ and a conjugate cyclic group stabilizes a point in the Λ-orbit of r. A Λ-orbit of a fixed point of a parabolic element is called a *cusp* so that a group of signature (1) has t cusps. The cusps of $\Gamma_0(N)$ are well-known. ([9]). Every parabolic fixed point $\Gamma_0(N)$ lies in the orbit of $\frac{a}{d}$ for some $d \mid N$. If we let $\begin{pmatrix} a_1 \\ d_1 \end{pmatrix}$ denote the orbit of $\frac{a_1}{d_1}$ then $\begin{pmatrix} a_1 \\ d_1 \end{pmatrix} = \begin{pmatrix} a_2 \\ d_2 \end{pmatrix}$ if and only if $d_1 = d_2 = d$ and $a_2 \equiv a_1 \bmod(d, N/d)$. This gives the well-known formula

$$\pi_0(N) = \sum_{d \mid N} \phi\left(d, \tfrac{N}{d}\right) \qquad (7)$$

for the cusp number of $\Gamma_0(N)$. We note that π_0 is a multiplicative function.

The hyperbolic measure of a fundamental region for a group Λ of signature (1) is given by $2\pi\mu(\Lambda)$ where

$$\mu(\Lambda) = 2g - 2 + \sum_{i=1}^{r} \left(1 - \frac{1}{m_i}\right) + t \tag{8}$$

Also, if Λ_0 is a subgroup of finite index in Λ then

$$|\Lambda : \Lambda_0| = \mu(\Lambda_0)/\mu(\Lambda) \tag{9}$$

4   The cusp number

In this section we compute the cusp number $\pi(N)$ of $\Gamma_B(N)$. We first need to review the method of obtaining the cusp number of a subgroup Λ_0 of index n in Λ in terms of the cusp number of Λ. This is done in generality in [11] but there are simplifications if $\Lambda_0 \triangleleft \Lambda$, [7, 12] and we just deal with this case. If P generates a parabolic cyclic group $\langle P \rangle$ in Λ then we have a parabolic conjugacy class $\{\langle P \rangle\}$. If P has exponent k mod Λ_0 then for all $T \in \Lambda$ $\{\langle TP^kT^{-1}\rangle\}$ is a parabolic class in Λ_0 which we say has been *induced* by P . The number of such classes turns out to be n/k so that if in (6), P_i has exponent k_i mod Λ_0 then the cusp number of Λ_0 is

$$n \sum_{i=1}^{t} \frac{1}{k_i} \tag{10}$$

Theorem 1 *If ρ is the number of prime factors of N/h^2 then*

$$\pi(N) = \begin{cases} \pi_0(N/h^2)/2^\rho & \text{if } 2^8 \nmid N \\ \pi_0(N/4h^2)/2^{\rho-2} & \text{if } 2^8 \parallel N \end{cases}$$

Proof. If the matrix (3) of determinant $e > 0$ represents a parabolic element of $\Gamma_B(N)$ then $(a+d)\sqrt{e} = \pm 2$ so that $e = 1$ or 4. As $e \parallel N/h^2$ it is easy to see that $e = 4$ can occur if and only if $2^8 \parallel N$. Thus if $2^8 \nmid N$ then $e = 1$ so that all parabolics in $\Gamma_B(N)$ lie in $\Gamma_C(N)$ and thus have exponent 1 mod $\Gamma_C(N)$. As $\Gamma_C(N)$ is conjugate to $\Gamma_0(N/h^2)$, and $|\Gamma_B(N) : \Gamma_C(N)| = 2^\rho$, (10) implies that $\pi(N) = \pi_0(N/h^2)/2^\rho$, if $2^8 \nmid N$.

From now on in this proof we assume that $2^8 \parallel N$.

Lemma 2 *If $\alpha \in \mathbf{Q} \cup \{\infty\}$ is fixed by a parabolic element of eterminant 4 then the denominator of α when expressed as a reduced fraction is exactly divisible by 16. Conversely, every fraction of this form is fixed by a parabolic element of eterminant 4.*

Proof. An element of $\Gamma_B(N)$ of eterminant 4 can be represented by the following matrix *of determinant 1* :

$$T = \begin{pmatrix} 2a & b/2h \\ N/2h & 2d \end{pmatrix}$$

and $a + d = \pm 1$ as T is parabolic. The fixed point α of T is then given by

$$\alpha = \frac{(a-d)}{cN/2h}.$$

As $4ad - bcN/4h^2 = 1$, c is odd and as $a + d = \pm 1$, $a - d$ is odd. As $2^8 \parallel N$ and $2^3 \mid h$, the denominator is exactly divisible by 16.

For the converse we note that if $N = 2^8 L$ (L odd) and $\alpha = \frac{x}{16y}$, x, y odd then

$$\begin{pmatrix} 4\frac{(1-Lxy)}{2} & \frac{Lx^2}{8} \\ \frac{-Ny^2}{8} & 4\frac{(1+Lxy)}{2} \end{pmatrix}$$

has determinant 4. If $3^2 \nmid N$ then $h = 8$ and so this matrix represents a parabolic element of $\Gamma_B(N)$ of eterminant 4. If $3^2 \mid N$ we just write $\frac{Lx^2}{8} = \frac{3Lx^2}{24}$, $\frac{Ny^2}{8} = \frac{3Ny^2}{24}$ and we still have an element of $\Gamma_B(N)$ of eterminant 4. Its fixed point is $\alpha = \frac{x}{16y}$.

Now if $\alpha \in \mathbf{Q} \cup \{\infty\}$ is fixed by $T \in \Gamma_B(N)$ then it is also fixed by $T^2 \in \Gamma_C(N)$ and as $\Gamma_C(N) \lhd \Gamma_B(N)$ we find that the cusps ($\Gamma_C(N)$ orbits of fixed points) are of two kinds. **Type 1 cusps** are cusps which are fixed by elements of $\Gamma_B(N)$ and **type 2 cusps** are cusps which are not fixed by elements of $\Gamma_B(N)$.

As $4 \parallel N/h^2$ we can write $N/h^2 = 4K$, K odd.

Lemma 3 *The number of type 1 cusps is $\pi_0(K)$. The number of type 2 cusps is $2\pi_0(K)$.*

Proof. By Lemma 2 a type 1 cusp is an orbit of a point of the form $\frac{x}{16y}$, x, y odd, $(x, y) = 1$, and all points in this orbit also have this form. By (4) $hx/16y$ is fixed by a parabolic element of $\Gamma_0(N/h^2)$ where $h = 24$ if $3^2 \mid N$, $h = 8$ otherwise. We deduce that the type 1 cusps are in one-to-one correspondence with the cusps of $\Gamma_0(N/h^2)$ of the form $\begin{pmatrix} a \\ 2d \end{pmatrix}$, d odd. By (7) their number is

$$\sum_{\substack{2d \mid N/h^2 \\ d \text{ odd}}} \phi\left(2d, \frac{N}{2dh^2}\right) = \sum_{\substack{2d \mid N/h^2 \\ d \text{ odd}}} \phi\left(2d, \frac{2K}{h^2}\right) = \sum_{d \mid K} \phi\left(d, \frac{K}{d}\right) = \pi_0(K).$$

The total number of cusps of $\Gamma_0(N/h^2)$ is

$$\pi_0(N/h^2) = \pi_0(4K) = \pi_0(4)\pi_0(K) = 3\pi_0(K).$$

Hence the number of type 2 cusps is $2\pi_0(K)$.

Now let u be the number of parabolic classes of eterminant 1 in $\Gamma_B(N)$ and v the number of classes of eterminant 4. If T has eterminant 1 then T has exponent 1 mod $\Gamma_C(N)$ and if T has eterminant 4 then it has exponent 2 mod $\Gamma_C(N)$.

By the first paragraph of this section we find, using $|\Gamma_B(N) : \Gamma_C(N)| = 2^\rho$ and lemma 3, that $2^\rho u = 2\pi_0(K)$, $2^\rho v/2 = \pi_0(K)$. Thus $u = v = \pi_0(K)/2^{\rho-1}$ and $\pi(N) = u + v = \pi_0(K)/2^{\rho-2}$.

Corollary 1 π is a multiplicative function.

Corollary 2 Let N have prime power decomposition $2^{\alpha_1}3^{\alpha_2}p_3^{\alpha_3}\ldots p_r^{\alpha_r}$. Then $\pi(N) = 1$ if and only if $\alpha_1 \leq 7$, $\alpha_2 \leq 3$ and $\alpha_i \leq 1$, $i = 3,\ldots r$.

Proof. By Corollary 1 we need only find the prime powers p^α for which $\pi(p^\alpha) = 1$. This is done easily using Theorem 1 and (7).

We note that $\pi(N) = 1$ means that $\Gamma_B(N)$ has one orbit in its action on $\mathbf{Q} \cup \{\infty\}$, i.e. $\Gamma_B(N)$ acts transitively on $\mathbf{Q} \cup \{\infty\}$. An alternative proof of Corollary 2, independent of Theorem 1, can be obtained using the ideas of Ogg [9]. There, for Galois-theoretic reasons, he calls a cusp $\begin{pmatrix} a \\ d \end{pmatrix}$ $(d \mid N)$ of $\Gamma_0(N)$ multiquadratic if $(d, N/d) \mid 24$ and shows that $\Gamma_B(N)$ acts transitively on the multiquadratic cusps. The values of N given in Corollary 2 are precisely those values for which all cusps are multiquadratic.

5   The finite periods of $\Gamma_B(N)$

The only elements of finite order in $PSL(2,\mathbf{R})$ are elliptic. We recall that the transformation (2) is elliptic if, when normalized so that $\alpha\delta - \beta\gamma = 1$, we have $\mid \alpha + \delta \mid < 2$. Thus the transformation T of $\Gamma_B(N)$ represented by (3) is elliptic if and only if $(a + d)\sqrt{e} < 2$ so that either $a + d = 0$ or $a + d = \pm 1$ and $e = 1, 2, 3$. Now $a + d = 0$ if and only if T has order 2 and it is easy to show that $e = 1, 2, 3$ corresponds to T having orders 3, 4, 6 respectively. Thus $\Gamma_B(N)$ can only have finite periods 2, 3, 4, 6. As we now see, we can calculate the values of N for which $\Gamma_B(N)$ has periods 3, 4, 6.

Theorem 2 *(i)* $\Gamma_B(N)$ *has at most one period of order 4.* $\Gamma_B(N)$ *has a period of order 4 if and only if $2 \parallel N/h^2$ and if p is an odd prime divisor of N/h^2 then $p \equiv 1 \bmod 4$.*

(ii) $\Gamma_B(N)$ *has at most one period of order 6.* $\Gamma_B(N)$ *has a period of order 6 if and only if $3 \parallel N/h^2$ and if p is a prime divisor of N/h^2, $(p \neq 3)$ then $p \equiv 1 \bmod 3$.*

*(iii) $\Gamma_B(N)$ has at most one period of order 3. $\Gamma_B(N)$ has a period of order
3 if and only if for each prime divisor p of N/h^2, $p \equiv 1 \bmod 3$.*

Proof. (i) If T given by (3) has order 4 then $e = 2$ and so $2 \parallel N/h^2$.
We may suppose that $a + d = +1$ so from $4ad - bcN/h^2 = 2$ we obtain
$(2a - 1)^2 \equiv -1 \bmod N/h^2$. Thus if $p \mid N/h^2$ then $p \equiv 1 \bmod 4$. Conversely,
if these conditions are satisfied we can construct a matrix (3) of eterminant
2 with $a + d = 1$.

To calculate the number of periods equal to 4 we need to know the number
of periods equal to 2 in the subgroup $\Gamma_C(N) = H^{-1}\Gamma_0(N/h^2)H$. From [10]
the number of periods equal to 2 in $\Gamma_0(M)$ is

$$\begin{cases} 0 \text{ if } 4 \mid M \\ \Pi_{p\mid M}\left(1 + \left(\frac{-1}{p}\right)\right) \text{ otherwise.} \end{cases}$$

Thus if N/h^2 has r prime divisors $\equiv 1 \bmod 4$ then $\Gamma_0(N/h^2)$ and hence $\Gamma_C(N)$
has 2^r periods equal to 2. Also $2 \parallel N/h^2$ so that $|\Gamma_B(N) : \Gamma_C(N)| = 2^{r+1}$.
In general, if Γ is a Fuchsian group and Λ is a normal subgroup of index n
then if $T \in \Gamma$ is elliptic of order m and if T has exponent k mod Λ then T
induces $\frac{n}{k}$ periods equal to $\frac{m}{k}$ on Λ ([6]). In our case $\Gamma = \Gamma_B(N)$, $\Lambda = \Gamma_C(N)$
so that $n = 2^{r+1}, m = 4, k = 2$ and thus T induces 2^r periods equal to 2
on $\Gamma_C(N)$. As $\Gamma_C(N)$ only has 2^r periods equal to 2 it follows that $\Gamma_B(N)$
has just one period equal to 4. (ii), (iii) are similar dealing with elements of
eterminant 3 and 1 respectively. See [7] for the analogous calculation when
N is square-free.

6   Calculation of the signatures

We have now calculated all the invariants of the signature of $\Gamma_B(N)$ except
for the genus g and the number of periods e_2 equal to 2. However, from (9) we
can compute $\mu(\Gamma_B(N))$ as follows. If Γ is the modular group then $\mu(\Gamma) = \frac{1}{3}$
and $|\Gamma : \Gamma_0(N)| = N \prod_{p\mid N}\left(1 + \frac{1}{p}\right)$ ([10]) so that using (5) and

$$\tau = \prod_{p\mid N}\left(1 + \frac{1}{p}\right) / \prod_{p\mid N/h^2}\left(1 + \frac{1}{p}\right)$$

we find that

$$\mu(\Gamma_B(N)) = \frac{N}{3.2^{\rho+1}h^2} \sum_{p\mid N/h^2}\left(1 + \frac{1}{p}\right) \tag{11}$$

We can now use (8) to give a relation between e_2 and g. If N is square-
free then the genus g has been calculated using class numbers of imaginary
quadratic fields in [4] where all values of g have been listed for square-free

$N \leq 163$. Also in [7], e_2 was calculated for square-free N and the signatures of $\Gamma_B(N)$ were tabulated for square-free $N \leq 47$. Even though we are as yet unable to compute e_2 and g for general N the following result is useful.

Theorem 3 $\Gamma_B(N)$ *contains at least one period equal to 2.*

Proof $\Gamma_B(N)$ contains the following elliptic element of order 2 and of eterminant N/h^2 :

$$W = \begin{pmatrix} 0 & -1/h \\ N/h & 0 \end{pmatrix}.$$

If $h^2 = N$ then $\Gamma_B(N) = \Gamma_C(N)$ (by §2) and by (4) we find that $\Gamma_B(N)$ is conjugate to the modular group Γ which has signature $(0; 2, 3; 1)$. Now suppose that $h^2 \neq N$. By Theorem 2 the possible periods of $\Gamma_B(N)$ are 2, 3, 4 and 6 so we need only show that W is not the square of an elliptic element of order 4 or the cube of an elliptic element of order 6. By §2, all squares lie in $\Gamma_C(N)$ so if the first possibility occurs then $W \in \Gamma_C(N)$, a contradiction as $N/h^2 \neq 1$. Now suppose that $W = \overline{W}^3$, where $\overline{W}$ has order 6. By Theorem 2, $\overline{W}$ has eterminant 3. As $\overline{W}^2$ has eterminant one, W also has eterminant 3. Thus $N/h^2 = 3$ and so by (11), $\mu(\Gamma_B(N)) = \frac{1}{3}$. By Theorem 2, $\Gamma_B(N)$ has no periods of order 3 or 4 and one period of order 6. Thus

$$2g - 2 + \frac{e_2}{2} + \frac{5}{6} + t = \frac{1}{3}$$

which shows that $e_2 > 0$.

We now compute some signatures for non square-free N .

Example 1. $N = 49$. By Theorem 1, $\pi(49) = 4$ by (7). By Theorem 2, $\Gamma_B(N)$ has one period equal to 3. From (11), $\mu(\Gamma_B(49)) = 14/3$ and by (8), $2g - 2 + \frac{e_2}{2} + \frac{2}{3} + 4 = \frac{14}{3}$. As $e_2 > 0$, by Theorem 3 we must have $g = 0$ and $e_2 = 4$. Thus the signature of $\Gamma_B(49)$ is $(0; 2, 2, 2, 2, 3; 4)$.

Similarly we can show that $\Gamma_B(50)$ has signature $(0; 2, 2, 2, 4; 3)$ and $\Gamma_B(25)$ has signature $(0; 2, 2, 2; 3)$.

If N is divisible by 2^2 or 3^2 and N/h^2 is square-free we can compute the signature by using Lemma 1 (ii) and [4] or [7]. In these ways it is possible to compute the signature of $\Gamma_B(N)$ for $1 \leq N \leq 74$.

We can also compute signatures for some much larger values of N .

Example 2. $N = 2^8$. By Theorem 1, $\pi(2^8) = \frac{1}{2^{-1}}\pi_0(1) = 2$. By (11), $\mu(\Gamma_B(2^8)) = \frac{1}{2}$ and hence the signature must be $(0; 2; 2)$. Similarly, $\Gamma_B(2^8.3^2)$ has signature $(0; 2; 2)$.

As a further illustration we find all values of N for which $\Gamma_B(N)$ is a triangle group. We now use the convention that a parabolic element is considered as an elliptic element of infinite order, so that, for example, the signature $(0; 2; 2)$ can be written as $(0; 2, \infty, \infty)$. A **triangle group** is a Fuchsian group of genus 0 and 3 periods (which may include ∞), and we may write $(0; \ell, m, n)$ as (ℓ, m, n). In some ways triangle groups are the simplest Fuchsian groups; in [5] it is shown that 'maps' (tessellations of orientable surfaces) can be parametrized by subgroups of Fuchsian groups containing a period 2 and that the regular maps correspond to normal subgroups. For these reasons it might be of interest to find the values of N for which $\Gamma_B(N)$ is a triangle group. As a Fuchsian group that contains a triangle group is itself a triangle group [12] this will give all values of N for which $\Gamma_0(N)$ can be imbedded as a normal subgroup of a triangle group.

By Theorem 3, $\Gamma_B(N)$ contains a period 2; also $\Gamma_B(N)$ contains parabolic elements so that ∞ must also be a period. Thus if $\Gamma_B(N)$ is a triangle group it must have one of the signatures $(2, \infty, \infty), (2, 3, \infty), (2, 4, \infty), (2, 6, \infty)$. $((2, 2, \infty)$ is not allowed as a signature of a Fuchsian group as $\mu = 0$ for such a signature by (8).)

We have seen in example 2 that $\Gamma_B(N)$ has signature $(2, \infty, \infty)$ for $N = 2^8$ or $2^8.3^2$. These are the only such values of N . For we must have $\mu(\Gamma_B(N)) = \frac{1}{2}$ by (8) and then by (11)

$$\frac{N}{3.2^{\rho+1}h^2} \prod_{p|N/h^2} \left(1 + \frac{1}{p}\right) = \frac{1}{2}$$

This implies that $N/h^2 = 4, 5$ or 6. We cannot have $N/h^2 = 5$ for then $N = 2^{\alpha_1}.3^{\alpha_2}.5, \alpha_1 \le 6, \alpha_2 \le 2$ and by Corollary 2 of Theorem $1, \pi(N) = 1$ and we require $\pi(N) = 2$. Similarly, $N/h^2 \ne 6$. If $N/h^2 = 4$ then $N = 2^8$ or $2^8.3^2$.

For the other triangle groups we have $\mu < \frac{1}{2}$ and then we find that $N = 2^a 3^b$ so that $N/h^2 = 2^\alpha 3^\beta$. As $\mu(\Gamma_B(N)) < \frac{1}{2}, (\alpha, \beta) = (0,0), (1,0), (0,1)$. Thus $a \le 7, b \le 3$ and the three possibilities for (α, β) correspond to (i) a, b both even, (ii) a odd, b even, (iii) a even, b odd. In case (i) $N/h^2 = 1$ and so $\Gamma_B(N)$ is isomorphic to the modular group and so has signature $(2, 3, \infty)$. In case (ii), $N/h^2 = 2$ so by Theorem 2 $\Gamma_B(N)$ has signature $(2, 4, \infty)$. In case (iii) $N/h^2 = 3$ and so by Theorem 2 $\Gamma_B(N)$ has signature $(2, 6, \infty)$. We sum up in

Theorem 4 $\Gamma_B(N)$ *is a triangle group for precisely 26 values of* N .
If $N = 1, 2^2, 2^4, 2^6, 3^2, 2^2.3^2, 2^4.3^2, 2^6.3^2$ *then* $\Gamma_B(N)$ *has signature* $(2, 3, \infty)$.
If $N = 2, 2^3, 2^5, 2^7, 2.3^2, 2^3.3^2, 2^5.3^2, 2^7.3^2$ *then* $\Gamma_B(N)$ *has signature* $(2, 4, \infty)$.
If $N = 3, 2^2.3, 2^4.3, 2^6.3, 3^3, 2^2.3^3, 2^4.3^3, 2^6.3^3$ *then* $\Gamma_B(N)$ *has signature* $(2, 6, \infty)$.

If $N = 2^8$, $2^8.3^3$ then $\Gamma_B(N)$ has signature $(2, \infty, \infty)$.

There is a connection between these signatures and Moonshine. Ogg has observed in [9] that for primes p , the genus of $\Gamma_B(p)$ is 0 precisely when p is a prime factor of the order of the Monster M. Also 119 is the largest square free N for which $\Gamma_B(N)$ has genus 0 and 119 is the largest order of an element of M . In the Conway-Norton Moonshine conjectures ([3, 8]) there is a near one-to-one correspondence of genus zero subgroups between $\Gamma_0(N)$ and $\Gamma_B(N)$ with Monster conjugacy classes. This breaks down for $N = 25, 49, 50$ and we have shown that these are the smallest values of N for which $\pi(N) \neq 1$. (See Corollary 2 to Theorem 1.) In fact these are the only such values of N corresponding to Monster conjugacy classes for which $\pi(N) > 1$ and if $\Gamma_B(N)$ has more than one cusp then so does every subgroup of $\Gamma_B(N)$ of finite index. It seems as though genus 0 *and* cusp number 1 is the important property in the Conway-Norton correspondence.

References

1.   M. Akbas and D. Singerman.   The normalizer of $\Gamma_0(N)$ in PSL$(2, \mathbf{R})$, to appear in Glasgow Math. J.

2.   A.O.L. Atkin and J. Lehner. Hecke operators on $\Gamma_0(m)$, Math. Ann. 185 (1970), 134-160.

3.   J.H. Conway and S.P. Norton. Monstrous moonshine, Bull. London Math. Soc., 11 (1979), 308-339.

4.   H. Helling. On the commensurability class of the rational modular group, J. London Math. Soc., (2) 2 (1970), 67-72.

5.   G.A. Jones and D. Singerman. Theory of maps on orientable surfaces, Proc. London Math. Soc. (3) 37 (1978) 273-307.

6.   C. Maclachlan. Maximal Normal Fuchsian groups, Illinois, J. Math. 15 (1971) 104-113.

7.   C. Maclachlan. Groups of units of zero ternary quadratic forms, Proc. of the Royal Soc. of Edinburgh, $88A, (1981), 141 - 157$.

8.   G. Mason. Finite groups and modular functions, Proc. Symposia pure math., Vol. 47, AMS, Providence $RI, (1985), 223 - 244$.

9.   A.P. Ogg. Modular Functions. Proc. Santa Cruz conf. on finite groups, Proc. symposia pure math., Vol. 37, AMS (1980).

10. B. Schoeneberg, Elliptic modular functions.   Springer-Verlag. (1974).

11. D. Singerman. Subgroups of Fuchsian groups and finite permutation groups, Bull. London Math. Soc., 2 (1970) 319-323.

12. D. Singerman. Finitely maximal Fuchsian groups, J. London Math. Soc. (2) 6 (1972), 29-38.

Completely Replicable Functions

D. ALEXANDER, C. CUMMINS, J. MCKAY, & C. SIMONS

Abstract. We find all completely replicable functions with integer coefficients, tabulate the new ones, and summarize the computations needed.

Monstrous moonshine. To each conjugacy class of cyclic subgroups, $\langle m \rangle$, of the Monster simple group, M, a modular function, $j_{\langle m \rangle}(z)$, was found empirically in [CN] for which the q-coefficients (Fourier coefficients) are the values of the trace in the so-called head representations. For the identity subgroup the function is the elliptic modular function $J(z) = j(z) - 744$. Here, and throughout, the computations are simplest to describe if we assume all our q-series to have constant term zero.

Replication. Replication enables us to associate with a formal q-series

$$f = \sum_{i=-1}^{\infty} a_i q^i, \quad a_{-1} = 1, \quad a_0 = 0, \tag{1}$$

$a_i \in \mathbf{C}$, certain functions of the same form, called the replicates of f. Although f is a formal q-series, it is useful to write $f = f(z)$, where $q = e^{2\pi i z}$, consistent with the properties of modular functions. We tacitly omit describing the Galois action [N], which is trivial when the q-series coefficients are rational integers.

The prototypical replication relation is that between the monstrous moonshine function $j_{\langle m \rangle}(z)$ for $\langle m \rangle \subset M$ and its pth replicate $j_{\langle m^p \rangle}(z)$ for $\langle m^p \rangle$. Conway and Norton [CN] note that monstrous moonshine functions satisfy identities involving f and its replicates which they call *replication formulae*. A replicable function is a function with a q-expansion of the form (1) for which replicates exist. Such functions also satisfy the replication formulae.

Norton [N] has conjectured that a function $q^{-1} + \sum_{i=1}^{\infty} a_i q^i, a_i \in \mathbf{Z}, i \geq 1$, is replicable if and only if either $a_i = 0$ for all $i > 1$ or it is the canonical Hauptmodul for a group of genus zero, containing $\Gamma_0(N)$ for some N and containing $z \to z + k$ precisely when k is an integer.

Partially supported by NSERC and FCAR grants.

Hecke Operators. Motivation for introducing the twisted Hecke operator $\widehat{T}_n$ derives from the action of the standard Hecke operator, T_n , on $J(z) = j(z) - 744$, given by

$$J|T_n = \frac{1}{n} \sum_{\substack{ad=n \\ 0 \le b < d}} J((az + b)/d) = P_n(J(z))/n \qquad (2)$$

which value is a polynomial in J since the sum is invariant under the modular group and J is a Hauptmodul holomorphic in the upper half-plane. Note that P_n is the unique polynomial such that $P_n(J(z))$ has a q-expansion q^{-n} mod $q\mathbf{Z}[q]$.

We introduce a twisted Hecke operator $\widehat{T}_n$ which, like T_n , acts linearly on q-coefficients yet takes certain functions $f(z)$ to $P_n(f(z))/n$.

More precisely, we call a function f replicable if there are replicate functions $\{f^{(a)}\}$ such that

$$P_n(f(z))/n = \frac{1}{n} \sum_{\substack{ad=n \\ 0 \le b < d}} f^{(a)}((az + b)/d), \qquad (3)$$

with $P_n(f(z)) = q^{-n}$ mod $q\mathbf{Z}[q]$, and we define $f|\widehat{T}_n$ to be the right side of (3).

The monic polynomial $P_n(t) \in \mathbf{Z}[a_1, a_2, \dots, a_{n-1}][t]$ is unique and we shall abuse notation by using P_n to denote the polynomial in each case. This definition of $\widehat{T}_n$ is provisional since we have not yet incorporated the Galois action.

Note that $J(z)$ of level $N = 1$ is the sole normalized modular function on which the Hecke operators T_n act as in (2) for all n, since $(N, n) = 1$ for all n. Replicable functions are defined so as to share this property under the action of the twisted Hecke operator $\widehat{T}_n$. In this case, however, the sum involves both f and its replicates.

From Norton [N] it follows that

$$\sum_{n=1}^{\infty} \frac{P_n(t)}{n} q^n = -\ln(q(f(z) - t)), \qquad (4)$$

and so $P_1(t) = t, P_2(t) = t^2 - 2a_1, P_3(t) = t^3 - 3a_1 t - 3a_2, \dots$.

We define coefficients $\{H_{m,n}\}$ by

$$f|\widehat{T}_n = \tfrac{1}{n} P_n(f(z)) = \tfrac{1}{n} q^{-n} + \sum_{m=1}^{\infty} H_{m,n} q^m, \qquad n \ge 1, \qquad (5)$$

so that $H_{m,n}$ is the coefficient of q^m in $f|\widehat{T}_n$ and $H_{m,1}$ is the coefficient of q^m in f (denoted H_m by Norton).

We find that $P_n(t)$ satisfies the recurrence relations:

$$P_0(t) = 1, \quad ra_{r-1} + \sum_{k=-1}^{r-2} a_k P_{r-k-1}(t) = tP_{r-1}(t), \quad r = 1, 2, \dots \quad (6)$$

while $\widehat{H}_{r,s} = (r+s)H_{r,s}$ satisfies

$$\widehat{H}_{r,s} = (r+s)H_{r+s-1} + \sum_{m=1}^{r-1}\sum_{n=1}^{s-1} H_{m+n-1}\widehat{H}_{r-m,s-n}. \quad (7)$$

Norton has another definition of replicability that is somewhat easier to use in practice.

A function f is replicable if $H_{m,n} = H_{r,s}$ whenever $mn = rs$ and $\gcd(m,n) = \gcd(r,s)$.

This is equivalent to the definition given above: assume f is replicable in Norton's sense, then set

$$f^{(k)}(z) = \sum_{i=-1}^{\infty} a_i^{(k)} q^i \quad (8)$$

where

$$a_i^{(k)} = k\sum_{d|k} \mu(d)H_{\frac{k}{d},dki}, \quad i > 0, \ a_{-1}^{(k)} = 1, \ a_0^{(k)} = 0 \quad (9)$$

and μ is the Möbius function. It follows that $f^{(1)} = f$. For any pair $r, s \in \mathbf{Z}^{>0}$, we find, by Möbius inversion, that

$$H_{r,rs} = \sum_{d|r} \frac{1}{d} a_{r^2 s/d^2}^{(d)} \quad (10)$$

and, since f is replicable under Norton's definition, this implies that

$$H_{m,n} = \sum_{d|(m,n)} \frac{1}{d} a_{mn/d^2}^{(d)} \quad (11)$$

which, from (5), gives (compare Serre [S, Chap.VII, §5.3])

$$f|\widehat{T}_n = \frac{1}{n} \sum_{\substack{ad=n \\ 0 \le b < d}} f^{(a)}((az+b)/d). \quad (12)$$

Conversely if f has replicates which satisfy (12) it follows that the $H_{m,n}$ of (5) satisfy (11) and so f is replicable as defined by Norton.

When $n = p$, a prime, we see that

$$pf|\widehat{T}_p = f^{(p)}(pz) + \sum_{k=0}^{p-1} f((z+k)/p). \tag{13}$$

In terms of the standard operators U_p and V_p where

$$U_p : a_n q^n \rightarrow a_{pn} q^n,$$
$$V_p : a_n q^n \rightarrow a_n q^{pn},$$

we have

$$pf|\widehat{T}_p = f|(\Psi^p V_p + pU_p) = P_p(f(z)) \tag{14}$$

where Ψ^p acts as an Adams operator (see Mason [**Mas**]); equivalently we may compute $f^{(p)}$ from

$$f^{(p)}(pz) = P_p(f(z)) - pf|U_p. \tag{15}$$

Complete replicability. A function is completely replicable if it and all its replicates are replicable. One would expect properties of the monstrous moonshine functions to be shared by the completely replicable functions (and they are). At the end we tabulate all non-monstrous completely replicable functions with rational integer coefficients. Complementary monstrous data are found in [**CN**] and [**MS**].

Method of Calculation. To find all completely replicable functions, we computed the larger class of all completely 2-replicable functions. These are functions whose iterated duplicates are replicable. Table 1 of [**N**] contains a list of all completely 2-replicable functions satisfying $f^{(2)} = f$. We call g a replication p^{th} root of f if $g^{(p)} = f$. With a small prime π, the replication square roots of these functions are found by first testing all choices of a_1, a_2, a_3 and a_5 mod 2π for replicability using replication identities and identities derived from them (see [**CN**]). Solutions are then lifted by π-adic approximation using identities up to $H_{145} = H_{5,29}$ so that the solutions found mod 2π, for some prime π, lift uniquely to $2\pi^k, k > 1$.

These calculations require further coefficients which are computed from a_1, a_2, a_3, a_5 and the coefficients of $f^{(2)}$ via the generalized Mahler recurrence relations [**Mah**] (compare [**B**]) derived from:

$$f(\gamma_0 z) + f(\gamma_1 z) + f^{(2)}(\gamma_2 z) = f(z)^2 - 2a_1,$$
$$\tag{16}$$

$$(f(\gamma_1 z) + f(\gamma_0 z))f^{(2)}(\gamma_2 z) + f(\gamma_0 z)f(\gamma_1 z) = 2a_2 f - f^{(2)} + 2(a_4 - a_1)$$

where
$$\gamma_0 = \begin{pmatrix} 1 & 0 \\ 0 & 2 \end{pmatrix}, \quad \gamma_1 = \begin{pmatrix} 1 & 1 \\ 0 & 2 \end{pmatrix}, \quad \gamma_2 = \begin{pmatrix} 2 & 0 \\ 0 & 1 \end{pmatrix};$$

namely, for $k \geq 1$:

$$a_{4k} = a_{2k+1} + \sum_{j=1}^{k-1} a_j a_{2k-j} + \tfrac{1}{2}(a_k^2 - a_k^{(2)}),$$

$$a_{4k+1} = a_{2k+3} + \sum_{j=1}^{k} a_j a_{2k+2-j} + \tfrac{1}{2}(a_{k+1}^2 - a_{k+1}^{(2)}) + \tfrac{1}{2}(a_{2k}^2 + a_{2k}^{(2)})$$

$$- a_2 a_{2k} + \sum_{j=1}^{k-1} a_j^{(2)} a_{4k-4j} + \sum_{j=1}^{2k-1} (-1)^j a_j a_{4k-j},$$

$$a_{4k+2} = a_{2k+2} + \sum_{j=1}^{k} a_j a_{2k+1-j}, \quad \text{and}$$

$$a_{4k+3} = a_{2k+4} + \sum_{j=1}^{k+1} a_j a_{2k+3-j} - \tfrac{1}{2}(a_{2k+1}^2 - a_{2k+1}^{(2)})$$

$$- a_2 a_{2k+1} + \sum_{j=1}^{k} a_j^{(2)} a_{4k+2-4j} + \sum_{j=1}^{2k} (-1)^j a_j a_{4k+2-j}.$$

$$(17)$$

Replication square roots are repeatedly extracted until functions which have no replication square roots mod 2π are found. In addition the prime power maps are calculated. In each case enough coefficients of the pth replicates of the non-monstrous functions are computed from (15) to reduce the number of candidate functions to at most one. A useful check is given by the congruence:

$$f^{(p)} \equiv f \pmod{p}.$$

Programs in Ford's language ALGEB [F] were written from procedures generated by Maple [M]. For the functions q^{-1} and $q^{-1} + q$ we found no prime for which the solutions mod 2π lifted uniquely to $2\pi^k$, $k > 1$. The function $q^{-1} - q$ is a root of $q^{-1} + q$ and we have assumed that no other roots of these functions exist. The recursive relations given here, together with the monstrous data in [CN] or [MS], determine the q-series.

Table. The table contains the initial coefficients a_1, a_2, a_3, and a_5 of 157 non-monstrous, completely replicable functions, which we believe to be the complete set. Each function is described by a number which is its "replication level", together with a small letter identifier; the prime power-maps follow. Capital letter identifiers indicate monstrous functions, for which

ATLAS notation is used as in [CN]. The ghosts [CN] 25Z, 49Z, and 50Z appear here as 25a, 49a, and 50a.

Non-monstrous completely replicable functions

f	Power maps		a_1	a_2	a_3	a_5
1a			1	0	0	0
1b			0	0	0	0
2a	1A		-492	0	-22590	-367400
2b	1a		-1	0	0	0
4a	2A		-76	0	-702	-5224
5a	1A		-6	20	15	0
6a	3A	2a	-33	0	-153	-713
6b	3A	2a	21	0	171	745
6c	3B	2a	-6	0	9	16
6d	3C	2A	16	-8	0	28
8a	4A		-20	0	-62	-216
8b	4B		8	0	-6	48
8c	4B		-8	0	-6	-48
9a	3A		0	14	0	65
9b	3A		9	-4	0	2
9c	3B		0	-4	0	2
9d	3C		-3	2	0	5
10a	5A	2a	8	0	35	100
10b	5a	2A	2	-4	7	0
10c	5a	2a	-2	0	-5	0
12a	6A	4B	-11	0	-21	-55
12b	6A	4a	5	0	27	41
12c	6C	4C	5	0	-5	9
12d	6C	4D	-3	0	3	-7
12e	6d	4B	4	0	0	-4
12f	6d	4a	-4	0	0	-4
14a	7A	2a	-9	0	-15	-33
14b	7B	2a	-2	0	-1	2
14c	7A	2a	5	0	13	37
15a	5A	3C	5	-2	0	-1
15b	5a	3A	3	2	-3	0
16a	8B		0	0	6	0
16b	8B		4	0	-2	8
16c	8B		-4	0	-2	-8
16d	8D		0	0	-2	0
16e	8C		2	0	-2	4
16f	8C		-2	0	-2	-4
16g	8b		2	0	2	-4
16h	8b		-2	0	2	4
18a	9b	6A	1	4	0	10
18b	9a	6b	0	0	0	7
18c	9A	6c	3	0	9	16
18d	9c	6B	0	4	0	10

18e	9a	6C		0	-2	0	1
18f	9b	6a		-3	0	0	-2
18g	9b	6b		3	0	0	-2
18h	9a	6A		4	-2	0	1
18i	9c	6c		0	0	0	-2
18j	9d	6d		1	-2	0	1
20a	10A	4a		-6	0	-7	-14
20b	10A	4a		4	0	3	16
20c	10C	4a		-1	0	-2	1
20d	10B	4C		0	0	3	-4
20e	10b	4B		2	0	-1	0
22a	11A	2a		3	0	4	11
24a	12A	8a		-5	0	-5	-9
24b	12A	8a		1	0	7	9
24c	12B	8a		-2	0	1	0
24d	12C	8b		-1	0	3	3
24e	12C	8c		1	0	3	-3
24f	12C	8C		-3	0	-1	-3
24g	12C	8C		3	0	-1	3
24h	12E	8E		1	0	-1	1
24i	12e	8b		2	0	0	0
24j	12e	8c		-2	0	0	0
25a	5B			-1	0	0	0
26a	13A	2a		2	0	4	6
27a	9A			3	-1	0	-1
27b	9A			0	2	0	5
27c	9B			0	-1	0	-1
27d	9b			0	2	0	-1
27e	9b			0	-1	0	2
28a	14A	4a		1	0	5	5
30a	15A	10a	6b	-4	0	-4	-5
30b	15a	10A	6d	1	2	0	3
30c	15B	10a	6c	-1	0	-1	1
30d	15A	10a	6a	2	0	2	7
30e	15b	10b	6A	-1	2	1	0
30f	15b	10c	6b	1	0	1	0
32a	16a			0	0	0	0
32b	16A			0	0	2	0
32c	16b			2	0	0	0
32d	16b			-2	0	0	0
32e	16d			0	0	0	0
34a	17A	2a		1	0	3	4
35a	7A	5a		1	-1	1	0
36a	18a	12b		-1	0	0	2
36b	18e	12A		0	2	0	1
36c	18h	12b		2	0	0	-1
36d	18a	12C		1	0	0	2
36e	18e	12d		0	0	0	-1
36f	18C	12I		-1	0	1	0
36g	18d	12F		0	0	0	2
36h	18h	12a		-2	0	0	-1

36i	18c	12f		-1	0	0	-1
38a	19A	2a		2	0	1	3
40a	20A	8a		0	0	3	4
40b	20B	8b		-2	0	-1	-2
40c	20B	8c		2	0	-1	2
40d	20D	8F		0	0	-1	0
40e	20e	8C		0	0	1	0
42a	21A	14a	6b	0	0	3	3
42b	21A	14c	6a	2	0	1	1
42c	21C	14A	6d	2	-1	0	0
42d	21D	14b	6c	1	0	2	2
44a	22A	4B		-3	0	-2	-3
44b	22B	4D		-1	0	0	-1
44c	22A	4a		1	0	2	1
45a	15A	9b		-1	1	0	2
45b	15C	9c		0	1	0	2
45c	15b	9a		0	-1	0	0
48a	24A	16b		1	0	1	-1
48b	24A	16c		-1	0	1	1
48c	24g	16e		-1	0	1	1
48d	24g	16f		1	0	1	-1
48e	24d	16g		-1	0	-1	-1
48f	24d	16h		1	0	-1	1
48g	24E	16a		0	0	0	0
48h	24H	16d		0	0	1	0
49a	7B			2	1	2	4
50a	25a	10E		1	2	2	4
52a	26A	4a		2	0	0	2
54a	27a	18B		1	1	0	1
54b	27c	18E		0	1	0	1
54c	27b	18c		0	0	0	1
54d	27c	18g		0	0	0	1
56a	28B	8a		1	0	1	1
56b	28A	8b		1	0	1	-1
56c	28A	8c		-1	0	1	1
58a	29A	2a		1	0	1	1
60a	30C	20d	12c	0	0	0	-1
60b	30B	20a	12b	0	0	2	1
60c	30b	20B	12e	-1	0	0	1
60d	30e	20e	12a	-1	0	-1	0
60e	30b	20b	12f	1	0	0	1
63a	21A	9a		0	0	0	2
64a	32b			0	0	0	0
66a	33B	22a	6a	0	0	1	2
70a	35A	14a	10a	1	0	0	2
72a	36A	24c		1	0	1	0
72b	36b	24A		0	0	0	1
72c	36d	24d		-1	0	0	0
72d	36d	24e		1	0	0	0
72e	36g	24F		0	0	0	0
76a	38A	4a		0	0	1	1

80a	40B	16a		0	0	1	0
82a	41A	2a		0	0	1	1
84a	42A	28a	12b	-2	0	-1	-1
90a	45a	30B	18a	1	-1	0	0
90b	45b	30A	18d	0	-1	0	0
96a	48g	32a		0	0	0	0
102a	51A	34a	6a	1	0	0	1
117a	39A	9a		0	1	0	0
120a	60B	40a	24a	0	0	0	1
126a	63a	42a	18b	0	0	0	0
132a	66A	44a	12a	0	0	1	0
140a	70A	28a	20a	1	0	0	0

We correct an error in [MS]: On page 265 class 29Z should read 25Z and signs should be inserted compatible with its sign pattern.

References

[B] R.E. Borcherds, *Monstrous moonshine and monstrous Lie superalgebras*, preprint (1989).

[CN] J.H. Conway & S.P. Norton, *Monstrous Moonshine*, Bull. Lond. Math. Soc. 11 (1979), 308-339.

[F] D.J. Ford, "On the Computation of the Maximal Order in a Dedekind Domain," Ph. D. Dissertation, Ohio State University, 1978.

[M] B.W. Char, K.O. Geddes, Gaston H. Gonnet, M.B. Monagan and S.M. Watt, "The Maple Reference Manual (5th edition)," Watcom, Waterloo, 1988.

[Mah] K. Mahler, *On a class of non-linear functional equations connected with modular functions*, J. Austral. Math. Soc. 22A (1976), 65–118.

[Mas] G. Mason, *Finite groups and Hecke operators*, Math. Ann. 283 (1989), 381–409.

[MS] J. McKay and H. Strauß, *The q-series of monstrous moonshine & the decomposition of the head characters*, Comm. in Alg. 18 (1990), 253–278.

[N] S.P. Norton, *More on Moonshine*, in "Computational Group Theory," edited by M. D. Atkinson, Academic Press, 1984, pp. 185–193.

[S] J-P. Serre, "A Course in Arithmetic," Springer-Verlag, 1973.

Appendix

The Monster & Moonshine — A Bibliography

[AS]   Akbas, M. & Singerman, D. The normalizer of $\Gamma_0(N)$ in $PSL(2, R)$, Glasgow Math. J., 32, (1990), 317-327.

[B1]   Borcherds, R. Vertex algebras, Kac-Moody algebras, & the Monster. Proc. Nat. Acad. Sci., 83, (1986), 3068-3071.

[B2]   Borcherds, R. Generalized Kac-Moody algebras. J. Alg., 115, (1988), 501-512.

[C1]   Conway, J.H. Monsters and Moonshine. Math. Intelligencer, 2, (1980), 164-171.

[C2]   Conway, J.H. A simple construction for the Fischer-Griess Monster group. Inv. Math., (1985), 513-540.

[CN]   Conway, J.H. & Norton, S.P. Monstrous Moonshine. Bull. Lond. Math. Soc.,11, (1979), 308-339.

[DVVV] Dijkgraaf, R., Vafa, C., Verlinde, E. & Verlinde, H. The Operator Algebra and Orbifold Models, Comm. Math. Physics, 123, (1989), 485-526.

[DGH]   Dixon, L., Ginsparg, P., Harvey, J., Beauty and the Beast: Superconformal symmetry in a Monster module. Comm. Math. Phys. (1988), 119.

[DGM1] Dolan, L., Goddard, P., and Montague, P., Conformal Field Theory of Twisted Vertex Operators. Nuclear Physics, B338, (1990), 529-601.

[DGM2] Dolan, L., Goddard, P., and Montague, P., Conformal Field Theory, Triality, and the Monster Group. Physics Letters, B236, (1990), 165-172.

[DKM]   Dummit, D., Kisilevsky, H., McKay, J. Multiplicative products of η-functions. Contemp. Math. 45, (1985), 89-98.

[F]   Fong, P. Characters arising in the monster-modular connection. A.M.S. Proc. Symp. Pure Math. 37, (1979), 557-559.

[FV]   Freed, D. & Vafa, C. Global anomalies on orbifolds, Comm. Math. Phys. 110, (1987), 349-389.

[FLM1] Frenkel, I., Lepowsky, J., Meurman, A. An E_8 approach to the Monster. Contemp. Math. 45, (1985), 99-120.

[FLM2] Frenkel, I., Lepowsky, J., Meurman, A. Vertex operator algebras& the Monster. Academic Press.(1989).

[G]   Griess, R.L. The Friendly Giant. Inv. Math. 69, (1982), 1-102.

[HL1]   Harada, K., Lang, M.-L. On a question of Conway-Norton, J. Alg. 125,(1989), 298-310.

[HL2]   Harada, K., Lang, M.-L. On some sublattices of the Leech lattice, Hokkaido Math. J. 19, (1990), 435-446.

[Koi1] Koike, M. On McKay's conjecture. Nagoya J. Math. 95, (1984), 85-89.

[Koi2]   Koike, M. The Mathieu group M_{24} and modular forms. Nagoya J. Math. 99, (1985), 147-157.

[Koi3]   Koike, M. Moonshines of $PSL_2(F_q)$ and the automorphism groupof the Leech lattice. Japanese J. Math. 12, (1986), 283-323.

[Koi4]   Koike, M. Moonshine for $PSL_2(F_7)$. In Automorphic forms and number theory (Sendai 1983), Adv. Studies in Pure Math.,7, North-Holland (1985), 103-111.

[Koi5]   Koike, M. Modular forms and the automorphism group of the Leech lattice. Nagoya Math.J. 112, (1988), 63-79.

[Kon1]   Kondo, T. Examples of multiplicative η-quotients. Sci. Papers College Arts-Sci. Univ. Tokyo 35, (1986), 133-149.

[Kon2]   Kondo, T. The automorphism group of the Leech lattice and elliptic modular functions. J.Math.Soc.Japan 37, (1985), 337-362.

[KonT]   Kondo, T., Tasaka T. The theta functions of the Leech lattice. Nagoya-Math.J. 101, (1986), 151-179.

[L]   Lang, M.-L. On a question raised by Conway-Norton. J. Math. Soc. ofJapan 41, (1989), 263-284.

[M1]   Mahler, K. On a class of non-linear functional equations connected with modular functions. J. Austral. Math. Soc. 22A, (1976), 65-118.

[M2]   Mahler, K. On a special nonlinear functional equation. Proc. R. Soc. Lond. A 378, (1981), 155-178.

[M3]   Mahler, K. On the analytic solution of certain functional and difference equations. Proc. R. Soc. Lond. A 389, (1983), 1-13.

[Ma1]   Mason, G. M_{24} and certain automorphic forms. Contemp. Math. 45, (1985), 223-244.

[Ma2]   Mason, G. Finite Groups and Modular Functions. Proc. Symp. Pure Math. 47, (1987), 181-210. (Norton, S.P. Appendix: Generalized Moonshine).

[Ma3]   Mason, G. Groups, discriminants, and the spinor norm.Bull. Lond. Math. Soc. 21, (1989), 51-56.

[Ma4]   Mason, G. Frame-shapes and rational characters of finite groups.J. Alg. 89 (1984), 237-246.

[Ma5]   Mason, G. Elliptic systems and the η-function. (to appear).

[Ma6]   Mason, G. Finite Groups and Hecke Operators. Math. Ann. 283, (1989), 381-409.

[Ma7]   Mason, G. On a system of elliptic modular forms attached to the large Mathieu group, Nagoya J. Math. 118, (1990), 175-195.

[McK]   McKay, J. Graphs, singularities, and finite groups. In Santa Cruz conference on finite groups, A.M.S. Proc. Symp. Pure Math. 37, (1979), 183-186.

[Nahm]   Nahm, W. Quantum field theories in one and two dimensions. Duke.J. Math. 54, (1987), 579 - 613.

[MN] Meyer, W., & Neutsch, W. Associative subalgebras of the Griess algebra. (1990) (to appear).

[N] Norton, S.P. More on Moonshine. In "Computational Group Theory" ed. Atkinson, M.D. Academic Press (1984), 185-193.

[O] Ogg, A.P. Modular functions, A.M.S. Proc. Symp. Pure. Math. 37, (1979), 521-532.

[Q1] Queen, L. Modular functions and finite simple groups, A.M.S. Proc. Symp. Pure Math. 37, (1979), 561-570.

[Q2] Queen, L. Some relations between finite groups, Lie groups and modular functions, Ph.D. thesis, Cambridge (1980).

[Q3] Queen, L. Modular functions arising from some finite groups, Math. Comp. 37, (1981), 547-580.

[S] Smit, J-D. Quantum Groups, Algebraic Geometry and Conformal Field theory, Ph.D. Thesis, Utrecht (1989).

[Th1] Thompson, J.G. Some numerology between the Fischer-Griess monster and the elliptic modular function. Bull. Lond. Math. Soc. 11, (1979), 352 - 353.

[Th2] Thompson, J.G. A finiteness theorem for subgroups of $PSL_2(R)$ commensurable with $PSL_2(Z)$. A.M.S. Proc. Symp. Pure Math. 37, (1979), 533-555.

[Ti1] Tits, J. Le Monstre. Seminaire Bourbaki 620, (1983/84). In Asterisque vol. 121-122, 105-122.

Introduction to the monster Lie algebra

RICHARD E. BORCHERDS

I would like to thank J. M. E. Hyland for suggesting many improvements to this paper.

The monster sporadic simple group, of order

$$2^{46}3^{20}5^{9}7^{6}11^{2}13^{3}17.19.23.29.31.41.47.59.71,$$

acts on a graded vector space $V = \oplus_{n \in Z} V_n$ constructed by Frenkel, Lepowsky and Meurman ([Fre]). The dimension of V_n is equal to the coefficient $c(n)$ of the elliptic modular function

$$j(\tau) - 744 = \sum_{n} c(n)q^{n} = q^{-1} + 196884q + 21493760q^{2} + \ldots$$

(where we write q for $e^{2\pi i \tau}$). The main problem is to describe V as a graded representation of the monster, or in other words to describe the trace $\mathrm{Tr}(g|V_n)$ of each element g of the monster on each space V_n. The best way to describe this information is to define the Thompson series

$$T_g(\tau) = \sum_{n \in Z} \mathrm{Tr}(g|V_n)q^{n}$$

for each element g of the monster, so that our problem is to work out what these Thompson series are. For example if 1 is the identity element of the monster then $\mathrm{Tr}(1|V_n) = \dim(V_n) = c_n$, so that the Thompson series $T_1(\tau) = j(\tau) - 744$ is the elliptic modular function. McKay, Thompson, Conway and Norton conjectured ([Con]) that the Thompson series $T_g(\tau)$ are always hauptmodules for certain modular groups of genus 0 (I will explain what this means in a moment). In this paper we describe the proof of this in [Bor] using an infinite dimensional Lie algebra acted on by the monster called the monster Lie algebra. These hauptmodules are known explicitly, so this gives a complete description of V as a representation of the monster.

We now recall the definition of a hauptmodule. The group $SL_2(Z)$ acts on the upper half plane $H = \{\tau \in C | \mathrm{im}(\tau) > 0\}$ by $\begin{pmatrix} a,b \\ c,d \end{pmatrix}(\tau) = \frac{a\tau+b}{c\tau+d}$. The elliptic modular function $j(\tau)$ is more or less the simplest function defined on H that is invariant under $SL_2(Z)$ in much the same way that the function $e^{2\pi i \tau}$ is the simplest function invariant under $\tau \to \tau + 1$. The element $\left(\begin{smallmatrix} 1,1 \\ 0,1 \end{smallmatrix}\right)$ of $SL_2(Z)$ takes τ to $\tau + 1$, so in particular $j(\tau)$ is periodic and can be written as a power series in $q = e^{2\pi i \tau}$. The exact expression for j is

$$j(\tau) = \frac{(1 + 240 \sum_{n>0} \sigma_3(n) q^n)^3}{q \prod_{n>0} (1 - q^n)^{24}}$$

where $\sigma_3(n) = \sum_{d|n} d^3$ is the sum of the cubes of the divisors of n; see any book on modular forms or elliptic functions, for example [Ser]. Another way of thinking about j is that is defines an isomorphism from the quotient space $H/SL_2(Z)$ to the complex plane, which can be thought of as the Riemann sphere minus the point at infinity.

There is nothing particularly special about the group $SL_2(Z)$ acting on H. We can consider functions invariant under any group G acting on H, for example some group G of finite index in $SL_2(Z)$. If G satisfies some mild conditions then the quotient H/G will again be a compact Riemann surface with a finite number of points removed. If this compact Riemann surface is a sphere, rather than something of higher genus, then we say that G is a genus 0 group. When this happens there is a more or less unique map from H/G to the complex plane C, which induces a function from H to C which is invariant under the group G acting on H. When this function is correctly normalized it is called a hauptmodule for the genus 0 group G. For example, $j(\tau) - 744$ is the hauptmodule for the genus 0 group $SL_2(Z)$.

For example, we could take G to be the group $\Gamma_0(2)$, where $\Gamma_0(N) = \{ \left(\begin{smallmatrix} a,b \\ c,d \end{smallmatrix}\right) \in SL_2(Z) | c \equiv 0 \bmod N \}$. The quotient H/G is then a sphere with 2 points removed, so that G is a genus 0 group. Its hauptmodule starts off $T_{2-}(\tau) = q^{-1} + 276q - 2048q^2 + \ldots$, and is equal to the Thompson series for a certain element of the monster of order 2 (of type 2B in atlas notation). Similarly $\Gamma_0(N)$ is a genus 0 subgroup for several other values of N which correspond to elements of the monster. (However the genus of $\Gamma_0(N)$ tends to infinity as N increases, so there are only a finite number of integers N for which it has genus 0; more generally Thompson has shown that there are only a finite number of conjugacy classes of genus 0 subgroups of $SL_2(R)$ which are commensurable with $SL_2(Z)$.) The first person to notice any connection between the monster and genus 0 subgroups was probably Ogg, who observed that the primes p for which the normalizer of $\Gamma_0(p)$ in $SL_2(R)$ has genus 0 are exactly those which divide the order of the monster.

So the problem we want to solve is to calculate the Thompson series $T_g(\tau)$ and show that they are hauptmodules of genus 0 subgroups of $SL_2(R)$. The difficulty with doing this is as follows. Frenkel, Lepowsky, and Meurman constructed V as the sum of two subspaces V^+ and V^-, which are the $+1$ and -1 eigenspaces of an element of order 2 in the monster. If an element g of the monster commutes with this element of order 2, then it is not difficult to to work out its Thompson series $T_g(q) = \sum_n \text{Tr}(g|V_n)q^n$ as the sum of two series given by its traces on V^+ and V^-, and it would be tedious but straightforward to check that these all gave hauptmodules. However if an element of the monster is not conjugate to something that commutes with this element of order 2 then there is no obvious direct way of working out its Thompson series, because it muddles up V^+ and V^- in a very complicated way.

We now give a bird's eye view of the rest of the paper. We first construct a Lie algebra from V, called the monster Lie algebra. This is a generalized Kac-Moody algebra, so we recall some facts about such algebras, and in particular show how each such algebra gives an identity called its denominator formula. A well known example of this is that the denominator formulas of affine Kac-Moody algebras are the Macdonald identities. The denominator formula for the monster Lie algebra turns out to be the product formula for the elliptic modular function j, and we use this to work out the structure of the monster Lie algebra, and in particular to find its "simple roots". Once we know the simple roots, we can write down a sort of twisted denominator formula for each element of the monster group. These twisted denominator formulas imply some relations between the coefficients of the Thompson series of the monster, which are strong enough to characterize them and verify that they are indeed hauptmodules.

We will now introduce the monster Lie algebra. This is a Z^2-graded Lie algebra, whose piece of degree $(m,n) \in Z^2$ is isomorphic as a module over the monster to V_{mn} if $(m,n) \neq (0,0)$ and to R^2 if $(m,n) = (0,0)$, so for small degrees it looks something like

$$
\begin{array}{ccccccccc}
 & \vdots & \vdots & \vdots & \vdots & \vdots & \vdots & \vdots & \\
\cdots & 0 & 0 & 0 & 0 & V_3 & V_6 & V_9 & \cdots \\
\cdots & 0 & 0 & 0 & 0 & V_2 & V_4 & V_6 & \cdots \\
\cdots & 0 & 0 & V_{-1} & 0 & V_1 & V_2 & V_3 & \cdots \\
\cdots & 0 & 0 & 0 & R^2 & 0 & 0 & 0 & \cdots \\
\cdots & V_3 & V_2 & V_1 & 0 & V_{-1} & 0 & 0 & \cdots \\
\cdots & V_6 & V_4 & V_2 & 0 & 0 & 0 & 0 & \cdots \\
\cdots & V_9 & V_6 & V_3 & 0 & 0 & 0 & 0 & \cdots \\
 & \vdots & \vdots & \vdots & \vdots & \vdots & \vdots & \vdots & \\
\end{array}
$$

This is a very big Lie algebra because the dimension of V_n increases exponentially fast; most infinite dimensional Lie algebras that occur in mathematics can be graded so that the dimensions of the graded pieces have only polynomial growth.

Very briefly, the construction of this Lie algebra goes as follows (this construction is not used later, so it does not matter if the reader finds this meaningless). The graded vector space V is a vertex algebra; roughly speaking this means that V has an infinite number of products on it which satisfy some rather complicated identities. (The Griess product on the 196884 dimensional space V_1 is a special case of one of the vertex algebra products.) We can also construct a vertex algebra from any even lattice, and the tensor product of two vertex algebras is a vertex algebra. We take the tensor product of the monster vertex algebra V with that of the two dimensional even unimodular Lorentzian lattice (defined by the matrix $\left(\begin{smallmatrix} 0,1 \\ 1,0 \end{smallmatrix} \right)$) to get a new vertex algebra W. The monster Lie algebra is then the space of physical states of the vertex algebra W; this is a subquotient of W, which is roughly a space of highest weight vectors for an action of the Virasoro algebra on W. This subquotient can be identified exactly using the no-ghost theorem from string theory, and turns out to be as described above.

We need to know what the structure of the monster Lie algebra is. It turns out to be something called a generalized Kac-Moody algebra, so we will now have a digression to explain what these are.

To motivate Kac-Moody algebras we first look at the structure of finite dimensional simple Lie algebras. A typical finite dimensional Lie algebra is $sl_4(R)$, the algebra of 4×4 matrices of trace 0 with bracket given by $[a, b] = ab - ba$. If we look at this Lie algebra we see that it has three subalgebras $sl_2(R)$ as indicated below (as well as many others):

$$\begin{pmatrix} * & * & & \\ * & * & & \\ & & & \\ & & & \end{pmatrix}, \begin{pmatrix} & & & \\ & * & * & \\ & * & * & \\ & & & \end{pmatrix}, \begin{pmatrix} & & & \\ & & & \\ & & * & * \\ & & * & * \end{pmatrix}.$$

The Dynkin diagram of $sl_4(R)$ is constructed by drawing a point for each of these sl_2's, and connecting them by various numbers of lines depending on how the sl_2's are arranged; for example, if the two sl_2's commute then we draw no lines between the corresponding points, and if the sl_2's overlap in one corner we draw a single line between the points. The Dynkin diagram of sl_4 is therefore $\bullet - \bullet - \bullet$, where the two outer dots are not joined because the corresponding sl_2's of sl_4 commute, and so on. Conversely we can reconstruct sl_4 from its Dynkin diagram by writing down some generators and relations from the Dynkin diagram. Roughly speaking, sl_4 is generated by an sl_2 for each point of the Dynkin diagram together with some relations depending on the lines between points; for example if two points are not joined we add relations saying that the corresponding sl_2's commute. We can construct all other finite dimensional split simple Lie algebras from their Dynkin diagrams (usually denoted by A_n, B_n, C_n, D_n, E_6, E_7, E_8, F_4, and G_2) in a similar way. If we are given some graph which is not one of these Dynkin diagrams of finite Lie algebras we can still construct a Lie algebra, called a Kac-Moody algebra, by writing down generators and relations; the main difference is that these Lie algebras will be infinite dimensional rather than finite dimensional. We can think of a Kac-Moody algebra as an algebra generated by a copy of sl_2 for each point of its Dynkin diagram.

The simplest example of an infinite dimensional Kac-Moody algebra is the Lie algebra $sl_2(R[z, z^{-1}])$ of 2×2 matrices whose entries are Laurent polynomials. (This is a slight simplification; the Kac-Moody algebra is really a central extension of this Lie algebra, but we will ignore this.) The Dynkin diagram is $\bullet = \bullet$, so this algebra is generated by 2 copies of $sl_2(R)$, which are the subalgebras of elements of the form $\begin{pmatrix} a,b \\ c,-a \end{pmatrix}$ and $\begin{pmatrix} a,bz \\ cz^{-1},-a \end{pmatrix}$. The algebra $sl_2(R[z, z^{-1}])$ can be Z^2-graded by letting $\begin{pmatrix} 0,z^n \\ 0,0 \end{pmatrix}$ have degree $(2n+1, 1)$, letting $\begin{pmatrix} z^n,0 \\ 0,-z^n \end{pmatrix}$ have degree $2n$, and letting $\begin{pmatrix} 0,0 \\ z^n,0 \end{pmatrix}$ have degree $(2n-1, 1)$. The picture we get looks like

$$\cdots \begin{pmatrix} 0 & 0 \\ z^{-2} & 0 \end{pmatrix} \qquad \begin{pmatrix} 0 & 0 \\ z^{-1} & 0 \end{pmatrix} \qquad \begin{pmatrix} 0 & 0 \\ 1 & 0 \end{pmatrix} \qquad \begin{pmatrix} 0 & 0 \\ z & 0 \end{pmatrix} \cdots$$

$$\cdots \begin{pmatrix} z^{-1} & 0 \\ 0 & -z^{-1} \end{pmatrix} \qquad \begin{pmatrix} 1 & 0 \\ 0 & -1 \end{pmatrix} \qquad \begin{pmatrix} z & 0 \\ 0 & -z \end{pmatrix} \cdots$$

$$\cdots \begin{pmatrix} 0 & z^{-1} \\ 0 & 0 \end{pmatrix} \qquad \begin{pmatrix} 0 & 1 \\ 0 & 0 \end{pmatrix} \qquad \begin{pmatrix} 0 & z \\ 0 & 0 \end{pmatrix} \qquad \begin{pmatrix} 0 & z^2 \\ 0 & 0 \end{pmatrix} \cdots$$

where on each point of Z^2 we have written something that spans the corresponding subspace of the Lie algebra.

We now describe the denominator formula of Kac-Moody algebras, and as an example show that for the Lie algebra $sl_2(R[z, z^{-1}])$ it becomes the Jacobi triple product identity. The characters $\text{Ch}(V)$ of finite dimensional irreducible representations V of finite dimensional simple Lie algebras are described by the Weyl character formula

$$\text{Ch}(V) = \frac{\sum_{w \in W} \det(w) e^{w(\lambda - \rho)}}{e^{-\rho} \prod_{\alpha > 0} (1 - e^\alpha)^{\text{mult}(\alpha)}}$$

and the same formula turns out to be true for the characters of lowest weight representations of Kac-Moody algebras. The only case we will use this is when V is the trivial one dimensional representation with character 1, when the character formula becomes the denominator formula

$$e^{-\rho} \prod_{\alpha > 0} (1 - e^\alpha)^{\text{mult}(\alpha)} = \sum_{w \in W} \det(w) e^{-w(\rho)}.$$

Rather than explain precisely what all the terms in this formula mean, we will just describe what happens in the case of $sl_2(R[z, z^{-1}])$. The product is over the "positive roots" α; for $sl_2(R[z, z^{-1}])$ these are the vectors $\alpha = (m, n)$ with $m > 0$ for which there is something in $sl_2(R[z, z^{-1}])$ of that degree, i.e., the vectors $(2m, 0)$, $(2m - 1, 1)$, and $(2m - 1, -1)$ for $m > 0$. The multiplicity $\text{mult}(\alpha)$ is the dimension of the subspace of that degree, which for $sl_2(R[z, z^{-1}])$ is always 1. If we put $e^{(m,n)} = x^m y^n$, then the product becomes

$$\prod_{m>0}(1 - x^{2m})(1 - x^{2m-1}y)(1 - x^{2m-1}y^{-1}).$$

The sum is a sum over all elements of the "Weyl group" W, which is a reflection group generated by one reflection for each point of the Dynkin diagram. In this case the Weyl group is the infinite dihedral group, which has an infinite cyclic group of index 2, so the sum over the Weyl group is essentially a sum over the integer s. If we work out explicitly what it is, the Weyl denominator formula becomes

$$\prod_{m>0}(1 - x^{2m})(1 - x^{2m-1}y)(1 - x^{2m-1}y^{-1}) = \sum_{n \in Z}(-1)^n x^{n^2} y^n,$$

which is the Jacobi triple product identity.

Not surprisingly, we can do the same thing with sl_2 replaced by any other simple finite dimensional Lie algebra, and we then obtain some identities called the Macdonald identities. (In fact there is more than one Macdonald identity for some finite dimensional Lie algebras, because there are sometimes ways of "twisting" this construction.)

Generalized Kac-Moody algebras are rather like Kac-Moody algebras except that we are allowed to glue together the sl_2's in more complicated ways, and are also allowed to use Heisenberg Lie algebras as well as sl_2's to generate the algebra. In technical terms, the roots of a Kac-Moody algebra may be either real (norm > 0) or imaginary (norm ≤ 0), but all the simple roots must be real. Generalized Kac-Moody algebras may also have imaginary simple roots. They have a denominator formula which is similar to the one above, except that it has some extra correction terms corresponding to the imaginary simple roots. (The simple roots of a generalized Kac-Moody algebra correspond to a minimal set of generators for the subalgebra corresponding to the positive roots; for example the simple roots of $sl_2(R[z, z^{-1}])$ are $(1, 1)$ and $(1, -1)$. For Kac-Moody algebras the simple roots also correspond to the points of the Dynkin diagram and to the generators of the Weyl group.)

We now return to the monster Lie algebra. This is a generalized Kac-Moody algebra, and we are going to write down its denominator formula, which says that a product over the positive roots is a sum over the Weyl group. The positive roots are the vectors (m, n) with $m > 0$, $n > 0$, and the vector $(1, -1)$, and the root (m, n) has multiplicity $c(mn)$. The Weyl group has order 2 and its nontrivial element maps (m,n) to (n,m), so it exchanges p and q. The denominator formula is the product formula for the j function

$$p^{-1} \prod_{m>0, n \in Z} (1 - p^m q^n)^{c(mn)} = j(p) - j(q).$$

(The left side is apparently not antisymmetric in p and q, but in fact it is because of the factor of $p^{-1}(1 - p^1 q^{-1})$ in the product.) The reason why we get $j(p)$ rather than just a monomial in p and q on the right hand side (as we would for ordinary Kac-Moody algebras) is because of the correction due to the imaginary simple roots of M. The simple roots of M correspond to a set of generators of the subalgebra E of the elements of M whose degree is to the right of the y axis (so the roots of E are the positive roots of M), and turn out to be the vectors $(1, n)$ each with multiplicity $c(n)$. In the picture of the monster Lie algebra given earlier, the simple roots are given by the column just to the right of the one containing R^2. The sum of the simple root spaces is isomorphic to the space V. The simple root $(1, -1)$ is real of norm 2, and the simple roots $(1, n)$ for $n > 0$ are imaginary of norm $-2n$ and have multiplicity $c(n) = \dim(V_n)$. As these multiplicities are exactly the coefficients of the j function, it is not surprising that j appears in the correction caused by the imaginary simple roots. This discussion is slightly misleading because we have implied that we obtain the product formula of the j function as the denominator formula of the monster Lie algebra by using our knowledge of the simple roots; in fact we really have to use this argument in reverse, using the product formula for the j function in order to work out what the simple roots of the monster Lie algebra are.

We will now extract information about the coefficients of the Thompson series $T_g(\tau)$ from a twisted denominator formula for the monster Lie algebra. For an arbitrary generalized Kac-Moody algebra there is a more high powered version of the Weyl character formula which states that

$$\Lambda(E) = H_*(E),$$

where E is the subalgebra corresponding to the positive roots. Here $\Lambda(E) = \Lambda^0(E) - \Lambda^1(E) + \Lambda^2(E)\ldots$ is the alternating sum of the exterior powers of E, and similarly $H_*(E)$ is the alternating sum of the homology groups $H_i(E)$ of the Lie algebra E (see [Car]). This identity is true for any Lie algebra because the H_i's are the homology groups of a complex whose terms are the Λ^i's. The left hand side corresponds to a product over the positive roots because $\Lambda(A \oplus B) = \Lambda(A) \otimes \Lambda(B)$, $\Lambda(A) = 1 - A$ if A is one dimensional, and E is the sum of $\text{mult}(\alpha)$ one dimensional spaces for each positive root α. It is more difficult to identify H_* with a sum over the Weyl group, and we do this roughly as follows. For Kac-Moody algebras H_i turns out to have dimension equal to the number of elements in the Weyl group of length i; for finite dimensional Lie algebras this was first observed by Bott, and was used by Kostant to give a homological proof of the Weyl character formula. The sum over the homology groups can therefore be identified with a sum over the Weyl group. For generalized Kac-Moody algebras things are a bit more complicated. The sum over the homology groups becomes a sum over the Weyl group, which is generated by an involution for each real simple root, and the thing we sum contains terms corresponding to the imaginary simple roots.

Anyway, we can work out the homology groups of E explicitly provided we know the simple roots of our Lie algebra; for example, the first homology group H_1 is the sum of the simple root spaces. For the monster Lie algebra we have worked out the simple roots using its denominator formula, which is the product formula for the j function, and the homology groups of E turn out to be $H_0 = R$, $H_1 = \sum_{n \in Z} V_n p q^n$, $H_2 = \sum_{m > 0} V_m p^{m+1}$, and all the higher homology groups are 0. Each homology group is a Z^2-graded representation of the monster, and we use the p's and q's to keep track of the grading. If we substitute these values into the formula $\Lambda(E) = H_*(E)$ we find that

$$\Lambda(\sum_{n \in Z, m > 0} V_{mn} p^m q^n) = \sum_m V_m p^{m+1} - \sum_n V_n p q^n.$$

Both sides of this are virtual graded representations of the monster. If we replace everything by its dimension we recover the product formula for the j function. More generally, we can take the trace of some element of the monster on both sides, which after some calculation gives the identity

$$p^{-1} \exp\left(-\sum_{i > 0} \sum_{m > 0, n \in Z} \mathrm{Tr}(g^i | V_{mn}) p^{mi} q^{ni} / i\right) = \sum_{m \in Z} \mathrm{Tr}(g | V_m) p^m - \sum_{n \in Z} \mathrm{Tr}(g | V_n) q^n$$

where $\mathrm{Tr}(g | V_n)$ is the trace of g on the vector space V_n.

These relations between the coefficients of the Thompson series turn out to be strong enough to characterize them and check that they are hauptmodules for genus 0 subgroups. Unfortunately this final step of the proof is rather messy. The relations above determine the Thompson series from their first few coefficie nts. Norton and Koike checked that certain modular functions of genus 0 also satisfy the same recursion relations. We can therefore prove that the Thompson series $T_g(q)$ are modular functions of genus 0 be checking that the first few coefficients of both functions are the same. Norton has conjectured that hauptmodules with integer coefficients are essentially the same as functions satisfying relations similar to the ones above, and a conceptual proof or explanation of this would be a big improvement to the final part of the proof. (It should be possible to prove this conjecture by a very long and tedious case by case check, because all functions which are either hauptmodules or which satisfy the relations above can be listed explicitly.)

Bibliography.

[Bor] R. E. Borcherds, Monstrous moonshine and monstrous Lie superalgebras, preprint.

[Car] H. Cartan, S. Eilenberg, Homological algebra, Princeton University Press 1956.

[Con] J. H. Conway, S. Norton, Monstrous moonshine, Bull. London. Math. Soc. 11 (1979) 308-339.

[Fre] I. B. Frenkel, J. Lepowsky, A. Meurman, Vertex operator algebras and the monster, Academic press 1988.

[Ser] J. P. Serre, A course in arithmetic. Graduate texts in mathematics 7, Springer-Verlag, 1973.

Remarks on Moonshine and Orbifolds

Geoffrey Mason[*]

1° Introduction. The celebrated paper of Conway-Norton [2] on connections between the Monster and certain modular functions opened up dramatic possibilities for the study of hitherto unrelated fields, but for a long time the foundations of this new subject remained obscure. Work of Borcherds [1] and Frenkel-Lepowsky-Meurman [4] showed that the origins of "moonshine" involved the theory of infinite-dimensional Lie algebras and that indeed the monster-modular connection could be understood in the context of two-dimensional conformal field theory. In an appendix to [6], Norton extended the Conway-Norton conjectures in a quite remarkable way. He expected that his new strengthened conjectures, as compelling as they were, were peculiar to the Monstrous situation and might help lead to an explanation of the genus zero problem. But these hopes were quickly dashed, in the sense that in [7] I showed that, roughly speaking, one can associate modular forms to the elements of *any* finite group in such a way that the axioms which Norton introduced were satisfied. Shortly thereafter I learned that the ideas of Norton were just those which axiomatize certain algebraic aspects of "conformal field theory on an orbifold."

Thus indeed Norton's ideas are susceptible to application to any finite group and it is my feeling that this approach will provide a suitable foundation to the subject of moonshine. In this note I want to discuss the notion of what I call an "elliptic system," which is just a mild extension of Norton's ideas in [6], and show how it can be thought of as a simultaneous generalization of a modular form and of a "moonshine module." As motivation a brief discussion of some aspects of orbifolds will be included, but a fuller discussion is omitted for lack of space. Nonetheless, it should be apparent that strong motivation can be gained from this point of view, and indeed it is mainly the physicists who have contributed to the group-theoretic aspects of the subject up till now. The reader may want to look at [3] in this regard.

---

[*] Supported by the National Science Foundation.

2° Group Actions. Let G be a group and X a left G-set, so that there is a left action on X by elements of G

$$g: x \to gx, \quad g \in G, \ x \in X \tag{2.1}$$

There is a right action of G on the space of functions $\mathcal{F} = \{f: X \to \mathbb{C}\}$ given by

$$f \cdot g : x \to f(gx), \quad f \in \mathcal{F} \tag{2.2}$$

Frequently one needs to modify this action by a 1-cocycle $c \in C^1(G, \mathcal{F})$. Here, $c: G \to \mathcal{F}$ is a map such that $c(g)$ is nowhere-vanishing on X for $g \in G$ and satisfies

$$c(g_1 g_2, x) = c(g_1, g_2 x)c(g_2, x) \tag{2.3}$$

where here, and below, we write $c(g, x)$ for the value $c(g)(x)$. Equation (2.3) ensures that the "stroke operator"

$$f|g: x \to c(g, x)f(gx) \tag{2.4}$$

is again a right action of G on $\mathcal{F}$.

What 1-cocycles are available to us? Clearly, this reduces to the case of a transitive G-set X, for which there is a pretty answer. For each $f \in \mathcal{F}$ nowhere-vanishing on X there is a "trivial" 1-cocycle, alias 1-coboundary, given by

$$c_f(g, x) = f(gx)/f(x) \tag{2.5}$$

Next, if we fix some elements $x_0 \in X$, with G_0 the stabilizer of x_0 in G, then (2.3) shows that the restriction of c to G_0, evaluated at x_0, is a character $\chi_c: G_0 \to \mathbb{C}^*$ of G_0. Thus we have a map

$$\phi : C^1(G, \mathcal{F}) \to \hat{G}_0 \ (= \text{character group})$$

$$c \to \chi_c \tag{2.6}$$

which is readily verified to be a homomorphism of abelian groups with kernel exactly the subgroup of 1-coboundaries. Furthermore this map is surjective: if $\chi \in \hat{G}_0$ and $G = \bigcup_i g_i G_0$ is a decomposition into left cosets of G_0, then the map $c_\chi : G \to \mathcal{F}$ defined by $c_\chi(g, g_i x) = \chi(h)$ if $gg_i = g_j h$ for $h \in G_0$ is a 1-cocycle satisfying $\phi(c_\chi) = \chi$. We have thus established the following, essentially a special case of the Eckmann-Shapiro lemma (cf. [5]):

Lemma: If X is a transitive G-set with G_0 the stabilizer of some fixed $x_0 \in X$, then there is a canonical isomorphism

$$H^1(G, \mathcal{F}) = C^1(G, \mathcal{F})/B^1(G, \mathcal{F}) \overset{\phi}{\underset{\cong}{\longrightarrow}} \hat{G}_0.$$

3° Modular Forms. We discuss modular forms from the above perspective. We take X to be the upper half-plane $H = \{z \in \mathbb{C} \mid imz > 0\}$ and G to be the group $SL_2(\mathbb{R})$. It operates on H in the usual way:

$$\gamma : \tau \to \frac{a\tau + b}{c\tau + d}, \ \gamma = \begin{pmatrix} a & b \\ c & d \end{pmatrix} \in SL_2(\mathbb{R}), \ \tau \in H \tag{3.1}$$

Note that $\{\pm I\} = Z(SL_2(\mathbb{R}))$ acts trivially, so that the action group is $PSL_2(\mathbb{R})$. The action is transitive, and the stabilizer of i is seen to be isomorphic to the group $SO_2(\mathbb{R})$, namely $K = \left\{ \begin{pmatrix} \cos\theta & -\sin\theta \\ \sin\theta & \cos\theta \end{pmatrix} \mid \theta \in \mathbb{R} \right\}$. Its (unitary) character group is generated by the morphism

$$\varepsilon : \begin{pmatrix} \cos\theta & -\sin\theta \\ \sin\theta & \cos\theta \end{pmatrix} \to e^{i\theta} \tag{3.2}$$

There is an obvious 1-cocycle associated with this situation which we learn about in high school! Namely, the derivative

$$\delta : \gamma \to \gamma' \tag{3.3}$$

which makes sense since (3.1) is analytic. The 1-cocycle property (2.3) is just the product rule for derivatives. Note that

$$\delta(\gamma, \tau) = \gamma'(\tau) = (c\tau + d)^{-2} \tag{3.4}$$

Then the recipe of section 2 shows that with respect to the point i, the corresponding character of K satisfies $\begin{pmatrix} \cos\theta & -\sin\theta \\ \sin\theta & \cos\theta \end{pmatrix} \to e^{2i\theta}$, that is

$$\phi(\delta) = \varepsilon^2 \tag{3.5}$$

Thus the image of δ under ϕ generates those characters of K which are trivial on $\{\pm I\}$ which is to be expected since, as we said, $\{\pm I\}$ is trivial on H.

The square root j of δ defined by

$$j(\gamma, \tau) = (c\tau + d)^{-1} \tag{3.6}$$

is also a 1-cocycle on $SL_2(\mathbb{R})$ and satisfies

$$\phi(j) = \varepsilon \tag{3.7}$$

One is usually interested in stroke operators of the form

$$f\mid_k \gamma(\tau) = j(\gamma, \tau)^k f(\gamma\tau) \tag{3.8}$$

for non-negative integers k. Since $SL_2(\mathbb{R})$ is transitive on H, however, the only $SL_2(\mathbb{R})$-invariants of the action (3.8) are uninteresting, and indeed we must restrict our attention to discrete subgroups of $SL_2(\mathbb{R})$ to get an interesting theory.

The prime example of a discrete subgroup of $SL_2(\mathbb{R})$ is $\Gamma = SL_2(\mathbb{Z})$. A (meremorphic) modular form of weight k on Γ is a function f which is an invariant of the kth stroke operator (3.8) for $\gamma \in \Gamma$ and which is meromorphic on $H \cup \{\infty\}$. "Meromorphic at ∞" means the following: by dint of (3.8) with $\gamma = \begin{pmatrix} 1 & 1 \\ 0 & 1 \end{pmatrix}$ we see that $f(\tau + 1) = f(\tau)$, so we have an identity of the form

$$f(\tau) = \sum_{n=-\infty}^{\infty} a_n\, e^{2\pi i n\tau}, \quad \tau \in H \tag{3.9}$$

We say that f is meromorphic at ∞ if $a_n = 0$ for $n \ll 0$. Thus if we set $q = q_\tau = e^{2\pi i \tau}$ (local parameter at ∞) then under the map $\tau \rightarrow q_\tau$, H maps to the interior of the unit disk D of the q-plane with origin removed and $f(q)$ is meromorphic in $D - \{0\}$. Then $f(\tau)$ is meromorphic at ∞ if $f(q)$ extends to a meromorphic function in D, in which case (3.9) is its Laurent expansion at 0.

A modular form f on Γ of weight zero will be called a modular function. According to (3.8) this just says that $f(\gamma\tau) = f(\tau)$ for $\gamma \in \Gamma$, together with meromorphy on $H \cup \{\infty\}$. The first assertion says that f lives as a (meromorphic) function on the orbit space $\Gamma\backslash H$, which is a punctured Riemann sphere. The second assertion is equivalent to the fact that f extends to a function on the compactification $\Gamma\backslash H \cup \{\infty\}$. Pictorially then, we have a sphere and a distinguished point ∞ together with the q-expansion of a meromorphic function at ∞:

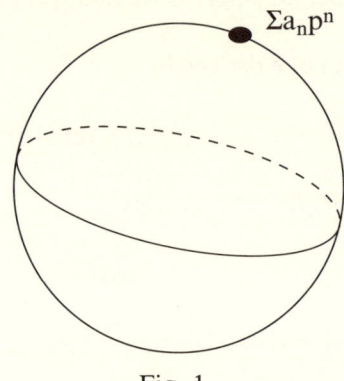

$$\Sigma a_n p^n$$

Fig. 1

For a form of weight k on Γ, say f, (3.8) becomes $f(\tau) = \gamma'(\tau)^{k/2} f(\gamma\tau)$ (we may as well assume k even for this discussion), that is it satisfies

$$f(\gamma\tau)d\gamma(\tau)^{k/2} = f(\tau)(d\tau)^{k/2}, \ \gamma \in \Gamma. \tag{3.10}$$

So f is just a (meromorphic) differential with the indicated Laurent expansion at ∞.

If we want to generalize this notion to modular forms of weight k on subgroups Δ of Γ, say of finite index, it is not enough to just restrict the preceding definitions to hold just for elements of Δ; one must do more. The following approach is somewhat unorthodox, but it points the way towards the generalizations we are heading for.

Let us therefore choose a left Γ-set, say S, which is finite. We will assume that we are also given a 1-cocycle $c \in C^1(\Gamma, \mathcal{F})$ where $\mathcal{F}$ is the space of functions on S, as in section 2. Then we can define a 1-cocycle on the left Γ-set $S \times H$ as follows: to each orbit O of Γ on S we assign a weight $k = k_O$, and then the 1-cocycle on $O \times H$ is given by (c, j^{k_O}). Denoting this 1-cocycle by J, we again have a Γ-action given by

$$f \| \gamma(s, \tau) = J(s, \tau)f(\gamma s, \gamma\tau) \tag{3.11}$$

where $f: S \times H \to \mathbb{C}$, and an invariant f of this action satisfies

$$f(s, \tau) = c(\gamma, s)j(\gamma, \tau)^{k_O}f(\gamma s, \gamma\tau), \tag{3.12}$$
$$\text{all } s \in O, \tau \in H, \gamma \in \Gamma.$$

We shall be interested in invariants which are also "meromorphic." More precisely, we require that for each $s \in S$, the function $f_s(\tau) = f(s, \tau)$ is meromorphic on $H \cup \{\infty\}$ in the sense discussed previously.

Consider first the case in which S is a transitive Γ-set. So there is a single orbit with weight k, say. Now fix $s \in S$. Then for $\gamma \in \Gamma_s$, (3.12) reads

$$f_s(\tau) = \chi_s(\gamma)j(\gamma, \tau)^k f_s(\gamma\tau) \tag{3.13}$$

where χ_s is the character of Γ_s associated with c (cf. (2.6)). This is similar to invariance under the stroke operator (3.8), except for the inclusion of the character χ_s, which is in fact a common feature of modular forms. Restricting further to the subgroup of Γ_s equal to $\ker \chi_s$, we can emulate our previous discussion and conclude that $f_s(\tau)$ "corresponds" to a form on the punctured Riemann surface $\ker \chi_s \backslash H$ together with its q-expansion at ∞. Of course, we can repeat this discussion for each $s \in S$. Pictorially:

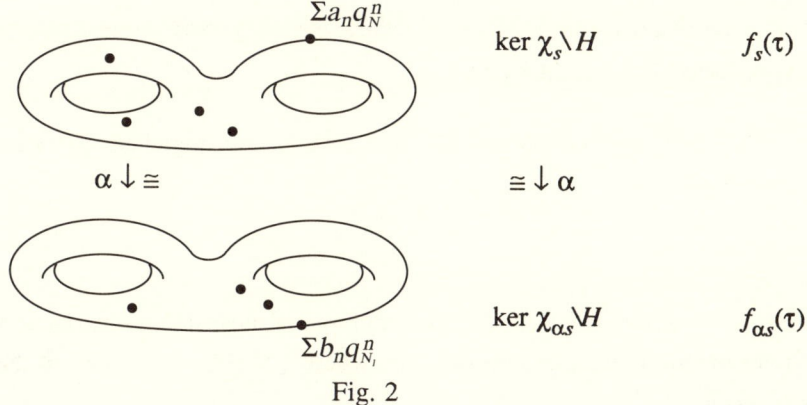

Fig. 2

Note than in general $\ker \chi_s \backslash H$ will have genus g for some g not necessarily equal to 0.

The vertical arrows in Fig. 2 correspond to conjugation by $\alpha \in \Gamma$. As α conjugates $\ker \chi_s$ to $\ker \chi_{\alpha s}$ it induces an (analytic) isomorphism of the corresponding Riemann surfaces. As long as we assume that there is $N \geq 1$ such that $\begin{pmatrix} 1 & N \\ 0 & 1 \end{pmatrix} \in \ker \chi_s$ then $f_s(\tau)$ will have a q-expansion $\Sigma a_n q_N^n$ at ∞ where $q_N = e^{2\pi i \tau / N}$. Note, however, that α generally does not fix ∞, and so if we identify all of the punctured surfaces what emerges is a single Riemann surface (punctured at the cusps) together with a collection of q-expansions at the cusps (there generally being several at each), the different q-expansion being related by the equation (3.12).

If, for example, $| \Gamma : \ker \chi_s |$ is finite, so that the integer N exists, this is precisely what we mean by a modular form of weight k on $\ker \chi_s$; alternatively we can regard $f_s(\tau)$ as a modular form of weight k on Γ_s with character χ_s.

For a 1-cocycle c on $S \times H$, let us denote by $\mathcal{M}(k, c)$ the set of modular forms satisfying the invariance property (3.12) (still assuming S to be a transitive Γ-set). Then $\mathcal{M}(k, c)$ is clearly a $\mathbb{C}$-vector space, and one verifies easily that if c_1 is another 1-cocycle cohomologous to c, say $c_1 = c \cdot c_h$ for some 1-coboundary c_h (cf. (2.5)), then multiplication by h induces a linear isomorphism $\mathcal{M}(k, c) \xrightarrow{\cong} \mathcal{M}(k, c_1)$. So for many purposes the nature of the space of functions f satisfying (3.12) depends only on the cohomology class of c.

Finally, let us turn to the case in which S is an arbitrary finite left Γ-set. We let $\mathcal{M}(k, c)$ be as before, where now k is regarded as a function $k : S \to \mathbb{Z}$ constant on Γ-orbits, so that $k(s) = k_O$ for $s \in O$. The preceding discussion applies to each Γ-orbit O on S, so that to an element of $f \in \mathcal{M}(k, c)$ corresponds a collection $\{X_O, f_O\}$ of punctured Riemann surfaces together with the q-expansions at the cusps of a modular form f_O of weight k_O on X_O.

It is convenient to collect all of the q-expansions together, that is one forms the q-expansion

$$F(\tau) = \Sigma \chi_n \, q^{n/N}$$

where now N is, say, the least common multiple of those denominators occuring in each of the individual q-expansions, and where $\chi : S \to \mathbb{C}$ is defined so that $f(s, \tau) = \Sigma \chi_n(s) q^{n/N}$.

A key point is the choice of the Γ-set S, and for moonshine it can be partially motivated by considering orbifolds. We turn to this aspect of the theory next.

4° Orbifolds. For the purposes of this note, an orbifold is simply the orbit space of a (finite) group G which acts on a manifold M. We write $M_1 = M/G$.

M_1 will be the "space-time" of some physical situation, and in string theory one considers a string propagating in M_1. It is convenient to consider maps

$$f : [0, 2\pi] \to M \qquad (4.1)$$

(For our purposes the nature of f, i.e., whether it is continuous or smooth, will be irrelevant.) Then f describes a *closed string* in M just when $f(0) = f(2\pi)$. But in M_1 there are additional closed strings; corresponding to an element $g \in G$ one can consider maps (4.1) which satisfy $f(2\pi) = gf(0)$. These are generally open in M, but obviously closed in M_1. We say that such an f *satisfies boundary conditions twisted by g.* Let us set

$$L_g = \{f \mid f \text{ satisfies boundary conditions twisted by } g\}.$$

Note that multiplication by $h \in G$ defines a canonical bijection $L_g \to L_{hgh^{-1}}$ and that this bijection respects the natural action of $C_G(g)$ on L_g.

In quantum theory one has to quantize the preceding situation. What results[*] is that for such $g \in G$ there is a (Hilbert) space of *twisted quantum states* H_g, corresponding to L_g, so that the preceding group-theoretic aspects are preserved. That is, H_g is naturally graded into finite-dimensional $C(g)$-modules; conjugation by $h \in G$ induces a natural isomorphism $H_g \overset{\to}{\cong} H_{hgh^{-1}}$.

For $g \in G$ and $h \in C_G(g)$ one can consider the graded trace

$$\mathrm{tr}_{H_g}(h) = \sum_n \mathrm{tr}_{H_g^{n/d}}(h)q^{n/d} \qquad (4.2)$$

where in (4.2) the space H_g is (fractionally) graded as $\bigoplus_n H_g^{n/d}$. We begin to make contact with earlier sections with the remark that $\mathrm{tr}_{H_g}(h)$ should be a modular function on some subgroup of $\Gamma = SL_2(\mathbb{Z})$.

Let us set

$$f(g, h; \tau) = \sum_n \mathrm{tr}_{H_g^{n/d}}(h) \, q^{n/d} \qquad (4.3)$$

Comparison of (4.3) and the remarks following (3.13) suggests that for the left Γ-set S we choose the set

$$P = P(G) = \{(g, h) \in G \times G \mid gh = hg\} \qquad (4.4)$$

How does Γ act on P? This can be motivated by considering the space of maps $X \xrightarrow{f} M_1$ where X is some fixed compact Riemann surface, which in fact is highly relevant in string theory. Consider in particular the case in which X has genus 1, so that $X \cong S^1 \times S^1$ is a torus. As in our discussion of the maps (4.1), a (continuous) map of X into M_1 corresponds to a map $f : [0, 2\pi] \times [0, 2\pi] \to M$ in which the two co-ordinate axes satisfy boundary conditions twisted by a pair of commuting elements g, h of G.

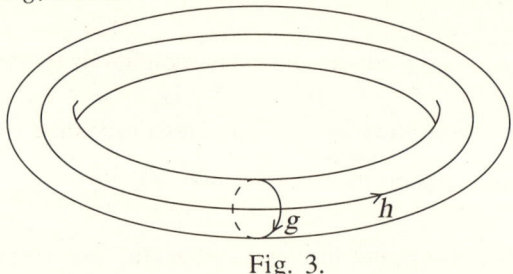

Fig. 3.

Alternatively, we can think of g and h as generators of a lattice $\mathbb{L} \cong \mathbb{Z}^2$ where $X \cong \mathbb{R}^2/\mathbb{L}$, or, what amounts to the same thing, we identify $\mathbb{L}$ with $\pi_1(X) = H_1(X; \mathbb{Z})$ and consider a group homomorphism $\pi_1(X) \to G$ in which generating cycles are mapped to g, h respectively.

There is a standard action of Γ on $H_1(X; \mathbb{Z}) \cong \mathbb{Z}^2$, namely the usual matrix multiplication

$$\begin{pmatrix} a & b \\ c & d \end{pmatrix} : (\overline{g}, \overline{h}) \longmapsto (\overline{g}^a \overline{h}^b, \overline{g}^c \overline{h}^d) \tag{4.5}$$

where in (4.5) we have used multiplicative notation for $\mathbb{Z}^2$. The induced action of Γ on P is simply (4.5), where now $(\overline{g}, \overline{h}) \in P$.

This concludes our brief foray into strings and orbifolds, although several issues have been left unattended—for example, how does the action (4.5) of Γ on P transfer to an action of Γ on the functions $f(g, h; \tau)$ of (4.3)? We pass over this, and turn immediately to elliptic systems.

---

* Naturally, this is a gross oversimplification of a complicated story. The main group-theoretic omission is that, in general, H_g realizes only a projective representation of $C_G(g)$.

5° Elliptic Systems. We abstract (and generalize) the situation of section 4. For a finite group G, an *elliptic system* is a family of (infinite-dimensional), virtual, complex vector spaces $\{H_g\}$ indexed by elements $g \in G$, which satisfy the following axioms:

1. Each H_g carries an integer grading as the direct sum of finite-dimensional, virtual $C_G(g)$-modules

$$H_g = \bigoplus_{n \in \mathbb{Z}} H_g^n .$$

2. For $h \in G$ there is a grade-preserving isomorphism of vector spaces

$$\phi_h : H_g \to H_{hgh^{-1}}$$

which is natural in the sense that for $v \in H_g^n$ and $c \in C_G(g)$ we have $\phi_h(cv) = (hch^{-1})\phi_h(v)$.

3. Let $P = P(G)$ (cf. (4.4)). Then there is a 1-cocycle $c \in C^1(\Gamma, \mathcal{F})$ ($\mathcal{F} =$ functions on P) and a Γ-invariant weight function $k: P \to \mathbb{Z}$ such that the values of c are roots of unity and such that the following holds: if $(g, h) \in P$ then there is an integer $N = N(g, h)$ such that the functions $f(g, h; \tau)$ defined by

$$f(g, h; \tau) = \sum_{n \in \mathbb{Z}} \text{tr}_{H_g^n} (h) q^{n/N} \tag{5.1}$$

satisfy (3.12). That is, we have

$$f(g, h; \tau) = c(\gamma; g, h) j(\gamma, \tau)^{k(g,h)} f((\gamma^\tau)^{-1} (g, h), \gamma\tau) \tag{5.2}$$

for $\gamma \in \Gamma$.

Remarks:

1. Axiom 2 says that in essence we have a single space $H_{\{g\}}$ for each conjugacy class $\{g\}$ of G.

2. The heart of the matter is, of course, axiom 3. The point of section 3 is that it tells us that, in essence, for each Γ-orbit O on $P(G)$ we have a modular form of some weight $k(O)$ and some character, on a certain congruence subgroup of Γ.

Moreover, the q-expansions at the various cusps are given, up to a root of unity, by certain trace functions (5.1).

3. As indicated in section 4, in string theory one encounters the situation in which k is identically zero, i.e., each $f(g, h; \tau)$ is a modular function. Moreover, in this case each H_g is a genuine, as opposed to virtual, vector space. Nevertheless it is useful to have the more general situation axiomatized.

Let us now discussion the relation of an elliptic system with the earlier notion of a (virtual) moonshine module. For simplicity, we deal with modular functions of weight zero.

We recall the following subgroup of Γ:

$$\Gamma_1(N) = \left\{ \begin{pmatrix} a & b \\ c & d \end{pmatrix} \equiv \begin{pmatrix} 1 & * \\ 0 & 1 \end{pmatrix} (\text{mod } N) \right\}$$

In the following we let Γ act on the right of $P(G)$ via

$$\begin{pmatrix} a & b \\ c & d \end{pmatrix} : (x, y) \rightarrow (x^a y^c, x^b y^d).$$

The following is an easy exercise.

Lemma: Let $g \in G$ have order N and let $P_0(g) = \{(x, y) \in P(G) \mid \langle x, y \rangle = \langle g \rangle\}$. Then Γ acts transitively on $P_0(g)$ and the stabilizer of $(1, g)$ is the subgroup $\Gamma_1(N)$.

From the lemma and the results of section 3 we see that an elliptic system attaches to each cyclic subgroup $\langle g \rangle$ of G a modular function (with character) on $\Gamma_1(|\langle g \rangle|)$.

If we now restrict ourselves to the "untwisted" space H_1, we have only the functions $f(1, h; \tau)$; we lose the critical relation (5.2). This is essentially the situation which pertains in a "moonshine" module.

Conversely, suppose that $\bigoplus_n H_n$ is a (virtual) moonshine module in the sense that for each $g \in G$ we know that $f_g(\tau) = \sum tr_{H_n} (g)q^n$ is a modular function on $\Gamma_1(|\langle g \rangle|)$ with character χ_g. We can then proceed as follows: fix a generator z for each cyclic subgrup of G, and for each $(x, y) \in P(G)$ with $\langle x, y \rangle = \langle z \rangle$ fix an element $\gamma_{(x,y)} \in \Gamma$ satisfying $(x, y) \gamma_{(x,y)} = (1, z)$. Such an element exists by the lemma. Now define, for $(x, y) \in P(G)$,

$$f(x, y; \tau) = \begin{cases} f_z(\gamma_{(x,y)}^{-1} \tau), & \langle x, y \rangle = \langle z \rangle \\ 0, & \langle x, y \rangle \text{ non-cyclic} \end{cases} \tag{5.3}$$

and define for $\alpha \in \Gamma$, $\langle x, y \rangle = \langle z \rangle$,

$$f \| \alpha(x, y; \tau) = \chi_z(\gamma_{(x,y)}^{-1} \alpha^{-1} \gamma_{(x,y)\alpha^{-1}}) f((x, y)\alpha^{-1}, \alpha\tau) \tag{5.4}$$

By the lemma we have $\gamma_{(x,y)}^{-1} \alpha^{-1} \gamma_{(x,y)\alpha^{-1}} \in \Gamma_1(|\langle z \rangle|)$, so (5.4) is meaningful. In fact following section 3 we find that f is an invariant in the sense that

$$f \| \alpha(x, y; \tau) = f(x, y; \tau), \qquad \alpha \in \Gamma.$$

Thus (5.2) is satisfied. What we are still missing are the spaces H_g for $1 \neq g \in G$ and the interpretation of $f(x, y; \tau)$ as $\mathrm{tr}_{H_x}(y)$.

There are some significant advantages in dealing with elliptic systems as opposed to moonshine modules. More precisely, because of the invariance property (5.2) (or (3.12)), we can deal with f as an ordinary modular form in many situations. For example, to construct an elliptic system (of zero weight) it is sufficient to take the quotient f_1/f_2 where f_1, f_2 are a pair of elliptic systems, possibly with different cocycles $c \in C^1(\Gamma, \mathcal{F})$ but with each $f(g, h; \tau)$ of the same (non-zero) weight. Examples can be found in [7] and [8]. Also, the theory of Hecke operators, as developed in [9], can be extended to elliptic systems where it is both more general and more conceptual. Details of this appear in [10] and [11], where we show how to construct for M_{24} an elliptic system of weight zero in which each $f(g, h; \tau)$ is either zero or a hauptmodul on the subgroup of $SL_2(\mathbb{R})$ (of genus zero) which fixes $f(g, h; \tau)$. That is, we have an affirmative solution to the genus zero problem for elliptic systems attached to M_{24}.

References

[1]   R. Borcherds, Vertex Algebras, Kac-Moody algebras, and the Monster, P.N.A.S. Vol. 83 (1986), 3068-3071.

[2]   J. Conway, S. Norton, Monstrous Moonshine, Bull. L.M.S. 11 (1979), 303-339.

[3]   R. Dijkraaf, C. Vafa, E. Verlinde, H. Verlinde, The operator algebra of orbifold models, preprint, 1988.

[4]   I. Frenkel, J. Lepowsky, A. Meurman, Vertex operator algebras and the Monster, Pure and Applied Math. Vol. 134, Academic Press, new York, 1988.

[5]   S. Maclane, Homology, Springer-Verlag, Heidelberg, 1963.

[6]   G. Mason, Finite groups and modular functions, Proc. Symp. Pure Math. Vol. 47, A.M.S., 1987.

[7]   G. Mason, Elliptic systems and the eta-function, Notas Soc. Mat. Chilé 8 (1989), 37-53.

[8]   G. Mason, On a system of elliptic modular forms attached to the large Mathieu group, Nagoya Math. J. Vol. 118 (1990), 177-193.

[9]   G. Mason, Finite groups and Hecke operators, Math. Ann. 283 (1989), 381-409.

[10]   G. Mason, Research Announcement: On G-elliptic systems and the genus zero problem for M_{24}, preprint, U.C.S.C. (1990).

[11]   G. Mason, Hecke operators and elliptic systems, in preparation.

The Classification of 3-Transposition Groups with Trivial Center

H. Cuypers

J.I. Hall*

§1. INTRODUCTION

A conjugacy class D of *3-transpositions* in the group G is a class of elements of order 2 such that, for all d and e in D, the order of the product de is 1, 2, or 3. If G is generated by the conjugacy class D of 3-transpositions, we say that (G, D) is a *3-transposition group* or (loosely) that G is a 3-transposition group. Such groups were introduced and studied by Bernd Fischer [10, 11] who classified all finite 3-transposition groups with no nontrivial normal, solvable subgroups. His work was of great importance in the classification of finite simple groups.

The basic example of a class of 3-transpositions is the class of transpositions in any symmetric group. This was the only class which Fischer originally considered [9], but Roger Carter pointed out that examples could be found in several of the classical groups as well. The transvections of symplectic groups over $GF(2)$ form a class of 3-transpositions, so additionally any subgroup of the symplectic group generated by a class of transvections is also a 3-transposition group. The symmetric groups arise in this way as do the orthogonal groups over $GF(2)$. Symplectic transvections over $GF(2)$ are special cases of unitary transvections over $GF(4)$, and this unitary class is still a class of 3-transpositions. The final classical examples are given by the reflection classes of orthogonal groups over $GF(3)$.

Fischer's principal result on 3-transpositions is the following theorem [10, 11].

(**1.1**) THEOREM. *Let G be a finite group generated by the conjugacy class D of 3-transpositions. Assume that G has no noncentral solvable normal subgroups. Then, up to a center, we may identify D with one of:*

(1) *The transposition class of a symmetric group;*

---

*partial support provided by the NSA(USA) and SERC(UK)

(2) *The transvection class of the isometry group of a nondegenerate orthogonal space over $GF(2)$;*

(3) *The transvection class of the isometry group of a nondegenerate symplectic space over $GF(2)$;*

(4) *A reflection class of the isometry group of a nondegenerate orthogonal space over $GF(3)$;*

(5) *The transvection class of the isometry group of a nondegenerate unitary space over $GF(4)$;*

(6) *A unique class of involutions in one of the five groups $\Omega_8^+(2){:}\Sigma_3$, $\Omega_8^+(3){:}\Sigma_3$, Fi_{22}, Fi_{23}, or Fi_{24}.*

In fact a special sort of 3-transposition group had been studied earlier than [9, 10, 11]. In 1960 M. Hall, while considering certain types of Steiner triple systems [16], looked at centerfree groups G generated by a conjugacy class D of involutions such that the product of any two had order 3. This is the 3-transposition case in which products of order two never occur. As Hall and Bruck [17] later noted, the study of this situation is also essentially equivalent to the examination of commutative Moufang loops of exponent 3 [3]. This is a very difficult and deep area which has little to do with the rest of our present work, and we avoid these "degenerate" 3-transposition groups whenever possible. For a survey on the Moufang loops, see [18]. See also [24].

The authors of this paper became interested in 3-transposition groups for geometric reasons [7, 12, 13], and in such situations restrictions on normal subgroups and finiteness conditions are not as natural as in the group theoretic context. This article gives a description of the complete classification of centerfree 3-transposition groups with the exception of the Moufang loop groups. Equivalently the classification gives, up to a central factor, all 3-transposition groups which contain a subgroup Σ_4.

To a large extent the conclusions are as expected. For groups with no nontrivial solvable normal subgroups, Fischer's classification remains valid, giving nondegenerate classical groups (but now on spaces whose dimension may be infinite) and the five sporadic examples of Theorem 1.1(6). For groups with solvable normal subgroups, the generic examples come from parabolic subgroups of larger nondegenerate classical groups. Equivalently, they come from the isometry groups of degenerate forms. Thus we expect the split extension by a classical group of a direct sum of natural modules. On rare occasions the module is not completely reducible or the extension does not split. Symmetric groups are sufficiently thin that they may act on solvable subgroups which are not elementary abelian. A complete and reasonably precise description of all the groups is given in the final section.

We mention one interesting consequence. In his deep work on Moufang loops, Bruck [3] proved that the corresponding 3-transposition groups are locally finite, that is, every finite subset generates a finite subgroup. This turns out to be true in general, confirming an earlier conjecture of the second author.

(**1.2**) THEOREM.   *Every 3-transposition group is locally finite.*

It is regretable that the only known proof of this theorem relies on the classification of the non-Moufang groups, observing that each is locally finite. An independent and direct proof would be of great interest and would also simplify the classification considerably.

In recent years several people (other than the present authors) have studied the classification problem for 3-transposition groups. M.-M. Virotte-Ducharme [19, 20] and Richard Weiss [21, 23] have given new proofs of Fischer's finite classification, and Weiss [22] considered the extension of the general theory of 3-transposition groups to the infinite case. François Zara [24, 25, 26] examined finite 3-transposition groups with solvable normal subgroups, starting from Fischer's classification.

The classification of 3-transposition groups with trivial center but not of Moufang type was completed in early 1991. The proof of the classification is nearly self-contained; and, in particular, we do not assume Fischer's classification. We do however appeal to a few special cases of the finite classification. So, for instance, we assume rather than prove that, for a finite 3-transposition group (G, D) and $d \in D$, if $\langle C_D(d) \rangle$ is a central extension of $PSU_6(2)$ then G is Fi_{22}. There are roughly ten such finite local classifications which are quoted from [10, 19, 21, 23]. Other than these, the only required results from the literature are of a more general nature, such as knowledge of various Schur multipliers.

This article is an expanded and updated report of the talk given by the second author at the Durham Group Theory Symposium 1990. The area of 3-transposition groups is perhaps singular in that much of the really significant work which has been done through the years remains unpublished to this day [4, 10, 25]. We do not plan on having our work suffer this same fate. Portions of it are already available [8, 12, 13, 14].

§2. SOME GENERAL THEORY

The 3-transposition property is nicely inductive. A quotient of a 3-transposition group is a 3-transposition group. If H is a subgroup of the 3-transposition group (G, D), then $H \cap D$ is a normal set of 3-transpositions within H, although it may no longer be a conjugacy class. In his classification Fischer's first task was to prove results which allowed him to do some inductive reductions.

(**2.1**) LEMMA. (Fischer [11, (2.1.1)].) *Let* $G = \langle D \rangle$ *with* D *a conjugacy class of* 3-*transpositions. Then* $Z(G) = Z_2(G)$, *and* $D \cap dZ(G) = d$, $d \in D$.

The message of this lemma is that, from a 3-transposition viewpoint, there is not much difference between G and $G/Z(G)$. (We shall see this more graphically in the next section.) In our classification the groups are usually only distinguished up to a center, although of course there are times in the proof where it is necessary to consider cental extensions. We say that two 3-transposition groups (G_1, D_1) and (G_2, D_2) have the same *central type* (usually abbreviated to *type*) if their central quotients $(G_1/Z(G_1), D_1Z(G_1)/Z(G_1))$ and $(G_2/Z(G_2), D_2Z(G_2)/Z(G_2))$ are isomorphic as 3-transposition groups.

Once we introduce some terminology, we may state Fischer's two main reductions for the finite case in a form which goes over to the general case.

In any 3-transposition group (G, D), for each $d \in D$, following Fischer we set $D_d = \{e \in D | \ |de| = 2\}$ and $A_d = \{e \in D | \ |de| = 3\}$. We next define two equivalence relations τ and θ on D. We write $d\tau e$ if $A_d = A_e$, and we write $d\theta e$ if $D_d = D_e$. We get two normal subgroups of G, the kernel of the action of G induced on each of the two equivalence relations: $\tau(G) = \langle\ de \mid d\tau e\ \rangle$ and $\theta(G) = \langle\ de \mid d\theta e\ \rangle$.

For finite G, Fischer proved that $\tau(G) = [O_2(G), G]$ and $\theta(G) = [O_3(G), G]$. It may be best to think of these groups as noncentral normal 2-subgroups and 3-subgroups of G, and the notation has been chosen to suggest this. The correspondence could be made precise even in the infinite case, but we prefer to leave the subgroups as described.

We also, following Fischer, define the normal subgroup

$$Q(G) = \langle\ de \mid d, e \in D, |de| = 2\ \rangle.$$

(**2.2**) THEOREM.   (NORMAL SUBGROUP THEOREM.) *Let* G *be generated by the conjugacy class* D *of* 3-*transpositions, and assume that* $Z(G) = 1$. *Then we have exactly one of:*
  (1) $\tau(G) \neq 1$;
  (2) $\theta(G) \neq 1$;
  (3) $Q(G)$ *is simple.*

(**2.3**) THEOREM.   (TRANSITIVITY THEOREM.) *Let* $G = \langle d \rangle Q(G)$ *be generated by the conjugacy class* D *of* 3-*transpositions, where* $d \in D$ *and* $Z(G) = \tau(G) = \theta(G) = 1$. *Then* $\langle C_D(d) \rangle$ *is transitive on* D_d *and* A_d.

The Normal Subgroup Theorem allowed Fischer to weaken the hypothesis for Theorem 1.1 to $O_2(G)O_3(G) \leq Z(G)$. The Transitivity Theorem is the basis for Fischer's inductive proof of the classification. Indeed transitivity on D_d is equivalent to the statement that D_d is a genuine conjugacy class of $C_G(d)$, not just a normal subset. Fischer considers each classical example as the possible type for the local subgroup $\langle D_d \rangle$ and finds only classical groups except in the case where $\langle D_d \rangle$ is of type $SU_6(2)$. This local characterization leads to the sporadic group Fi_{22}, which in turn leads to Fi_{23} and finally to Fi_{24}, where the progression and proof finish.

The proof of both of these theorems in the general case depends upon the following lemma, a stronger version of which will be discussed in a later section.

(2.4) LEMMA.   *If the 3-transposition group G has a quotient Σ_4, then $\tau(G)\theta(G) \neq 1$.*

The Normal Subgroup Theorem is proved in a similar manner to Fischer's original proof for finite groups [11, (3.2.7)] with (2.4) and (7.1) below providing the required version of Burnside's $p^\alpha q^\beta$ theorem.

It is clear from Fischer's proof [11] of the finite Transitivity Theorem and Aschbacher's elementary treatment of it [1] that the theorem derives from the Normal Subgroup Theorem. The proof of (2.3) follows Weiss' approach [22] for extending these methods and the result from the finite to the infinite case.

§3. FISCHER SPACES AND PRESENTATIONS

For the general classification we do not possess the strong inductive hold which Fischer had and must find something to replace it. In addition to general group theoretic techniques, there are two main tools in our proof — Fischer spaces and presentations.

Aside from the sporadic groups, all the conclusions to Fischer's classification have a geometric (or linear algebraic) description, and it is in that manner that we recover them. To do so we must extract their geometry from their group theoretic properties. As the 3-transpositions are transvections and reflections, each 3-transposition t is naturally associated with a projective point or 1-dimensional subspace, its *center*, the range of $t - 1$. This is isotropic for the symplectic and unitary examples and nonsingular for the orthogonal examples. The two generator subgroups therefore correspond to projective lines. The Σ_3 subgroups of the orthogonal examples over $GF(3)$ correspond to lines tangent to the quadric, while in the other examples the lines are hyperbolic.

This suggests associating to any 3-transposition group (G, D) a partial linear space $\Pi(G, D)$ whose points are the 3-transpositions and whose lines are the corresponding subgroups Σ_3. In fact the very first study of special types of 3-transposition groups occured in exactly this context. M. Hall [16, 17] looked at Steiner triple systems, all of whose planes are affine over $GF(3)$. He noted that these arose exactly as $\Pi(G, D)$ for centerfree 3-transposition groups G in which no pair of distinct 3-transpositions commute. This is the "degenerate" case of Moufang 3-transposition groups which was mentioned in the introduction.

Francis Buekenhout [4] observed that the approaches of Hall and Fischer could be combined. Let us call a *Fischer space* a partial linear space in which all lines have exactly three points, and any pair of intersecting lines lies in a subspace which is either an affine plane over $GF(3)$ or a dual affine plane over $GF(2)$.

(**3.1**) THEOREM. (Buekenhout [4].) *There is a correspondence between connected Fisher spaces and 3-transposition groups with trivial center. More precisely, for each 3-transposition group (G, D), the space $\Pi(G, D)$ is a connected Fischer space; and every connected Fischer space is isomorphic to $\Pi(G, D)$ for some 3-transposition group (G, D). Two 3-transposition groups give rise to the same Fischer space if and only if they are of the same central type.*

The proof is elementary (see, for instance, [12, §3]).

3-Transposition groups are defined in terms of their two generator subgroups. Among other things, the theorem (3.1) states that the three generator subgroups of 3-transposition groups are also restricted in shape. (In this form the result can be found in Fischer's work, [11, (1.6)].) To see why this is the case, we study presentations.

If (G, D) is a 3-transposition group, and Δ is a subset of D, the *diagram* of Δ is the graph with the members of Δ as nodes, two adjacent precisely when their product has order 3. It is important to notice that $D \cap \langle \Delta \rangle$ is a conjugacy class of $\langle \Delta \rangle$ if and only if Δ is connected and that distinct connected components of Δ generate subgroups which commute.

It is immediate that the subgroup $\langle \Delta \rangle$ of G is a homomorphic image of the Coxeter group with Dynkin diagram Δ. So, for instance, the diagram A_n, a path with n nodes, gives rise to a presentation of Σ_{n+1} in terms of its transposition class, revealing Σ_{n+1} as the Weyl group $W(A_n)$.

We may consider other than spherical diagrams. In particular, the only connected diagrams on three nodes are the path A_3, giving Σ_4, and the triangle, the affine diagram $\tilde{A}_2$. The corresponding Weyl group $W(\tilde{A}_2)$ is the extension of the 2-dimensional root lattice Λ by $W(A_2) \simeq \Sigma_3$. To get a 3-transposition group we must factor by two or three times the lattice Λ, giving 3-transposition quotients $W_2(\tilde{A}_2) \simeq 2^2{:}\Sigma_3 \simeq \Sigma_4$ again and $W_3(\tilde{A}_2) \simeq 3^2{:}\Sigma_3 \simeq SU_3(2)'$. The corresponding Fischer spaces are, respectively, the dual affine plane over $GF(2)$ and the affine plane over $GF(3)$, as in

(3.1). Of course there are more direct ways of proving this, but the Coxeter diagram point of view is helpful to maintain (see, for instance, (5.3) below.)

Although induction is not available to us, we should remember that the groups we are hunting are all locally finite. Having found all three generator 3-transposition groups we can look at subgroups with four generators, five generators, and so on, to see if useful presentations occur.

Here we let (G, D) be a 3-transposition group. A D-subgroup of G is any subgroup H with $H = \langle D \cap H \rangle$.

(**3.2**) PROPOSITION. (See [8, 14, 19, 25].) *Let H be a D-subgroup of G of type $SU_3(2)'$, and suppose that $d \in D - H$ with $|C_D(d) \cap H| = 1$. Then $\langle H, d \rangle$ is of type $2^{1+6}:SU_3(2)'$.*

(**3.3**) PROPOSITION. *Let I be a D-subgroup of G of type $2^{1+6}:SU_3(2)'$; and, for some $d \in (D \cap I) - Z(I)$, let $C = d^{O_2(I)}$ of size 4. Let $e \in D - I$. If $|C_D(e) \cap C|$ is 2, then $\langle I, c \rangle$ is of type $\Omega_8^+(2):\Sigma_3$. If $|C_D(e) \cap C|$ is 3, then $\langle I, e \rangle$ is of type $(3^8 \oplus 3^8):2^{1+6}:SU_3(2)'$ or its quotient $3^8:2^{1+6}:SU_3(2)'$.*

This was proven with the help of Leonard Soicher and his coset enumeration program *Enum* as part of a census of all five generator groups $\langle I, e \rangle$. Zara [27] has found and verified by hand related presentations for the two triality groups.

§4. CLASSIFICATION OF REDUCTIVE GROUPS

In the introduction it was observed that the subgroups $\tau(G)$ and $\theta(G)$ of the 3-transposition group G generically are normal unipotent subgroups. It is thus natural, following the terminology of algebraic groups and [27], to call G *reductive* if $\tau(G) = \theta(G) = 1$. In this section we shall trace through the classification of 3-transposition groups in the reductive case. We find that Fischer's classification theorem (1.1) still holds with the finiteness assumption removed.

Assume throughout this section that $G = \langle D \rangle$ with D a conjugacy class of 3-transpositions. Assume additionally that $Z(G) = \tau(G) = \theta(G) = 1$.

§4.1. STEP 1. MOUFANG TYPE

Suppose that the 3-transposition group (H, E) has no E-subgroup isomorphic to Σ_4. In that case the diagram of E is connected with no subdiagram A_3, and so must be complete. Therefore we find that $H = \langle e \rangle \theta(H)$, for any $e \in E$. This is the case considered by M. Hall and Bruck [3, 16, 17]. Our assumptions on G preclude this.

§4.2. Step 2. Symplectic type

Let us say that G is of *symplectic type* if it has D-subgroups Σ_4 but no D-subgroup of central type $SU_3(2)'$. (We hope that having two uses for the word "type" does not cause confusion.) Not only is $SU_3(2)'$ a unitary group, but it is also appears within a Sylow 3-normalizer of $O_5(3)$. Therefore the lack of D-subgroups $SU_3(2)'$ effectively keeps G from being orthogonal over $GF(3)$ or unitary over $GF(4)$ (and so sporadic as well). That is, the denial of $SU_3(2)'$ forces G to look like a symplectic group or one of its subgroups, hence the name.

This case is considered in [12, 13].

(4.1) Theorem.   *The group G of symplectic type is symmetric, orthogonal over $GF(2)$, or symplectic over $GF(2)$.*

The class D is identified uniquely as the transposition class for the symmetric groups and the transvection class for the orthogonal and symplectic groups, except in the case $\Sigma_6 \simeq Sp_4(2)$ which has two generating classes of 3-transpositions, exchanged by an outer automorphism.

The proof is elementary [12]. To each $d \in D$ we associate the vector $\tilde{d} \in GF(2)^D$ given by $\tilde{d}_e = 0$ if d and e commute and $\tilde{d}_e = 1$ if de has order 3. Then it can be easily checked that the elements of D induce transvections on the subspace $\tilde{V}$ of $GF(2)^D$ spanned by the various $\tilde{d}$. The crucial point is that, because there are no D-subgroups of type $SU_3(2)'$, if d, e, f are the three members of D in a subgroup Σ_3, then $\tilde{d} + \tilde{e} + \tilde{f} = 0$ in $\tilde{V}$. So $\tilde{V}$ gives an embedding of the Fischer space for G via transvection centers in a projective space over $GF(2)$.

§4.3. Step 3. Orthogonal type

Now say that G is of *orthogonal type* if it has D-subgroups of type $SU_3(2)'$, but no D-subgroup of type $2^{1+6}{:}SU_3(2)'$. The subgroup $2^{1+6}{:}SU_3(2)'$ is the transvection generated part of a transvection centralizer in $SU_5(2)$. Therefore this denial keeps G from being unitary and so forces it to look like an orthogonal group over $GF(3)$.

This is the case considered in [8].

(4.2) Theorem.   *The group G of orthogonal type is an orthogonal group over $GF(3)$.*

Some of these groups have two generating classes of 3-transpositions, exchanged by an outer automorphism [8, 19].

The group theoretic restriction on subgroup type is equivalent, using (3.2), to the geometric property **(P)** studied by the first author [7]: for any affine plane π of the Fischer space there is no point external to π and collinear with all but one of the points of π. This allows the construction of a partial parallel relation on the lines of the Fischer space of G. Unlike the symplectic case, it seems difficult to reconstruct the entire vector space over $GF(3)$ on which G acts (3 is a lot bigger than 2), but it is possible to construct all the tangent lines through a given point on the quadric. That is, the parallel relation can be used to reconstruct the parabolic subgroup which is the stabilizer of a singular point in the orthogonal geometry. This subgroup is then uniquely extended to G.

§4.4. STEP 4. UNITARY TYPE

Next say that G is of *unitary type* if it has D-subgroups of type $2^{1+6}:SU_3(2)'$ but no D-subgroup of type

$$\Omega_8^+(2):\Sigma_3 \text{ or } 3^{1+8}:2^{1+6}:SU_3(2)'.$$

The group $\Omega_8^+(2):\Sigma_3$ is a D-subgroup of each of the five sporadic groups in Theorem 1.1(6); and $3^{1+8}:2^{1+6}:SU_3(2)'$ occurs within a 3-local subgroup of Fi_{23} [6]. Therefore the denial forces G to look like a unitary group. (At the Durham Symposium it was suggested to characterize the unitary groups by the absence of subgroups $\Omega_8^+(2):\Sigma_3$ alone, and in fact that can be done. But it suits the proof better to include a ban on subgroups of type $3^{1+8}:2^{1+6}:SU_3(2)'$ as well.)

(4.3) THEOREM.   *The group G of unitary type is a unitary group over $GF(4)$.*

The class is identified uniquely.

(4.4) THEOREM.   *Suppose that, for some $d \in D$, we have $\tau(C_G(d)) \neq 1$. Then either G is a symplectic group over $GF(2)$ or G is a unitary group over $GF(4)$.*

Theorem 4.4 is proved by finding the lines of the associated polar space as commuting subsets of members of D, following Weiss' approach [21].

For (4.3) let the D-subgroup I have type $2^{1+6}:SU_3(2)'$, and choose $d \in (D \cap I) - Z(I)$. Let $C = d^{O_2(I)}$, of order 4. We claim that the elements of $C - \{d\}$ are all τ-equivalent within $C_G(d)$; so G of unitary type satisfies the hypotheses of (4.4). The claim is a consequence of the presentation result (3.3).

§4.5. Step 5. Sporadic type

Finally the group G is of *sporadic type* if it is of none of the preceding types. Equivalently, it is of sporadic type if it has a D-subgroup isomorphic to $\Omega_8^+(2){:}\Sigma_3$ or of type $3^{1+8}{:}2^{1+6}{:}SU_3(2)'$.

(4.5) Theorem.   *The group G of sporadic type is one of the five groups* $\Omega_8^+(2){:}\Sigma_3$, $\Omega_8^+(3){:}\Sigma_3$, Fi_{22}, Fi_{23}, *or* Fi_{24}.

The analysis is rather complicated. The main step is

(4.6) Proposition.   *If G is of sporadic type and a, b, c are three distinct commuting elements of D, then the subgroup $H = \langle D \cap C_G(a,b,c) \rangle$ is not of sporadic type.*

So, for instance, in Fi_{24} the subgroup H is of type $SU_6(2)$. Theorem 4.5 is then proved by returning to Fischer's inductive approach and considering the various local characterizations. In most cases these characterizations can be easily reduced to earlier cases. In particular the judicious use of (4.4) allows many cases to be avoided entirely. However we do not actually construct the sporadic Fischer groups, but instead reduce to some of the local characterization problems which were solved by Fischer [10], Virotte-Ducharme [19, 20], and Weiss [23], and then quote the earlier work.

§4.6. Classification

The previous steps in this section now combine to give a proof of the reductive classification:

(4.7) Theorem.   *Fischer's classification, Theorem 1.1, remains valid for infinite groups.*

Several corollaries are helpful in considering situations where solvable normal subgroups appear.

(4.8) Corollary.   *Let (H, E) be a 3-transposition group with $H/Z(H)$ one of the conclusions to Theorems 1.1 and 4.7. Then $C_H(e)/\langle D_e \rangle$ is cyclic of order dividing 6 or H is $SU_5(2)$ or $Z_2 \times SU_5(2)$ and $C_H(e)/\langle D_e \rangle$ is $\mathcal{A}_4$ or $SL_2(3)$, respectively.*

(4.9) Corollary.   *Let (H, E) be a 3-transposition group with $H/Z(H)$ one of the conclusions to Theorems 1.1 and 4.7, and let e, f be a noncommuting pair from E. Then either $H = \langle D_e, D_f \rangle$ or, up to isomorphism, $H = \Sigma_\Omega$ and $\langle D_e, D_f \rangle = \Sigma_{\Omega - \omega}$, for some $\omega \in \Omega$.*

As can be seen in the $SU_5(2)$ case, a central extension may behave differently from the original group in important ways. Another special case is that of $Z_3 \cdot {}^+\Omega_6^-(3)$, the nonsplit central extension of Z_3 by ${}^+\Omega_6^-(3)$. The Z_3 splits off in the centralizer of a 3-transposition, so the quotient of (4.9) has order 6 in this case.

§5. THE NORMAL SUBGROUP THEOREM REVISITED

Now we consider a 3-transposition group (G, D) with trivial center and containing the nontrivial normal subgroup $N < G'$. Let $\overline{G} = G/N$.

(5.1) PROPOSITION. *Suppose that, for each pair $d, e \in D$ with $|\bar{d}\bar{e}| = 3$, we have $G = \langle D_d, D_e \rangle$. Then $[N, G]$ is elementary abelian.*

As G is generated by $D - d^N$, to prove the proposition it is enough to check that $[N, d]$ commutes with $[N, e]$ for each $\bar{e} \neq \bar{d}$. This is clearly the case if d and e commute. The commutator $[[N, d], [N, e]]$ lies in $[N, d] \cap [N, e]$ because $[N, d]$ ($\leq \langle d^N \rangle$) and $[N, e]$ normalize each other. Also $[N, d] \leq \langle d^N \rangle$ is centralized by $\langle D_d \rangle$ and $[N, e]$ is centralized by $\langle D_e \rangle$. If d and e do not commute, then the intersection is central in G and so trivial, proving the proposition.

The next result can be thought of as a refinement to the Normal Subgroup Theorem.

(5.2) THEOREM. *Either*
(1) *the quotient $\overline{G}/\theta(\overline{G})$ is symmetric, or*
(2) *$[N, d]$ is elementary abelian and $[N, G] \leq \tau(G)\theta(G)$.*

The proposition and (4.9) can be used to prove the theorem, but it is better not to use the weight of the reductive classification. It is enough to observe that for $\overline{G} \simeq W(D_4)$ the proposition is valid, and then to characterize the symmetric groups by the absence of subgroups with type $W(D_4)$, extending the finite characterization of [10, (6.2)].

(5.3) COROLLARY. (F. Zara [25, 26].) *If $\overline{G}$ contains a subgroup of type $W_p(\widetilde{D}_4)$, then N can not be an elementary abelian q-group, $\{p, q\} = \{2, 3\}$.*

§6. MODULES

From (4.9), (5.1), and Theorem 5.2 we find that when G with trivial center has no symmetric quotient (of order bigger than 2) any nontrivial normal subgroup $N = [N, G]$ is a $GF(2)$- or $GF(3)$-module for the nearly simple

3-transposition group $\overline{G} = G/N$. In this section we discuss how to identify the module N and G under these assumptions.

The result of Zara (5.3) allows the immediate rejection of various possiblities for the pair $(\overline{G}, N)$. For instance,

(**6.1**) PROPOSITION. *If $\overline{G}$ has type $SU_n(2)$, $n \geq 4$, and N is an elementary abelian 3-group, then $n \leq 5$.*

This is true because, for $G \simeq SU_6(2)$, the subgroup $\langle C_D(d) \rangle \simeq 2^{1+8}:SU_4(2)$ contains a D-subgroup $W_2(\widetilde{D_4})$.

The main tool for construction and identification of the modules is the following.

(**6.2**) LEMMA. *If U is a $C_G(d)$-submodule of $[N, d]$, then $\|\langle \langle U^G \rangle, d]\| \leq |U|$.*

This is powerful because the subgroup $P = \langle D_d \rangle$ of $C = C_G(d)$ is very large by (4.8) but must act trivially on $U \leq \langle d^N \rangle$. Any cyclic module N must arise as a quotient of the induced module $U \uparrow_C^G / \langle [U \uparrow_C^G, d, P]^G \rangle$, a module which can usually be identified as the expected natural module [8, 14].

The most complicated case is that of $SU_5(2)$ where $\overline{C}/\overline{P}$ is isomorphic to $\mathcal{A}_4$ or $SL_2(3)$.

(**6.3**) LEMMA. *A nontrivial cyclic and indecomposable module $N = [N, G]$ for $\overline{G}$ of type $SU_5(2)$ is one of:*
   (1) The natural module $GF(4)^5$;
   (2) An extension $GF(4)^5 \cdot GF(4)^5$ of a natural module by a natural module;
   (3) An irreducible module $GF(3)^{10}$.

Here (2) and (3) may be thought of as the *mod 2* and *mod 3* reductions of the 5-dimensional quaternionic lattice for $Z_2 \times SU_5(2)$ [5] which can be constructed as above by "inducing" the 1-dimensional quaternionic representation of $SL_2(3)$ up to $Z_2 \times SU_5(2)$.

For the most part, the modules N must be completely reducible with all irreducible factors the anticipated natural modules. The next question which must be answered is whether the extension of N by $\overline{G}$ splits. Generically it does. Splitting is found using Gaschütz' theorem [2, (10.4)] within a 3-transposition centralizer. Induction and the complete reducibility of modules for smaller groups allow the construction of a complement for the centralizer in most of the finite cases. Finite splitting then implies infinite splitting. The infrequent occurence of indecomposable but reducible modules such as (6.3)(2) above allows a few nonsplit groups to occur. For instance, consider an extension $GF(4)^7 . SU_7(2)$. The transvection centralizer in $SU_7(2)$ is $2^{1+10}:SU_5(2)$; so (6.3)(2) obstructs the construction of a complement to $GF(4)^7$ in the subgroup $GF(4)^7 . (2^{1+10}:SU_5(2))$. In fact a nonsplit extension of $GF(4)^7$ by $SU_7(2)$ does exist, apparently first constructed in [14].

§7. Groups with symmetric quotient

We now consider 3-transposition groups (G, D) with a symmetric quotient of degree at least 3.

It is easy to see that within the wreathed product $K \wr \Sigma_\Omega$ the transposition class can not be a class of 3-transpositions unless each element of K has order 1, 2, or 3. Zara [25, 26] observed that conversely if K has this property, then the transposition class of $K \wr \Sigma_\Omega$ is indeed a class of 3-transpositions. Denote the transposition generated subgroup by $Wr(K, \Omega)$.

Let us call a group a *strong $\{2,3\}$-group* if each element has order 1, 2, or 3.

(**7.1**) PROPOSITION.   *A strong $\{2,3\}$-group is one of:*

   (1) *an elementary abelian 2-group;*

   (2) *a group of exponent 3;*

   (3) *(Generalized Σ_3) the extension of an elementary abelian 3-group by an involution acting without fixed points;*

   (4) *(Generalized $\mathcal{A}_4$) the extension of an elementary abelian 2-group by an element of order 3 acting without fixed points.*

This is elementary. The groups of (2) are well understood [15, 18.2] and, in particular, are nilpotent of class at most 3.

(**7.2**) THEOREM.   *Let G have $Z(G) = 1$ and possess a homomorphic image Σ_Ω with $|\Omega| > 3$. Then G is a quotient of $Wr(K, \Omega)$, for some strong $\{2,3\}$-group K.*

The earlier (2.4) is a weak version of this result. Notice that when K is a "generalized Σ_3" the group G is essentially the extension by the orthogonal monomial group over $GF(3)$ of a direct sum of natural modules. Similarly "generalized $\mathcal{A}_4$" is associated with the unitary monomial group over $GF(4)$.

The proof is in the spirit of that of (5.1). We can no longer force the intersections $[N, d] \cap [N, e]$ to be central, but (see (4.9)) they are centralized by large subgroups covering $\Sigma_{\Omega-\omega}$. This is the beginning of a construction of the wreathed product.

The groups which have a quotient Σ_3 (but not a quotient Σ_4) are more difficult to deal with.

(**7.3**) THEOREM.   *Suppose G has trivial center and a quotient Σ_3 but no quotient Σ_4. Then we have one of:*

   (1) $G = \langle d \rangle \theta(G)$;

   (2) $G/\tau(G) \simeq SU_3(2)'$ *and $\tau(G)$ has class at most 2;*

(3) G *has type* $\oplus_{i \in I}(GF(3)^8):(2^{1+6}:SU_3(2)')$, *for some nonempty index set I;*

(4) $Q(G)$ *is simple and* $G/Q(G) \simeq \Sigma_3$.

Assume that we do not have (1). The cases where $\theta(G/\tau(G)) \neq \tau(G)$ and $\tau(G/\theta(G)) \neq \theta(G)$ are handled by elementary techniques. In the first case a result of Fischer [11, (2.2.7)] implies that $G/\tau(G) \simeq SU_3(2)'$. One then proves, following (5.1) and (7.2), that commutators

$$[[\tau(G), d], [\tau(G), e], [\tau(G), f]]$$

are trivial, giving (2). (A more precise description of the groups in (2) is given in the next section. It was observed by Zara [25] that groups of class two can occur here.) In the second case the conclusion (3) follows using (2), (3.3), and (5.2).

This leaves the case where either $Q(G/\tau(G))$ or $Q(G/\theta(G))$ is simple. Here one (regretably) resorts to the reductive classification, Theorem 4.7. The only possibility is that $G/\tau(G)\theta(G)$ is either $\Omega_8^+(2):\Sigma_3$ or $\Omega_8^+(3):\Sigma_3$. Then $\tau(G)\theta(G) = 1$ follows relatively easily [14, 25].

§8. THE LIST

In this section we describe, up to central type, all 3-transposition groups G not satisfying $G = \langle d \rangle \theta(G)$, for $d \in G$. Except where otherwise stated, there is a unique generating class of 3-transpositions, and that class should be reasonably apparent from the description. Examples of 3-transposition groups of each sort do exist. (There is some duplication in the list but not much.)

Throughout I is an indexing set, possibly empty and possibly infinite.

(1) $\oplus_{i \in I} V_i : FO(V, q)$.

Here (V, q) is a nondegenerate orthogonal space over $GF(2)$ of dimension at least 4 but not of dimension 4 and Witt index 2. Each V_i is an isometric copy of V. The group G is the split extension of the direct sum $\oplus_{i \in I} V_i$ by that subgroup $FO(V, q)$ of $O(V, q)$ which is generated by all transvections.

(2) $\oplus_{i \in I} V_i : FSp(V, f)$.

(V, f) is a nondegenerate symplectic space over $GF(2)$ of dimension at least 6, and each V_i is a copy of V. $FSp(V, f)$ is that subgroup of the full isometry group $Sp(V, f)$ which is generated by the symplectic transvections.

(3) $\oplus_{i \in I} V_i : {}^+\Omega(V, q)$.

(V, q) is a nondegenerate orthogonal space over $GF(3)$ of dimension at least 5, and each V_i is a copy of V. $^+\Omega(V, q)$ is that subgroup of the full isometry group $O(V, q)$ which is generated by the reflections of +-type. D is either the reflection class of +-type or I is empty, V has finite even dimension, q has discriminant $-$, and D is the class of negative reflections of $-$-type. (See [8, (7.3)] or [19].)

(4) $\oplus_{i \in I} V_i : FSU(V, f)$.

(V, f) is a nondegenerate unitary space over $GF(4)$ of dimension at least 4, and each V_i is a copy of V. $FSU(V, f)$ is that subgroup of the full isometry group $U(V, f)$ which is generated by the unitary transvections.

(5) $\oplus_{i \in I}(GF(4)^6):(Z_3 \cdot {}^+\Omega_6^-(3))$, $I \neq \emptyset$.

This is the subgroup of $\oplus_{i \in I}(GF(4)^6):SU_6(2)$ of (4) corresponding to the transvection subgroup $Z_3 \cdot {}^+\Omega_6^-(3)$ of $SU_6(2)$, [6, 14].

(6) $\oplus_{i \in I}(GF(3)^7):(Z_2 \times Sp_6(2))$, $I \neq \emptyset$.

$GF(3)^7$ is the *mod 3* root lattice for $W(E_7) \simeq Z_2 \times Sp_6(2)$.

(7) $\oplus_{i \in I}(GF(3)^8):(2 \cdot O_8^+(2))$, $I \neq \emptyset$.

$GF(3)^8$ is the *mod 3* root lattice for $W(E_8) \simeq 2 \cdot O_8^+(2)$.

(8) $\oplus_{i \in I}(GF(3)^{10}):(Z_2 \times SU_5(2))$, $I \neq \emptyset$.

See (6.3). A group of this type with $|I| = 1$ occurs in a 3-local subgroup of Fi_{24} [6].

(9) A quotient of $\oplus_{i \in I}(GF(3)^{10}):{}^+\Omega_5^-(3)$, $I \neq \emptyset$.

$GF(3)^{10}$ is a nonsplit module extension of a natural module by a natural module for the group $^+\Omega_5^-(3) \simeq Z_2 \times SU_4(2)$. A subgroup of (8).

(10) A quotient of $\oplus_{i \in I}(GF(3)^{12}):(Z_3 \cdot {}^+\Omega_6^-(3))$, $I \neq \emptyset$.

Nonsplit module extension of a natural module by a natural module.

(11) $\oplus_{i \in I}(GF(3)^7) \cdot {}^+\Omega_7^-(3)$, $I \neq \emptyset$.

Nonsplit extension; see [8]. Contains a subgroup of type (9). For $|I| = 1$ this is a 3-local subgroup of Fi_{24} [6].

(12) $\oplus_{i \in I}(GF(3)^8) \cdot {}^+\Omega_8^-(3)$, $I \neq \emptyset$.

Nonsplit extension; see [8]. Contains a subgroup of type (10).

(13) A quotient of $\oplus_{i \in I}(GF(4)^{10}):SU_5(2)$, $I \neq \emptyset$.

Nonsplit module extension. See (6.3).

(14) $\oplus_{i \in I}(GF(4)^7) \cdot SU_7(2)$, $I \neq \emptyset$.

Nonsplit extension; see [14]. Contains a subgroup of type (13).

(15) Fischer groups of sporadic type $\Omega_8^+(2){:}\Sigma_3$, $\Omega_8^+(3){:}\Sigma_3$, Fi_{22}, Fi_{23}, Fi_{24}.

(16) A quotient of $Wr(K, \Omega)$, $|\Omega| \geq 4$, where K is a strong $\{2,3\}$-group.

See (7.1) and (7.2). There are two generating classes of 3-transpositions if and only if $|\Omega| = 6$ and $K = 1$.

(17) $\oplus_{i \in I}(GF(3)^8){:}(2^{1+6}{:}SU_3(2)')$, $I \neq \emptyset$.

See (3.3) and (7.3). Subgroup of a group of type (8) corresponding to the subgroup $2^{1+6}{:}SU_3(2)'$ of $Z_2 \times SU_5(2)$. For $|I| = 1$ a group of this type is in a 3-local subgroup of Fi_{23} [6].

(18) A quotient of $T_I{:}SU_3(2)'$, where the 2-group T_I is described below, $I \neq \emptyset$.

For each $i \in I$, let U_i be a copy of the natural module $V = GF(4)^3$ for $SU_3(2)'$. Also, for each distinct pair $i, j \in I$, let $W_{i,j}$ be a copy of V. Set $U = \oplus_{i \in I} U_i$ and $W = \oplus_{i,j \in I} W_{i,j}$. Then T_I is a uniquely determined central extension (of class 2 if $|I| > 1$)

$$0 \longrightarrow W \longrightarrow T_I \longrightarrow U \longrightarrow 0$$

which contains an abelian subgroup $V_i \simeq U_i$ representing U_i, for each $i \in I$. These subgroups satisfy $[V_i, V_j] = W_{i,j}$, for all distinct pairs $i, j \in I$.

REFERENCES

[1] M. Aschbacher, A homomorphism theorem for finite graphs, Proc. Amer. Math. Soc. **54** (1976), 468-470.

[2] M. Aschbacher, 'Finite Group Theory,' Cambridge Univ. Press, London, 1986.

[3] R.H. Bruck, 'A Survey of Binary Systems,' Ergebnisse der Mathematik und ihrer Grenzgebiete **20**, Springer-Verlag, Berlin, 1958.

[4] F. Buekenhout, 'La géométrie des groupes des Fischer,' unpublished notes, Free University of Brussels, 1974.

[5] A.M. Cohen, Finite quaternionic reflection groups, J. Algebra **64** (1980), 293-324.

[6] J.H. Conway, R.T. Curtis, S.P. Norton, R.A. Parker, R.A. Wilson, 'Atlas of Finite Groups,' Clarendon Press, Oxford 1985.

[7] H. Cuypers, On a generalization of Fischer spaces, Geom. Dedicata, **34** (1990), 67-87.

[8] H. Cuypers and J.I. Hall, 3-Transposition groups of orthogonal type, J. Algebra, to appear.

[9] B. Fischer, A characterization of the symmetric groups on 4 and 5 letters, J. Algebra **3** (1966), 88-98.

[10] B. Fischer, Finite groups generated by 3-transpositions, University of Warwick lecture notes, 1969.

[11] B. Fischer, Finite groups generated by 3-transpositions I, Invent. Math. **13** (1971), 232-246.

[12] J.I. Hall, Graphs, geometry, 3-transpositions, and symplectic F_2-transvection groups, Proc. London Math. Soc. (Ser. 3) **58** (1989), 89-111.

[13] J.I. Hall, Some 3-transposition groups with normal 2-subgroups, Proc. London Math. Soc. (Ser. 3) **58** (1989), 112-136.

[14] J.I. Hall, 3-Transposition groups with non-central normal 2-subgroups, J. Algebra, to appear.

[15] M. Hall, Jr., 'The Theory of Groups,' Macmillan, New York, 1959.

[16] M. Hall, Jr., Automorphisms of Steiner triple systems, Proc. Symp. in Pure Math. **6** (1962), 47-66.

[17] M. Hall, Jr., Group theory and block designs, in: 'Proceedings of the International Conference on the Theory of Groups at Canberra 1965,' eds.: L.G. Kovács and B.H. Neumann, Gordon and Breach, New York, 1967.

[18] J.D.H. Smith, Commutative Moufang loops: the first 50 years, Algebras, Groups and Geom., **3** (1985), 209-234.

[19] M.-M. Virotte-Ducharme, Couples fischériens presque simple, Thése Paris 7, 1985.

[20] M.-M. Virotte-Ducharme, Une construction du groupe de Fischer $Fi(24)$, Mém. de la Soc. Math. de France **27** (1987).

[21] R. Weiss, On Fischer's characterization of $Sp_{2n}(2)$ and $U_n(2)$, Comm. Algebra **11** (1983), 2527-2554.

[22] R. Weiss, 3-transpositions in infinite groups, Math. Proc. Cambridge Philos. Soc. **96** (1984), 371-377.

[23] R. Weiss, A uniqueness lemma for groups generated by 3-transpositions, Math. Proc. Cambridge Philos. Soc. **97** (1985), 421-431.

[24] F. Zara, Sur les couples Fischeriens de largeur 1, Europ. J. Combin. **4** (1983), 185-199.

[25] F. Zara, Classification des couples fischeriens, Thése Université de Picardie, 1984.

[26] F. Zara, A first step toward the classification of Fischer groups, Geom. Dedicata **25** (1988), 503-512.

[27] F. Zara, Propriétés des groupes de Fischer $D_4(2){:}S_3$ et $D_4(3){:}S_3$, preprint 1990.

(S_3, S_6)-Amalgams

W. Lempken, C. Parker, P. Rowley

This article summarises the results of [LPR1] and [LPR2] in which (S_3, S_6)-amalgams are classified. We begin with the following

Hypothesis 1 *Let p be a prime, and G^* be a group containing proper finite subgroups P_1^* and P_2^* which satisfy*

i) $G^* =< P_1^*, P_2^* >$;

ii) no non-trivial normal subgroup of G^* is contained in $P_1^* \cap P_2^*$; and

iii) $Syl_p(P_1^* \cap P_2^*) \subseteq Syl_p(P_1^*) \cap Syl_p(P_2^*)$.

When Hypothesis 1 holds we have

$$P_1^* \xleftarrow{\phi_1} P_1^* \cap P_2^* \xrightarrow{\phi_2} P_2^*$$

where the ϕ_i are group monomorphisms, and so we may form the free amalgamated product $G = P_1 *_{(P_1^* \cap P_2^*)} P_2^*$. Taking Γ to be a certain coset graph we may reinterpret Hypothesis 1 as (see [DS] for more details):

Hypothesis 1' *Let p be a prime, Γ a tree and G a subgroup of $\mathrm{Aut}\Gamma$ for which*

i) G_γ is finite for each vertex γ of Γ;

ii) G has two orbits on the vertices and is transitive on the edges of Γ; and

iii) $Syl_p(G_{\gamma\tau}) \subseteq Syl_p(G_\gamma) \cap Syl_p(G_\tau)$ *for each edge $\{\gamma, \tau\}$ in Γ.*

Let X_1 and X_2 be two given groups. We call G an (X_1, X_2)-amalgam if G satisfies Hypothesis 1' and additionally

i) $G_\delta/O_p(G_\delta) \cong X_1$ and $G_\tau/O_p(G_\tau) \cong X_2$ for all edges $\{\delta, \tau\}$ in Γ; and

ii) $C_{G_\gamma}(O_p(G_\gamma)) \leq O_p(G_\gamma)$ for all vertices γ of Γ.

Here we shall only consider classifying amalgams when $p = 2$. By a classification of an (X_1, X_2)-amalgam we mean the determination of the possible structures for G_δ and G_τ (where $\{\delta, \tau\}$ is an edge of Γ).

Goldschmidt [G] investigated groups satisfying Hypothesis $1'$ with the further restriction that $[G_\delta : G_{\delta\tau}] = 3 = [G_\tau : G_{\delta\tau}]$ for any, and hence all, edges $\{\delta, \tau\}$ of Γ. There he lists all the possible pairs (G_δ, G_τ), fifteen in all, up to isomorphism. One interesting consequence is that $\mid G_{\delta\tau} \mid$ divides 2^7. Of course Goldschmidt's result includes the classification of all the (S_3, S_3)-amalgams. Following in the footsteps of [G] many other amalgams have been studied. We do not attempt to survey this recent work and henceforth just concentrate on (S_3, S_6)-amalgams. We first recall some notation related to the action of G on the tree Γ. Let $V(\Gamma)$ denote the set of vertices of Γ. For each $\lambda \in V(\Gamma)$ we define

$$\Delta(\lambda) = \{\tau \in \Gamma \mid \{\tau, \lambda\} \text{ is an edge of } \Gamma\};$$
$$Q_\lambda = O_2(G_\lambda) = \text{Stab}_G(\Delta(\lambda)) \text{ (pointwise); and}$$
$$Z_\lambda = < \Omega_1(Z(G_{\lambda\tau})) \mid \tau \in \Delta(\lambda) > .$$

Because $G_{\lambda\tau}$ is a finite non trivial 2-group, Z_λ is non-trivial. Since $C_{G_\lambda}(Q_\lambda) \leq Q_\lambda$ by hypothesis, Z_λ is a subgroup of $\Omega_1(Z(Q_\lambda))$ and hence fixes all the vertices in $\Delta(\lambda)$. Because G acts as a group of automorphisms of Γ, there exist vertices in Γ which are not fixed by Z_λ. The vertices that are closest to λ, under the usual distance function $d(.,.)$, and moved by Z_λ are of central interest. More precisely we define a parameter b as follows.

$b = \min_{\lambda \in V(\Gamma)} b_\lambda$ where
$b_\lambda = \min_{\tau \in V(\Gamma)} \{d(\tau, \lambda) \mid Z_\lambda \not\leq Q_\tau\}$.

b is referred to as the critical distance. The set of pairs $(\alpha, \alpha') \in V(\Gamma) \times V(\Gamma)$ such that $d(\alpha, \alpha') = b$ and $Z_\alpha \not\leq Q_{\alpha'}$ is denoted by $\mathcal{C}$. The elements of $\mathcal{C}$ are called critical pairs and appear at every turn in [LPR1] and [LPR2] The parameter b, in some sense, measures the complexity of the amalgam - the larger b is the more non-central chief factors G_λ has in Q_λ. Thus our first objective is to bound b. This is achieved in [LPR1] where it is shown that

Theorem A ([LPR1]) *Suppose that G is an (S_3, S_6)-amalgam. Then $b \in \{1, 2, 3, 5\}$.*

The main difficulties in the proof of Theorem A arise because of the following facts:

1) S_6 is a rank 2 Lie-type group;

2) The natural $GF(2)S_6$-modules exhibit a large degree of FF-ness, that is, there are many offending subgroups;

3) If V is a natural $GF(2)S_6$-module and $S \in \mathrm{Syl}_2(S_6)$, then there exists an involution centralising the unique S-invariant hyperplane of V(the central transvection); and

4) the 2-rank of S_6 is 3.

The proof of Theorem A involves a very detailed inspection of groups related to critical pairs. We begin by supposing that $(\alpha, \alpha') \in \mathcal{C}$. Then our investigation proceeds down five different paths:

A1. $Z_\tau \not\leq Z(G_\tau)$ for all $\tau \in V(\Gamma)$.

A2. $[Z_\alpha, Z_{\alpha'}] \neq 1$ and $G_\alpha/Q_\alpha \cong S_6$.

A3. $[Z_\alpha, Z_{\alpha'}] \neq 1$ and $G_\alpha/Q_\alpha \cong S_3$.

A4. $[Z_\alpha, Z_{\alpha'}] = 1$ and $G_\alpha/Q_\alpha \cong S_6$.

A5. $[Z_\alpha, Z_{\alpha'}] = 1$ and $G_\alpha/Q_\alpha \cong S_3$.

Case A1 is almost always reduced to a pushing-up problem which has been solved by Meierfrankenfeld. We mention that in a dissertation, [W], T. Westerhoff investigated (S_3, S_6)-amalgams satisfying A1 and A2. We have revised his work. The complexity of the analysis necessary in order to bound the size of b climbs steadily through each of the cases, culminating in A5. In case A5 we are forced into an investigation of the normal subgroups V_δ and W_δ, of G_δ, where $G_\delta/Q_\delta \cong S_6$. Here $V_\delta =< Z_\tau \mid \tau \in \Delta(\delta) >$ and $W_\delta =< Z_\tau \mid d(\tau, \delta) \leq 3 >$. The interest in these groups is prompted by the fact that $\eta(G_\delta, V_\delta/Z_\delta) \neq 0 \neq \eta(G_\delta, W_\delta/V_\delta)$ ($\eta(K, L)$ is the number of non-central chief factors that K has in L), and that V_δ/Z_δ is a module for $G_\delta/Q_\delta \cong S_6$. In [§12;LPR1] the structure of V_δ/Z_δ is determined. In fact, it is shown that V_δ/Z_δ is a quotient of $\binom{4}{1} \oplus 1$, where $\binom{4}{1}$ denotes one of the orthogonal $GF(2)S_6$-modules. From this position the work divides into three mainline cases and two offshoots. If it were not for the central transvection (see 3), then the work in case A5 would be reduced by at least two thirds. Once armed with Theorem A, the structures of the (S_3, S_6)-amalgams are at hand.

Theorem B ([LPR2]) *Suppose that G is an (S_3, S_6)-amalgam. Then $b = 1$ or 2 and, for an edge $\{\delta, \tau\}$ of Γ, $\mid G_{\delta\tau} \mid$ divides 2^{15}; moreover, the structure of G_δ and G_τ is as summarised in Table 1.*

		(S_3, S_6)-Amalgams		
	b	G_δ	$\eta(G_\tau, O_2(G_\tau))$	Example(s)
S_1	1	$2^{(1+4)}S_6$	1	$A_{12}, O_7(3)$
S_1^1	1	$2^{(1+4+1)}S_6$	1	$S_{12}, SO_7(3)$
S_2	1	$2^{(1+4)}S_6$	3	$Sp_6(2)$
S_3	1	$2^4 S_6$	2	$U_4(3).2$
S_3^1	1	$2^{(4+1)}S_6$	2	$U_4(3).2^2$
S_4	2	$2^4 S_6$	3	$L_6(3)$
S_4^1	2	$(2^4 \times 2)S_6$	3	—
S_4^2	2	$2^{(4+1)}S_6$	3	$PGL_6(3)$
S_4^3	2	$(2^{(4+1)} \times 2)S_6$	3	—
S_5	2	$(4^4 \times 2)S_6$	1	$L_6(5), U_6(5)$
S_5^1	2	$(4^4 \times 2)(S_6 \times 2)$	1	—
S_5^2	2	$(4^5)S_6$	1	—
S_5^3	2	$(4^5)(S_6 \times 2)$	1	—
S_6	2	$(4 * 2_+^{1+4})S_6$	3	Co_3

Table 1.

The horizontal line in the last column indicates that no examples are known to the authors.

The proof of Theorem B starts with the information given in Theorem A (that is, b is 1,2,3 or 5). Then each configuration is analysed separately. We remark that while $b = 1$ and 2 give their examples comparitively easily, it is not until the groups G_δ and G_τ have been constructed that contradictions to the (almost existent)$b = 3$ cases reveal themselves. The reason that the $b = 5$ case is not contradicted by the general arguments of [LPR1] is that for $b \geq 6, W_\delta$ is elementary abelian, while when $b = 5, W_\delta$ is non-abelian. Thus in [LPR2] the $b = 5$ configuration yields to a combination of the big b methods together with particular equalities caused by b being 5.

Finally, we would like to briefly discuss the amalgams listed in table 1. Firstly, recalling the case divison A1, A2,...,A5 given earlier; A1 yields the amalgams $S_1, S_1^1, S_2, S_4, S_4^1, S_4^2$ and S_4^3. A2 gives S_5, S_5^1, S_5^2 and S_5^3. Case A3, S_6 and case A4, S_3 and S_3^1. The resilient case A5 provides no examples.

Secondly, we recall the Timmesfeld criterion for an amalgam to be connected. We associate with each amalgam a diagram, Δ, whose nodes are the subgroups $P_1, P_2, \ldots, P_n$. Two nodes, $i, j \in \Delta$, are connected if and only if $O_p(P_i) \cap O_p(P_j)$ is not normal in P_i and is not normal in P_j. Thus Timmesfeld says that an amalgam is connected whenever the associated diagram is connected. Put $P_{ij} =< P_i, P_j >, B = \bigcap_{i=1}^{n} P_i$ and $K_{ij} = \text{Core}_{P_{ij}}(B)$. In [ST] Stellmacher and Timmesfeld classify, knowing P_{ij}/K_{ij} only locally, a collection of connected rank-3 amalgams showing, amongst other things, that not all the vertices are joined (that is, Δ is not a triangle). Our work on (S_3, S_6)-amalgams may be viewed in the context of rank 3-amalgams in which we have $P_i/O_2(P_i) \cong S_3$ for $i = 1, 2, 3$ and $P_{12}/K_{12} \cong S_6$. Note that we impose no conditions upon P_{13}/K_{13} and P_{23}/K_{23}. We find that the amalgams $S_1, S_1^1, S_5, S_5^1, S_5^2$ and S_5^3 provide examples for which the associated diagram is not connected.

References.

[DS]   A. Delgado and B. Stellmacher, Weak (B, N)-pairs of rank 2, in: A. Delgado, D. Goldschmidt and B. Stellmacher, Groups and graphs: new results and methods, DMV seminar Bd 6, Basel-Boston-Stuttgart, 1985.

[G]   D. Goldschmidt, Automorphisms of Trivalent Graphs, Annals of Mathematics, 111, 1980, 377-406.

[LPR1]   W. Lempken, C. Parker, P. Rowley, (S_3, S_6)-Amalgams I, II, III, IV, V and VI, preprints and manuscripts, University of Manchester Institute of Science and Technology.

[LPR2]   W. Lempken, C. Parker, P. Rowley, The Structure of (S_3, S_6)-Amalgams I and II, manuscripts, Justus-Liebig-Universität Giessen.

[ST]   B. Stellmacher, F. Timmesfeld, Rank-3 amalgams, preprint.

[W]   T. Westerhoff, Eine Kennzeichnung einiger (Σ_6, Σ_3)-Amalgame, Dissertation, Universität Bielefeld.

Acknowledgements

Lempken was funded by SERC GR/D/69655

Parker was funded by NATO B/RFN/8269

Rowley's three trips to Germany to advance work on this project were supported by SERC GR/F71768

Pushing Down Minimal Parabolic Systems

Peter Rowley

The finite groups of Lie type have a number of structural features in common. Of particular note are the systems of parabolic subgroups. These systems of subgroups have a rich structure and have had a significant impact on the study of groups of Lie type. So, not surprisingly, this has prompted a number of attempts to broaden the concept of a parabolic system and related ideas. For a small sample of such endeavours see [7] and [8]. Here we consider a particular type of (generalized) minimal parabolic system. For the remainder of this paper G will be a group which contains finite subgroups $P_1, \ldots, P_n$ and satisfies the following:-

1. $G = \langle P_i \mid i \in I \rangle \neq \langle P_j \mid j \in J \subsetneq I \rangle$ (where $I = \{1, \ldots, n\}$);

2. for each $k \in I, S := \bigcap_{i \in I} P_i$ is a Sylow 2-subgroup of P_k; and

3. $P_i/O_2(P_i) \cong L_2(2)$ for each $i \in I$.

For $J = \{j_1, \ldots, j_r\} \subseteq I$ we put $P_J = P_{j_1 \ldots j_r} = \langle P_j \mid j \in J \rangle$ and $S_{j_1 \ldots j_r} = \mathrm{core}_{P_J} S$. Also we set $S_o = \mathrm{core}_G S$.

The above situation occurs in the familiar example $L_{n+1}(2)$. Such an example comes equipped with a Dynkin diagram and we now seek, by analogy, to define a diagram for our minimal parabolic system. We take I to be the set of nodes with the following rules (and notation) for the bonds. Let $i, j \in I, i \neq j$.

$$
\begin{array}{ll}
\underset{i}{\circ} \qquad \underset{j}{\circ} & \text{if and only if } P_{ij}/S_{ij} \cong L_2(2) \times L_2(2) \\[2mm]
\underset{i}{\circ}\!\!-\!\!\!-\!\!\!-\!\!\underset{j}{\circ} & \text{if and only if } P_{ij}/S_{ij} \cong L_3(2) \\[2mm]
\underset{i}{\circ}\!\!=\!\!\!\overset{\sim}{=}\!\!\!=\!\!\underset{j}{\circ} & \text{if and only if } P_{ij}/S_{ij} \cong \hat{S}_6
\end{array}
$$

($\hat{S}_6$ denotes the group H which is uniquely specified by requiring H' to be the non-split triple cover of A_6, $H/O_3(H') \cong S_6$ and $H' = C_H(O_3(H'))$.)

We are drawn to the above type of bonds, rather as a moth is to a light, because of the following remarkable series of examples.

Diagram	G
$\circ\!\!-\!\!-\!\!-\!\!\circ\!\!=\!\!\overset{\sim}{=}\!\!=\!\!\circ$	M_{24} or He
$\circ\!\!-\!\!-\!\!-\!\!\circ\!\!-\!\!-\!\!-\!\!\circ\!\!=\!\!\overset{\sim}{=}\!\!=\!\!\circ$	$.1$
$\circ\!\!-\!\!-\!\!\circ\!\!-\!\!-\!\!\circ\!\!-\!\!-\!\!\circ\!\!=\!\!\overset{\sim}{=}\!\!=\!\!\circ$	M

Our interest here is in what can be deduced given that our minimal parabolic system has a prescribed diagram. For example Timmesfeld [13] uses a result of Tits to show that if $\{P_i \mid i \in I\}$ has diagram o——o——.. .o——o, then $G/S_o \cong L_{n+1}(2)$ with $\{P_i/S_o \mid i \in I\}$ being the "standard" minimal parabolic subgroups containing S/S_o. We now survey other results of this nature when the diagram contains the bond o══o. Since we are only examining G/S_o, there is no loss in generality in assuming $S_o = 1$.

Theorem 1 *(Rowley [9]) If $\{P_i \mid i \in I\}$ has diagram* o——o══o, *then* $|S| = 2^9$ *or* 2^{10}.

Theorem 1 is established by investigating the intersections of subgroups such as $O_2(P_i)$ and $O_2(P_{ij})$ with suitably chosen conjugates so as to produce other subgroups of S. The ultimate aim is to locate subgroups of S which are simultaneously normalized by P_1, P_2 and P_3; such subgroups are then contained in S_o at which point the proof of Theorem 1 is speedily concluded. The philosophy of this type of approach is to use information about the top of S to construct additional subgroups of S which are lower down in S. These subgroups are used in turn to build yet more subgroups of S further down the subgroup lattice of $S-$ pushing down seems an apt description for this process. Pushing down type arguments also appear in [2] and [1]. For studying parabolic systems there is another body of technique - the so-called amalgam method (see, for example, [3]). This works in the opposite direction to pushing down by considering the normal closure of certain subgroups of S and working upwards. S. Heiss [4] and, independently, A. Ivanov [5] have studied further the two outcomes of Theorem 1. With the aid of a computer (for Todd Coxeter enumerations) they show that $G \cong M_{24}$ or He in the case $|S| = 2^{10}$ and that $G \cong 3^7.Sp_6(2)$ when $|S| = 2^9$.

We now assume that when we have the following diagram it is labelled as indicated

$$\underset{n}{o}\text{——}\underset{n-1}{o}\cdots\underset{4}{o}\text{——}\underset{3}{o}\text{——}\underset{2}{o}\text{══}\underset{1}{o}$$

and that $\mid S/S_{123}\mid = 2^{10}$.

Theorem 2 *(Rowley [10]) If $\{P_i \mid i \in I\}$ has diagram*

$$o\text{——}o\text{——}o\text{══}o$$

and $|S| \neq 2^{10}$, *then* $|S| = 2^{21}$ *or* 2^{25}.

This result is also proved using pushing down arguments. A related result which employs the amalgam method is

Theorem 3 *(C. Parker [6]) Suppose $n = 4$ and put $H = \langle P_1, P_2, P_3 \rangle$. If $\{P_i \mid i \in \{1,2,3\}\}$ has diagram* o——o══o, $C_H(S_{123}) \leq S_{123}$ *and* $C_{P_4}(O_2(P_4)) \leq O_2(P_4)$, *then* $|S| = 2^{21}, 2^{22}$ *or* 2^{25}.

We remark that, invoking the work of Heiss and Ivanov mentioned above, we may eliminate the 2^{25} possibility in Theorems 2 and 3.

Theorem 4 *(Rowley [11]) If $\{P_i \mid i \in I\}$ has diagram*

$$\circ\!\!-\!\!\!-\!\!\!-\!\!\!-\!\!\circ\!\!-\!\!\!-\!\!\!-\!\!\!-\!\!\circ\!\!-\!\!\!-\!\!\!-\!\!\!-\!\!\circ\!\!=\!\!\!=\!\!\!=\!\!\circ$$

and $2^{10} \neq |S/S_{1234}| \neq 2^{25}$, then $|S| = 2^{46}$.

The obvious question as to whether we may extend the minimal parabolic system in Theorem 4 once more is answered in

Theorem 5 *(Rowley [12]) There is no minimal parabolic system $\{P_i \mid i \in I\}$ with diagram* $\circ\!\!-\!\!\!-\!\!\!-\!\!\circ\!\!-\!\!\!-\!\!\!-\!\!\circ\!\!-\!\!\!-\!\!\!-\!\!\circ\!\!-\!\!\!-\!\!\!-\!\!\circ\!\!=\!\!\!=\!\!\circ$ *and $2^{10} \neq |S/S_{1234}| \neq 2^{25}$.*

Theorems 4 and 5 also use the pushing down approach. We devote the remainder of this paper to highlighting various parts of the proof of Theorem 5. When $n \geq 4$ we have at our disposal an important, albeit elementary, tool– the Replication Lemma ([Lemma 3.4; [11]]). We just consider a special case, the general features being readily apparent. Suppose that $n = 6$ and that our minimal parabolic subgroups are as follows:-

$$\underset{F}{\circ}\!\!-\!\!\!-\!\!\!-\!\!\underset{E}{\circ}\!\!-\!\!\!-\!\!\!-\!\!\underset{D}{\circ}\!\!-\!\!\!-\!\!\!-\!\!\underset{C}{\circ}\ \!\!-\!\!\!-\!\!\!-\!\!\underset{B}{\circ}\!\!=\!\!\!=\!\!\underset{A}{\circ}$$

We must pause for a moment to establish some notation. By hypothesis $\langle A, E\rangle/O_2(\langle A, E\rangle) \cong L_2(2) \times L_2(2)$, and A contains a subgroup A_E of index 2 which projects onto one of the $L_2(2)$-direct factors. So $A_E \trianglelefteq \langle A, E\rangle$. Likewise B and C possess analogous subgroups B_E and C_E. Put $K = \langle A, B, C\rangle$ and $K_E = \langle A_E, B_E, C_E\rangle$; observe that E normalizes K_E.

Now suppose that $X \leq O_2(E)$ and that $X \trianglelefteq K$. Let $e \in E\backslash S$ be such that $e^2 \in S$. So, as $X \leq O_2(E), XX^e$ is a subgroup of S and K_E normalizes X and X^e. Hence K_E normalizes XXe and $X \cap X^e$ and

$$XX^e/X \cong_{K_E} X^e/X \cap X^e$$

(meaning they are isomorphic as K_E-operator groups). So what's new? Now let Y be a subgroup of XXe which contains X and suppose, say, that $Y \trianglelefteq \langle A, B\rangle$.So $Y/X \cong_{\langle A_E, B_E\rangle} Y \cap X^e/X \cap X^e$. Since $X \trianglelefteq S$ and $e^2 \in S, e$ normalizes $X \cap X^e$ and therefore $Y^{e^{-1}} \cap X$ is a subgroup of X containing $X \cap X^e$ which is normalized by $\langle A_E, B_E\rangle$. Hence the section Y/X has been replicated in the section $X/X \cap X^e$. Often with some additional pleading we may obtain $Y^{e^{-1}} \cap X \trianglelefteq \langle A, B\rangle$ and $X \cap X^e \trianglelefteq K$ and so, among other things, we can deduce properties of the $\langle A, B\rangle$-chief series of $X/X \cap X^e$ from knowledge of the subgroup lattice of S/X.

We investigate a minimal parabolic system of the type described in Theorem 5 with the aim of deriving a contradiction. Put $T_{13} = S_{12345}$ (to match up

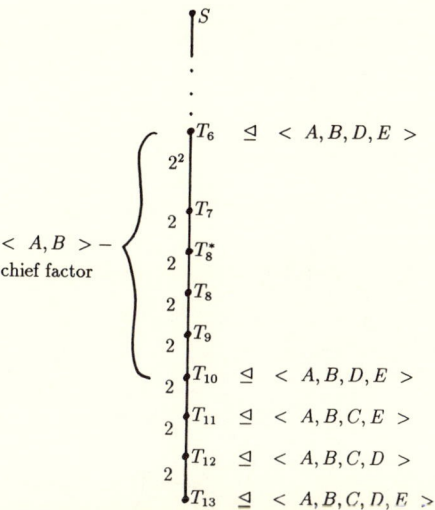

Figure 1.

with the notation in [11]). By Theorem 4 $|S/T_{13}| = 2^{46}$ and from [11] we have the subgroups at the bottom of S/T_{13} as in Figure 1.

Setting $Z_1 = S_{23456}$ we note that, by a result mentioned earlier, $\langle B, C, D, E, F\rangle/Z_1 \cong L_6(2)$ with $\{B/Z_1, C/Z_1, D/Z_1, E/Z_1, F/Z_1\}$ the minimal parabolic subgroups containing S/Z_1. In particular, $\langle D, E, F\rangle/S_{456} \cong L_4(2)$ and $\langle C, D, E, F\rangle/S_{3456} \cong L_5(2)$.

In the course of establishing Theorem 5 we make the acquaintance of a large number of subgroups of S. Here are a few of the cast:-

$$K_{13} := \mathrm{core}_F T_{13} \; ; \qquad K_i = T_i \cap T_i^f \, (i \le 12) \qquad f \in F\backslash S$$
$$K_8^* = T_8^* \cap T_8^{*f}$$
$$i \le 13 \qquad L_i = K_i \cap K_i^e \qquad L_8^* = K_8^* \cap K_8^{*e} \qquad e \in E\backslash S$$
$$M_i = L_i \cap L_i^d \qquad M_8^* = L_8^* \cap L_8^{*d} \qquad d \in D\backslash S$$
$$N_i = M_i \cap M_i^c \qquad N_8^* = M_8^* \cap M_8^{*c} \qquad c \in C\backslash S$$
$$P_8^* = N_8^* \cap N_8^{*b} \qquad b \in B\backslash S$$

It turns out that the above subgroups are independent of the element chosen from $P_i\backslash S$. Several of these groups are displayed in Figure 2. Note that in the sequence of elements chosen we are moving left to right along the diagram.

We pick up the proof of Theorem 5 at the point where the groups K_i have just entered the picture. A relatively short argument together with the Replication Lemma shows that if $K_{13} = T_{13} \cap T_{13}^f$ for some $f \in F\backslash S$, then T_{13}/K_{13}

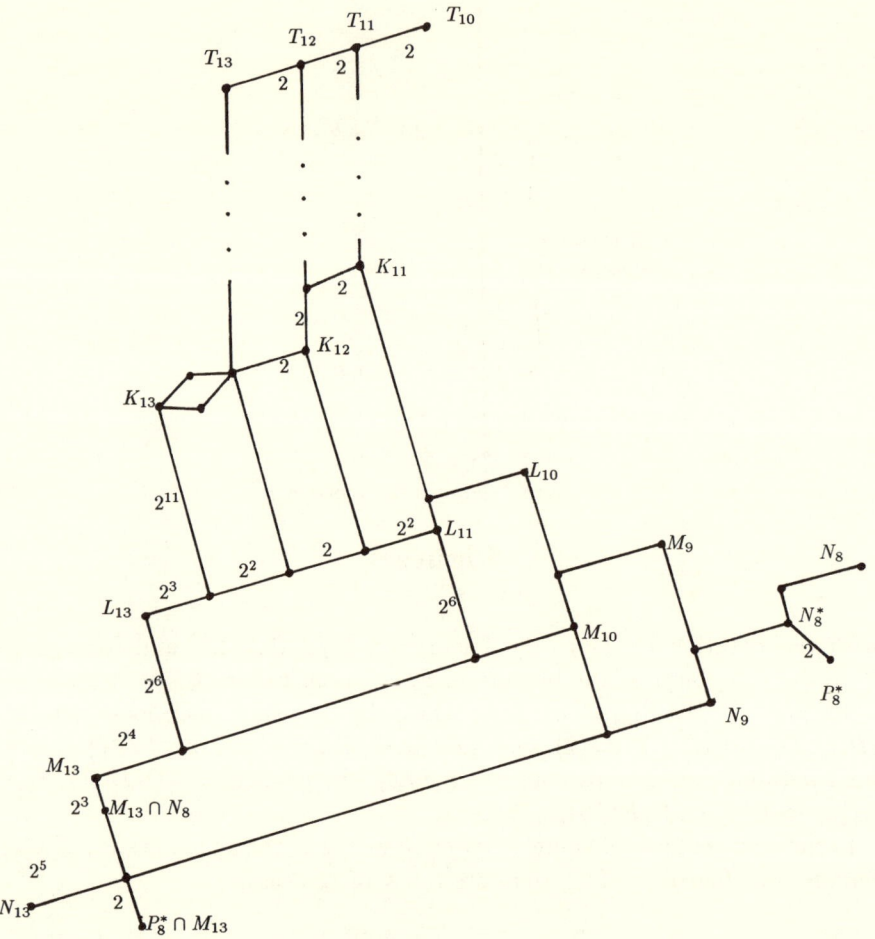

(Note that $K_{11}/K_{12} \cong L_{11} \cap K_{12}$, $L_{10}/L_{11} \cong M_{10}/M_{10} \cap L_{11}$ and $M_9/M_{10} \cong N_9 \cap M_{10}$.)

Figure 2.

must be an extra special group of order 2^{26}! So $K_{13} \neq T_{13} \cap T_{13}^f$ for $f \in F \backslash S$; in fact $[T_{13} \cap T_{13}^f : K_{13}] = 2$ and $[T_{13} : K_{13}] = 2^{27}$. Having $K_{13} \neq T_{13} \cap T_{13}^f$ for any $f \in F \backslash S$ throws quite a spanner in the works; apart from denying us the use of the Replication Lemma it also hampers our investigation of L_{13}(via [Lemma 3.2; [11]]). Luckily were able to weave our way round this problem and show that $L_{13} \trianglelefteq \langle A, B, C, E, F \rangle$ with L_{11}/L_{13} being an 8-dimensional chief factor for $\langle E, F \rangle$. So $L_{11}/L_{13} \cong_{\langle E,F \rangle} s\ell_3(2)$, the simple adjoint (or Steinberg) module for $L_3(2)$. This looks a most promising development since many of the $\langle E, F \rangle$-chief factors known to us at this stage are isomorphic to a natural 3-dimensional $L_3(2)$-module (for example L_{10}/L_{11}). Such small chief factors don't normally reveal many secrets. However our joy is short-lived as we discover that any information in the subgroup lattice when projected onto L_{11}/L_{13} is entirely consistent. So we must travel yet deeper into the bowels of S!

Now $[K_{13} : K_{13} \cap L_6] = 2^4$ with $K_{13} \cap L_6 \trianglelefteq D$. Hence $[K_{13} \cap L_6 : L_{13}] = 2^{10}$, and so $[L_{13} : M_{13}] \leq 2^{10}$. Also we are able to pinpoint the location of M_{10} and thus deduce that $[M_{10} : M_{13}] \leq 2^{15}$. Because M_{10}/M_{13} admits the group $\langle D, E, F \rangle$ an inspection of the irreducible $GF(2)L_4(2)$ modules and the fact that L_{11}/L_{13} projects into M_{10}/M_{13} to give an irreducible $\langle E, F \rangle 8$-dimensional subquotient reveals that

$$M_{10}/M_{13} \cong_{\langle D,E,F \rangle} V \text{ or } V/\langle I_4 \rangle.$$

Here V is the set of 4×4 trace zero matrices over $GF(2)$ and I_4 is the 4×4 identity matrix. If the former possibility holds, then we obtain $|N_9/N_{13}| \leq 2^{24}$ while $|N_9/N_{13}| \leq 2^{23}$ if the latter holds. Since M_{10}/M_{13} projects into N_9/N_{13} and N_9/N_{13} admits $\langle C, D, E, F \rangle$, the irreducible $GF(2)L_5(2)$-modules force $|M_{10}/M_{13}| = 2^{15}$ and N_9/N_{13} to be isomorphic to the simple adjoint $L_5(2)$-module.

Somewhat like a fifth column we have also been building another chain of subgroups which culminates in P_8^*. There we have that N_8/P_8^* is a chief factor for $\langle B, C, D, E, F \rangle$ of order 2^6. Also we can show that $M_{13} \cap N_8/M_{13} \cap P_8^*$ has order 2^3. Since M_{13} is normal in $\langle D, E, F \rangle, \langle D, E, F \rangle$ acts upon $M_{13} \cap N_8/M_{13} \cap P_8^*$ whence $O^2(\langle D, E, F \rangle)$ centralizes this section. But then $O^2(\langle D, E, F \rangle)$ centralizes a 2^3-section of N_8/P_8^* which is impossible and gives us our desired contradiction.

References

[1] N. CHEAITO, S_3-complexes and the McLaughlin group, Ph.D. Thesis, University of Manchester Institute of Science and Technology 1985.

[2] A. CHERMAK, On certain groups with parabolic-type subgroups over $\mathbf{Z}_2$, J. London Math. Soc. (2) 23 (1981) 265-279.

[3] A. DELGADO and B. STELLMACHER, Weak (B, N)-pairs of rank 2, in: A. Delgado, D. Goldschmidt and B. Stellmacher, Groups and graphs: new results and methods, DMV Seminar Bd. 6, Basel-Boston- Stuttgart, 1985.

[4] S. HEISS, A class of parabolic systems of rank 3 containing a $\hat{S}_6$ subsystem, preprint, Freie Universität, Berlin 1989.

[5] A. IVANOV, unpublished.

[6] C. PARKER, Groups containing a subdiagram o———o══o, preprint, University of Giessen 1989.

[7] M. RONAN, and S. SMITH, 2-local geometries for some sporadic groups, AMS Symposia in Pure Mathematics. 37(Finite Groups) *American Math. Soc.*, 1980, 283-289.

[8] M. RONAN, and G. STROTH, Minimal parabolic geometries for the sporadic groups. *Europ. J. Combinatorics*, (1984)5, 59-91.

[9] P. ROWLEY, On the minimal parabolic system related to M_{24}, *J. London Math. Soc.* (2) 40 (1989), 40-57 .

[10] P. ROWLEY, Minimal parabolic systems with diagram o———o———o══o, to appear in *J. of Algebra*.

[11] P. ROWLEY, On the minimal parabolic system related to the monster simple group, preprint UMIST.

[12] P. ROWLEY, The non-existence of certain minimal parabolic systems, in preparation.

[13] F. TIMMESFELD, Tits Geometries and Parabolic Systems in Finitely Generated Groups. I, *Math. Z.* 184(1983) 377-396.

Nonspherical spheres

G. Stroth

Let G be a finite simple group, p a prime, S a Sylow p-subgroup of G and let $\{P_1, \ldots, P_n\}$ be a minimal parabolic system of G in the sense of [RoStr]. Then this parabolic system does not determine G uniquely. In fact in most cases there is even an infinite group having the same parabolic system. In this paper we will make an additional assumption which will hopefully make it possible to determine G in most cases. For this approach we will use the geometry Γ associated to the parabolic system (The objects are the cosets of the maximal parabolics $G_i = \langle P_j \mid j \neq i \rangle$, $i = 1, \ldots, n$ and incidence iff two different cosets have nontrivial intersection).

We will consider a geometry Γ (firm, residually connected) over a finite type set I. Furthermore let G be a subgroup of $Aut(\Gamma)$ acting flag transitively on Γ. Then we define

Definition *A thin subgeometry Δ of Γ will be called an apartment in Γ with respect to G iff*

(1) I is also the type set of Δ.

(2) Δ is residually connected.

(3) $N = G_\Delta$ (the setwise stabilizer of Δ in G) acts flag transitively on Δ.

We make the following observation: Suppose Γ possesses an apartment Δ with $G_\Delta = N$. Let us fix a maximal flag F of Δ. Set $B = G_F$. Then we may choose notation such that $P_i = G_{F_i}$, $i = 1, \ldots, n$, where F_i are the comaximal flags in F. As N acts flag transitively on Δ we get that for any i there is a reflection x_i on F_i. Now obviously $\langle x_1, \ldots, x_n \rangle$ acts flag transitively on Δ. So by (2) we get that $N = \langle x_1, \ldots, x_n \rangle (B \cap N)$. Furthermore as Δ is thin we get that $B \cap N \trianglelefteq N$. Finally as Δ is a subgeometry of Γ we have $P_i \cap N = (B \cap N)\langle x_i \rangle$, $i = 1, \ldots, n$. This gives motivation for the following definition for our original parabolic system

Definition. *Let $\mathcal{P} = \{P_1, \ldots, P_n\}$ be a minimal parabolic system for G, $B = P_1 \cap \ldots \cap P_n$. A subgroup N of G is called a Weyl group for $\mathcal{P}$ iff*

(1) $N = \langle x_1, \ldots, x_n \rangle$, $x_i \in P_i - B$, $x_i^2 \in B$.

(2) $B \cap N \trianglelefteq N$.

(3) $N \cap P_i = (B \cap N)\langle x_i \rangle$, $i = 1, \ldots, n$.

We have just seen that for an apartment Δ of a geometry Γ we have that G_Δ is always a Weyl group of the parabolic system built by G_{F_i}, $i = 1, \ldots, n$, where the F_i are the comaximal flags of a maximal flag F of Δ.

Suppose that a parabolic system possesses a Weyl group N. If additionally we have

(A) $G = BNB$ and

(B) $BgBhB \subset (BgB) \cup (BghB)$ for all $g, h \in N$

then we have a group with BN-pair. If the rank is at least three (and some irreducible property), then by the work of J. Tits [Tits] we get that G has a normal subgroup which is a group of Lietype over some finite field. In this paper we will basically investigate the following situation

(BNB) We have a geometry Γ with flag transitive automorphism group G. Assume that Γ possesses an apartment Δ with respect to G. Let $N = G_\Delta$. Then $G = BNB$, where $B = G_F$ for some maximal flag F of Δ. (this means that for any two maximal flags F^1, F^2 there is some $g \in G$ such that F^1, $F^2 \in \Delta^g$).

The geometries we are going to investigate should be close to the parabolic systems in the sense of [RoStr]. Hence we will ask for

(p-closed) If F is a maximal flag of Γ, then $G_F = B$ is p-closed.

First results

(1.1) Lemma *Assume (BNB). Let F_i be a comaximal flag in Δ, then G_{F_i} acts 2-transitively on the residue of F_i in Γ.*

Proof. The residue of F_i in Δ consists of two elements. Let $a, b \in res_\Gamma(F_i)$. Then $F_1 \cup \{a\}$, $F_1 \cup \{b\}$ are maximal flags. By (BNB) there is some $g \in G$ such that both belong to Δ^g. Hence $g \in G_{F_i}$. So G_{F_i} is 2-homogeneous. But as G_{F_i} contains some reflection on F_i we get the assertion.

By the classification of 2-transitive groups we now get that if we have (p-closed) then basically G_{F_i} induces a rank one Lietype group over a field of characteristic p on $res(F_i)$. So at least in the case that all G_{F_i} induce nonsolvable groups we have what is called a weak BN - pair [DeSte]. For convenience of the reader we recall the definition in our terminology.

Definition *Let Γ be a geometry over I, $|I| = n$, with flag transitive group G. Let F be a maximal flag and F_i, $i = 1, \ldots, n$, be the comaximal subflags. We call G an irreducible weak BN-pair of rank n if the following hold*

(1) $B = G_F$ is p-closed for some prime p.

(2) $P_i = G_{F_i}$ induces a rank one Liegroup over a finite field of characteristic p on $res(F_i)$.

(3) (irreducible) Let K_i be the kernel of the representation of P_i on $res(F_i)$, $i = 1, \ldots, n$. Then either $\langle P_i, P_j \rangle = P_i P_j$ or $K_i \cap K_j$ is neither normal in P_i nor in P_j, $i, j = 1, \ldots, n$.

Now in what follows we try to classify some irreducible weak BN-pairs satisfying (BNB). The idea in doing this is to investigate the possibilities of a Weyl group in the corresponding parabolic system. While the stabilizer of an apartment always is a Weyl group the contrary is not true. The point is that the geometry defined by N (the objects are the cosets of $\langle x_i \rangle (B \cap N)$, $i = 1, \ldots, n$, and two different cosets are incident iff they have nontrivial intersection) is not neccessarily a subgeometry of Γ.

(1.2) Lemma *Let Γ be of rank two (points, lines) with incidence graph a complete bipartite graph $K_{3,3}$. Let G be a flag transitive automorphism group on Γ, $G \simeq (\mathbb{Z}_3 \times \mathbb{Z}_3) \cdot 2$. Then Γ possesses no apartment with respect to G.*

Proof. Suppose false. Then $N = G_\Delta$ would be a Weyl group. This yields $N \simeq \Sigma_3$. Hence Δ consists of three points and three lines. This means that $\Delta = \Gamma$, contradicting Δ to be thin.

Remark. The corresponding chamber system in (1.2) possesses an apartment with six chambers.

2. Results on irreducible weak BN - pairs with (BNB)

(2.1) Theorem *[StrWe]. Let G be an irreducible weak BN - pair of rank 2 satisfying (BNB). Then one of the following holds*

(i) $G = P_1 P_2$

(ii) G *is a rank two group of Lietype (Γ is the corresponding building)*

(iii) $G \simeq A_6$ *or* $^2F_4(2)'$ *(Γ is the corresponding building)*

(iv) $G \simeq 3 \cdot S_6$ *(Γ is the 3 - fold cover of the $Sp(4,2)$ 4 - gon)*

(2.2) Corollary. *Let G be an irreducible weak BN - pair of rank n satisfying (BNB). Then G is a semiclassical parabolic system. If for no flag F_{ij} of corank two $G_{F_{ij}}$ induces $3 \cdot S_6$ on $res(F_{ij})$ then we even have a classical Tits system. (For definitions see [Ti] where semiclassical is called semiparabolic).*

We now use the classification of classical Tits systems (see [Str], [Mei]).

(2.3) Lemma. *Let G be a classical Tits system of rank three*

(i) *If the diagram is of spherical type and Γ possesses an apartment with respect to G, then G is a Liegroup and Γ is the corresponding building.*

(ii) *If the diagram is of type* ◯—◯══◯ *or* ◯══◯══◯ *and G is locally isomorphic to $G_2(3)$ then Γ possesses no appartment with respect to G.*

Proof. If the diagram is of spherical type then we have that G is a group of Lietype or $G \simeq A_7$. In the case of A_7 and also in (ii) we always have a system $\{P_1, P_2, P_3\}$ of rank three where $\langle P_1, P_3 \rangle / O_2(\langle P_1, P_3 \rangle) \simeq (\mathbb{Z}_8 \times \mathbb{Z}_8) \cdot 2$. Now application of (1.2) gives the assertion.

Remark. In (ii) one can show that in the second case G does not possesses a Weyl group at all. In the first case G possesses a Weyl group. In fact one can show

(2.4) Lemma. *Let G be locally isomorphic to $G_2(3)$ with diagram* ◯—◯══◯. *If G possesses a Weyl group N, then there is a normal subgroup K of G contained in N such that one of the following holds:*

(i) $G/K \simeq G_2(3)$ and $N/K \simeq G_2(2)$ or $3^4 \cdot 16$

(ii) $G/K \simeq 3^{14} \cdot G_2(3)$ and $N/K \simeq 3^{13} \cdot 3^4 \cdot 16$

Proof. We first give generators and relations for the parabolic system. We have three minimal parabolics P_1, P_2, P_3, where $P_i/O_2(P_i) \simeq \Sigma_3$, $i = 1, 2, 3$. Further we have $\langle P_1, P_2 \rangle \simeq E_8 L_3(2)$, $\langle P_2, P_3 \rangle \simeq G_2(2)$ and $\langle P_1, P_3 \rangle \simeq 2^{1+4}(\mathbb{Z}_3 \times \mathbb{Z}_3) \cdot 2$. This gives the following generators and relations for the parabolic system.

$a^4 = b^4 = c^2 = d^2 = [a, b] = (ac)^2 = (bc)^2 = (cd)^2 = dadb^{-1} = 1$ (relations for B).
$P_1 = \langle B, w \rangle \; : \; w^3 = (ab)^w b^{-1} adc = (cda^{-1}b)^w dcb^2 = (b^2 cd)^w a^{-1} b^{-1} = (ca^2 bw)^2 = (a^{-1}b)^w da^{-1} b^{-1} = 1$.
$P_2 = \langle B, x \rangle : x^3 = a^x b^{-1} = b^x ab = c^x c = (xd)^2 = 1$.
$P_3 = \langle B, y \rangle : y^3 = (a^{-1}b)^y da^{-1} b^{-1} = (cda^{-1}b)^y b^{-1} adc = adyda^{-1} y = 1$.

Set $u = xd$, $z = yca^2 b$ and $t = wcadb^2 c$. Then we have the following further relations:

$$(uz)^6 = (tu)^3 = [w, y].$$

Now using CAYLEY we can check the possibilities for the group N. This yields that basically (up to conjugation) there are two possibilities

$N_1 : x_1 = ta^2 b^2$, $x_2 = uc$, $x_3 = za^2 b^2$ and
$N_2 : x_1 = a^2 b^{-1} cy$, $x_2 = dx$, $x_3 = a^2 b^{-1} cw$.

Let $K_i = \text{Core}_G(N_i)$. Then we get with CAYLEY that in the first case $G/K_1 \simeq G_2(3)$ while in the second case $G/K_2 \simeq 3^{14} \cdot G_2(3)$.

(It is an easy counting argument that (BNB) does not hold if $G/K \not\simeq G_2(3)$. Whether it holds if $G/K \simeq G_2(3)$ is open).

Let G be an automorhism group of $U_4(3)$. Then G possesses a parabolic system belonging to $\bigcirc\!\!=\!\!\bigcirc\!\!=\!\!\bigcirc$.

(2.5) Proposition. *(i) [Glau] The classical system for $U_4(3)$ with diagram $\bigcirc\!\!=\!\!\bigcirc\!\!=\!\!\bigcirc$ possesses two types of apartment. One with 16 points and one with 18 points.*

(ii) Let G be a classical system with diagram $\bigcirc\!\!=\!\!\bigcirc\!\!=\!\!\bigcirc$ which is locally isomorphic to $U_4(3)$. If Γ possesses an apartment with 16 or 18 points then $G \simeq U_4(3) \cdot 2 \cdot 2$.

Proof. We just have to prove (ii). First we give generators and relations for the parabolic system. We have three maximal parabolics G_1, G_2, G_3: We get the following relations:

$G_1 = \langle a, b, v, w, u \rangle : v^4 = w^4 = v^2 w^2 = u^2 = (vu)^2 = w^v w = (uw)^4 = a^3 = b^3 = [a, b] = a^v a = b^v b = [w, a] = a^u b^{-1} = b^u a^{-1} = (wb)^3 = 1.$
Set $t = w^b$. Then we have: $t^2 w^2 = (tw)^4 = [w^u, w] = [w^u, t] = [t^u, w] = [t^u, t] = 1.$
$G_2 = \langle v, w, u, p, y \rangle,\ y = a^{-1}b : p^3 = (w^2)^p (wu)^2 = u^p w^2 u (wu)^2 (tu)^2 = ((tu)^2)^p uw^2 = ((wu)^2)^p (wu)^2 w^2 = w^p v^{-1} w^{-1} (tu)^2 (wu)^2 = (ptyu(tu)^2)^4 = v^p w^{-1} (wu)^2 w^2 u = tpt^{-1} pv^{-1} (wu)^2 w^2 u = (py)^5 = 1.$
$G_3 = \langle v, w, u, p, z \rangle,\ z = ab : (ptzv)^8 = (pz)^5 = 1.$

As automorphisms of this system we get m and r with the following relations:
$m^2 = r^2 v^2 = [r, w] = rr^u = [r, v] = r^m r = v^m v = [m, w] = [m, uv] = [r, a] = [r, b] = [m, a] = [m, b] = m^p m v^{-1} (wu)^2 u = r^p v^{-1} r^{-1} (tu)^2 = 1.$

A search for Weyl groups using CAYLEY now implies that we have to add the automorphisms m and r. The we get basically two possible Weyl groups

$N_1 : x_1 = muz^{-1}v,\ x_2 = w^{-1}turvp,\ x_3 = u(tu)^2 r^{-1} t^{-1} w^{-1} y$ and
$N_2 : x_1 = muz^{-1}v,\ x_2 = w^{-1}turvp,\ x_3 = tuwmry.$

Case 1: Set $t_1 = x_3 x_1 (x_2 x_1)^2$ and $s_1 = x_2 t_1 x_1$ and add the following relations:
$x_1^2 = x_2^2 = x_3^2 = (x_1 x_2)^4 = (x_2 x_3)^4 = (x_1 x_3)^2 = t_1^4 = [s_1, t_1] = (x_1 s_1)^2 = 1.$

Case 2: Set $t_1 = (x_1 x_2)^2 x_1 x_3$ and $s_1 = t_1^3$ and add the following relations:
$x_1^2 = x_2^2 = x_3^2 = (x_1 x_2)^4 = (x_2 x_3)^4 = (x_1 x_3)^2 = t_1^6 = [x_1, s_1] = [x_2, s_1] = [x_3, s_1] = [(t_1^2)^{x_2}, t_1^2] = (t_1^2)^{x_2 x_1} (t_1^2)^{x_2} = 1.$

Then we get that N_1 is of order 128 while N_2 is of order 144. Adding this relations to the relations of the parabolic system shows that we define $U_4(3) \cdot 2 \cdot 2$ in both cases.

Remark. N does not satisfy (BNB).

(2.6) Proposition. *Let $G = 3^4 \cdot S_6$. Then G possesses a classical system Γ with diagram $\bigcirc\!\!=\!\!\bigcirc\!\!-\!\!\bigcirc$. This system possesses a Weyl group N isomorphic to $3^3 \cdot (\Sigma_4 \times \mathbb{Z}_2)$ such that $G = BNB$ holds. This N dos not come from an apartment of the corresponding geometry.*

Proof. First we give the relations for the parabolic system.

$B : a^2 = b^2 = c^2 = u^2 = [u, a] = [u, c] = [a, b] = 1,\ b^c = ba.$

$P_1 = \langle B, r \rangle : r^3 = 1 = [u, r], \ a^r = b, \ b^r = ab, \ r^c = r^{-1}.$
$P_2 = \langle B, t \rangle : t^3 = 1 = [au, t], \ a^t = c, \ c^t = ac, \ t^b = t^{-1}.$
$P_3 = \langle B, s \rangle : s^3 = 1 = [u, s], \ a^s = b, \ b^s = ab, \ s^c = s^{-1}.$

Furthermore we have: $(rctb)^5 = (tas^{-1}c)^5 = [r, s] = 1 = (ucruabt)^4 = (ucsuabt)^4$. Set $H = \langle rc, tb, abs^{-1}c \rangle$ and $K = \mathrm{core}_G(H)$. Then we get with CAYLEY that $L = G/K \simeq 3^4 \cdot S_6$.

Set $m = rcu$, $n = tbau$ and $e = abs^{-1}cu$. Then we get $N = \langle m, n, e \rangle \simeq 3^3 \cdot (\mathbb{Z}_2 \times \Sigma_4)$ as a subgroup of L. Furthermore one can check that N satisfies the conditions of a Weyl group and using Cayley one can show that $L = BNB$.

Let Γ be the corresponding geometry. Then Γ possesses exactly 81 points. The point stabilizer in N is of order 16, so as a subgeometry an apartment belonging to N would contain all the points of Γ, which is impossible as an apartment has to be thin.

One can show that N is the only Weylgroup in G satisfying BNB. So there are no apartments with (BNB) at all.

(2.7) Proposition. *Let G be semiclassical with diagram* $\bigcirc\!\!=\!\!\bigcirc\!\!\overset{\sim}{=}\!\!\bigcirc$. *Then* $G \simeq M_{24}$, *He or $3^7 \cdot Sp_6(2)$. None of these possess apartments with (BNB).*

Proof. [IvSh], [Hei].

Remark. Let G be a classical Tits system. Assume that any rank three residue is a building. Then S. Heiss is going to show that G has to be an automorphism group of a Liegroup if there are apartments with (BNB). It is very likely that the same conclusion holds without any restriction on the rank three residues. Here the hardest case is the coverings of the geometry belonging to LyS.

References

[DeSte]   A. Delgado, B. Stellmacher, Weak (B, N) - pairs of rank 2, in Groups and Graphs: New Results and Methods, Birkhäuser, Basel, Boston 1985.

[Glau]   private communication.

[Hei]   S. Heiss, On a parabolic system of type M_{24}, J. Algebra, to appear.

[IvSh]   A.A. Ivanov, S.V. Shpectorov, The P-geometry for M_{23} has no nontrivial 2-covering, Europ. J. Combinatorics 11 (1990), 373 - 381.

[Mei]   Th. Meixner, Parabolic systems: The $GF(3)$ - case, preprint.

[RoStr]   M. Ronan, G. Stroth, Minimal parabolic geometries for the sporadic groups, Europ. J. Comb. 5 (1984), 59 - 82.

[Str]   G. Stroth, A local classification of some finite classical Tits geometries and chamber systems of characteristic $\neq 3$, Geom. Dedicata 28 (1988), 93 - 106.

[StrWe]   G. Stroth, R. Weiss, Groups with the BNB - property, Geom. Dedicata 35 (1990), 251 - 282.

[Ti]   F. Timmesfeld, Classical locally finite Tits chamber systems of rank 3, J. Algebra 124, (1989), 9 - 59.

[Tits]   J. Tits, Buildings of spherical type and finite BN - pairs, Lecture notes in Math. 386 (1974), Springer Verlag.

On the 2-local structure of finite groups

Bernd Stellmacher

1. Let H be a finite group of even order, $S \in \mathrm{Syl}_2(H)$ and $\mathrm{Loc}(S) = \{N_H(D) \mid 1 \neq D \leq S\}$. H is said to be of **characteristic 2 type** if

$$C_P(O_2(P)) \leq O_2(P) \text{ for every } P \in \mathrm{Loc}(S).$$

Being of characteristic 2 type is a local property of H; i.e. a property which only concerns the elements of $\mathrm{Loc}(S)$. Yet this local property also restricts the global structure of H. It is easy to see that one of the following holds:

(1) $H \in \mathrm{Loc}(S)$; i.e. $O_2(H) \neq 1$.

(2) S is cyclic or a (generalized) quaternion group, and $O^{2'}(H) = SA$ where A is an abelian normal subgroup of H.

(3) $F^*(H)$, the generalized Fitting subgroup of H, is a simple group of characteristic 2 type.

Case 3 is the most interesting one. In this case, one may ask whether this global property of H in turn yields restrictions on the structure of the elements in $\mathrm{Loc}(S)$. Indeed, using the classification of the finite simple groups, one gets one of the following cases for $P \in \mathrm{Loc}(S)$:

(i) P is solvable.

(ii) $F^*(P/O_2(P))$ is a central product of the Fitting subgroup of $P/O_2(P)$ and quasisimple groups E such that $E/Z(E)$ is a group of characteristic 2 type.

(iii) $H \cong M_{23}$ and $P/O_2(P) \cong A_7$, $H \cong F_3$ and $P/O_2(P) \cong A_9$, $H \cong \mathrm{Co}_2$ and $P/O_2(P) \cong \mathrm{Aut}(M_{22})$, or $H \cong J_4$ and $P/O_2(P) \cong 3\mathrm{Aut}(M_{22})$.

In this paper, we want to study the structure of the elements of $\mathrm{Loc}(S)$ for a group of characteristic 2 type without using this classification. This will be done from a special point of view which was inspired by the so-called amalgam method. More precisely, we are interested in the structure of certain subgroups P_1 and P_2 of H both containing S such that

$$(*) \quad O_2(P_1) \neq 1 \neq O_2(P_2) \text{ and } O_2(\langle P_1, P_2 \rangle) = 1.$$

Note that P_1 and P_2 are contained in elements of $\mathrm{Loc}(S)$, and so P_1 and P_2 are also of characteristic 2 type. The amalgam method is a tool to investigate the chief factors of P_i in $O_2(P_i)$, $i = 1, 2$, provided one has some information about $P_i/O_2(P_i)$. The purpose of this paper is to give this information under certain additional assumptions.

In many of the arguments the prime 2 can be replaced by any prime. Moreover, some of the arguments do not require any global hypothesis on H, in particular neither finiteness nor simplicity. We will therefore start with a very general hypothesis and later successively carry on to more specialized hypotheses.

The most general set-up we are investigating is

Hypothesis I *Let H be a group, p a prime and S a non-trivial finite p-subgroup of H.*

Assume Hypothesis I. We first have to define the "local" subgroups of H we are interested in. Let $S \leq U \leq H$. We define :

$O_S(U)$ is the largest subgroup of S which is normal in U,

$O^S(U)$ is the intersection of all normal subgroups Y of U with $YS = U$,

U is S-**constrained**, if $C_S(O_S(U)) \leq O_S(U)$,

$\mathcal{L}(S) = \{P \leq H \mid S \leq P \text{ and } S \neq O_S(P) \neq 1\}$,

$\mathcal{FL}(S) = \{P \in \mathcal{L}(S) \mid P \text{ is finite}\}$,

$\mathcal{ML}(S) = \{P \in \mathcal{FL}(S) \mid P = \langle S^P \rangle$ and S is contained in a unique maximal subgroup of $P\}$,

$\mathcal{CL}(S) = \{P \in \mathcal{L}(S) \mid P = \langle S^P \rangle$ and $O^S(P) \leq \langle T^P \rangle$ for every $T \leq S$ with $T \not\leq O_S(P)\}$.

If U is finite and $S \in \mathrm{Syl}_p(U)$, then $O_S(U) = O_p(U)$ and $O^S(U) = O^p(U)$, and U S-constrained implies U p-constrained.

The elements of $\mathcal{ML}(S)$ and $\mathcal{CL}(S)$ very vaguely resemble "minimal parabolic subgroups" resp. "parabolic subgroups with a connected Dynkin diagram" in the context of finite groups of Lie type. At least the analogy carries far enough to introduce a graph for the elements in $\mathcal{CL}(S)$ resp. $\mathcal{ML}(S)$ which generalizes the Dynkin diagram for groups of Lie type.

Let $\mathcal{P}$ be a non-empty subset of $\mathcal{CL}(S)$. The **Timmesfeld graph** $\Lambda(\mathcal{P})$ is a graph whose vertices are the elements of $\mathcal{P}$ and whose edges are the sets $\{P_1, P_2\}$, $P_1, P_2 \in \mathcal{P}$ satisfying

$$(*) \quad O_S(P_1) \cap O_S(P_2) \text{ is neither normal in } P_1 \text{ nor in } P_2.$$

Hypothesis II *Assume Hypothesis I. In addition, let H be finite and $S \in Syl_p(H)$, and suppose the following:*

(i) *Every element in $\mathcal{L}(S)$ is of characteristic p type.*

(ii) *S is contained in at least two maximal p-local subgroups of H.*

The following result, the Thompson Replacement Theorem, will be used in the next sections. The proof is elementary and can be found in [1].

(1.1) *Let X be a finite group, p a prime and V a faithful finite $GF(p)X$-module. Suppose that there exists an elementary abelian p-subgroup A of X such that $|V/C_V(A)| \leq |A| \neq 1$. Then there exists a subgroup $1 \neq A^*$ in A such that $[V, A^*, A^*] = 1$ and $|A^*||C_V(A^*)| \geq |V|$.*

2. In this section we assume Hypothesis I.

(2.1) *Let $P \in \mathcal{L}(S)$ and $P = \langle S^P \rangle$. Then the following hold:*

(a) $P = O^S(P)S$.

(b) *If N is normal in $O^S(P)$ and $O^S(P)/N$ is a finite p-group, then $O^S(P) = N$.*

(c) $[O_S(P), O^S(P), O^S(P)] = 1$ *implies* $[O_S(P), O^S(P)] = 1$.

Proof. Let Y_1 and Y_2 be normal subgroups of P such that $Y_i S = P$ for $i = 1, 2$. Then $P/Y_1 \cap Y_2$ is a finite p-group, and $P = \langle S^P \rangle$ implies $P = (Y_1 \cap Y_2)S$. Since S is finite we get that $O^S(P)$ is of finite index in P and $P = O^S(P)S$.

Let N be a normal subgroup of $O^S(P)$ such that $O^S(P)/N$ is a finite p-group. Then P/N is a finite p-subgroup , and $P = \langle S^P \rangle$ implies that $P = SN$ and $N = O^S(P)$.

Assume that $[O_S(P), O^S(P), O^S(P)] = 1$. Then $O^S(P)/C_{O^S(P)}(O_S(P))$ is a finite p-group, and (b) implies (c).

(2.2) *Let $\emptyset \neq \mathcal{P} \subseteq C\mathcal{L}(S)$ and $L = \langle P \mid P \in \mathcal{P} \rangle$. Suppose that $L \in \mathcal{L}(S)$ and $\Lambda(\mathcal{P})$ is connected. Then $L \in C\mathcal{L}(S)$.*

Proof. Let $T \leq S$ and $L_0 = \langle T^L \rangle$, and let $\mathcal{P}_0 = \{P \in \mathcal{P} \mid O^S(P) \leq L_0\}$ and $\mathcal{P}_1 = \mathcal{P} \backslash \mathcal{P}_0$.

Suppose first that $\mathcal{P}_0 = \emptyset$. Then $L_0 \cap S = O_S(P) \cap L_0$ for every $P \in \mathcal{P}$ since $\mathcal{P} \subseteq C\mathcal{L}(S)$, and $L_0 \cap S$ is normal in L. Hence $T \leq L_0 \cap S \leq O_S(L)$.

Suppose now that $\mathcal{P}_0 \neq \emptyset$. Then clearly $T \not\leq O_S(L)$. Assume that $\mathcal{P}_1 \neq \emptyset$, and let $P_1 \in \mathcal{P}_1$ and $P_2 \in \mathcal{P}_0$. Then $S \cap L_0 \leq O_S(P_1)$, and so $O_S(P_1) \cap O_S(P_2)$

is normal in P_2. Thus, P_1 and P_2 are not adjacent in $\Lambda(\mathcal{P})$. But then $\mathcal{P}_1$ and $\mathcal{P}_0$ are in different connected components of $\Lambda(\mathcal{P})$, a contradiction.

We have shown that $\mathcal{P}_0 = \mathcal{P}$. This implies by (2.1) that $L_0 S = L$ and $O^S(L) \leq L_0$; i.e. $L \in \mathcal{CL}(S)$.

(2.3) *Let H be a finite group and $S \in Syl_p(H)$. Then $\mathcal{L}(S) \neq \emptyset$ and $\mathcal{ML}(S) \neq \emptyset$, or $N_H(S)$ is the unique maximal p-local subgroup of H containing S.*

Proof. Let L be a maximal p-local subgroup of H containing S. Suppose that $L \neq N_H(S)$. Then $L \in \mathcal{L}(S)$ and $\mathcal{L}(S) \neq \emptyset$.

Assume now that $\mathcal{L}(S) \neq \emptyset$. Choose $M \in \mathcal{L}(S)$ such that $|M|$ is minimal. Then $M \in \mathcal{ML}(S)$.

(2.4) *Let $P \in \mathcal{ML}(S)$ and $\overline{P} = P/O_p(P)$. Then the following hold:*

(a) $O^p(\overline{P}/\Phi(\overline{P}))$ *is a minimal normal subgroup of* $\overline{P}/\Phi(\overline{P})$.

(b) $\Phi(\overline{P})$ *is a p'-group.*

(c) *If P is solvable, then $O^p(\overline{P})$ is a q-group, q a prime.*

Proof. Let M be the unique maximal subgroup of P containing S, and let N be the largest normal subgroup of P which is in M. Note that $O_p(P) \leq N$ since $P = \langle S^P \rangle$. By a Frattini argument, $\overline{N}$ is a nilpotent p'-group.

Assume that there is a maximal subgroup $\overline{L}$ of $\overline{P}$ such that $\overline{N} \not\leq \overline{L}$. Then $\overline{P} = \overline{L}\,\overline{N}$ and $\overline{S}^{\overline{x}} \leq \overline{L}$ for some $\overline{x} \in \overline{P}$. It follows that $\overline{L} = \overline{M}^{\overline{x}}$ and $\overline{N} \leq \overline{L}$, a contradiction. Hence $\overline{N} = \Phi(\overline{P})$, and (a) and (b) are easy to verify.

Suppose that P is solvable. Then the existence of Hall subgroups in P yields that P is a $\{p, q\}$-group.

Definition.   Let $P \in \mathcal{FL}(S)$. A subnormal subgroup K of P is a **p-component** of P, if either

(i) K is perfect and $K/O_p(K)$ is quasisimple, or

(ii) $K/O_p(K)$ is a q-group, q a prime, $K = O^p(K)$ and $KS \in \mathcal{ML}(S)$.

$\mathcal{P}(S)$ is the set of all $P \in \mathcal{FL}(S)$ which contain a p-component K such that $P = \langle K, S \rangle = \langle S^P \rangle$.

(2.5) *The elements of $\mathcal{P}(S)$ satisfy the conclusion of (2.4).*

Proof. Let $P \in \mathcal{P}(S)$ and $P = \langle K, S \rangle$, K a p-component of P. We may assume that K is not solvable. Let $P_0 = \langle K^P \rangle$ and $\overline{P} = P/O_p(P)$. Then $\overline{P}_0$ is a central product of the $\overline{P}$-conjugates of $\overline{K}$, and $\overline{P} = \overline{P}_0 \overline{S}$. Hence, it suffices to show that $Z(\overline{P}_0) \leq \Phi(\overline{P})$. Assume that $\overline{L}$ is a maximal subgroup of $\overline{P}$ such that $Z(\overline{P}_0) \not\leq \overline{L}$. Then $\overline{P} = \overline{L}Z(\overline{P}_0)$ and $O^p(\overline{P}) = \overline{P}_0 = O^p(\overline{L})Z(\overline{P}_0)$. Since $Z(\overline{P}_0) \leq \Phi(\overline{P}_0)$ we conclude that $\overline{P}_0 = O^p(\overline{L})$ and $Z(\overline{P}_0) \leq \overline{L}$, a contradiction.

One of the reasons to focus on the elements P of $\mathcal{P}(S)$ (and $\mathcal{ML}(S)$) rather than those of $\mathcal{L}(S)$ is their "easy" normal subgroup structure. As (2.5) indicates there exists only one "interesting" chief factor of P outside $O_p(P)$. The next three lemmata rephrase this fact in other terms.

(2.6) $\mathcal{P}(S) \cup \mathcal{ML}(S) \subseteq \mathcal{CL}(S)$.

(2.7) *Let $P \in \mathcal{CL}(S)$. Suppose that V is a normal subgroup of P in $O_S(P)$. Then either $O^S(P) \leq C_P(V)$ or $C_S(V) \leq O_S(P)$.*

(2.8) *Let $P \in \mathcal{CL}(S)$. Suppose that Z is a normal subgroup of S and $V = \langle Z^P \rangle$. Then either $V = Z$ or $C_S(V) \leq O_S(P)$.*

The next lemma is based on an idea of Gomi in [5] which was brought to my attention by A. Delgado.

(2.9) *Assume Hypothesis II. Then there exist elements $P_1 \in \mathcal{ML}(S)$ and $P_2 \in \mathcal{P}(S)$ such that $O_p(\langle P_1, P_2 \rangle) = 1$.*

Proof. Set $D = \bigcap_{X \in \mathcal{P}(S)} O_p(X)$. Then $D \neq 1$ and $N_H(D) \in \mathcal{L}(S)$ since $Z(S) \leq D$, and $N_H(S) \leq N_H(D)$ since $N_H(S)$ operates on $\mathcal{P}(S)$ by conjugation.

By our hypothesis there exists $P_1 \in \mathcal{L}(S)$ such that $P_1 \not\leq N_H(D)$. Thus, choosing $|P_1|$ minimal with that property we get $P_1 \in \mathcal{ML}(S)$.

Let $\mathcal{P}(S) = \{L_1, ..., L_n\}$, $Y_i = \langle P_1, L_i \rangle$ and $\overline{Y}_i = Y_i/O_p(Y_i)$, $i = 1, ..., n$. We may assume that $O_p(Y_i) \neq 1$. Let $\Lambda_i = \{F \in \mathcal{P}(S) \mid O^p(F)$ is subnormal in $Y_i\}$. Then $[F^*(\overline{Y}_i), S] = \langle O^p(\overline{F}) \mid F \in \Lambda_i \rangle$ and

$$O_p(Y_i) = \bigcap_{F \in \Lambda_i} O_p(F).$$

We conclude that $D \leq O_p(Y_i)$ and

$$D \leq \bigcap_{i=1}^n O_p(Y_i) \leq \bigcap_{i=1}^n O_p(L_i) = D,$$

and so $D = \bigcap_{i=1}^n O_p(Y_i)$ and $P_1 \leq N_H(D)$, a contradiction.

3.   Assume that H satisfies Hypothesis II with $p = 2$. Then (2.9) provides us with a pair of subgroups $P_1 \in \mathcal{ML}(S)$ and $P_2 \in \mathcal{P}(S)$ such that $O_p(\langle P_1, P_2 \rangle) = 1$. The subgroup $P_1/O_2(P_1)$ has the more restrictive structure (see (4.2)). We are therefore mainly interested in the structure of $P_2/O_2(P_2)$. The general idea is to investigate this structure using the action of P_2 on certain chieffactors of P_2 in $O_2(P_2)$. Our aim will be to show, under additional hypotheses and allowing certain exceptions,

      (FF1) P_2 possesses a failure-of-factorization module V such that $C_S(V) = O_2(P_2)$.

Here V is a failure-of-factorization module for P_2 provided V is a $GF(2)P_2$-module and there exists an elementary abelian subgroup $A \neq 1$ in $P_2/C_{P_2}(V)$ such that

      $(*)$ $|A| \geq |V/C_V(A)|$.

Cooperstein [3] and Aschbacher [1] have determined all finite groups G having a failure-of-factorization module in characteristic 2 provided $F^*(G)$ is a known quasisimple group. It turns out that only certain quasisimple groups of Lie type in characteristic 2 and the alternating groups (allowing $3A_6$) occur as $F^*(G)$.

Set $Z_i = \langle \Omega_1(Z(S))^{P_i} \rangle$, $i = 1, 2$. Since $O_2(\langle P_1, P_2 \rangle) = 1$ we have $Z_k \not\leq Z(P_k)$ for some $k \in \{1, 2\}$. We can distinguish the following two cases:

(I) $Z_1 \not\leq Z(P_1)$.

(II) $Z_1 \leq Z(P_1)$.

In case (I) the subdivision can be continued by

(Ia) $Z_2 \not\leq Z(P_2)$, $J(S) \not\leq O_2(P_2)$.

(Ib) $Z_2 \not\leq Z(P_2)$, $J(S) \leq O_2(P_2)$.

(Ic) $Z_2 \leq Z(P_2)$.

Similarly, case (II) can be subdivided into:

(IIa) $Z_2 \not\leq Z(P_2)$, $J(S) \not\leq O_2(P_2)$.

(IIb) $Z_2 \not\leq Z(P_2)$, $J(S) \leq O_2(P_2)$.

In the cases (Ia) and (IIa) Z_2 is a failure-of-factorization module for P_2. Moreover, (2.6) and (2.8) yield $C_S(Z_2) = O_2(P_2)$. Hence, we are done in these cases.

In this section, we will deal with case (IIb) (see (3.5)).

Notation. Let P be a group and V a finite $GF(p)P$-module, and let $1 = V_0 \leq V_1 \leq ... \leq V_n = V$ be a chief series of $GF(p)P$-submodules of V (we write the modules multiplicatively). We define:

$\mathrm{cf}(P,V)$ is the number of non-central chief factors in the chief series $V_0 \leq \dots \leq V_n$;

$$\mathcal{Q}(S,V) = \{A \leq S \mid [V,A,A] = 1 \neq [V,A]\};$$

$q(S,V) = \min\{\log_{|A/C_A(V)|}(|V/C_V(A)|) \mid A \in \mathcal{Q}(S,V)\}$ if $\mathcal{Q}(S,V) \neq \emptyset$, and $q(S,V) = 0$ if $\mathcal{Q}(S,V) = \emptyset$.

$$\mathcal{Q}^*(S,V) = \{A \in \mathcal{Q}(S,V) \mid \log_{|A/C_A(V)|}(|V/C_V(A)|) = q(S,V)\}.$$

$r(S,V) = \min\{q(S,V_i/V_{i-1}) \mid 1 \leq i \leq n$ and $q(S,V_i/V_{i-1}) \neq 0\}$ if there exists $k \geq 1$ such that $q(S,V_k/V_{k-1}) \neq 0$, and $r(S,V) = 0$ if $q(S,V_k/V_{k-1}) = 0$ for every $k = 1,\dots,n$.

Hypothesis III *Assume Hypothesis I, and let $P \in \mathcal{CL}(S)$ and V be a finite $GF(p)P$-module . Suppose that there exists a $GF(p)O_S(P)$-submodule W in V such that*

(i) $V = \langle W^P \rangle$ *and* $V \neq W$,

(ii) $P = N_P(W)O^S(P)$.

Set $q = q(O_S(P),W)$, $c = \mathrm{cf}(P,V)$ and $r = r(S,V)$.

(3.1) *Assume Hypothesis III. Suppose that there exists $A \in \mathcal{Q}(S,V)$ such that $A \leq C_P(W)$, $A \not\leq O_S(P)$ and*

$$|V/C_V(A)| \leq |A/C_A(V)|.$$

Then the following hold:

(a) $|A/A \cap O_S(P)|^{rc-1} \leq |A \cap O_S(P)/C_A(V)|.$

(b) $|A \cap O_S(P)/C_A(V)|^{q-1} \leq |A/A \cap O_S(P)|.$

(c) $(q-1)(rc-1) \leq 1$, *if* $q-1 \geq 0$ *or* $rc-1 \geq 0$.

Proof. Let $A_0 = A \cap O_S(P)$, $A^* = C_A(V)$ and $\overline{A} = A/A_0$. Since $V \neq W$ there exists a non-central $GF(p)P$ chief factor Y in V; i.e. $\mathrm{cf}(P,V) \geq 1$. Moreover, $C_S(Y) = O_S(P)$ since $P \in \mathcal{CL}(S)$; in particular $C_A(Y) \leq O_S(P)$. Since $A \not\leq O_S(P)$ this implies that $q(S,Y) \neq 0 \neq r$. We conclude that

$$|\overline{A}|^{cr} \leq |V/C_V(A)| \leq |\overline{A}||A_0/A^*|.$$

This shows (a).

Define subgroups $A_0 \geq A_1 \geq ... \geq A_s = A^*$ and conjugates $W_i = W^{x_i}$, $x_i \in P$, such that $A_i = C_{A_{i-1}}(W_i)$, $i = 1, ..., s$. Then

$$|A_{i-1}/A_i|^q \leq |W_i/C_{W_i}(A_{i-1})|$$

and

$$|A_0/A^*|^q \leq \prod_{i=1}^{s} |W_i/C_{W_i}(A_{i-1})|.$$

Moreover,

$$|W_i/C_{W_i}(A_{i-1})| = |W_i C_V(A_{i-1})/C_V(A_{i-1})| \leq |C_V(A_i)/C_V(A_{i-1})|$$

and

$$|C_V(A_s)/C_V(A_{s-1})|...|C_V(A_1)/C_V(A_0)| = |V/C_V(A_0)|$$
$$\geq \prod_{i=1}^{s} |W_i/C_{W_i}(A_{i-1})| \geq |A_0/A^*|^q.$$

Now (b) follows from $|V/C_V(A)| \geq |V/C_V(A_0)|$, and (a) and (b) imply (c).

(3.2) *Assume the hypothesis of (3.1), and let*

$$h = \log_{|A/A \cap O_s(P)|}(A \cap O_S(P)/C_A(V)).$$

Then $h(q-1) \leq 1$ and $rc \leq h+1$.

Proof. Note that $|A/A \cap O_S(P)| \neq 1$. Then (3.1)(b) implies $h(q-1) \leq 1$, and (3.1)(a) gives $rc - 1 \leq h$.

(3.3) *Assume Hypothesis III, and let $Q = [O_S(P), O^S(P)]$. Suppose that $c = 1$, $C_S(W) \not\leq O_S(P)$ and $[V, Q] \neq 1$. Then $Q \in \mathcal{Q}(O_S(P), W)$ and $|[W, Q]| = |Q/C_Q(W)|$.*

Proof. Let $U = O^S(P)$ and $V_0 = C_V(U)$. Note that $V = [V, U]W$ and $[V, U, Q] \leq V_0$. Note further that $U \leq \langle C_S(W)^P \rangle$ since $P \in \mathcal{CL}(S)$ and so $Q = \langle C_Q(W)^P \rangle$ by (2.1). It follows that $[V, Q] = [V, U, Q] \leq V_0$ and $[V, Q, Q] = 1$; in particular $Q \in \mathcal{Q}(O_S(P), W)$.

Pick $x \in W \backslash V_0$. Then $[V, U] \leq \langle x^U \rangle V_0$ and $V = \langle x^U \rangle V_0 W$; i.e.

$$[V, Q] = [V, U, Q] = [x, Q]$$

and $|[W, Q]| \not\leq |Q/C_Q(x)|$. Now the assertion follows.

We now return to the discussion at the beginning of this section. Let P_i and Z_i, $i = 1, 2$, be defined as there, and let $V_1 = \langle Z_2^{P_1} \rangle$.

(3.4) *Assume Hypothesis II with $p = 2$. Suppose that $Z_1 \leq Z(P_1)$ and $J(S) \leq O_2(P_2)$. Then one of the following holds:*

(a) $Z_2 \not\leq O_2(P_1)$.

(b) $0 \neq q(S, Z_2) \leq 2$.

(c) $0 \neq q(S, Z_2^*) \leq 1$, *where Z_2^* is the $GF(2)P_2$-module dual to Z_2.*

(d) V *is abelian and* $0 \neq r(S, V) < 1$.

(e) $[V_1 \cap Z(O_2(P_1)), O^2(P_1)] \neq 1$ *and* $0 \neq q(S, V_1 \cap Z(O_2(P_1))) \leq 1$.

Proof. Note that $J(S) \not\leq O_2(P_1)$ since $O_2(\langle P_1, P_2 \rangle) = 1$. Suppose that (P_1, P_2) is a counterexample. Then $V_1 \leq O_2(P_1)$.

Assume first that V_1 is non-abelian. Then there exists $x \in P_1$ such that $[Z_2, Z_2^x] \neq 1$. Now either $q(Z_2^x, Z_2) \leq 1$ or $q(Z_2, Z_2^x) \leq 1$ and (b) holds, a contradiction.

Hence, V_1 is abelian. Assume next that $\mathrm{cf}(S, V_1) \leq 1$. Since (c) does not hold, we get from (3.3) applied to $W = Z_2$, $V = V_1$ and $P = P_1$ that either $C_S(Z_2) \leq O_2(P_1)$ or $[V_1, O_2(P_1) \cap O^2(P_1)] = 1$. In the first case $J(S) \leq O_2(P_1)$, a contradiction. In the second case let $Q = O_2(P_1) \cap O^2(P_1)$, and let V_0 in V_1 be a normal subgroup of P_1 which is minimal with $[V_0, O^2(P_1)] \neq 1$. Then $[V_0, O_2(P_1), O^2(P_1)] = 1$ and $[O_2(P_1), O^2(P_1), V_0] = 1$. Hence, the 3-subgroup lemma implies that $[V_0, O^2(P_1), O_2(P_1)] = 1$; i.e.

$$[V_1 \cap Z(O_2(P_1)), O^2(P_1)] \neq 1.$$

Now (e) follows from (1.1), a contradiction.

Assume finally that $\mathrm{cf}(P_1, V_1) \geq 2$. We apply (3.1) to $W = Z_2, V = V_1$ and $P = P_1$. Since (b) does not hold there exists by (1.1) $A \leq O_2(P_2)$ such that $0 \neq q(A, V_1) \leq 1$ and $A \not\leq O_2(P_1)$. For the same reason $q - 1 > 2$, and $rc - 1 \geq 1$ since (d) does not hold. But now (3.1)(c) gives

$$q - 1 \leq (q - 1)(rc - 1) \leq 1$$

and $q \leq 2$, a contradiction.

Remark. Under the additional hypothesis that the components of $P_1/C_{P_1}(Z_1)$ are "known" cases (d) and (e) of (3.4) do not occur; see (4.7) and use that $Z_1 \leq Z(P_1)$.

4. In this section H is a finite group, V is a finite faithful $GF(2)H$-module, $O_2(H) = 1$ and S is a non-trivial 2-subgroup of H.

The purpose of this chapter is to describe certain failure-of-factorization modules of H. We will start with the solvable case. The result should be well known.

(4.1) *Let* $J = \langle A^h | A \in Q^*(S,V), h \in H \rangle$ *and* $N = J \cap O_{2'}(H)$. *Suppose that* $0 \neq q(S,V) < 2$ *and* $C_H(O_{2'}(H)) \leq O_{2'}(H)$. *Then the following hold:*

(a) *J/N is an elementary abelian 2-group.*

(b) *$N = E_1 \times ... \times E_n$ and $V = V_0 \times V_1 \times ... \times V_n$ such that $V_0 = C_V(N)$, $E_i \cong C_3$, $V_i = [V, E_i]$ and $|V_i| = 4$ for $i = 1, ..., n$.*

(c) *If $q(S,V) \leq 1$, then $q(S,V) = 1$, $V_0 = C_V(J)$ and $J \cong L_2(2) \times ... \times L_2(2)$.*

We now continue with the component case. Some of the results can be found in [1], others can be calculated with the help of [7] and [8]. We will write $E \cong (S)L_n(q)$, $(S)U_n(q)$ etc., if E is a quasisimple factor-group of $SL_n(q)$, $SU_n(q)$ etc..

We now define the following classes of finite quasisimple groups E with $|Z(E)|$ odd :

$\mathcal{K}_{\textbf{even}}$: *$E/Z(E)$ is isomorphic to a group of Lie type in characteristic 2 (including $Sp_4(2)'$, $G_2(2)'$, ${}^2F_4(2)'$).*

$\mathcal{K}_{\textbf{odd}}$: *$E/Z(E)$ is isomorphic to a group of Lie type in characteristic $p \neq 2$.*

$\mathcal{K}_{\textbf{alt}}$: *$E/Z(E) \cong A_n$, $n \geq 5$.*

$\mathcal{K}_{\textbf{spor}}$: *$E/Z(E)$ is isomorphic to one of the 26 sporadic groups.*

$\mathcal{K} = \mathcal{K}_{\text{even}} \cup \mathcal{K}_{\text{odd}} \cup \mathcal{K}_{\text{alt}} \cup \mathcal{K}_{\text{spor}}$.

(4.2) *Suppose that S is contained in a unique maximal subgroup of H and $F^*(H) \in \mathcal{K}$. Then one of the following holds:*

(a) *$F^*(H) \in \mathcal{K}_{odd}$.*

(b) *$F^*(H) \cong A_{2^n+1}$.*

(c) *$F^*(H) \cong L_2(2^n)$, $Sz(2^n)$ or $(S)U_3(2^n)$.*

(d) *$F^*(H) \cong (S)L_3(2^n)$, $Sp_4(2^n)$, A_6 or $3A_6$, and there exists an element in S which induces a diagram automorphism in $F^*(H)$.*

(4.3) *Suppose that S is contained in a unique maximal subgroup of H and $0 \neq q(S,V) < 2$. Let K be a component of H and $K \in \mathcal{K}$. Then there exists $A \in Q^*(S,V)$ such that $[K,A] \neq 1$, and for $L = \langle K^A \rangle$ and $W = [V,L]$ the following hold:*

(a) *$1 \leq q(A,W) < 2$.*

(b) $L \cong L_2(2^n)$, $L_2(2^n) \times L_2(2^n)$, A_{2^n+1}, or $(S)L_3(2^n)$.

(4.4) *Let* $C_V(F^*(H)) = 1$. *Suppose that* $F^*(H) \cong L_2(2^n)$, $n \geq 2$, *and* $0 \neq q(S, V) \leq 1$. *Then either*

(a) $q(S \cap F^*(H), V) = 1$, *and* V *is a natural* $L_2(2^n)$-*module, or*

(b) $q(S \cap F^*(H), V) = 2$, $n = 2$, *and* V *is a natural* $\Omega_4^-(2)$-*module.*

(4.5) *Let* K *be a component of* H, $A \in \mathcal{Q}^*(S, V)$ *and* $L = \langle K^A \rangle$. *Suppose that* $L \neq K$ *and* $0 \neq q(S, V) < 2$. *Then* $L \cong L_2(2^n) \times L_2(2^n)$, *and* $V/C_V(L)$ *is the direct product of two natural* $L_2(2^n)$-*modules for* K. *In particular* $q(S, V) \geq \frac{3}{2}$.

(4.6) *Let* $F^*(H) \cong A_{2n+1}$, $n \geq 2$, *and* $V = [V, F^*(H)]$. *Suppose that* $0 \neq q(S, V) < 2$. *Then one of the following holds:*

(a) $V/C_V(F^*(H))$ *is a natural* A_{2n+1}-*module.*

(b) $F^*(H) \cong L_2(4)$ *and* $V/C_V(F^*(H))$ *is a natural* $L_2(4)$-*module.*

(c) $F^*(H) \cong A_7$ *and* $|V/C_V(F^*(H))| = 2^4$.

(4.7) *Suppose that the following hold:*

(i) S *is contained in a unique maximal subgroup of* H.

(ii) *The components of* H *are in* $\mathcal{K}$.

(iii) $0 \neq q(S, V) \leq 1$.

Then $q(S, V) = 1$ *and* $C_V(S) \not\leq C_V(O^2(H))$, *and there exists a subnormal subgroup* K^* *in* H *such that for* $A \in \mathcal{Q}^*(S, V)$:

(a) $\langle K^{*S} \rangle S = H$.

(b) $K^* \cong L_2(2^n)$ *or* Σ_{2^n+1}.

(c) $A = (A \cap K^*) \times C_A(K^*)$.

(d) $A \cap K^* \in Syl_2(K^*)$, *or* $K^* \cong \Sigma_{2^n+1}$ *and* $A \cap K^*$ *is generated by trans-positions.*

5.   The purpose of this section is to treat case (Ib) of section 3. We assume Hypothesis I with $p = 2$. Let $P \in \mathcal{FL}(S)$ such that

(i) $S \in Syl_2(P)$,

(ii) P is of characteristic 2 type.

Notation. Let T be a finite 2-group. Then $\mathfrak{U}(T)$ is the set of all elementary abelian subgroups of maximal order in T (i.e. $J(T) = \langle \mathfrak{U}(T) \rangle$),

$$B(T) = C_T(\Omega_1(Z(J(T)))),$$

$$V = \langle \Omega_1(Z(S))^P \rangle.$$

(5.1) *Let* $\overline{P} = P/O_2(P)$. *Suppose that* $\overline{P}/\Phi(\overline{P}) \cong L_2(2^n)$, $n \geq 1$, *and that neither* $J(S)$ *nor* $\Omega_1(Z(S))$ *is normal in* P. *Then* $B(S) \in Syl_2(O^2(P)B(S))$.

Proof. In [2] this was shown for $\overline{P} \cong L_2(2^n)$, and the proof is the same for $\overline{P}/\Phi(\overline{P}) \cong L_2(2^n)$.

(5.2) *Let* $\overline{P} = P/O_2(P)$ *and* $\overline{P}/\Phi(\overline{P}) \cong L_2(2^n)$, $n \geq 1$. *Suppose that no non-trivial characteristic subgroup of* S *is normal in* P. *Then* $[O_2(P), O^2(P)] = W \leq \Omega_1(Z(O_2(P)))$, $\Phi(\overline{P}) = 1$, *and* $W/C_W(P)$ *is a natural* $L_2(2^n)$-*module.*

Proof. As for (5.1) this is essentially [2] where the case $\overline{P} \cong L_2(2^n)$ was treated. An elementary and short proof of (5.2) can be found in [6].

(5.3) *Let* H *be finite and* $P \in \mathcal{ML}(S)$ *with* $[P, \Omega_1(Z(S))] \neq 1$, *and let* $1 \neq L$ *be a normal subgroup of* P. *Suppose that every component of* $P/C_P(V)$ *is in* $\mathcal{K}$ *and that for every 2-local subgroup* Y *of* H *containing* $B(S)$:

(i) Y *is of characteristic 2 type.*

(ii) $B(S) = B(S_0)$ *for* $B(S) \leq S_0 \in Syl_2(Y)$.

Then for $T = S \cap L$ *and* $L_0 = \langle B(T)^L \rangle$ *either* $O_2(\langle N_H(B(T)), P \rangle) \neq 1$ *or the following hold:*

(a) $P = L_0 S$ *and* $O_2(L_0) = V$.

(b) $L_0/O_2(L_0) \cong L_2(2^n)$, $n \geq 1$, *and* $O_2(L_0)/Z(L_0)$ *is a natural* $L_2(2^n)$-*module.*

(c) $|Z(L_0)| \leq 2^n$.

Proof. Let $\overline{P} = P/C_P(V)$, $B = B(T)$ and $N = \langle N_H(B), P \rangle$. If $B \leq O_2(L)$, then B is normal in P, and $O_2(N) \neq 1$. Assume that $B \nleq O_2(L)$. Then $B \nleq O_2(P)$ and $P = SL_0$ by (2.6).

Let C be a characteristic subgroup of B which is normal in L_0. Then $N \leq N_H(C)$ since $P = L_0 N_P(B)$. Hence, we may assume:

(1)   No non-trivial characteristic subgroup of B is normal in L_0.

In particular, $J(T)$ is not normal in L_0. In addition, $V = \langle \Omega_1(Z(S))^{L_0} \rangle$ since $P = SL_0$ and so by (2.8) $C_T(V) = O_2(L)$. We conclude that $F^*(\overline{L}_0) = F^*(\overline{P})$ and

(2) $[F^*(\overline{L}_0), \overline{B}] = F^*(\overline{L}_0)$.

Next we want to show:

(3)   There exists a subgroup $\overline{E}$ in $\overline{L}_0$ such that
  (i) $\overline{B} \in \text{Syl}_2(\overline{E})$,
  (ii) $\overline{E}/O_2(\overline{E}) \cong L_2(2^n)$, $n \geq 1$,
  (iii) $\langle \overline{E}, \overline{S} \rangle = \overline{P}$,
  (iv) $\overline{B} = \overline{B(S)}$.

Since $J(T)$ is not in $O_2(L_0)$ we get from (1.1) that $0 \neq q(B, V) \leq 1$. Suppose first that $F^*(\overline{L}_0)$ is solvable. Then by (4.1)

$$F^*(\overline{L}_0)\overline{B} = L_2(2) \times \ldots \times L_2(2),$$

and (3) follows.

Suppose now that $F^*(\overline{L}_0)$ is not solvable. Then there exists a component K of $\overline{L}_0$ such that $F^*(\overline{L}_0) = \langle K^{\overline{P}} \rangle$. We apply (4.7). We get $K \cong L_2(2^n)$ or A_{2^n+1}. Now we apply (4.6). Then either (3) holds for $K = \overline{E}$ or $K \leq E_1$ and $E_1 \cong \Sigma_{2^n+1}$. In the latter case $\overline{B} = (\overline{B} \cap E_1) \times C_{\overline{B}}(E_1)$, and $\overline{B} \cap E_1$ is generated by transpositions. Hence, there exists a subgroup $\overline{E} \leq E_1$ so that (3) holds with $\overline{E}/O_2(\overline{E}) \cong \Sigma_3 \cong L_2(2)$. We also get $q(B, V) = 1$ and that

(4) $V \leq B$.

Let $\overline{E}$ be as in (3), and let $F \leq L_0$ such that $\overline{E} = \overline{F}$ and $B \leq F$, and let $|F|$ be minimal with these properties. Then $F \cap T \in \text{Syl}_2(F)$ since $O_2(L_0) \in \text{Syl}_2(C_{L_0}(V))$, and $N_F(T \cap F)C_F(V)$ is the unique maximal subgroup of F containing $T \cap F$, i.e. $F \in \mathcal{ML}(T \cap F)$. It follows from (2.4) that

$$\hat{F}/\Phi(\hat{F}) \cong L_2(2^n) \text{ where } \hat{F} = F/O_2(F).$$

Hence (5.1) yields $B \in \text{Syl}_2(F)$. Moreover, from (3)(iii) and (1) we get that no non-trivial characteristic subgroup of of B is normal in F.

Let $X = [O_2(F), O^2(F)]$. Then (5.2) implies that $X \leq V$ and $X/C_X(F)$ is a natural $L_2(2^n)$-module. We conclude that $F/O_2(F) \cong L_2(2^n)$ and $[O_2(P), O^2(F)] = X$ since B is normal in $BO_2(P)$. By (3)(iv) $B(S) \leq BO_2(P)$ and thus

(5) $B(S) \leq N_P(F)$ and $O^2(F) = O^2(B(S)F)$.

Let $X_0 = \langle X^g | g \in N_H(B) \rangle$. If $X_0 \leq O_2(F)$, then X_0 is normal in N. Thus, we may assume that there exists $g \in N_H(B)$ such that $X^g \not\leq O_2(F)$; i.e. $B = O_2(F)X^g$. Note that $F = O_2(F)\langle X^g, X^{gf} \rangle$ for $f \in F\backslash N_F(B)$. Set $B_0 = C_B(\langle X^g, X^{gf} \rangle)$. Then $O_2(F) = B_0 X$ and $O_2(F^g) = B_0 X^g$. Hence

$$\Phi(O_2(F)) = \Phi(B_0) = \Phi(O_2(F^g)).$$

Let $M = N_H(\Phi(B_0))$ and assume that $\Phi(B_0) \neq 1$. Then $\langle F, F^g \rangle \leq M$. Note that by (5) $B(S) \leq M$, and so M is of characteristic 2 type. In addition, $[O_2(M), B(S)] \leq O_2(M) \cap B(S)$ since $B(S) = B(S_0)$ for $B(S) \leq S_0 \in \mathrm{Syl}_2(M)$.

Let $F_0 = B(S)F$. Then $[O_2(M), B(S)] \leq O_2(M) \cap O_2(F_0)$ and so by (5) $[O_2(M), O^2(F)] \leq O_2(M) \cap X$. Hence either $[O_2(M), O^2(F)] = 1$ or X. In the first case, M is not of characteristic 2 type, a contradiction. In the second case $X \leq O_2(M) \cap F^g \leq O_2(F^g)$ and $[X, X^g] = 1$, a contradiction since $X^g \not\leq O_2(F)$. We have shown:

(6) $\quad \Phi(O_2(F)) = 1$.

From (4) we get that $O_2(F) \leq C_B(V) \leq O_2(L_0)$. Now our choice of F shows that $\overline{L}_0 \cong L_2(2^n)$ and $L_0 = F$. In particular, (4.4) yields (a) and (b). Hence, it remains to prove (c).

According to (b) there are exactly two maximal elementary abelian subgroups in B one of which is $O_2(F)$. Hence we may assume that $|N_H(B)/N_H(B) \cap N_H(O_2(F))| = 2$. Since $B(S) \leq N_H(B) \cap N_H(O_2(F))$ there exists $g \in N_H(B) \cap N_H(B(S))$ such that $[X^g, X] \neq 1$. Hence $B(S)$ normalizes $Z(F) \cap Z(F)^g = D$. Since $|Z(B)/Z(F)| = 2^n$ we have $|Z(F)| \leq 2^n$ or $D \neq 1$. Assume that $D \neq 1$, and let $M = N_H(D)$. Then $\langle B(S), F, F^g \rangle \leq M$, and a contradiction follows as above.

(5.4) *Assume Hypothesis II with $p = 2$, and let $P_1 \in \mathcal{ML}(S)$ and $P_2 \in \mathcal{P}(S)$ be as in (2.9). Suppose that the following hold:*

(i) $[\Omega_1(Z(S)), P_1] \neq 1$.

(ii) *The non-solvable composition factors of P_1 are in $\mathcal{K}$.*

(iii) *All 2-local subgroups of H containing $B(S)$ are of characteristic 2 type.*

Then $B(S)$ is not normal in P_2.

Proof. Assume that $B(S)$ is normal in P_2. We apply (5.3) with $P_1 = P = L$. Then either $O_2(\langle N_H(B(S)), P_1 \rangle) \neq 1$ or $B(S)$ contains exactly two maximal elementary abelian subgroups, one of which is normal in P_1.

The first case gives $O_2(\langle P_1, P_2 \rangle) \neq 1$ since $P_2 \leq N_H(B(S))$, a contradiction. In the second case $O^2(P_2)$ normalizes these two abelian subgroups and so again $O_2(\langle P_1, P_2 \rangle) \neq 1$, a contradiction.

We now return to the dicussion started in section 3. Suppose that we are in case (Ib). Then by (2.6) and (2.8) $C_S(Z_2) = O_2(P_2)$ and thus $B(S) \leq O_2(P_2)$. Under the hypotheses of (5.4) this rules out case (Ib).

6.   In this section we assume Hypothesis II with $p = 2$.

Definition. Let $\mathcal{T}$ be the set of all 2-subgroups of H which contain $B(S)$.

Let U be a subgroup of H with $O_2(U) \neq 1$. Then U is a **uniqueness subgroup** of H, if for every $T \in \mathrm{Syl}_2(H)$ and $P = \langle B(T), U \rangle$ one of the following holds:

(i) U is not subnormal in P.

(ii) U is subnormal in every $L \in \mathcal{L}(T)$ with $P \leq L$.

Let U be a uniqueness subgroup of H and $T \in \mathrm{Syl}_2(H)$. Then (U, T) is an **exceptional pair**, if the following hold:

(1) U is a 2-component of $\langle U, T \rangle$.

(2) $U \not\leq N_H(B(T))$.

(3) There exists a 2-group T_0 and $L \in \mathcal{L}(T_0)$ such that $\langle B(T), U \rangle \leq L$, and U is not subnormal in L.

Remark. Let (U, T) be an exceptional pair and $P = \langle U, B(T) \rangle$. By (1) $\langle U, T \rangle \in \mathcal{P}(T)$, and U is subnormal in $\langle U, T \rangle$. Hence, U is also subnormal in P, and $U \leq [U, B(T)]$.

Since U is a uniqueness subgroup of H it follows that U is subnormal in every $M \in \mathcal{L}(\tilde{T})$ with $\tilde{T} \in \mathrm{Syl}_2(H)$ and $P \leq M$. In particular, the subgroup T_0 in (3) is not a Sylow 2-subgroup of H.

In some of the proofs in sections 6 and 7 we will have to treat uniqueness subgroups U satisfying (1) and (2). It is then convenient to distinguish whether (U, T) is an exceptional pair or not.

The first case will be treated in (6.4). It turns out that this case is directly related to the concept of pushing up. In the second case we get that $\langle U, B(T) \rangle$ is subnormal in every 2-local subgroup it is contained in . This is much stronger than condition (ii) in the definition of a uniqueness subgroup.

(6.1) *The set of uniqueness subgroups of H is closed under conjugation.*

Proof. Let U be a uniqueness subgroup of H and $h \in H$, $T \in \mathrm{Syl}_2(H)$ and $P = \langle B(T), U^h \rangle$. Suppose that there exists $L \in \mathcal{L}(T)$ with $P \leq L$ such that U^h is subnormal in P but not in L. Then U is subnormal in $P^{h^{-1}} = \langle B(T^{h^{-1}}), U \rangle$ but not in $L^{h^{-1}}$, a contradiction.

(6.2) *Let U be a uniqueness subgroup of H. Suppose that U is subnormal in $\langle U, S \rangle$. Then U is contained in a unique maximal element of $\mathcal{L}(S)$.*

Proof. Let Ω be the set of all conjugates X of U which are subnormal in $\langle X, S \rangle$. Since $O_2(U) \neq 1$ we also have $O_2(\langle U, S \rangle) \neq 1$. Hence, there exists a non-empty subset $\Omega_0 \subseteq \Omega$ such that $\langle \Omega_0 \rangle$ is contained in an element of $\mathcal{L}(S)$.

Let Ω_0 be maximal (with respect to inclusion) with that property and $\langle\Omega_0\rangle \leq$ $Y \in \mathcal{L}(S)$. Note that by (6.1) every element of Ω_0 is subnormal in Y. Hence $O_2(\langle\Omega_0\rangle) \neq 1$, and the maximality of Ω_0 implies that $Y \leq N_H(\Omega_0) \in \mathcal{L}(\mathcal{S})$.

Let Ω_1 be another subset of Ω which is maximal with $\langle\Omega_1\rangle \leq N_H(\Omega_1) \in$ $\mathcal{L}(S)$. In addition, we choose Ω_1 so that $\Omega_2 = \Omega_0 \cap \Omega_1$ is maximal. Assume that $\Omega_2 \neq \emptyset$. Then $S \leq N_H(\Omega_2) \in \mathcal{L}(S)$, and $\langle\Omega_2\rangle$ is subnormal in $\langle\Omega_0\rangle$ and $\langle\Omega_1\rangle$. It follows that $\Omega_2 \subset N_{\Omega_0}(\Omega_2)$ and $\Omega_2 \subset N_{\Omega_1}(\Omega_2)$. But the maximality of Ω_2 yields $N_\Omega(\Omega_2) \subseteq \Omega_0 \cap \Omega_1 = \Omega_2$, a contradiction. We have shown that $\Omega_0 \cap \Omega_1 = \emptyset$.

Now let $U \in \Omega_0$. Then $N_H(\Omega_0)$ is the unique maximal element of $\mathcal{L}(S)$ containing U.

(6.3) *Let $P_2 \in \mathcal{P}(S)$ and U be a 2-component of P_2 such that $P_2 = \langle U, S\rangle$. Suppose that U is a uniqueness subgroup of H. Then there exists $P_1 \in$ $\mathcal{ML}(S)$ and a maximal subgroup B of P_1 such that*

(i) $O_2(\langle P_1, P_2\rangle) = 1$, *and*

(ii) $\langle B, P_2\rangle \in \mathcal{L}(S)$.

Proof. By (6.2) there exists a unique maximal element $P_2^* \in \mathcal{L}(\mathcal{S})$ with $P_2 \leq P_2^*$, and by our hypothesis, P_2^* is not the only maximal element of $\mathcal{L}(S)$. Hence, there exists $P_1 \in \mathcal{L}(\mathcal{S})$ such that $P_1 \not\leq P_2^*$; i.e. $O_2(\langle P_1, P_2\rangle) = 1$. Choose P_1 minimal with that property. Then the assertion follows.

Notation. In the following, let U, B, P_1 and P_2 be as in (6.3), and let P_2^* be the unique maximal element of $\mathcal{L}(S)$ containing P_2. In particular $\langle B, P_2\rangle \leq P_2^*$.

(6.4) *Let (U, S) be an exceptional pair. Suppose that every 2-local subgroup of H containing $B(S)$ is of characteristic 2 type. Then there exists $T \in \mathcal{T}$ and $P = \langle T, U\rangle$ such that U is subnormal in P and either*

(a) $T = S$, $P \in \mathcal{P}(S)$, $0 \neq q(S, V) \leq 1$ *and* $C_S(V) = O_2(P)$ *where* $V = \Omega_1(Z(O_2(P)))$, *or*

(b) $[\Omega_1(Z(T)), P] = 1$, *and there exists* $L \in \mathcal{L}(T)$ *and a non-solvable 2-component E of L such that*

(b1) $L = \langle E, P\rangle$,

(b2) *U is not subnormal in L,*

(b3) *U is subnormal in every proper subgroup of L containing P,*

(b4) *no non-trivial characteristic subgroup of T is normal in L,* (b5) $T \notin Syl_2(H)$.

Proof. Since (U, S) is an exceptional pair, there exists $T_0 \in \mathcal{T}$, and $L \in \mathcal{L}(T_0)$ with $U \leq L$ such that

(i) U is subnormal in $\langle U, B(S) \rangle$.

(ii) U is not subnormal in L.

(iii) $T_0 \notin \mathrm{Syl}_2(H)$ and $B(S) \not\leq O_2(\langle U, B(S) \rangle)$.

Among all T_0 and L with (i)–(iii) we first choose $|T_0|$ maximal and then $|L|$ minimal. Let $P_0 = \langle U, B(S) \rangle$ and $S_0 \in \mathrm{Syl}_2(P_0)$. We may assume that $S_0 \leq T_0$. Let $V_0 = \Omega_1((Z(O_2(P_0))))$ and $W = \langle \Omega_1(Z(T_0))^L \rangle$.

Suppose first that $[V_0, O^2(P_0)] \neq 1$. Then also $[V_0, U] \neq 1$ and thus $[\Omega_1(Z(O_2(U))), U] \neq 1$. Let $P = \langle U, S \rangle$ and $V = \Omega_1(Z(O_2(P)))$. Note that $P \in \mathcal{P}(S)$ and so by (2.6) $P \in \mathcal{CL}(S)$. From (2.7) we get that $C_S(V) = O_2(P)$.

If $J(S) \leq O_2(P)$, then also $B(S) \leq O_2(P)$ contradicting $U \not\leq N_H(B(S))$. Hence, $J(S) \not\leq O_2(P)$ and by (1.1) $0 \neq q(S, V) \leq 1$, and (a) holds.

Suppose now that $[V_0, P_0] = 1$. Note that $\Omega_1(Z(T_0)) \leq B(T_0) \leq S_0$ and thus $[\Omega_1(Z(T_0)), P_0] = 1$. Now (iii) and the maximality of $|T_0|$ show that U is subnormal in $\langle U, T_0 \rangle$. Set $T = T_0$ and $P = \langle U, T \rangle$. The minimality of $|L|$ yields (b3) and the maximality of $|T|$ (b4). Hence, it remains to prove (b1).

Define $N = C_L(W)$, $\overline{L} = L/O_2(L)$ and $C = C_L(\overline{N})$. Then U is subnormal in PN since $PN \leq C_H(\Omega_1(Z(T)))$. Hence

$$(*) \quad UN/N \text{ is not subnormal in } L/N.$$

Assume first that $L = CP$. Let $\overline{E}$ be a component of $\overline{C}$. We may assume $\overline{L} \neq \langle \overline{P}, \overline{E} \rangle$. Hence (b3) implies that U is subnormal in $\langle P, E \rangle$, and so either $\overline{U}$ is a component of $\overline{C}$ or $[\overline{E}, U] = 1$. The first case contradicts $(*)$. In the second case let $\overline{E}_0 = \langle \overline{E}^L \rangle$. Then $\overline{U} \leq C_{\overline{L}}(\overline{E}_0)$ and $\overline{P} C_{\overline{L}}(\overline{E}_0) \neq \overline{L}$. Hence, $\overline{U}$ is subnormal in $C_{\overline{L}}(\overline{E}_0)$ and thus in $\overline{L}$, a contradiction. Thus, we may assume that $F^*(\overline{C}) = F(\overline{C})$.

Let $\tilde{L} = L/N$ and $\tilde{N}_1 = O_2(\tilde{L})$. Since $U \not\leq N_1$ we have that $T \cap N_1 = O_2(P) \cap N_1$ and so $U \leq N_L(T \cap N_1)$. Now (b3) and $(*)$ imply $T \cap N_1 = O_2(L)$. Hence $O_2(\tilde{L}) = 1$.

Assume that U is not solvable. Then $[\overline{U}, \overline{N}] = 1$ since $\overline{U} \not\leq \overline{N}$; i.e. $\overline{U} \leq \overline{C}$ and $L = CT$. In particular $F^*(\overline{L}) = F^*(\overline{C}) = F(\overline{C})$. By (b4) $J(T) \not\leq O_2(L)$. Let $L_0 = \langle B(T)^L \rangle$. Then (1.1) and (4.1) show that $\tilde{L}_0$ is solvable. But $U \not\leq N_H(B(T))$ since $B(T) = B(S)$ and so $\tilde{U} \leq \tilde{L}_0$, a contradiction.

Assume that U is solvable. Let $Q = O^2(P) \cap O_2(P)$. If $\overline{Q} = 1$, then $W = C_W(O^2(P)) \times [W, O^2(P)]$ and $[\Omega_1(Z(T)), O^2(P)] \neq 1$ since $U \not\leq N$, a contradiction. Hence $\overline{Q} \neq 1$. Since U is subnormal in PN we get that

$[\overline{Q}, \overline{N}] \leq O_2(\overline{N}) = 1$ and $\overline{Q} \leq \overline{C}$. It follows that $\overline{U}$ is not subnormal in $\overline{PF(C)}$ since $[\overline{Q}, F(\overline{C})] \neq 1$, and (b3) yields $\overline{L} = \overline{PF(C)}$.

Since $O_2(\tilde{L}) = 1$ and $\tilde{L}$ is solvable we get $C_{\tilde{L}}(O_{2'}(\tilde{L})) \leq O_{2'}(\tilde{L})$. Now as above (4.1) yields $U \leq F(\tilde{L})$ contradicting $(*)$.

We have shown that $L \neq CP$. Hence, (b3) implies that U is subnormal in PC. Assume that U is not solvable. Then $[\overline{N}, \overline{U}] = 1$ since $U \not\leq N$. Hence $\overline{U} \leq \overline{C}$, and U is subnormal in L, a contradiction.

Assume that U is solvable. Let Q be as above. Then we get as there that $\overline{Q} \neq 1$ and $\overline{Q} \leq \overline{C}$. But $O^2(\overline{P}) = \langle \overline{U}^{\overline{P}} \rangle$ is subnormal in $\overline{PC}$ and so $\overline{Q} \leq O_2(\overline{C}) = 1$, a contradiction.

Remark. Let L be as in (6.4)(b) and $\overline{L} = L/O_2(L)$. Then $F^*(\overline{L})$ is generated by the conjugates of $\overline{E}$, and $\overline{L} = F^*(\overline{L})\overline{P}$. Moreover, if $W = \langle \Omega_1(Z(T))^L \rangle$, then $C_T(W) = O_2(L)$ since E does not centralize W.

By (b4) W is a failure-of-factorization module for $L/C_L(W)$. Suppose that $\overline{E} \in \mathcal{K}$. Then $B(T) \not\leq O_2(P)$ and [3] imply that $O^2(P) \leq F^*(\overline{L})$. Hence, the structure of $O^2(\overline{P})$ (and thus $\overline{U}$) can be deduced from the structure of $\overline{E}$; in fact, $\overline{U}$ is subnormal in a maximal 2-local subgroup of $F^*(\overline{L})$ containing $\overline{T} \cap F^*(\overline{L})$.

7. In this section we will finish the discussion started at the beginning of section 3.

Let L be a finite group and $T \in \mathrm{Syl}_2(L)$. We will say that L satisfies (X), if (X) holds in L and (X) is one of the following properties:

($\mathbf{FF_1}$) L possesses a $GF(2)L$-module V such that $C_T(V) = O_2(L)$ and $0 \neq q(T, V) \leq 1$.

($\mathbf{FF_2}$) L possesses a $GF(2)L$-module V such that $C_T(V) = O_2(L)$ and $0 \neq q(T, V) \leq 2$.

($\mathbf{PU}$) $L = \langle B(T)^L \rangle T$, $C_T(\Omega_1(Z(O_2(L)))) = O_2(L)$, and no nontrivial characteristic subgroup of T is normal in L.

($\mathbf{L_2N}$) There exists a normal subgroup L^* in L containing $O_2(L)$ such that for $V = \langle \Omega_1(Z(T))^L \rangle$

(i) $C_L(V) = O_2(L)$, and

(ii) $L^*/O_2(L) \cong L_2(2^n)$ and V is a natural $L_2(2^n)$-module.

Note that (PU) implies ($\mathrm{FF_1}$). In this section we assume:

Hypothesis IV *Hypothesis II holds with $p = 2$, and U, P_1 and P_2 are defined as in section 6 (below (6.3)). We further assume:*

(1) *Every 2-local subgroup of H containing $B(S)$ is of characteristic 2 type.*

(2) *The non-solvable composition factors of P_1 are in $\mathcal{K}$.*

Recall from section 6 that $P_1 \in \mathcal{ML}(S)$, $P_2 \in \mathcal{P}(S)$ and $O_2(\langle P_1, P_2 \rangle) = 1$. In particular, (P_1, P_2) has the same properties as the pair used in the beginning of section 3. In addition, P_2 contains the uniqueness subgroup U. We will now discuss the cases listed in section 3.

Case (Ia) and (IIa). As already mentioned in section 3, we get:

      (a$_1$) P_2 satisfies (FF$_1$).

Case (IIb). Here (3.4) applies. Let W be any $GF(2)P_1$-module such that $0 \neq q(S, W) \leq 1$ and $C_S(W) = O_2(P_1)$. Then (4.7) shows:

$$q(S, W) = 1 \text{ and } C_W(S) \not\leq C_W(P_1).$$

We conclude that the cases (3.4)(d) and (e) do not occur. Hence, either (a$_1$) holds or

      (a$_2$) P_2 satisfies (FF$_2$)

or

      (b$_1$) $Z_1 \not\leq O_2(P_2)$.

Case (Ib). This case is ruled out by (5.4); see also the remark below (5.4).

Case (Ic). This is the only case where we will use the fact that P_2 contains a uniqueness subgroup. We distinguish the two cases whether (U, S) is an exceptional pair or not.

Assume first that (U, S) is an exceptional pair. Then (6.4) applies. There exists a 2-subgroup T of H containing $B(S)$ and $P \in \mathcal{L}(T)$ such that $P = \langle U, T \rangle$ and U is subnormal in P, and either

      (a$_3$) $T = S$, $P \in \mathcal{P}(S)$, and P satisfies (FF$_1$),

or

      (a$_4$) $P \leq L \in \mathcal{L}(T)$ such that $L = \langle E, P \rangle$ for some non-solvable 2-component E of L and

(i) L satisfies (PU),

(ii) U is not subnormal in L,

(iii) U is subnormal in every proper subgroup of L containing P,

(iv) $T \not\in \mathrm{Syl}_2(H)$ and $[\Omega_1(Z(T)), P] = 1$.

Assume now that (U, S) is not an exceptional pair. This case is covered by the following result where $V_2 = \langle Z_1^{P_2} \rangle$ and $W_1 = \langle V_2^{P_1} \rangle$.

(7.1) *Assume Hypothesis IV. Suppose that $Z_2 \leq Z(P_2)$ and that (U, S) is not an exceptional pair. Then one of the following holds:*

(a_1) P_2 *satisfies* *(FF$_1$), and P_1 satisfies (L$_2$N).*

(b_1) $Z_1 \not\leq O_2(P_2)$.

(b_2) $V_2 \not\leq O_2(P_1)$, *and P_1 satisfies (L$_2$N).*

(b_3) $W_1 \not\leq O_2(P_2)$, *and P_1 satisfies (L$_2$N).*

The proof of (7.1) uses the amalgam method. It is rather long and technical and thus omitted here.

We have found that one of the cases (a_1)–(a_4), (b_1)–(b_3) holds. Assume, in addition, that $U/O_2(U) \in \mathcal{K}$.

In the first four cases the results of Cooperstein [3] and Aschbacher [1] and Meierfrankenfeld-Stroth, [7] and [8], show that $U/O_2(U)$ is an alternating group or a group of Lie type in characteristic 2, allowing some exceptions in case (a_2).

In the last three cases information about $U/O_2(U)$ can be derived by a different method. First one gets information about the chieffactors of P_2 inside $O_2(P_2)$. This then yields information about $P_2/O_2(P_2)$ (and thus $U/O_2(U)$) possibly after a case by case discussion.

8.   This section is concerned with the existence of suitable uniqueness subgroups. We assume Hypothesis II with $p = 2$. For any subgroup X of H let $E(X/O_2(X))$ be the subgroup generated by all of the components of $X/O_2(X)$.

Let $\mathcal{E}(S)$ be the set of all subgroups E of H with

(∗) E is a non-solvable 2-component of $\langle E, S \rangle$.

An element $E \in \mathcal{E}(S)$ is called **smooth**, if for every $T \in \text{Syl}_2(H)$ and $M \in \mathcal{L}(T)$

(∗∗) $\langle B(S), E \rangle \leq M$ implies that $\overline{E} \leq E(\overline{M})$ where $\overline{M} = M/O_2(M)$.

Let $\mathcal{E}_s(S)$ be the set of all smooth elements of $\mathcal{E}(S)$, and let $\mathcal{E}_s^*(S)$ be the set of all maximal elements of $\mathcal{E}_s(S)$ (with respect to inclusion).

(8.1) *Let $E \in \mathcal{E}_s(S)$ and $B(S) \leq T \in \text{Syl}_2(H)$. Suppose that $F \in \mathcal{E}(T)$ and $E \leq F$. Then $F \in \mathcal{E}_s(T)$.*

Proof. Note that $B(S) = B(T)$ and $\langle E^F \rangle = F$. Now the assertion follows directly from the definition of $\mathcal{E}_s(T)$.

(8.2) *Let $T \in \text{Syl}_2(H)$, $M \in \mathcal{L}(T)$ and $\overline{M} = M/O_2(M)$. Suppose that $\overline{F}_1, ..., \overline{F}_n$ are the components of $\overline{M}$ and $\overline{E}$ is a subgroup of $E(\overline{M})$. Then there exists a subgroup $\overline{L}_i \leq \overline{F}_i$, $i = 1, ..., n$, such that for*

$$\overline{L} = Z(E(\overline{M})) \prod_{i=1}^{n} \overline{L}_i$$

the following hold:

(a) $\overline{E} \leq \overline{L}$ and $\overline{E}C_{E(\overline{M})}(\overline{F}_i) = \overline{L}_i C_{E(\overline{M})}(\overline{F}_i)$.

(b) $O_2(\overline{E}) \leq O_2(\overline{L})$.

(c) $N_{\overline{M}}(\overline{E}) \leq N_{\overline{M}}(\overline{L})$.

(d) $[\overline{L}_i, \overline{E}] = 1$ if and only if $\overline{L}_i \leq Z(\overline{L})$.

(e) If $\overline{E}$ is perfect and subnormal in $\overline{L}$, then $\overline{L}_i = \overline{E} \cap \overline{F}_i$.

Proof. We choose $\overline{E} \cap \overline{F}_i \leq \overline{L}_i \leq \overline{F}_i$ and $|\overline{L}_i|$ minimal such that (a) holds. Note that $\overline{L}_i$ is normal in $\overline{L}$.

From (a) we get that $[O_2(\overline{E}), \overline{L}_i] \leq O_2(\overline{L}_i) \leq O_2(\overline{L})$. Hence (b) holds.

Let $x \in N_{\overline{M}}(\overline{E})$. Then $\overline{L}_i^x \leq \overline{F}_j$ for some j and thus by (a) $(\overline{L}_i Z(\overline{F}_i))^x = \overline{L}_j Z(\overline{F}_j) \leq \overline{L}$. This gives (c). Assertion (d) is easy to calculate.

Suppose that $\overline{E}$ is perfect and subnormal in $\overline{L}$. Let $\overline{E}_i = \overline{E} \cap \overline{F}_i$. Then $\overline{E}_i$ is subnormal in $\overline{L}_i$. Assume that $\overline{E}_i \neq \overline{L}_i$. Then there exists a maximal normal subgroup $\overline{N}$ of $\overline{L}_i$ such that $\overline{E}_i \leq \overline{N}$ and $[\overline{E}, \overline{L}_I] \leq \overline{N}$. Now (a) implies $\overline{L}_i' = [\overline{E}, \overline{L}_i] \leq \overline{N}$. On the other hand, $\overline{L}_i' C_{E(\overline{M})}(\overline{F}_i) = \overline{E}C_{E(\overline{M})}(\overline{F}_i)$ since $\overline{E}$ is perfect. It follows that $\overline{L}_i = \overline{L}_i'\overline{E}_i$ by the minimality of $\overline{L}_i$ and so $\overline{L}_i \leq \overline{N}$, a contradiction.

(8.3) Let $E \in \mathcal{E}_s^*(S)$ and $T \in Syl_2(H)$. Suppose that $\langle B(S), E \rangle \leq M \in \mathcal{L}(T)$. Then each of the following conditions implies that E is contained in some 2-component of M.

(a) E is subnormal in $\langle E, T \rangle$.

(b) $E \leq N_H(B(S))$.

Proof. Let $\overline{M} = M/O_2(M)$, and let $\overline{F}_i$ be a component of $\overline{M}$ such that $[\overline{E}, \overline{F}_i] \neq 1$.

Suppose that (a) holds. Then it suffices to show that $\overline{E} \leq \overline{F}_i$. We apply (8.2) with the notation given there. Since E is subnormal in $\langle E, T \rangle$ we get that $\overline{E} \leq [\overline{F}_i \cap \overline{T}, \overline{E}]$ or $[\overline{F}_i \cap \overline{T}, \overline{E}] \leq O_2(\overline{E})$. The first case gives $\overline{E} \leq \overline{F}_i$. In the second case, $\overline{F}_i \cap \overline{T} \leq N_{\overline{M}}(\overline{E})$ and so by (8.2)(c) $\overline{F}_i \cap \overline{T} \leq N_{\overline{M}}(\overline{L} \cap \overline{F}_i)$. Since $F_i \cap T \in Syl_2(F_i)$ also $\overline{L}_i \cap \overline{T} \in Syl_2(\overline{L}_i)$. Hence $[\overline{L}_i \cap \overline{T}, \overline{E}] \leq O_2(\overline{E})$ and (8.2)(a) imply that $\overline{L}_i \cap \overline{T} = O_2(\overline{L}_i)$, i.e. $\overline{L}_i/O_2(\overline{L}_i)$ has odd order. Now by the Odd-Order-Theorem [4] $\overline{L}_i$ is solvable while $\overline{E}$ is perfect. This contradicts (8.2)(a).

Suppose that (b) holds. Let $N = N_H(B(S))$. We may assume that $B(S) = B(T)$. Then $T \leq N$. Since E is subnormal in $\langle E, S \rangle$ we get with condition (a) that $E \leq F$ for some 2-component F of N. By (8.1) $F \in \mathcal{E}_s(S)$ and so $F = E$ by the maximality of E. Hence E is also subnormal in $\langle E, T \rangle$, and the assertion follows with condition (a).

(8.4) *Suppose that $E \in \mathcal{E}_s^*(S)$ and $E \leq N_H(B(S))$. Then there exists $T \in Syl_2(H)$ and $F \in \mathcal{E}_s^*(T)$ such that*

(a) $B(T) = B(S)$,

(b) $E \leq F$, *and F is a uniqueness subgroup of H,*

(c) E *is subnormal in $\langle E, T \rangle$.*

Proof. Among all $T \in Syl_2(H)$ with $B(T) = B(S)$ and all $F \in \mathcal{E}(T)$ with $E \leq F$ we choose $|F|$ maximal. By (8.1) $F \in \mathcal{E}_s(T)$, and by (8.3) E is subnormal in $N_H(B(S))$. Hence E is also subnormal in $\langle T, E \rangle$. In particular, (c) holds.

Suppose that $\tilde{T} \in Syl_2(H)$ and $M \in \mathcal{L}(\tilde{T})$ such that $\langle B(T), F \rangle \leq M$. We may assume that $B(T) = B(\tilde{T})$. Hence, by (8.3) $E \leq \tilde{F}$ for some 2-component $\tilde{F}$ of M. Since $F/O_2(F)$ is quasisimple we also have $F \leq \tilde{F}$. The maximality of $|F|$ gives that $F = \tilde{F}$. Hence, F is a uniqueness subgroup.

(8.5) *Let $E \in \mathcal{E}_s^*(S)$, $E_0 = \langle E^{B(S)} \rangle$ and $N = N_H(E_0)$. Suppose that $T \in Syl_2(H)$ with $T \cap N \in Syl_2(N)$ and $E \not\leq N_H(B(S))$. Then one of the following holds:*

(a) $0 \neq q(B(S), V) \leq 1$ *where* $V = \Omega_1(Z(O_2(E_0)))$, *and* $C_{S \cap E_0}(V) = O_2(E_0)$.

(b) *There exists $F \in \mathcal{E}_s(T)$ with $E \leq F$ such that for $F_0 = \langle F^{B(T)} \rangle$:*

(b1) $0 \neq q(B(T), W) \leq 1$ *where* $W = \Omega_1(Z(O_2(F_0)))$, *and* $C_{F_0 \cap T}(W) = O_2(F_0)$.

(b2) *There exists $O_2(F) < D \leq T \cap F$ such that E is subnormal in $N_F(D)$.*

(c) E *is subnormal in $C_H(\Omega_1(Z(T)))$.*

Proof. Suppose that the assertion holds for E and T. Then it also holds for E and T^x where $x \in N_N(E)$. Hence, we may assume that $S \cap N \leq T \cap N$ since $N = N_N(E)B(S)$.

Let E be a counterexample, and let V be as in (a). Assume that $[V, E] \neq 1$. Then $C_{E_0 \cap S}(V) = O_2(E_0)$. If $[J(S), V] = 1$, then also $[B(S), V] = 1$ and $E \leq N_H(B(S))$, a contradiction. Hence, we have $[J(S), V] \neq 1$, and (1.1) implies (a). But then E is not a counterexample, a contradiction.

We have shown that $[V, E] = 1$. Let $E^* = \langle E^S \rangle$, and note that $E^* \leq N$. Then $[\Omega_1(Z(O_2(E^*))), E] \leq V$ and thus also $[\Omega_1(Z(O_2(E^*))), E^*] = 1$. Since $O_2(E^*) \leq N \cap S$ we get that $[\Omega_1(Z(T)), E^*] = 1$.

Let $C = C_H(\Omega_1(Z(T)))$; i.e. $E^* \leq C$. We apply (8.2) with $\overline{C} = C/O_2(C)$. Then there exists $\overline{L} \leq E(\overline{C})$ so that $\overline{E}^* \leq \overline{L}$ and $O_2(E^*) \leq O_2(L)$. Moreover, by (8.2)(e) $\overline{E}$ is in a component of $\overline{C}$, if E^* is subnormal in L.

Since $E^* \not\leq N_H(B(S))$ we get that $E^* \leq \langle E^{*B(S)} \rangle$. Now $[O_2(L), B(S)] \leq B(S)$ implies $[O_2(L), E^*] \leq O_2(E^*)$; i.e. $O_2(L) \leq X \leq N_H(E^*)$ where X is the largest subgroup of $N_H(E^*)$ with $[E^*, X] \leq O_2(E^*)$. There exists $x \in X$ such that $O_2(L) \leq S^x$. Let $A = \Omega_1(Z(S^x))$ and $W_0 = \Omega_1(Z(O_2(L)))$. Then $A \leq W_0$ and $[E^*, W_0] = 1$ since $[\Omega_1(Z(O_2(E^*))), E^*] = 1$. Hence, $E^* \leq C_L(W_0) \leq C_H(A)$. But $E^{*x} = E^*$, and so (8.3) and the maximality of E imply that E^* is subnormal in $C_H(A)$. Now E^* is subnormal in $C_L(W_0)$ and thus also in L, and E is contained in a 2-component F of C.

If $[F, A] = 1$, then $E = F$ and (c) holds. Thus we have that $[F, A] \neq 1$ and so $[F, W] \neq 1$, W as in (b). Now (8.1) implies (b1).

Let $D = O_2(L) \cap F$. If $D = O_2(F)$, then $[W, E] \neq 1$ implies (a), a contradiction. Hence $O_2(F) < D$. Moreover, $O_2(E) \leq D$ and so, again since (a) does not hold, $[A, N_F(D), E] = 1$. Since E is subnormal in $C_H(A)$ we also get that E is subnormal in $N_F(D)$, and (b2) holds. Hence, E is not a counterexample.

(8.6) *Suppose that $E \in \mathcal{E}_s^*(S)$. Then there exists $T \in Syl_2(H)$ and $F \in \mathcal{E}_s^*(T)$ such that $B(T) = B(S)$ and $E \leq F$, and one of the following holds:*

(a) *F is a uniqueness subgroup of H, and E is subnormal in $\langle E, T \rangle$.*

(b) *Let $F_0 = \langle F^{B(S)} \rangle$ and $V = \Omega_1(Z(O_2(F_0)))$. Then*

(b1) *$0 \neq q(B(S), V) \leq 1$ and $C_{F_0 \cap T}(V) = O_2(F_0)$, and*

(b2) *there exists $O_2(F) \leq D \leq T \cap F$ such that E is subnormal in $N_F(D)$.*

Proof. If $E \leq N_H(B(S))$, then (a) follows from (8.4). Suppose that $E \not\leq N_H(B(S))$.

Let $\tilde{T} \in Syl_2(H)$ and $M \in \mathcal{L}(\tilde{T})$ with $\langle E, B(S) \rangle \leq M$, and let $E_0 = \langle E^{B(S)} \rangle$. We may assume that $B(S) = B(\tilde{T})$.

We now apply (8.2) to E_0 and M. Then there exists a subgroup L of M with $E_0 \leq L$ which has the properties given in (8.2). In particular $B(S) \leq N_M(O_2(L))$. Since $E_0 = [E_0, B(S)]$ we get that $O_2(L) \leq N_H(E_0)$.

Let $T \in Syl_2(H)$ with $B(S)O_2(L) \leq N_T(E_0) \in Syl_2(N_H(E_0))$, and let $W_0 = \Omega_1(Z(O_2(L)))$. Then $\Omega_1(Z(T)) \leq W_0$. By (8.5) either (b) holds, or E is subnormal in $C_H(\Omega_1(Z(T)))$.

Assume the latter case. If $[W_0, E] \neq 1$, then again (b) holds since $[B(S), W_0] \neq 1$. Hence, we may also assume that $[W_0, E] = 1$; i.e. E is subnormal in $C_L(W_0)$ and so E is subnormal in L. By (8.2) $E \leq F$ for some 2-component

F of M. Again, either (b) holds or $F \leq C_H(\Omega_1(Z(T)))$ and $F = E$. Hence, if (b) does not hold, then (a) follows for $F = E$.

Remark. Let E be any element in $\mathcal{E}_s(S)$. Then $E \leq E^* \in \mathcal{E}_s^*(S)$ and

$(*)$ E is subnormal in $\langle E, S \rangle$.

By (8.6) there exists $T \in \mathrm{Syl}_2(H)$ and $F \in \mathcal{E}_s^*(T)$ with $B(S) = B(T)$ and $E \leq E^* \leq F$ such that

$(**)$ E^* is subnormal in $\langle E^*, T \rangle$, or E^* is subnormal in $N_F(D)$

for some $O_2(F) \leq D \leq T$.

Now information about $F/O_2(F)$ can be derived as in the discussion of sections 6 and 7. Via $(*)$ and $(**)$ this information also yields information about $E/O_2(E)$. For example, if $F/O_2(F) \in \mathcal{K}_{\mathrm{even}}$, then also $E/O_2(E) \in \mathcal{K}_{\mathrm{even}}$.

A final word about the elements in $\mathcal{E}(S) \backslash \mathcal{E}_s(S)$. Let $E \in \mathcal{E}(S) \backslash \mathcal{E}_s(S)$. Then there exists $T \in \mathrm{Syl}_2(H)$ and $M \in \mathcal{L}(T)$ such that $\langle B(S), E \rangle \leq M$ but $\overline{E} \not\leq E(\overline{M})$ where $\overline{M} = M/O_2(M)$.

Using Schreier's conjecture either $\overline{E}$ operates non-trivially on the components of $\overline{M}$ or there exists an odd prime p such that $[O_p(\overline{M}), \overline{E}] \neq 1$. In the first case for a suitable odd prime q the q-rank of $E(\overline{M})$ is at least 5. In the second case the p-rank of $O_p(\overline{M})$ is at least 2, and for small p-rank, e.g. 2 or 3, only very particular possibilities for $E/O_2(E)$ occur . For example, if the p-rank of $\overline{M}$ is 2, then $E/O_2(E) \cong L_2(p)$ or $L_2(4)$.

References

1. M. Aschbacher, $GF(2)$-Representations of finite groups, Am. J. Math. 104 (1982), 683-771.

2. B. Baumann, Über endliche Gruppen mit einer zu $L_2(2^n)$ isomorphen Faktorgruppe, Proc. Am. Math. Soc. 74 (1979), 215-222.

3. B. Cooperstein, An enemies list for factorization theorems, Comm. Alg. 6 (1978), 1239-1288.

4. W. Feit, J. G. Thompson, Solvability of groups of odd order, Pac. J. Math. 13 (1963), 775-1029.

5. K. Gomi, On the 2-local structure of groups of characteristic 2 type, J. Alg. 108 (1987), 492-502.

6. U. Meierfrankenfeld, Eine Lösung des Pushing-Up Problems für eine Klasse endlicher Gruppen, Dissertation Bielefeld (1986).

7. U. Meierfrankenfeld, G. Stroth, On quadratic $GF(2)$-modules for Chevalley groups over fields of odd order, preprint.

8. U. Meierfrankenfeld, G. Stroth, Quadratic $GF(2)$-modules for sporadic simple groups , preprint.

Groups generated by k-root subgroups - a survey.

F.G.Timmesfeld

1 Statement of the Theorems.

Let k be a field (commutative), G a group and Σ a class of abelian subgroups generating G. Then Σ is a class of k-root subgroups of G, if the following holds:

1. If $A, B \in \Sigma$, then one of the following holds:

   (a) $[A, B] = 1$

   (b) $\langle A, B \rangle \simeq (P)SL_2(k)$ and A, B are full unipotent subgroups of $\langle A, B \rangle$.

   (c) $\langle A, B \rangle' \le Z(\langle A, B \rangle)$ and $[a, B] = [A, b] = [A, B] \in \Sigma$ for all $a \in A^{\#}, b \in B^{\#}$

2. G satisfies the maximality condition for Σ subgroups. [1] (A Σ-subgroup U is a subgroup $U = \langle U \cap \Sigma \rangle$ where $U \cap \Sigma = \{A \in \Sigma \mid A \le U\}$. Maximality condition means that all ascending chains of Σ-subgroups are finite.)

If (1)(c) never occurs we call Σ a *degenerate* class of k-root subgroups or a class of $k - transvections$ of G. Otherwise Σ is $non - degenerate$. If (1)(b) never occurs, it has been shown in [16], extending a well-known theorem of Baer, that G is nilpotent. So this case is of no interest.

---

[1] The minimality condition has been weakened in the meantime in [19] in so far, that one only needs to demand it for "unipotent" Σ-subgroups. Here a Σ-subgroup U is unipotent, if there exist no $A, B \in U \cap \Sigma$ with $\langle A, B \rangle \simeq (P)SL_2(k)$.

$X \simeq (P)SL_2(k)$ means that $X \simeq SL_2(k)$ or $PSL_2(k)$. A full unipotent subgroup of $X = SL_2(k)$ is a conjugate of $\{(\begin{smallmatrix} 1 & 0 \\ a & 1 \end{smallmatrix}) \mid a \in k\}$, while a full unipotent subgroup of $\bar{X} = PSL_2(k)$ is just the image of a full unipotent subgroup of X. It is known, see [9], that the class Δ (resp.$\bar{\Delta}$) of full unipotent subgroups of X (resp.$\bar{X}$) is invariant under automorphisms of X (resp.$\bar{X}$). Thus if we have two isomorphisms:

$$\sigma, \varphi : X(\text{resp.}\bar{X}) \to \langle A, B \rangle$$

then $\sigma(\Delta) = \varphi(\Delta)$ (resp. $\sigma(\bar{\Delta}) = \varphi(\bar{\Delta})$). So we can speak of the class of full unipotent subgroups of the "abstract" group $\langle A, B \rangle$.

If $|k| < \infty$ then the condition $[a, B] = [A, b] = [A, B]$ in (1)(c) follows from the others. This is due to the fact that the commutator map:

$$\begin{aligned} B &\to [A, B] \\ b &\to [a, b], \, a \in A^{\#} \quad \text{fixed} \end{aligned}$$

is a monomorphism and thus must be an isomorphism, since $|B| = |k| = |[A, B]|$.

Typical example of groups generated by a class of k-root subgroups are Lie-type and classical groups over k (resp. extension skew fields of k) with Σ the class of centers of the "long root subgroups". Here it follows immediately from the Chevalley commutator relations and the action of the group on its spherical building that (1) holds. But it is not at all obvious that (2) is also satisfied. This is true if k is algebraically closed (and, of course, if G is finite), since then subgroups generated by elements in Σ are algebraic subgroups and thus have a dimension attached to them. In general it follows from a non-trivial theorem on fix-point subgroups under automorphisms of a group generated by k-root subgroups.

The origin of the notion comes from B. Fischer's well-known paper [11] "Groups generated by 3-transpositions", which is just the very special case G finite, $k = GF(2)$ and Σ degenerate. But, as the following theorems will show, it is possible to obtain a complete classification in this far greater generality.

This paper is a survey mainly of the results of [17], [19] and [20].

1.1 Theorem

Let k be a field and G a group generated by a class Σ of k-transvections. Suppose there exist $A \neq B \in \Sigma$ with $[A, B] = 1$ and one of the following two conditions is satisfied:

(i) $|k| > 3$ and G has no nilpotent normal subgroup;

(ii) $|k| \leq 3$, G' is simple, $Z(G) = 1$ and $Z(\langle C_\Sigma(AB)\rangle) \cap \Sigma \neq \{A, B\}$. (This means "lines" are thick!)

Then one of the following holds:

1. $G \simeq PSp(2n, k)$, $n \geq 2$ and either Σ is the class of symplectic transvection groups, or k is non-perfect of char 2 and there exists a field $\bar{k}$ with $k^2 \subseteq \bar{k} \subseteq k$ such that

   $$\Sigma = \{\bar{T}_v \mid v \neq 0 \quad \text{a vector of the natural (projective) module} \quad V \text{ of } G\}$$

   and $\bar{T}_v = \{t_c : w \to w + (w, v)cv \quad \text{or} \quad w \in V \mid c \in \bar{k}\}$.

2. $G \simeq PSU_n(\bar{k}, f)$ where $\bar{k}$ is a divisionring containing k, f a nondegenerate σ-hermitian form of Witt-index ≥ 2, σ an involutory antiautomorphism of $\bar{k}$ with $\bar{k}_\sigma = \{c \in \bar{k} \mid c = c^\sigma\} = k$. Further, Σ is the uniquely determined class of "unitary transvection groups". Moreover, if $\bar{k}$ is noncommutative, then char $k \neq 2$ and $\bar{k}$ is a quaternion algebra over k.

3. k is infinite of char.2. There exists a quaternion divisionring $\bar{k}$ over k, a finite dimensional left $\bar{k}$-vector space V, a nondegenerate pseudo-quadratic form $q : V \to \bar{k}/k$ of Witt-index ≥ 2 with associated σ-hermitian form f, where σ is the standard antiautomorphism of $\bar{k}$ (satisfying $\bar{k}_{\sigma,1} = \{c + c^\sigma \mid c \in \bar{k}\} = k$) such that the following holds:

   (a) There exists a monomorphism $\varphi : G \to PO(V, q)$ with $\varphi(G) \trianglelefteq PO(V, q)$,

   (b) $\varphi(\Sigma)$ is the set (of images under $P : O(V, q) \to PO(V, q)$) of orthogonal transvection groups:

   $$T_p = \{t_c : v \to v + f(v, p)cp \mid c \in k\} \quad \text{where}$$
   $$0 \neq p \in V \quad \text{with} \quad q(p) = 0.$$

4. There exists a field $\bar{k}$ of char.2 with $\bar{k}^2 \subseteq k \subset \bar{k}$, a finite dimensional k-vector space V, a non-degenerate quadratic form $q : V \to \bar{k}$ of Witt-index ≥ 2 such that the following holds:

(a) $\dim_{\bar{k}^2} k = \dim_{\bar{k}} V^\perp$. ($V^\perp$ defined with resp. to the associated bilinear form f).

(b) There exists a monomorphism $\varphi : G \to PO(V, q)$ such that $\varphi(G) \trianglelefteq PO(V, q)$ and $\varphi(\Sigma)$ is the set (of images under $P : O(V, q) \to PO(V, q)$) of orthogonal transvection groups:

$$T_p = \left\{ t_c : w \to w + f(w, p)c \left(p + \frac{1}{c} v_c \right) \,\middle|\, c \in k, v_c \in V^\perp \right\}$$

where $0 \neq p \in V$ with $q(p) = 0$.
($G \simeq C_n(k, \bar{k})$ in the notation of $[21, p.204]$.)

5. k is infinite. There exists a Cayley-division algebra K over k, such that the set of subspaces of Σ (as defined in section 3) is isomorphic to the uniquely determined polar space $\wp$ of rank 3 over K. (The planes of which are non-desarguesion.) Moreover, G is isomorphic to the uniquely determined non-trivial simple normal subgroup of $\mathrm{Aut}(\wp)$. ($G \simeq E_7^K$ in the notation of $[21]$.)

6. k is infinite; $(\Sigma, \mathcal{L}, \in)$ is a non-embeddable Moufang generalized quadrangle (in the sense of §8 of $[21]$) on which the elements of Σ act as "transvection groups" (i.e. fix all lines through a given point pointwise); where

$$\mathcal{L} = \{ C_\Sigma(AB) \mid A \neq B \in \Sigma \ \text{ with } \ [A, B] = 1 \}.$$

In case $\bar{k}$ is non-commutative, SU is just the normal subgroup of U generated by transvections. Examples of type (6) are the orthogonal groups of Witt-index 2, in which case our generalized quadrangle is the dual of the natural. But since the classification of Moufang generalized quadrangles still does not seem to be complete (for example in $[10]$ the case of fields of char2 is not treated at all), I have to desist from it.

The second possibility for Σ in case 1 of (1.1) when k is non-perfect of characteristic 2 did not appear in $[17]$. The main reason for this little gap is that the polar space obtained from Σ is the same for each subfield $\bar{k}$ of k containing k^2. We will show in §2 how to close this gap and complete the determination of Σ in this case.

1.2   Theorem.

Suppose G is a group generated by a non-degenerate class Σ of k-root subgroups. Then G has a nilpotent normal "radical" $R(G)$ such that the following hold:

1. $\tilde{R} = R(G)/R(G) \cap Z(G)$ is abelian.

2. $G^* = G/R(G)$ is quasisimple.

3. $C_{\tilde{R}}(G^*) = 1, \tilde{R} = [\tilde{R}, G^*]$ and $[\tilde{R}, A^*, B^*] = 1$
   for all commuting pairs $A, B \in \Sigma$.

4. For $\bar{G} = G^*/Z(G^*)$ and $\bar{\Sigma}$ one of the following holds:

   (a) $\bar{G} \simeq PSL_n(k), n \geq 3$ and $\bar{\Sigma}$
       is the class of root groups of transvections of $\bar{G}$.

   (b) $\bar{G} \simeq P\Omega(V, q)$, V a finite dimensional k-vectorspace, $q : V \to k$
       a non-degenerate quadratic form of Witt-index ≥ 3 and $\bar{\Sigma}$ is the
       class of "Siegel-transvection groups" of $\bar{G}$. (I.e. the elements of
       $\bar{\Sigma}$ are in 1-1-correspondence with the singular lines of V.)

   (c) $\bar{G} \simeq G_2(k)$ (resp. $G_2(2)'$), $^3D_4(\bar{k}), ^6D_4(\bar{k}), \bar{k}$ a separable extension of
       degree 3 of k, $^1E_{6,2}^{16}(k)$, $^2E_{6,2}^{16''}(k)$, $E_{8,2}^{78}(k)$ in the notation of table II
       of [23] (here k is infinite) or char $k = 3$ and $\bar{G} \simeq G_2(k, K)$ where
       $k \subset K \subset \bar{k}, \bar{k}$ a field with $\bar{k}^3 = k$ (for definition see [21,(10.3.2)].
       Moreover, except when char $k = 3$ and $\bar{G} \simeq G_2(k)$ or $G_2(k, K), \bar{\Sigma}$
       is the class of long root subgroups of $\bar{G}$. (In the latter case $\bar{\Sigma}$
       might be the class of long or short root subgroups.)

   (d) $\bar{G} \simeq F_4(k), ^2E_6(\bar{k}), \bar{k}$ a separable extension of degree 2 of k, $E_{7,4}^9(k)$,
       $E_{8,4}^{28}(k)$ in the notation of table II of [23] or char $k = 2$ and $\bar{G} \simeq$
       $F_4(k, K)$ where $k \subset K \subset \bar{k}$ and $\bar{k}$ a field with $\bar{k}^2 = k$. (See
       [21,(10.3.2)].) Moreover, except when char $k = 2$ and $\bar{G} \simeq F_4(k)$ or
       $F_4(k, K), \bar{\Sigma}$ is the class of long root subgroups of $\bar{G}$. (In the latter
       case $\bar{\Sigma}$ might be again the class of long or short root subgroups.)

   (e) $\bar{G} \simeq E_6(k), E_7(k), E_8(k)$ and $\bar{\Sigma}$ is the class of long root subgroups
       of $\bar{G}$.

5. $\tilde{R} = 1$ except in cases (4)(a),(4)(c) or in case (4)(b) when the Witt-
   index of q is 3.

Theorems (1.1) and (1.2) together give an "abstract" classification of al-
most all simple algebraic groups of relative rank ≥ 2. If $|k| \geq 4$ in Theorem
(1.1) one does not need the condition that G has no nilpotent normal subgroup. Namely, as in Theorem (1.2), one can defined in this case a nilpotent
radical $R(G)$ and show that conditions (1), (2) and (3) of Theorem (1.2) are
satisfied and then apply Theorem (1.1) to the simple group $\bar{G} = G^*/Z(G^*)$
generated by the class $\bar{\Sigma}$ of k-transvections.

If $\bar{G}$ is symplectic, unitary over fields or linear, it has been shown in [18] that $\tilde{R}$ is the direct sum of natural (resp. natural and dual) modules for G^*. If $\bar{G}$ is orthogonal of Witt-index 3 in Theorem (1.2), then $\tilde{R}$ should be the direct sum of spin modules and if $\bar{G}$ is "of type G_2" $\tilde{R}$ should be the direct sum of natural modules. (These statements are easy to show if G is finite.)

There are three main applications of Theorem (1.1) and (1.2) I can see at the moment.

1. Uniform treatment of the corresponding classifications for finite groups. ([1],[2], [3], [11], [13],[14]).
   Although not in all these papers our hypothesis is satisfied from the start, it is very often shown to hold in the course of the proof.

2. Determination of subgroups generated by long-root subgroups of algebraic groups over k. (The corresponding problem has been solved by Cooperstein [8], for finite exceptional Lie-type groups.)
   It is clear that such a subgroup is a group generated by a set of $k - root$ subgroups in our sense. Hence the results of [17] and [19] apply. (So for example under some irreducibility condition only the question, which of the groups of Theorem (1.1) and Theorem (1.2) occur in the different cases, remains.)
   Indeed, as Theorem (2.1) will show, the theory of groups generated by k-root subgroups which provides a large number of intermediate results and structural properties of the groups classified, seems to be a good tool to obtain information about the simple algebraic groups, without using algebraic geometry.

3. Quadratic action.
   In his paper on quadratic pairs [12], J. Thompson shows, roughly after one third of the proof, that there is a class of quadratically acting subgroups which satisfy our condition (1) and (2). (k a finite field of char $p \geq 5$.) So from this point on our results might take over. (Condition (3) is of course satisfied since G is finite.) It is my hope that one might be able to prove something similar, perhaps only in more restricted situations, for quadratic pairs (G, V) with V a finite dimensional kG-module and k an arbitrary field of char 0 or char $p \geq 5$.

In fact many of the results obtained are stronger than stated in Theorems (1.1) and (1.2). So for example for induction purposes one has to treat the case of groups generated by a "set of k-root subgroups" and then prove some central product theorem, i.e. prove that such a group is mod some nilpotent normal subgroup a central product of finitely many groups generated by classes of k-root subgroups. Moreover, in some sense the case $k = GF(2)$, is treated completely in [17]. Namely one introduces a rank function on Σ. If now rk $\Sigma = 1$, i.e. there don't exist different commuting elements, it follows from a theorem of Bruck [4] on commutative Moufang-loops of exponent 3, that G is finite. From this point on, one can show for arbitrary rank by induction on rk Σ using the results of [17], that G is finite and then use the finite classification [11], which has been revised by R.Weiss [25], [26].

Thus only the complete classification in case $k = GF(3)$ in Theorem (1.1) remains open. Here a theorem like Bruck's seems difficult, since one has to consider a non-solvable 3-generator group, namely $U_3(3)$.

2 Automorphisms of groups generated by k-root subgroups.

The fundamental question, whether the class of centers of the long-root subgroups of a simple Lie-type group over a field k is a class of k-root subgroups in our sense is answered by the following theorem, which is the main theorem of [20].

2.1 Theorem

Let G be a quasisimple group generated by the class Σ of k-root subgroups. If Σ is degenerate, assume that Σ satisfies the additional conditions of (1.1) (i.e. there exist different commuting elements A, B in Σ and $Z(\langle C_\Sigma(AB)\rangle) \cap \Sigma \neq \{A, B\}$ if $|k| \leq 3$). Let X be a set of automorphisms of G, $\Sigma^\circ = \{C_A(X) \mid A \in \Sigma, C_A(X) \neq 1\}$ and $R = \langle \Sigma^\circ \rangle$. Define a graph $\mathcal{F}(\Sigma^\circ)$ by joining two elements $A^\circ, B^\circ \in \Sigma^\circ$ where $A^\circ = C_A(X)$ if and only if $\langle A^\circ, B^\circ \rangle$ is not nilpotent (i.e. is a subgroup of $\langle A, B \rangle \simeq SL_2(k)!$) and assume that $\mathcal{F}(\Sigma^\circ)$ is connected. Suppose further that $|A^\circ| > 2$ for some $A^\circ \in \Sigma^\circ$. Then there exists a subfield ℓ of k such that Σ° is a class of ℓ-root subgroups of R.

It is clear that the above theorem answers the question raised before. Namely let $G(k)$ be a quasisimple Lie-type group over k, k a perfect field,[2] and let $\bar{k}$ be the algebraic closure of k, and $G(\bar{k})$ the Lie-type group of the same type over $\bar{k}$. Then, by the remarks in section 1, the class $\bar{\Sigma}$ of

---

[2]If k is not perfect one has to take the separable closure $\bar{k}$ of k.

centers of the long-root subgroups of $G(\bar{k})$ is a class of $\bar{k}$-root subgroups. Let $X = Aut(\bar{k} : k)$. Then X can be considered as a subgroup of the automorphism group of $G(\bar{k})$ with fix-point-group $G(k)$. Further

$$\Sigma = \{C_A(X) \mid A \in \bar{\Sigma}, C_A(X) \neq 1\}.$$

Hence Σ is a class of k-root subgroups of $G(k)$ by (2.1) .

2.2

It is straightforward to show that the connectedness of $\mathcal{F}(\Sigma^\circ)$ in (2.1) implies that there exists a subfield ℓ of k such that

$$\langle A^\circ, B^\circ \rangle \simeq (P)SL_2(\ell)$$

with A°, B° full unipotent subgroups of $\langle A^\circ, B^\circ \rangle$ for each pair $A^\circ, B^\circ \in \Sigma^\circ$ for which $\langle A^\circ, B^\circ \rangle$ is not nilpotent. Further, if $A, B \in \Sigma$ with $C = [A, B] \in \Sigma$ and $A^\circ = C_A(X) \neq 1$, $B^\circ = C_B(X) \neq 1$, then $C^\circ = C_C(X) \neq 1$ and $[A^\circ, B^\circ] \leq C^\circ$. For $a \in (A^\circ)^\#$ let $\chi_a : B \to C$ given by $b \to [a, b]$. Then by the definition of k-root subgroups, χ_a is an isomorphism. Let $B_\circ = \chi_a^{-1}(C^\circ)$. Then $B^\circ \leq B_\circ$ and, to show that condition (1)(c) of the definition of ℓ-root subgroups holds for Σ°, we need to show $B^\circ = B_\circ$, i.e. $B_\circ = C_B(X)$.

Now for $\alpha \in X$ and $b \in B_\circ$ we have

$$\chi_a(b) = [a, b] = [a, b]^\alpha = [a^\alpha, b^\alpha] = [a, b^\alpha] = \chi_a(b^\alpha).$$

Hence $b = b^\alpha$ for each $\alpha \in X$, since χ_a is an isomorphism , and thus $B_\circ \leq B^\circ$.

Hence only the proof of the maximality condition for Σ°-subgroups of R remains. Now this proof depends on the following three tools:

1. Induction on the rank of G.
   Namely let $A, B \in \Sigma$ with $\langle A, B \rangle \simeq (P)SL_2(k)$ and $A^\circ = C_A(X) \neq 1$, $B^\circ = C_B(X) \neq 1$. Let $C_A = \langle C_\Sigma(A) \rangle$, $\Delta = C_\Sigma(A) \cap C_\Sigma(B)$ and $V = \langle \Delta \rangle$. Then $C_A = M_A \cdot V$, $M_A \trianglelefteq C_A$, $M_A \cap V = 1$ and M_A/A abelian. Now, by induction on rank G, which is defined differently in the degenerate and non-degenerate case, we may assume that (2.1) holds for the action of X on V.

2. Let $(U_i)_{i \in I}, I = \{1, 2, 3, ...\}$ be an ascending chain of Σ°-subgroups of R with $A^\circ \in \Sigma^\circ \cap U_1$. Then there exists a number r, depending only on G, with the property that

   $$(U_{j+1} - U_j) \cap \Omega_{A^\circ} \neq \emptyset \quad \text{for at most} \quad r \quad \text{elements} \quad j \in I$$

where

$$\Omega_{A^\circ} = \{D^\circ \in \Sigma^\circ \mid \langle A^\circ, D^\circ \rangle \simeq (P)SL_2(\ell)\}.$$

Namely, first one shows that M_A/A carries the structure of a finite dimensional kV-module, where the scalar action is given by $H = N_Y(A) \cap N_Y(B), Y = \langle A, B \rangle$. Let $s = \dim_k M_A/A$ and $M_A^\circ = C_{M_A}(X)$. Then one shows that M_A°/A° carries the structure of a finite dimensional ℓV°-module, where $V^\circ = \langle \Delta^\circ \rangle, \Delta^\circ = \{C_D(X) \mid D \in \Delta\}$ with scalar action given by $H^\circ = C_H(X)$ and

$$r = \dim M_A^\circ/A^\circ \le s.$$

Then one shows that (2) holds for the so defined r.

3. Now the final aim of the proof is to construct from the infinite ascending chain $(U_i)_{i \in \mathbb{N}}$, $A^\circ \in \Sigma^\circ \cap U_1$ of Σ°-subgroups of R with the help of (2) an infinite ascending chain $(V_i)_{i \in I}$ of Σ°-subgroups of $C_{A^\circ} = \langle C_{\Sigma^\circ}(A^\circ) \rangle$ and then use (1) and the finite dimensionality of M_{A°/A° to show that such a chain cannot exist. Let $U = \bigcup_{i \in \mathbb{N}} U_i$. Then one has as an extreme case to consider the possibility

$$\Omega_{D^\circ} \cap U = \emptyset \quad \text{for each} \quad D^\circ \in \Sigma^\circ \cap U.$$

In this case one wants to show that U is nilpotent and $\Sigma^\circ \cap Z(U) \ne \emptyset$ to embed the chain $(U_i)_{i \in \mathbb{N}}$ into C_{A°, $A^\circ \in \Sigma^\circ \cap Z(U)$. For this one uses a nilpotence criterion, which we state since it might be of independent interest.

2.3 Proposition

Let $\mathcal{W}$ be a set of nilpotent subgroups of the group G satisfying the following conditions:

1. G satisfies the maximality condition for $\mathcal{W}$-subgroups. ($\mathcal{W}$-subgroups are subgroups $U = \langle U \cap \mathcal{W} \rangle, U \cap \mathcal{W} = \{A \in \mathcal{W} \mid A \le U\}$.)

2. If $A, B \in \mathcal{W}$ then $\langle A, B \rangle$ is nilpotent.

3. If $A, B \in \mathcal{W}$ then $[A, B] \in \mathcal{W}$.

Then $\langle \mathcal{W} \rangle$ is nilpotent.

Notice that (1) is always satisfied if $|\mathcal{W}| < \infty$. In the above situation we apply (2.3) for

$$\mathcal{W} = \{A \in \Sigma \mid A^\circ = C_A(X) \in \Sigma^\circ \cap U\}.$$

The following corollary, which ought to be well-known, is an immediate consequence of (2.1).

2.4 Corollary

Let G be a Lie-type group over the arbitrary field k and Σ the conjugacy class of centers of the long root-subgroups of G. Call a Σ-subgroup U unipotent, if $\langle A, B \rangle$ is nilpotent for all $A, B \in U \cap \Sigma$. Then unipotent Σ-subgroups are nilpotent.

Namely by (2.1) G satisfies the maximality condition for Σ-subgroups. Hence (2.4) is a consequence of [16].

3 Proof of Theorem (1.1).

Let in this section Σ be a degenerate class of k-root subgroups of G, satisfying the hypothesis of (1.1). If $\Lambda \subset \Sigma$ with $\langle \Lambda \rangle$ abelian let

$$\underline{\Lambda} = Z(\langle C_\Sigma(\Lambda) \rangle) \cap \Sigma.$$

Then it is easy to see that $\underline{\underline{\Lambda}} = \underline{\Lambda}$. Abelian subsets Λ of Σ with $\Lambda = \underline{\Lambda}$ are called *subspaces*. The final aim for the proof of (1.1) is to show that the set of subspaces of Σ is a non-degenerate thick polar space of finite rank ≥ 2 in the sense of Tits-Veldkamp [21]. To be able to do this one has to perform a certain amount of group-theory, which I will describe in this section.

3.1 The radical $R(G)$

For $A, B \in \Sigma$ with $X = \langle A, B \rangle \simeq (P)SL_2(k)$ let $N_A = N_{\langle A \rangle}(\underline{B}), N_B = N_{\langle B \rangle}(\underline{A}), N = N_A N_B$ and $Y = \langle \underline{A}, \underline{B} \rangle$. Then the following hold:

1. $Y = N \cdot X$, $N \lhd Y$ and $N \cap X = 1$.

2. $N' \leq N_A \cap N_B \leq Z(Y)$.

3. $N/N_A \cap N_B$ is the direct sum of natural kX-modules.

4. N_A is independent of the choice of B.

To prove (3) one needs a criterion for an abelian group to be a direct sum of natural $kSL_2(k)$-modules, see [17,(2.7)], which is one of the main tools for the proof of (1.1).

Now one sets

$$R(G) = \langle N_A \mid A \in \Sigma \rangle$$

Then $R(G)$ is by (4) well defined and one has

5. $R(G) \triangleleft G$.

6. $\tilde{R} = R(G)/R(G) \cap Z(G)$ is abelian.

7. If $|k| \geq 4$, then $\bar{G} = G/R(G)$ is quasisimple.

8. $C_{\tilde{R}}(\bar{G}) = 0, \tilde{R} = [\tilde{R}, \bar{G}]$ and $[\tilde{R}, \bar{A}, \bar{B}] = 0$ for commuting $A, B \in \Sigma$.

As mentioned in Section 1, (8) determines the structure of $\tilde{R}$ as $k\bar{G}$-module, in case $\bar{G}$ is symplectic or unitary over fields.

3.2 Triangles.

Let $A, B, C \in \Sigma$ with $\langle A, B \rangle \simeq \langle B, C \rangle \simeq (P)SL_2(k)$ and $[A, C] = 1, A \neq C$. Then the triple (A, B, C) is called a *triangle* and the group $Y = \langle A, B, C \rangle$ a *triangle group*. The determination of the structure of a triangle group is another important ingredient of the proof of (1.1). One has:

Let $Y = \langle A, B, C \rangle$ be a triangle group, $N = R(Y)$ defined with respect to $\Delta = A^Y$, $X = \langle A, B \rangle$ and $Z = N \cap Z(Y)$. Then the following hold:

1. $Y = N \cdot X, N \cap X = 1$.

2. $N' \leq Z$ and N/Z is a natural kX-module.

3. Let $A_1 \neq A_2 \in A^N$ with $A_1 \neq A \neq A_2$. Then $\langle A^N \rangle = AA_1A_2 = AA_1Z$.

4. If $R(G) = 1$ and $|k| \geq 4$, then $A_2 \cap AA_1 = 1$ and $AA_1A_2 = AA_1(AA_1A_2 \cap Z)$.

5. If $R(G) = 1$ and $|k| \geq 4$, then $A^N \subseteq \{A, C\}$. Especially "lines" are thick.

3.3 The permutation action of G on Σ

Suppose from now on that G satisfies the additional assumptions of (1.1). For $A \in \Sigma$ let

$$
\begin{aligned}
\Sigma_A &= C_\Sigma(A) - A \\
\Omega_A &= \Sigma - C_\Sigma(A) \\
C_A &= \langle C_\Sigma(A) \rangle.
\end{aligned}
$$

Then one shows:

1. Σ_A is a class of k-transvections of $\langle \Sigma_A \rangle$.

2. C_A acts transitively on Ω_A.

3. C_A is Σ-maximal (i.e. maximal among Σ-subgroups!).

(1) and (2) of course imply that G acts as a rank 3 permutation group on Σ. The proof of (1)-(3) is similar to [11]. But instead of using subsets of maximal cardinality one uses subsets of maximal "rank", where the rank is defined via chains of subspaces contained in the subset. Moreover, if $|k| \geq 4$, the structure of the triangle groups shows in general that the lines are thick. This leads to the following conjecture.

3.4   Conjecture [3]

Suppose G is a group generated by a class Σ satisfying part (1) of the definition of k-transvections with $|k| \geq 4$. Suppose further that G satisfies (i) of (1.1), there exist commuting elements $A \neq B \in \Sigma$ and all ascending chains of subspaces of Σ are finite. Then the set of subspaces of Σ is a thick, non-degenerate polar space of finite rank ≥ 2.

By Tits's classification of polar spaces of finite rank ≥ 3 this should again determine G in case rank $\Sigma \geq 3$. But in contrast to (1.1) it seems impossible to prove the Moufang-condition in the rank 2-case, since this proof uses the global maximality condition.

3.5   Shult spaces

For commuting elements $A \neq B \in \Sigma$ let

$$\ell_{A,B} := \{\underline{A, B}\}.$$

The sets $\ell_{A,B}$ are called lines, while the elements of Σ are called points. One shows:

(+) The set of points and lines of Σ form a thick, non-degenerate Shult space of finite singular rank ≥ 2. (In the sense of [5]!).

The proof of (+) is based on (3.2) and (3.3)(1)-(3). Now it follows immediately from (+) and [5] that the set of subspaces of Σ is a polar space in the sense of [21]. Now (1.1) follows from the Tits-Veldkamp classification of polar spaces of finite singular rank ≥ 3.

---

[3]This conjecture has been proved in the meantime by my student Anja Steinbach.

3.6

In the rest of this section I will show, how one obtains the second possibility in (1.1)(1).

By (3.5) and section 7 of [17] we may from now on assume that the set $\mathcal{S}(\Sigma)$ of subspaces of Σ is a thick, non-degenerate polar space of finite singular rank ≥ 2, which satisfies the Moufang condition if rank $\mathcal{S}(\Sigma) = 2$. Thus we may assume rank $\mathcal{S}(\Sigma) \geq 3$. (Since in the rank 2-case we do not have a complete classification in (1.1)!) Now by [17] either case (5) of (1.1) holds or $\mathcal{S}(\Sigma)$ is embeddable in the sense of [21]. Let $X = \langle A, B \rangle$; A, B non-commuting elements of Σ. Then in various places of section 8 of [17] the following quotation appears:

$(*)$ $\qquad X^{\perp\perp} := C_{\Sigma}(C_{\Sigma}(X)) = X \cap \Sigma$

(by section 7 of [17]!). Now for $(*)$ one would need that $X \cap \Sigma$ is a full "hyperbolic" line, which has not been proved and which is wrong in the symplectic groups over non-perfect fields of char 2, when Σ is not the class of full transvection groups. One can close this gap as follows:

For $A \in \Sigma$ let

$$
\begin{aligned}
M_A &= R(C_A) \\
\bar{A} &= Z(C_A) \cap M_A.
\end{aligned}
$$

Then the following holds:

3.7

1. $(3.6)(*)$ holds if $A = \bar{A}$ for $A \in \Sigma$.

2. Suppose $\mathcal{S}(\Sigma)$ is an embeddable polar space of rank ≥ 3 and $A \neq \bar{A}$ for $A \in \Sigma$. Then $\bar{\Sigma} = \{\bar{A} \mid A \in \Sigma\}$ is a class of $\bar{k}$-transvections of G, where $\bar{k}$ is a field with $k \subseteq \bar{k}$.

3. Either $k = \bar{k}$ or $\bar{k}$ is non-perfect of characteristic 2, $\bar{k}^2 \subseteq k \subseteq \bar{k}$ and G is a symplectic group over $\bar{k}$.

We will prove (3.7)(1) and sketch the proof of (2) and (3).
Let X be as in (3.6)$(*)$ and $\Delta = C_{\Sigma}(X)$, $V = \langle \Delta \rangle$. Then by [17,(7.4)] $C_A = M_A V$. Hence $C_{\Sigma}(\Delta) \subseteq A \cup \Omega_A$. Let $C \in C_{\Sigma}(\Delta)$, $C \notin X \cap \Sigma$. Then $C \in \Omega_A$ and by [17,(7.4)] there exists a $m \in M_A$ with $C = B^m$. We show

$m \in \bar{A}$, which obviously proves (1).

Now

$$\Delta^m = C_{\Sigma_A}(B)^m = C_{\Sigma_A}(C) \supseteq \Delta.$$

Hence $V^m = V$ and $\Delta^m = \Delta$, since $\Delta = \Sigma \cap V$, by (3.3)(*).
Now $M_A \cap V \leq C_{M_A}(V)$, since Δ is the set of points of Σ_A perpendicular to B, and M_A fixes all lines through A by definition. Further, since $\Sigma_A = \Delta^{M_A}$ and since $M'_A \leq A$ by [17,(7.4)], we obtain $C_{M_A}(V) \leq \bar{A}$. Hence $M_A \cap V = 1$ by (3.3)(3). But then

$$[m, V] \leq V \cap M_A = 1 \quad \text{and} \quad m \in C_{M_A}(V) = \bar{A}$$

which is to show.

Now the proof of (1) shows that (3.6)(*) holds for $\bar{\Sigma} = \{\bar{A} \mid A \in \Sigma\}$, i.e.

$$(+) \qquad C_{\bar{\Sigma}}(C_{\bar{\Sigma}}(\bar{X})) = \bar{X} \cap \bar{\Sigma}, \bar{X} = \langle \bar{A}, \bar{B} \rangle, \text{ since } A \leftrightarrow \bar{A}, A \in \Sigma$$
is a $1 - 1$-correspondence.

Let now W be the natural module over some skew-field L on which a perfect central extension $\hat{G}$ of G acts as a group generated by transvections. (I.e. there is a class $\hat{\Sigma}$ of transvection groups generating $\hat{G}$, which projects onto Σ. See[17, sect.8].) Identify, being slighty incorrect, G with $\hat{G}$. Let v, w be vectors of W such that A consists of transvections corresponding to Lv and B of transvections corresponding to Lw. Then $\bar{A}$ centralizes $v^\perp$, and $\bar{B}$ centralizes $w^\perp$, since $\bar{A}$ centralizes $C_{\bar{\Sigma}}(A)$. Hence it is easy to see that $\bar{A}$ and $\bar{B}$ consist also of transvections corresponding to Lv resp. Lw. Let $H = Lv + Lw$. Then $W = H \oplus H^\perp$ and H is a hyperbolic plane. Now $(+)$ implies that for each singular point P of H there exists a $\bar{D} \in \bar{\Sigma} \cap \bar{X}$, such that $\bar{D}$ consists of transvections corresponding to P.

By the proof of (1) $\bar{A}$ acts transitively on $(\bar{\Sigma} \cap \bar{X}) - \bar{A}$. Let T_v, T_w be as in [17](8.1),(8.5) or (8.6). Then $\bar{A} \leq T_v$, $\bar{B} \leq T_w$. Suppose there exists a $t \in T_v - \bar{A}$. Then $(T_w)^t = T_{w^t}$ and w^t is a singular vector of H. Hence, by the above, there is a $\bar{D} \in \bar{\Sigma} \cap \bar{X}$ with $\bar{D} \leq T_w^t$. Thus, there exists an $\alpha \in \bar{A}$ with $\bar{B}^\alpha = \bar{D} \leq T_w^t$. This implies $T_w^\alpha = T_w^t$ and $t\alpha^{-1} \in N_{T_v}(T_w) = 1$, a contradiction to $t \notin \bar{A}$. This shows $\bar{A} = T_v$, $\bar{B} = T_w$ and by [17,sec.8] $\bar{X} \simeq SL_2(\bar{k})$ where $\bar{k}$ is a subfield of L. (In fact $\bar{k}$ is as k in Theorem (1.1) and L as $\bar{k}$ in that theorem!) Especially $\bar{\Sigma}$ is a class of $\bar{k}$-transvections of G.

This proves (2). Now we have

$$SL_2(k) \simeq X \leq \bar{X} \simeq SL_2(\bar{k}) \quad \text{and} \quad N_{\bar{X}}(\bar{A}) \leq N_G(A).$$

Hence $[17,(2.4)]$ implies either $A = \bar{A}$ and $k = \bar{k}$ or $\bar{k}$ is non-perfect of characteristic 2 and $\bar{k} \supseteq k \supseteq \bar{k}^2$. (The latter is not stated explicitly in $[17,(2.4)]$ but follows easily from the proof.)

Thus it remains to show that G is symplectic, when $k \neq \bar{k}$. Since (1) holds for $\bar{\Sigma}$, Theorem (1.1) holds with $\bar{\Sigma}$ and $\bar{k}$ in place of Σ and k. Obviously $C_A = C_{\bar{A}}$ and so $M_A = M_{\bar{A}}$. Now in all cases except in the symplectic groups of characteristic 2, we have $M'_{\bar{A}} = \bar{A}$, since $\bar{A}$ is a full root group of transvections and $M_{\bar{A}}$ the unipotent radical of $N_G(\bar{A})$. But then by $[17,(7.4)(3)]$

$$\bar{A} = M'_{\bar{A}} = M'_A \leq A$$

and thus $\bar{A} = A$ and $\bar{k} = k$. (In the symplectic groups in char 2 one has $M'_{\bar{A}} = 1$!)

4   Proof of Theorem 1.2.

We assume in this section that Σ is a non-degenerate class of k-root subgroups of G. Similary as in the degenerate case one can define a nilpotent normal "radical" $R(G)$ satisfying

(a) $G/R(G)$ is quasisimple.

(b) $\tilde{R} = R(G)/R(G) \cap Z(G)$ is abelian.

(c) $\tilde{R} = [\tilde{R}, G]$ and $C_{\tilde{R}}(G) = 1$.

(d) $[\tilde{R}, A, B] = 1$ for all commuting $A, B \in \Sigma$.

Here one does not need to distinguish between $|k| \leq 3$ and $|k| > 3$, since G is perfect anyway.

The aim of the proof is, as in section 3, to construct the spherical building $\mathcal{B}$ of G and show that Σ is the class of "centers of the long-root subgroups on $\mathcal{B}$". But in contrast to section 3 the construction of $\mathcal{B}$ differs in the different cases of (1.2), which explains the greater length and complexity of the proof. We will describe these constructions in this section. To be able to do this we need some notation.

For $A \in \Sigma$ let

$$
\begin{aligned}
\Sigma_A &= C_\Sigma(A) - A \\
\Omega_A &= \{B \in \Sigma \mid \langle A, B \rangle \simeq (P)SL_2(k)\} \\
\Psi_A &= \{B \in \Sigma \mid \langle A, B \rangle \ \text{special with} \ [A, B] \in \Sigma\} \\
\Lambda_A &= \{B \in \Sigma \mid B \in \Sigma_A \ \text{and} \ \Sigma \cap AB \ \text{is a partition of} \ AB\}.
\end{aligned}
$$

It is easy to see that, if $C \in \Psi_A$, $B = [A, C] \in \Lambda_A$. We use the following pictorial notation:

$$
\begin{array}{cc}
A & B \\
\circ & \circ
\end{array}
\qquad \text{if and only if} \qquad B \in \Sigma_A
$$

$$
\begin{array}{cc}
A & B \\
\circ\!\!-\!\!-\!\!-\!\!\circ
\end{array}
\qquad \text{if and only if} \qquad B \in \Omega_A
$$

$$
\begin{array}{cc}
A & B \\
\circ\!\!=\!\!=\!\!\circ
\end{array}
\qquad \text{if and only if} \qquad B \in \Psi_A.
$$

As in the case of the triangle groups in the degenerate case, the determination of certain 3-generator groups is important.

4.1

Let $A, B, C \in \Sigma$. Then

(a) If
$$
\begin{array}{ccc}
A & B & C \\
\circ\!\!=\!\!=\!\!\circ\!\!=\!\!=\!\!\circ
\end{array}
$$
, then $\langle A, B, C \rangle$ is nilpotent.

(b) If
$$
\begin{array}{ccc}
C & B & A \\
\circ\!\!-\!\!-\!\!\circ\!\!=\!\!=\!\!\circ
\end{array}
$$
and $X = \langle B, C \rangle, N = \langle A^X \rangle$ then the following
hold:

(i) $\langle A, B, C \rangle = N \cdot X, \quad N \cap X = 1$.

(ii) $N = C_N(X) \times [N, X]$ with $[N, X]$ the natural kX-module.

(iii) If $R(G) = 1$, then $\Sigma \cap N$ is a partition of N. Especially either $C_N(X) = 1$ or $C_N(X) \in \Sigma$.

By the next lemma one can define a rank function on G, which is closely connected but not equal to the Lie-rank of G.

4.2

Let N be a subgroup of G with $N' = 1$ and $\Sigma \cap N$ a partition of N. Then the following hold:

(a) N is a finite dimensional k-vector space with $\Sigma \cap N$ the point set.

(b) $\langle N_\Sigma(N)\rangle$ induces the $SL_n(k)$ on N, $n = \dim_k N$.

(c) If $A \in N_\Sigma(N) - C_\Sigma(N)$, then $[N, A] = N \cap \Lambda_A$ and A induces a root group of transvections on N.

From now on we assume that $R(G) = 1 = Z(G)$. The analysis of so called weak TI-subsets, gives a natural distinction between linear groups and certain groups of small rank and the other groups in (1.2). Here Λ is a *weak* TI-*subset* of Σ, if the following hold:

(i) $\Lambda \neq \Sigma$ and $|\Lambda| \neq 1$.

(ii) $\Lambda \cap \Lambda^d = \begin{cases} \emptyset \\ \Lambda \end{cases}$ for each $d \in D(\Sigma)$ where $D(\Sigma) = \{t \in A^\# \mid A \in \Sigma\}$.

Λ is a TI-subset if (ii) holds for all $g \in G$. One has:

4.3

The following are equivalent:

(a) $\langle C_\Sigma(A)\rangle$ is not Σ-maximal,

(b) There exists a weak TI-subset Λ of Σ.

4.4

Suppose case (b) of (4.3) holds and let Λ be a weak TI-subset of Σ and $N = \langle \Lambda \rangle$. Then:

(a) $N' = 1$ and Λ is a partition of N.

(b) Either Λ is a TI-subset of Σ or $\dim_k N \le 3$.

(c) If Λ is a TI-subset, then $G \simeq PSL_n(k)$, $n = \dim_k N + 1$.

(c) is proved by constructing a projective space with Λ^G as point set.

4.5

Suppose now Λ is a weak TI-set, but not a TI-set. Then one of the following holds:

(a) $\dim_k N = 2$ and case (c) of (1.2) holds.

(b) $\dim_k N = 3$ and G is orthogonal of Witt-index 3.

In (a) one shows that $(\Sigma, \Lambda^G, \in)$ is a thick generalized hexagon satisfying the Moufang condition. In (b) the construction of the polar space is more complicated, since Σ corresponds to the singular lines, and Λ^G to the maximal singular subspaces. So the points have to be "rediscovered".

Assume from now on that $C_A = \langle C_\Sigma(A) \rangle$ is Σ-maximal for $A \in \Sigma$. Let $B \in \Omega_A$ be fixed and

$$
\begin{aligned}
M_A &= \langle \Lambda_A \rangle \\
\Delta &= \Sigma_A \cap \Sigma_B \\
V &= \langle \Delta \rangle
\end{aligned}
$$

We have

4.6

The following hold:

(a) $M'_A \leq A \leq Z(M_A)$

(b) $C_A = M_A \cdot V$, $M_A \cap V = 1$ and Δ is a set of k-root subgroups of V.

(c) M_A acts regularly on Ω_A.

(4.6)(a) is one of the main tools for the proof of (1.2). It follows directly from the definition of M_A and [16] that M_A is nilpotent. But the proof that the class of M_A is equal to 2 is long and complicated. I am not able to prove (c) in greater generality (i.e. without assuming that C_A is Σ-maximal). The reason is that there is a counterexample, namely $U_3(3) = G_2(2)'$, in which case $|M_A| = 16$ but $|\Omega_A| = 32$.

4.7

Suppose that Δ is not a conjugacy class in V. Then G is an orthogonal group of Witt-index ≥ 4 and Σ is the class of Siegel transvection groups.

The proof of (4.7) acutally splits into two cases. Namely one shows that one of the following holds:

(a) $V = V_1 * V_2$, $V_1 \simeq SL_2(k)$, $V_2 \not\simeq SL_2(k)$ $\qquad$ or

(b) $V = V_1 * V_2 * V_3$, $V_i \simeq SL_2(k)$ for $i = 1, 2, 3$.

Now in case (a) one constructs a certain V_2-invariant maximal commuting subset of Σ and shows that the conjugacy class of these subsets taken as points, and the elements of Σ taken as lines form a polar space.

In case (b) one could argue similarly, constructing a certain $V_2 * V_3 \simeq \Omega^+(4, k)$-invariant maximal commuting subset of Σ, if one would know that $V_2 * V_3$ is invariant under $N(A) \cap N(B)$. To show this amounts to showing that the triality automorphism which permutes the V_i is not inner, for which I don't have any tool. Thus I use instead the theory of parabolic systems and classical Tits chamber systems [15], [22] to determine G in case (b). This is to my knowledge the first application of this theory to "arbitrary" groups.

From now on we assume that Δ is a conjugacy class in V. This final case obviously splits into two subcases:

1. Δ is a degenerate class of k-root subgroups of V.

2. Δ is non-degenerate.

We have

4.8

Suppose 1. holds. Then case (d) of Theorem (1.2) is satisfied.

Let $\mathcal{L} = \{CD \cap \Sigma \mid C \in \Sigma, D \in \Lambda_C\}$. Then one shows, to prove (4.8), that $(\Sigma, \mathcal{L}, \in)$ satisfies the point-line axiom system of [6] for metasymplectic spaces. From this point (4.8) is essentially a consequence of [21]. (One has to show that the elements of Σ act as long- root subgroups on the corresponding building of type F_4.)

Finally one shows:

4.9

Suppose 2. holds. Then case (e) of Theorem (1.2) is satisfied.
Here one shows that $(\Sigma, \mathcal{L}, \in)$, $\mathcal{L}$ as above, satisfies the hypothesis of Theorem 2 of [7]. From this point the proof that case (e) of (1.2) holds is similar as in (4.8).

5   Some remarks.

There are certain natural questions which arise with the proofs of (1.1) and (1.2), on which I like to comment in this final section.

5.1   Generalization to skew fields?

Here the question is: Is it possible to allow k to be a skew field in (1.1) and (1.2)? This is a natural question since the main tool of the proof a property of $SL_2(k) = X$; namely:

   1. Let $A \neq B$ be two full unipotent subgroups of X. Then for each $a \in A$ there exists a unique $b = b(a) \in B$ with $a^b = b^{-a}$.

also holds for skew fields. On the other hand many properties of the classical groups and linear algebra is used, which then would have to be generalized to skew fields.

5.2   Complete the classification in case (6) of (1.1).

If the classification of Moufang generalized quadrangles announced in [24] is written down, this would amount to show that the only automorphism groups of quadrangles of "mixed type" arising in our situation occur already in case (4) of (1.1). On the other hand, there might be a direct proof for this, since our conditions are stronger than the Moufang condition.

5.3   Alternative classification of spherical buildings of rank ≥ 3?

Here the question is: Given a thick spherical building $\mathcal{B}$ of rank ≥ 3. Does there necessarily exist a field k such that the class Σ of centers of the long root subgroups of $Aut(\mathcal{B})$ is a class of k-root subgroups of $\langle \Sigma \rangle$ in our sense? If this would be the case, it would be worthwhile to make the proofs of (1.1) and (1.2) independent of the classification of spherical buildings by using instead certain graph methods as in [14]. This is certainly possible, but the proof would be much lengthier.

On the other hand the answer to the question is certainly no, since for example in the case of polar spaces of infinite dimension but finite Witt index the maximality condition is not satisfied. (Also there exist examples in which k is a skew field.) But still, when $\mathcal{B}$ is finite, the answer is yes and this should not be too difficult to show independently of the classification of spherical buildings. Thus the question should be specified to: Do there exist natural conditions on $\mathcal{B}$, more general than finiteness, such that Σ is a class of k-root subgroups of $\langle \Sigma \rangle$ under these conditions?

5.4 Replace the maximality condition by some handier condition

In full generality this seems very difficult. But for example as in (3.4) a finite rank condition might suffice.

References

[1] Aschbacher, M.: Finite groups generated by odd-transpositions I, Math.Z.127, 45-56; (1972), II, III, IV J. Algebra 26, (1973).

[2] Aschbacher, M.: A characterization of the unitary and symplectic groups over finite fields of characteristic at least 5. Pac. J. Math.47, 5-26, (1973).

[3] Aschbacher, M. and Hall, M.: Groups Generated by a Class of Elements of order 3. J. Algebra 24, 591-612, (1973).

[4] Bruck, R. H.: A Survey of Binary Systems. Ergebnisberichte Berlin, Heidelberg, New York, Springer (1958).

[5] Buekenhout, F. and Shult, E.E.: On the Foundations of Polar Geometry. Geom.Dedicata 3 , 155-170, (1974).

[6] Cohen, A. M.: An Axiom System for Metasymplectic Spaces. Geom.Dedicata 12, 417-433, (1982).

[7] Cohen, A. M. and Cooperstein, B.N.: A Characterizion of some Geometries of Lie-type. Geom. Dedicata 15, 73-105, (1983).

[8] Cooperstein, B. N.: The Geometry of root subgroups in exceptional groups I. Geom. Dedicata 8, 317-381, (1979). II Geom. Dedicata 15, 1-45 (1983).

[9] Dieudonné, J.: La géométrie des groupes classiques. Ergebnisberichte 14, Berlin, Heidelberg, New York, Springer (1957).

[10] Faulkner, J. R.: Groups with Steinberg relations and coordinatization of polygonal geometries. Mem. Am. Math. Soc. 185 (1977).

[11] Fischer, B.: Finite groups generated by 3-transpositions. Mim. Notes. Univ. of Warwick (1969) (unpublished) and Inv. Math. 13, 232-246, (1971).

[12] Thompson, J.: Quadratic pairs (unpublished)

[13] Timmesfeld, F. G.: A Characterization of Chevalley and Steinberg groups over F_2. Geom. Dedicata 1, 269-323, (1973).

[14] Timmesfeld, F. G.: Groups generated by Root-Involutions I. J. Algebra 33, 75-134 (1975), II J.Algebra 35, 367-441, (1975).

[15] Timmesfeld, F. G.: Tits geometries and parabolic systems in finitely generated groups I, II. MZ.184, 377-396, 449-487, (1983).

[16] Timmesfeld, F. G.: A remark on a theorem of Baer. Arch. Math.54, 1-3, (1990).

[17] Timmesfeld, F. G.: Groups generated by k-transvections. Inv. Math. 100, 167-206, (1990).

[18] Timmesfeld, F. G.: On the Identification of natural modules for symplectic and linear groups defined over arbitrary fields. Geom. Dedicata 35, 127-142, (1990).

[19] Timmesfeld, F. G.: Groups generated by k-root subgroups. To appear in Inv. Math.

[20] Timmesfeld, F. G.: Automorphisms of groups generated by k-root subgroups. Preprint.

[21] Tits, J.: Buildings of spherical Type and finite BN-pairs. Lecture Notes in Math. 386, Springer (1974).

[22] Tits, J.: "A local Approach to Buildings" in The Geometric Vein, The Coxeter Festschrift, Springer, 519-547, (1981).

[23] Tits, J.: Classification of Algebraic Semisimple Groups. Proc. Symp. Pure Math. IX, 33-62, (1966).

[24] Tits, J.: Classification of buildings of spherical type and Moufang polygons: A survey Atti. de Conv.Linea 17. In "Theorie Combinatorie", 230-246, (1976).

[25] Weiss, R.: On Fischer's characterisations of $SP_{2n}(2)$ and $U_n(2)$. Com. Alg.11, 2527-2554, (1983).

[26] Weiss, R.: A uniqueness lemma for groups generated by 3-transpositions. Proc. Comb.Phil.Soc.97, 421-431, (1985).

Finiteness questions for geometries

PETER J. CAMERON

This article discusses some results and open questions on the general theme: if we assume that some or all residues of small rank of a Buekenhout geometry are finite, under what conditions may we conclude that the geometry is finite?

1. RANK 2 GEOMETRIES

A rank 2 geometry is simply an incidence structure consisting of a set of "points" and a set of "lines", with a relation of "incidence" between them. The *incidence graph* of a rank 2 geometry is the bipartite graph whose vertices are the points and lines, a point and line being adjacent if they are incident.

In the remainder of this paper, most of the rank 2 geometries I consider are drawn from several important classes:

1. Digons. A digon is a geometry in which any point is incident with any line. Such a geometry is completely trivial; but, as we shall see, digons play an important rôle in the theory of geometries of higher rank.

2. Partial geometries. The definition is due to Bose (1963), though Bose's notation is different. A partial geometry is characterised by the properties

    (i) any line is incident with $s + 1$ points, and any point with $t + 1$ lines;

    (ii) two points are on at most one line;

    (iii) if the point p is not incident with the line l, then p is collinear with exactly α points of l.

Here s, t, α are usually positive integers, though at the end of this section we shall briefly allow one of them to be infinite. Various specialisations of partial geometries are important:

A *linear space* is a partial geometry with $\alpha = s + 1$. It is characterised by the properties that any line has $s + 1$ points, and any two points lie on a unique line. (In another language, it is a 2-(v, k, λ) design, where $v = 1 + s(t + 1)$.) Special cases of linear spaces include *projective planes* ($\alpha = s + 1 = t + 1$: two lines meet in a unique point) and *complete graphs* ($s = 1, \alpha = 2$: any line has two points).

A *generalized quadrangle* is a partial geometry with $\alpha = 1$.

3. Generalized polygons. A generalized n-gon is a rank 2 geometry whose incidence graph has diameter n and girth $2n$, and in which, given any element x (point or line), there is an element y opposite x (at distance n from x in the incidence graph). For $n = 2, 3, 4$ we obtain digons, projective planes, and generalized quadrangles respectively. We assume the existence of parameters s and t such that any line has $s + 1$ points and any point lies on $t + 1$ lines. (This condition automatically holds provided that the geometry is *thick*, i.e., any line is incident with at least three points and dually.)

Recent work on geometries related to the Monster (as discussed at the Symposium by Sasha Ivanov, Peter Rowley and others) suggests that we add two more specific geometries to this list, namely

(i) the Petersen graph (whose points and lines are the vertices and edges of this celebrated graph) with $s = 1, t = 2$, and

(ii) a geometry with 45 points and 45 lines, having $s = t = 2$, which is a triple cover of the unique generalized quadrangle with $s = t = 2$, and admits the group $3.S_6$.

Some time ago, I raised the question:

Problem 1.1. Is there a generalized n-gon ($n > 2$) with s finite ($s > 1$) and t infinite?

An easy argument shows that, if n is odd, then necessarily $s = t$. I showed that there is no generalized quadrangle with $s = 2$ and t infinite; and the same conclusion for $s = 3$ was established by Bill Kantor and simplified by Andries Brouwer. And there matters rest. The problem (for quadrangles with $s = 2$) originally arose in the context of infinite permutation groups. The next case required for this application is not a polygon, but a partial geometry. So I propose:

Problem 1.2. Is there a partial geometry with s finite, $\alpha < s$, and t infinite?

Note that partial geometries with $\alpha = s + 1$ are linear spaces, and exist for any finite s and infinite t; and partial geometries with $\alpha = s$ are "transversal designs" or "dual nets" (whose existence is equivalent to that of $s-1$ mutually orthogonal Latin squares of order $t + 1$), and these also exist for all finite s and infinite t. So the first open case, namely $s = 3, \alpha = 2$, is precisely what is required for the application.

Another problem, not really relevant to finiteness questions but important in the study of flag-transitive geometries, is the following:

Problem 1.3. Determine the pairs $(\mathcal{G}, G)$, where $\mathcal{G}$ is a finite generalized quadrangle and G a group of automorphisms of $\mathcal{G}$ acting transitively on flags (incident point-line pairs).

2. GEOMETRIES OF FINITE RANK

I begin with a brief summary of the definition of a Buekenhout geometry (Buekenhout 1979). The elements (called *varieties*) are of several different types, described by a *type map* from the set X of vertices to the finite set Δ of *types*. (The *rank* is the number of types.) There is also a reflexive and symmetric *incidence relation* I on X. A *flag* is a set of mutually incident varieties. We assume *transversality*, i.e. a maximal flag contains one variety of each type. (This implies that incident varieties of the same type are equal.) The *type* of a flag is its image under the type map, and its *cotype* is the complement of its type in Δ. *Rank* and *corank* are the cardinalities of type and cotype.

The *residue* $\mathcal{R}(F)$ of a non-maximal flag F is the set of varieties not in F which are incident with every variety in F. The *type* of $\mathcal{R}(F)$ is the cotype of F, and the *rank* its cardinality. We assume *residual connectedness*, i.e. any residue of rank at least 2 (regarded as a graph, with varieties as vertices and incidence as adjacency) is connected. All the above assumptions are inductive, i.e. they hold in every residue of rank at least 2. (In order to have any chance of deducing finiteness from purely local hypotheses, some connectivity condition is obviously necessary; residual connectedness is convenient because it is inductive, but it is possible to get away with less in some cases.)

A *diagram* is a labelling of the edges of the complete digraph on Δ with classes of rank 2 geometries. (We assume that the geometries labelling (j, i) are the duals of those labelling (i, j).) A geometry *belongs to* the diagram if any rank 2 residue with points of type i and blocks of type j belongs to the class labelling (i, j).

It is customary to represent a diagram pictorially, by associating a symbol with each class of rank 2 geometries. For most of the classes in the last section, there are conventional symbols: a digon is represented by the absence of an edge; a projective plane by a single edge o——o, a generalized quadrangle by a double edge o═══o, and other classes by using an ornament to label the edge. Thus a linear space, circle, or partial geometry is represented by o—L—o, o—c—o, or o—pg—o respectively; a generalized n-gon by o—(n)—o, and the Petersen graph and the triple cover of the 15-point quadrangle by o—P—o and o═$\sim$═o respectively. The dual of a geometry is denoted by reversing, or affixing an asterisk to, the symbol for the geometry; thus o—L^*—o denotes the class of dual linear spaces. (For self-dual classes, this is not necessary.) Sometimes we write the names of the types (point, line, ... or $0, 1, 2, \ldots$) above the corresponding nodes of the diagram.

For example,

$$\underset{\text{point}}{\circ} \overset{\mathcal{K}}{\text{———}} \underset{\text{line}}{\circ} \overset{\mathcal{L}}{\text{———}} \underset{\text{plane}}{\circ}$$

is the class of rank 3 geometries in which the points and lines incident with any plane form a $\mathcal{K}$-geometry, the lines and planes incident with any point form a $\mathcal{L}$-geometry, and any point and plane incident with a common line are incident with one another. (Note the rôle of the digon here.)

We usually assume the existence of *parameters* or *orders*, integers s_i $(i \in \Delta)$ such that the residue of a flag of cotype i has cardinality $s_i + 1$. The geometry is called *firm* (resp. *thick*) if $s_i \geq 1$ (resp. $s_i > 1$) for all $i \in \Delta$. We avoid the difficulties of the last section by assuming that all parameters are finite. Note, however, that Problems 1.1 and 1.2 can be generalised:

Problem 2.1. When can it happen that all but one of the parameters of a geometry are finite and the remaining parameter is infinite?

In the pictorial representation, parameters are written underneath the corresponding node. Thus the Petersen graph is

$$\underset{1}{\circ} \overset{P}{\text{———}} \underset{2}{\circ} \, .$$

3. STRONG FINITENESS

One very important question concerning Buekenhout geometries is the following.

Problem 3.1. For which diagrams does it hold that any geometry belonging to the diagram and having all its rank 2 residues finite is itself finite?

Of course, this question is not sensible in complete generality; we should restrict the diagrams to familiar ones which have been studied (for example, let the classes of rank 2 geometries used be generalized n-gons for various n, linear spaces, partial geometries, etc.).

Most finiteness theorems which are known are "strong" in the sense that they assert that some diameter of a geometry belonging to a particular diagram is bounded by a function of the diagram alone. For a simple example, consider geometries belonging to the diagram

$$\underset{1}{\circ}\!\!\overset{L}{\underline{\qquad}}\!\!\underset{2}{\circ}\!\!\overset{L}{\underline{\qquad}}\!\!\underset{3}{\circ}\ \ldots\ \underset{n-1}{\circ}\!\!\overset{L}{\underline{\qquad}}\!\!\underset{n}{\circ}$$

It is well known that these are matroids of rank $n + 1$. So, if the left-most varieties are points and the next are lines, two points lie on a unique line (the diameter with respect to points and lines is 1). Suppose that all rank 2 residues are finite. Then (by induction) a point lies on finitely many lines, each of which contains finitely many more points; so the geometry is finite.

This is trivial and well-known. Also familiar is the fact that *Tits geometries* (Buekenhout geometries in which all rank 2 residues are generalized polygons) belonging to spherical diagrams have bounded diameter, and so are finite if all rank 2 residues are finite. (See Ott and Ronan (1981).)

Recently, Sergei Tsaranov (to appear) has provided a general framework in which results of this sort can be proved. In order to outline his method, I must describe the relation between geometries and chamber systems.

A *chamber* in a geometry is a maximal flag. Let $\mathcal{C}$ be the set of chambers. On $\mathcal{C}$, we can define equivalence relations R_i for $i \in \Delta$, by the rule

$$R_i = \{(C, C') : C, C' \in \mathcal{C}, \quad (C)_j = (C')_j \text{ for all } j \neq i\},$$

where $(C)_j$ is the unique variety of type j in C. If $(C, C') \in R_i$, we say that C and C' are *i-adjacent*, and sometimes write $C \sim_i C'$. The set $\mathcal{C}$, equipped with the relations R_i for $i \in \Delta$, is a fairly typical example of a *chamber system*. If the geometry has parameters $(s_i : i \in \Delta)$, then each equivalence class of R_i has size $s_i + 1$.

Remarks: 1. Not every chamber system comes from a geometry. Another important type of chamber system consists of Latin squares, where the chambers are the cells, and the three equivalence relations are "same row", "same column", and "same entry".

2. There is a canonical way to try to recover a geometry from a chamber system: the varieties of type i are the equivalence classes of the transitive closure of $\bigcup_{j \neq i} R_j$, and incidence is defined by non-empty intersection. If a chamber system comes from a geometry $\mathcal{G}$, this procedure gives a geometry $\hat{\mathcal{G}}$, which may not be the same as $\mathcal{G}$ (but is at least as "well-connected", and gives rise to the same chamber system as $\mathcal{G}$ does).

3. Residual connectedness of a geometry implies connectedness of its chamber system (but not conversely). The latter condition can often be used in place of the former.

Returning to the chamber system of a geometry: Let M be the monoid generated by the relations R_i $(i \in \Delta)$, under the operation of composition of relations.

Observation: If M is finite, and if the geometry has finite parameters, then it is finite. (For, if the geometry has finite parameters, then an easy induction shows that, for any $X \in M$ and $C \in \mathcal{C}$, the set

$$\{C' \in \mathcal{C} : (C, C') \in X\}$$

is finite.)

In order to exploit this observation, we note that certain equations are satisfied by the R_i as a consequence of the diagram. So we define a monoid M^* by the presentation

$$M^* = \langle r_i \, (i \in \Delta) \; : \; \Phi \rangle,$$

where Φ is a convenient set of monoid equations $u(r_1, r_2, \ldots) = v(r_1, r_2, \ldots)$ for which $u(R_1, R_2, \ldots) = v(R_1, R_2, \ldots)$ holds in consequence of the diagram. Then there is a surjective homomorphism from M^* to M. Hence, if M^* is finite, then so is M.

In fact, there is a more subtle way to show that M is finite. An *attractor* in M is an element X such that $R_i X = X R_i = X$ for all $i \in \Delta$. Note that if M is finite then it has an attractor, namely the universal relation on $\mathcal{C}$. (This follows from the connectedness of $\mathcal{C}$, which, as we remarked, follows from the residual connectedness of the geometry.) Conversely, the universal relation is the only possible non-empty attractor. Using this, it can be shown that, if the monoid generated by every subset of the R_i has an attractor, then M is finite. Furthermore, if M^* has an attractor, then so does M. So the existence of attractors in M^* suffices to show finiteness of M; and this condition is genuinely weaker than finiteness of M^*. (*Remark*: If the geometry has finite parameters, then the existence of an attractor in M^* forces finiteness of the geometry. In general chamber systems, it is possible for M to have an attractor but a submonoid generated by a subset of the relations to fail to have one; I do not know whether this is possible in a residually connected geometry. Clearly, there is more to discover here!)

The first obvious class of relations consists of the equations

$$r_i^2 = r_i$$

for all $i \in \Delta$. These hold because the R_i are equivalence relations. I will call these the *idempotence relations*.

For the next class of equations, we consider the rank 2 residues. If $\mathcal{K}$ is a class of rank 2 geometries, we define

$$d_p(\mathcal{K}) = \max\{d(x,y) : x, y \text{ varieties of some } \mathcal{G} \in \mathcal{K}, \quad x \text{ is a point}\},$$

and

$$d_l(\mathcal{K}) = \max\{d(x,y) : x, y \text{ varieties of some } \mathcal{G} \in \mathcal{K}, \quad x \text{ is a line}\},$$

where d denotes the distance in the incidence graph of $\mathcal{G}$. (These diameters were first introduced by Buekenhout (1983).) Then, for example, generalized n-gons have $d_p = d_l = n$; linear spaces which are not projective planes have $d_p = 3, d_l = 4$; and partial geometries which are not linear spaces or dual linear spaces have $d_p = d_l = 4$. Note that we always have either $d_p = d_l$ or $\{d_p, d_l\} = \{2m - 1, 2m\}$ for some integer m. The pair (d_p, d_l) is the *diameter pair* of $\mathcal{K}$.

Now we can imagine that each ordered edge of the diagram carries the ordered pair of numbers (d_p, d_l).

If a rank 2 geometry $\mathcal{G}$ has $d_p = d$, then $R_1 R_2 R_1 \ldots$ (d terms) is the universal relation, and so is equal to each of the four expressions obtained by multiplying on the left or right by R_1 or R_2 (though two of these four equations are consequences of the idempotence relations). Similarly, if $d_l = e$, then we obtain five more expressions all equal to the universal relation and hence equal to the previous ones. These equations, with r_1 and r_2 substituted for R_1 and R_2, are the *rank 2 diameter relations* for $\mathcal{G}$. Now the rank 2 diameter relations for a diagram of arbitrary rank consist of all such relations arising from the diameter pairs of the classes of rank 2 residues.

Tsaranov determined all cases of rank at least 3 where the monoid M^* defined by the idempotence and rank 2 diameter relations is finite or has an attractor.

In the special case where $d_p = d_l$ holds for each edge of the diagram, the monoid M^* has the same order as the *Coxeter group* of the corresponding diagram; so it is finite if and only if the diagram is one of the familiar *spherical Coxeter diagrams* describing finite Euclidean reflection groups, viz. $A_n, C_n, D_n, E_6, E_7, E_8, F_4, H_3$ and H_4 (Coxeter and Moser 1957). This generalises the result of Ott and Ronan cited earlier. Consider, for example, the diagram

$$\circ\!\!-\!\!\!-\!\!\!-\!\!\circ\overset{pg}{-\!\!\!-\!\!\!-}\circ$$

investigated by Dan Hughes (personal communication). Any geometry with this diagram must have point-diameter at most 2, since the Coxeter complex has point-diameter 2. (This means that, given any two points, either they are collinear, or there is a point collinear with both; this can be interpreted in the chamber system, and hence in the monoid.)

Apart from this, there are very few cases where finiteness can be proved in this way. One of these is that of a "linear" diagram in which each stroke has $d_p = 3$ and $d_l = 4$ (as in our matroid diagram earlier).

Problem. Is it possible to prove further finiteness theorems by adding further relations, for example, relations in 3 variables derived from properties of the rank 3 residues?

For example, geometries with diagram

$$\overset{P}{\underset{1\quad\quad 2\quad\quad 2}{\circ\!-\!\!-\!\!-\!\circ\!-\!\!-\!\!-\!\circ}}$$

have been determined by Ivanov and Shpectorov (1988). All are finite. The finiteness here definitely does *not* follow from the Tsaranov relations. Nevertheless, if we consider the diagram

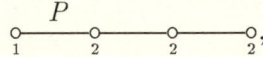

it might be possible to prove finiteness by adjoining the entire multiplication table of each (in turn) of the monoids for the left-hand subdiagram, as relations on the first three generators.

I am grateful to Sasha Ivanov, Leonard Soicher and Graham Weetman for their comments, which have significantly improved this section.

4. WEAK FINITENESS

There are a few known cases where finiteness can be proved, but no diameter bound can be established since infinite geometries in the class may have infinite diameter. These are cases in which there is a bound for some diameter associated with the geometry in terms of the parameters. No general theory is available, but results of this sort have been proved for some specific diagrams.

For example, consider geometries with diagram

$$\overset{c}{\circ\!\!-\!\!-\!\!\circ}\overset{c^*}{-\!\!-\!\!\circ},$$

where as usual $\circ\!\!\overset{c}{-\!\!-\!\!}\circ$ is the geometry whose lines are all pairs of points, and $\circ\!\!\overset{c^*}{-\!\!-\!\!}\circ$ is its dual. These geometries are the *semibiplanes*, connected incidence structures in which two points lie in 0 or 2 blocks and dually. (The varieties are points, lines and blocks, where a line can be identified with two points incident with two blocks.) It is well known that the incidence graph of a semibiplane has constant valency $s + 2$, so that the parameters are

$$\underset{1}{\overset{c}{\circ}}\!\!-\!\!-\!\!\underset{s}{\overset{}{\circ}}\!\!-\!\!-\!\!\underset{1}{\overset{c^*}{\circ}},$$

and that the diameter of the incidence graph is at most $s + 2$. So a semibiplane with finite parameters is finite, but the diameter bound involves the parameters.

I would like to make a little digression here. Following Neumaier (1982), a *rectagraph* is a connected graph without triangles, in which any path of length 2 is contained in a unique 4-cycle. A rectagraph gives rise to a geometry with diagram $\underset{1}{\overset{}{\circ}}\!\!=\!\!=\!\!\underset{2}{\overset{}{\circ}}\!\!\overset{c}{-\!\!-\!\!}\underset{3}{\overset{}{\circ}}$, in which varieties of type 1, 2, 3 are vertices, edges, and 4-cycles respectively. The incidence graph of a semibiplane is the same thing as a bipartite rectagraph. Now it is readily shown that a rectagraph is regular; if its valency is n, then its diameter is at most n and the number of vertices at most 2^n. The extremal graph for both bounds is the n-cube Q_n.

This observation was further generalized in the following theorem, proved by Peter Frankl and Richard Wilson in 1984 but (I believe) not published by them:

Theorem 4.1. *Let Γ be a connected, locally finite graph having the property that, for all positive integers t, there do not exist t vertices having precisely $t - 1$ common neighbours. Then Γ is regular (of valency n, say); its diameter is at most n and it has at most 2^n vertices, with equality (in either bound) if and only if it is the n-cube.*

Another family of geometries for which such a theorem holds are the "extended generalized quadrangles" or *EGQs*, discussed by Cameron, Hughes and Pasini (to appear) and others. These have the diagram $\circ\!\!\overset{c}{\rule{1.2cm}{0.4pt}}\!\!\circ\!\!=\!\!=\!\!\circ$. (Note the tantalising similarity with the last case.) The parameters exist, and are $(1, s, t)$, where (s, t) are the parameters of a GQ. In an EGQ, varieties of type 2 can be identified with the edges of a graph on the set of points, and varieties of type 3 correspond to certain cliques in this *point graph*; the *point diameter* is the diameter of the point graph. Cameron *et al.* and Del Fra and Ghinelli (to appear) showed:

Theorem 4.2. *(i) An EGQ with parameters $(1, s, t)$ has point diameter at most $s + 1$.*

(ii) Suppose that $s \geq 3$ and that the point diameter is $s + 1$. Then there are at most $\binom{2s+2}{s+1}$ points, with equality only for the Johnson geometry.

The Johnson geometry has as points the $(s + 1)$-subsets of a $(2s + 2)$-set, two points adjacent if their intersection has cardinality s; the blocks are the two types of maximal cliques in the graph.

The geometries meeting the bound for small s are easily found. If $s = 2$ then $t = 1, 2$ or 4, and there is a unique geometry of diameter 3 for each value. These geometries are affine polar spaces (note that $\circ\!\!\overset{c}{\underset{1}{\rule{1.2cm}{0.4pt}}}\!\!\underset{2}{\circ}$ is an affine plane), with 20, 32 and 56 points respectively. They are double covers of one-point extensions of GQs. See Cameron *et al.* for details.

Any EGQ with $s = 1$ is complete tripartite on $3t + 3$ points: that is, the points fall into three classes of size $t + 1$, and the blocks are all transversals of these classes.

Note that, apart from small cases, the bound can only be attained when $t = 1$. Better bounds are available for $t > 1$, but they are complicated to state; see the cited papers.

Both of these results can be extended. Dan Hughes has shown recently that weak finiteness theorems hold for the diagrams

$$\circ \!\!\!\!\xrightarrow{\quad L \quad} \!\!\!\!\circ \!\!\!\!\xrightarrow{\quad pg \quad} \!\!\!\!\circ$$

and

$$\circ \!\!\!\!\xrightarrow{\quad pg \quad} \!\!\!\!\circ \!\!\!\!\xrightarrow{\quad L \quad} \!\!\!\!\circ \ .$$

On the other hand, the tessellation of the Euclidean plane by squares, and numerous affine buildings, show that no finiteness theorem holds for the diagram $\circ \!\!=\!\!\!=\!\! \circ \!\!=\!\!\!=\!\! \circ$.

Problem 4.3. Does weak finiteness hold for the diagram

$$\circ \!\!\!\!\xrightarrow{\quad pg \quad} \!\!\!\!\circ \!\!\!\!\xrightarrow{\quad pg \quad} \!\!\!\!\circ$$

if at least one (or perhaps both) of the strokes are restricted to partial geometries with $\alpha > 1$?

It seems very likely that, at least in the case of nets ($\alpha = t$) or dual nets ($\alpha = s$), such a theorem should be true.

ORBIT THEOREMS

It is a well-known consequence of Block's Lemma (1967) that a group of automorphisms of a finite projective plane has equally many orbits on points and lines. Some years ago, Bill Kantor raised the question:

Problem 5.1. Does the same conclusion hold for infinite projective planes?

I see little hope of proving this in general; indeed, it may well be false, even for quite restricted classes of planes. There seem to be two possible techniques which might work in some cases.

The first technique would attempt to mimic the finite proof, replacing counting by integration, and the adjacency matrix by a Radon transform. (For example, the method might work for projective planes over a compact, or possibly locally compact, field.) But this method has a built-in limitation:

only measurable orbits could be handled (so very wild groups or automorphisms, if such exist, could not be dealt with), and orbits which are null sets would be invisible.

In connection with the last point, Buekenhout (1968) gives an example (due to G. Valette) of a linear space with a group of automorphisms having more point-orbits than line-orbits. (This cannot happen in the finite case, by Block's Lemma.) The example consists of the real hyperbolic plane with its boundary adjoined, with the usual lines. He shows that real points and points at infinity can be distinguished geometrically, and so lie in different orbits (in fact there are just two orbits.) On the other hand, the group is transitive on lines. In accordance with the speculations above, note that one of the point orbits is a null set.

The other possible line of attack would be to approximate infinite structures by finite ones. For example, the projective plane over a local field could be approximated by Hjelmslev planes over finite local rings, and we can do counting arguments in these finite planes.

REFERENCES

Block, R. E. (1967), On the orbits of collineation groups, *Math. Z.* **96**, 33–49.

Buekenhout, F. (1968), Remarques sur l'homogénéité des espaces linéaires et des systèmes de blocs, *Math. Z.* **104**, 144–146.

Buekenhout, F. (1979), Diagrams for geometries and groups, *J. Combinatorial Theory* (A) **27**, 121–151.

Buekenhout, F. (1983), (g, d^*, d)-gons, *Finite Geometries* (ed. N. L. Johnson, M. J. Kallaher & C. T. Long), pp. 93–111, *Lecture Notes Pure Appl. Math.* **82**, Marcel Dekker, New York.

Cameron, P. J., Hughes, D. R. & Pasini, A. (to appear), Extended generalized quadrangles, *Geometriae Dedicata*.

Coxeter, H. S. M. & Moser, W. O. J. (1957), *Generators and Relations for Discrete Groups*, Springer, Berlin.

Del Fra, A. & Ghinelli, D. (to appear), A classification of extended generalized quadrangles with maximum diameter.

Ivanov, A. A. & Shpectorov, S. V. (1988), Geometries for sporadic groups related to the Petersen graph, I, *Commun. Algebra* **16**, 925–953.

Neumaier, A. (1982), Rectagraphs, diagrams, and Suzuki's sporadic simple group, *Algebraic and Geometric Combinatorics* (ed. E. Mendelsohn), 305–318, *Ann. Discrete Math.* **15**, North-Holland, Amsterdam.

Ott, U. & Ronan, M. A. (1981), On buildings and locally finite Tits geometries, *Finite Geometries and Designs* (ed. P. J. Cameron, J. W. P. Hirschfeld & D. R. Hughes), 272–274, *London Math. Soc. Lecture Notes* **49**, Cambridge Univ. Press, Cambridge.

Tsaranov, S. V. (to appear), Representations and classification of Coxeter monoids, *Europ. J. Combinatorics*.

Kac-Moody groups and their automorphisms.

R.W. Carter

1.   The theory of Kac–Moody algebras, which generalises the theory of finite dimensional semisimple Lie algebras over the complex field, was introduced independently by V.G. Kac and R.V. Moody in 1967. [K1] [M].

Each finite dimensional semisimple Lie algebra g   over $\mathbb{C}$ has a Cartan matrix $A = (A_{ij})$ associated to it, where

$$A_{ij} \in \mathbb{Z}, \quad A_{ii} = 2, \quad A_{ij} \leq 0 \text{ if } i \neq j,$$

$A_{ij} = 0$ if and only if $A_{ji} = 0$, and all the leading minors

$$A_{11}, \quad \begin{vmatrix} A_{11} & A_{12} \\ A_{21} & A_{22} \end{vmatrix}, \quad \ldots \quad \det A$$

of A are positive. Conversely the Lie algebra g is uniquely determined by its Cartan matrix A . g may be described in terms of generators and relations as follows:

g is generated by elements $e_1 \ldots e_\ell$, $h_1 \ldots h_\ell$, $f_1 ,\ldots f_\ell$ subject to relations

$$[h_i h_j] = 0$$
$$[e_i f_i] = -h_i$$
$$[e_i f_j] = 0 \quad \text{if } i \neq j$$
$$[h_i e_j] = A_{ij} e_j$$
$$[h_i f_j] = -A_{ij} f_j$$
$$(ad e_i)^{1-A_{ij}} e_j = 0 \quad \text{if } i \neq j$$
$$(ad f_i)^{1-A_{ij}} f_j = 0 \quad \text{if } i \neq j.$$

These are called Serre's relations.

In order to generalise the theory of finite dimensional semisimple Lie algebras to that of Kac–Moody algebras we begin with a generalised Cartan matrix (G.C.M.) rather than an ordinary Cartan matrix. A matrix $A = (A_{ij})$ is a G.C.M. if it satisfies the conditions

$$A_{ij} \in \mathbb{Z}, \quad A_{ii} = 2, \quad A_{ij} \leq 0 \text{ if } i \neq j,$$
$$A_{ij} = 0 \text{ if and only if } A_{ji} = 0.$$

Given any G.C.M. A we may define the corresponding Kac–Moody Lie algebra $g(A)$ as the Lie algebra over $\mathbb{C}$ generated by elements e_1 .... e_ℓ, h_1 ... h_ℓ, f_1 ... f_ℓ subject to Serre's relations.

The Kac–Moody algebras can be divided naturally into three classes. The G.C.M. A is said to have finite type if its leading minors are all positive. A has affine type if all its proper leading minors are positive but $\det A = 0$. A has indefinite type otherwise.

The Kac–Moody algebra $g(A)$ is finite dimensional if and only if A has finite type. If A has affine type $g(A)$ is infinite dimensional. These Kac–Moody algebras of affine type have been investigated in great detail. [K2]. Many results on the structure and representation theory of finite dimensional semisimple Lie algebras over $\mathbb{C}$ carry over to the infinite dimensional situation. In addition the theory of Kac–Moody algebras of affine type has had many and varied applications in pure mathematics, applied mathematics and theoretical physics. In contrast, relatively little is known about Kac–Moody algebras of indefinite type.

2.   If A is an ordinary Cartan matrix and $g(A)$ the corresponding finite dimensional semisimple Lie algebra over $\mathbb{C}$ there is a corresponding simply-connected algebraic group $G(A)$ over $\mathbb{C}$ whose Lie algebra is $g(A)$. It is natural to try to construct a Kac–Moody group $G(A)$ over $\mathbb{C}$ for any G.C.M. A which reduces to the simply-connected algebraic group when A is of finite type. There is a series of papers of Kac and Peterson in which such groups $G(A)$ are defined and investigated. The groups $G(A)$ are obtained by exponentiating the action of $g(A)$ on its highest weight modules. Many of the properties of the simply-connected algebraic group $G(A)$ are shown by Kac and Peterson to generalise to the case where A is a G.C.M. In particular $G(A)$ has subgroups which form what is called a 'refined Tits system', generalising the properties of a Tits system or BN–pair in the theory of algebraic groups.

3.   There is an alternative approach to the theory of Kac–Moody groups due to J. Tits. The construction of the above group $G(A)$ is generalised by Tits in

two respects. In the first place one wishes to define a Kac–Moody group corresponding to A over any field k , not just the complex field. More generally one would like to define such a Kac–Moody group over any commutative ring R . In fact one would like to have a functor

$$R \to G(R)$$

from commutative rings to groups, depending on the G.C.M. A , which reduces to the Chevalley-Demazure functor in the case when A has finite type.

The second type of generalisation introduced by Tits is that one need not consider only the analogue of the simply–connected algebraic group $G(A)$. There are in general many different connected reductive algebraic groups associated with a given Cartan matrix A . They correspond, up to isomorphism, to the different root data with Cartan matrix A . A root datum with Cartan matrix A is a quadruple

$$D = (X , \Pi , Y , \Pi^{\vee})$$

where X, Y are free abelian groups of the same finite rank n with a non-degenerate map $X \times Y \to \mathbb{Z}$ which induces isomorphisms

$$\mathrm{Hom}(Y, \mathbb{Z}) \cong X , \quad \mathrm{Hom}(X, \mathbb{Z}) \cong Y .$$

Π is a finite subset of X , $\Pi^{\vee}$ a finite subset of Y and there is a bijection $\alpha_i \to \alpha_i^{\vee}$ between Π and $\Pi^{\vee}$ satisfying

$$\langle \alpha_j , \alpha_i^{\vee} \rangle = A_{ij} .$$

If we are given a G.C.M. A instead of an ordinary Cartan matrix we can define a root datum D with G.C.M. A in the same way as above. One would therefore like to define a functor

$$R \to G_D(R)$$

from commutative rings to groups for any such root datum D with G.C.M. A . This is done by Tits in his paper [T].

The question then arises as to whether this group functor G_D defined by Tits is the 'most natural one'. Tits argues convincingly that, at least when k is a field, the group $G_D(k)$ obtained by his method is indeed the most natural one. Tits shows that any group functor satisfying a few basic conditions (which a Kac-Moody group functor certainly ought to satisfy) has the property that $G_D(k)$ is determined up to a unique isomorphism when k is a field.

4.   The Kac–Moody group $G_D(k)$ has subgroups satisfying an axiom system rather like that of a split BN-pair, which one has in the theory of Chevalley groups. However in the situation of Kac-Moody groups one has a more special situation than that of a split BN-pair. In fact one has subgroups B, B^-, N in $G = G_D(k)$ which satisfy a system of axioms which might be called a 'split double BN-pair'. Both pairs (B,N) and (B^-,N) form split BN-pairs in G. Of course when the G.C.M. A has finite type its Weyl groups $W = N/B \cap N$ is finite and its element w_0 of maximal length transforms B into B^-. However in the general case W has no element of maximal length and we need the axioms of a 'split double BN-pair' rather than of a split BN-pair to describe the situation. These axioms are closely related to the axioms of a 'refined Tits system' introduced by Kac and Peterson.

Just as groups with a BN-pair act on the geometric structures called buildings, so groups with a double BN-pair act on double buildings, which are described in Tits' article in the present volume.

5.   We next describe the Kac-Moody group $G_D(k)$ by generators and relations, following the ideas of Tits. We begin with a field k and with the root datum $D = (X, \Pi, Y, \Pi^v)$ where X, Y are free abelian groups of rank n,

$$\Pi = \{\alpha_1, \dots \alpha_\ell\} \subset X \text{ and } \Pi^v = \{\alpha_1^v, \dots \alpha_\ell^v\} \subset Y \text{ with } \langle \alpha_j, \alpha_i^v \rangle = A_{ij}.$$

Now the G.C.M. A may be a singular matrix and $\alpha_1, \dots \alpha_\ell$ or $\alpha_1^v, \dots \alpha_\ell^v$ or both may be linearly dependent. In order to introduce the Weyl group and the system of real roots we thus introduce a set $\tilde\Pi = \{\tilde\alpha_1, \dots \tilde\alpha_\ell\}$ and let $Z\tilde\Pi$ be the free abelian group with the elements of $\tilde\Pi$ as basis. We define $s_i : Z\tilde\Pi \to Z\tilde\Pi$ by

$$s_i(\tilde\alpha_j) = \tilde\alpha_j - A_{ij}\tilde\alpha_i$$

and let W be the group of transformations of $Z\tilde\Pi$ generated by s_i, $i = 1, \dots \ell$. We define $\Phi = W(\tilde\Pi)$ to be the set of real roots. Then each element of Φ is a Z-combination of $\tilde\alpha_1, \dots \tilde\alpha_\ell$ in which all coefficients are ≥ 0 or all coefficients are ≤ 0. We define Φ^+, Φ^- to be the subsets in which the coefficients are ≥ 0 or ≤ 0 respectively.

W can be made to act on X, Y by means of the formulae

$$s_i(\chi) = \chi - \langle \chi, \alpha_i^v \rangle \alpha_i \qquad \chi \in X$$

$$s_i(\gamma) = \gamma - \langle \alpha_i, \gamma \rangle \alpha_i^v \qquad \gamma \in Y$$

although these actions need not be faithful.

We next define $H = Y \otimes_{\mathbb{Z}} k^*$. Thus $H \cong k^* \times ... \times k^*$ with n factors. We may define an action of W on H by

$$s_i(\gamma \otimes \lambda) = s_i(\gamma) \otimes \lambda \qquad \gamma \in Y, \ \lambda \in k^*.$$

We may also define an action of the simple roots $\tilde{\alpha}_i$ as maps from H to k^*. We define

$$\tilde{\alpha}_i : H \to k^*$$

by $\qquad \tilde{\alpha}_i(\gamma \otimes \lambda) = \lambda^{\langle \alpha_i, \gamma \rangle} \qquad \gamma \in Y, \ \lambda \in k^*.$

For each $\alpha \in \Phi$ we may write $\alpha = w(\tilde{\alpha}_i)$ for some $w \in W$ and some i and then define $\alpha : H \to k^*$ unambiguously by $\alpha(h) = \tilde{\alpha}(w^{-1}h)$.

We now come to the definition of $G_D(k)$ by generators and relations. For each $\alpha \in \Phi$ and each $\lambda \in k$ we introduce a symbol $x_\alpha(\lambda)$. The Kac–Moody group $G_D(k)$ will be generated by the elements $x_\alpha(\lambda)$, $\alpha \in \Phi$, $\lambda \in k$ and the elements $h \in H$. In order to describe the relations between these generators we define

$$x_i(\lambda) = x_{\tilde{\alpha}_i}(\lambda)$$
$$x_{-i}(\lambda) = x_{-\tilde{\alpha}_i}(\lambda)$$
$$n_i(\lambda) = x_i(\lambda)x_{-i}(\lambda^{-1})x_i(\lambda)$$
$$n_i = n_i(1)$$

$$h_i(\lambda) = \alpha_i^v \otimes \lambda \in H.$$

Then $G_D(k)$ is the group generated by elements $x_\alpha(\lambda)$, $\alpha \in \Phi$, $\lambda \in k$ and h $\in H$ subject to the relations:

    (i)   relations in H

(ii) $\quad h_i(\lambda) = n_i(\lambda)\, n_i^{-1}$

(iii) $\quad x_\alpha(\lambda) x_\alpha(\mu) = x_\alpha(\lambda+\mu)$

(iv) $\quad h x_i(\lambda) h^{-1} = x_i(\tilde{\alpha}_i(h)\lambda)$

(v) $\quad n_i\, h\, n_i^{-1} = s_i(h)$

(vi) $\quad n_i\, x_\alpha(\lambda) n_i^{-1} = x_{s_i(\alpha)}(\eta_{i,\alpha}\lambda)$ where $\eta_{i,\alpha} \in \{1,-1\}$.

(vii) Suppose $\alpha, \beta, \in \Phi$ with $\alpha \neq \pm\beta$ and suppose there exist w, w' $\in$ W with $w(\alpha) \in \Phi^+$, $w(\beta) \in \Phi^+$, $w'(\alpha) \in \Phi^-$, $w'(\beta) \in \Phi^-$. Then

$$\left[x_\alpha(\lambda),\, x_\beta(\mu)\right] = \prod_{\substack{i,j \in \mathbb{Z} \\ i,j>0 \\ i\alpha+j\beta \in \Phi}} x_{i\alpha+j\beta}\!\left(C_{ij\alpha\beta}\,\lambda^i\mu^j\right)$$

for certain $C_{ij\alpha\beta} \in \mathbb{Z}$ uniquely determined by i,j,α,β and the root datum D , and the ordering of the terms on the right hand side.

A pair of roots α,β satisfying the hypothesis of (vii) is called by Tits a prenilpotent pair.

6. The root datum D is called simply-connected if $\alpha_1^\vee \dots \alpha_\ell^\vee$ form a $\mathbb{Z}$-basis for Y .

There is, up to isomorphism, a unique simply-connected root datum D with a given G.C.M. A . In the corresponding simply connected Kac-Moody group $G_D(k)$ the elements $h_i(\lambda)$ generate H , since the elements $\alpha_i^\vee$ generate Y .

However by relation (ii) above $h_i(\lambda)$ can be expressed as a product of generators of the form $x_\alpha(\lambda)$ for $\alpha \in \Phi$, $\lambda \in k$. Thus the simply connected Kac-Moody group $G_D(k)$ is generated by the elements $x_\alpha(\lambda)$ for $\alpha \in \Phi$, $\lambda \in k$.

We define

$$x_i(\lambda) = x_{\tilde{\alpha}_i}(\lambda)$$
$$x_{-i}(\lambda) = x_{-\tilde{\alpha}_i}(\lambda)$$
$$n_i(\lambda) = x_i(\lambda) x_{-i}(\lambda^{-1}) x_i(\lambda)$$
$$n_i = n_i(1)$$

$$h_i(\lambda) = n_i(\lambda) n_i^{-1} .$$

Then $G_D(k)$ is generated by the elements $x_\alpha(\lambda)$, $\alpha \in \Phi$, $\lambda \in k$ subject to relations

(i)　$h_i(\lambda)h_i(\mu) = h_i(\lambda\mu)$

(ii)　$h_i(\lambda)h_j(\mu) = h_j(\mu)h_i(\lambda)$

(iii)　$x_\alpha(\lambda)x_\alpha(\mu) = x_\alpha(\lambda+\mu)$

(iv)　$h_j(\mu)x_i(\lambda)h_j(\mu)^{-1} = x_i(\mu^{A_{ji}}\lambda)$

(v)　$n_i h_j(\mu)n_i^{-1} = h_j(\mu)h_i(\mu^{-A_{ji}})$

(vi)　$n_i x_\alpha(\lambda)n_i^{-1} = x_{s_i(\alpha)}(\eta_{i,\alpha}\lambda)$

(vii)　Suppose $\alpha,\beta \in \Phi$ is a prenilpotent pair of roots. Then

$$\left[x_\alpha(\lambda), x_\beta(\mu)\right] = \prod_{\substack{i,j \in \mathbb{Z} \\ i>0, j>0 \\ i\alpha+j\beta \in \Phi}} x_{i\alpha+j\beta}\left(C_{ij\alpha\beta}\,\lambda^i\mu^j\right).$$

These relations are derived from the relations for an arbitrary Kac-Moody group using the fact that the elements $h_i(\lambda)$ generate H and that H is isomorphic to $k^* \times ... \times k^*$.

7.　　If A is an ordinary Cartan matrix we may construct a corresponding extended Cartan matrix $\tilde{A}$ as follows. Let $\alpha_1, ... \alpha_\ell$ be the set of simple roots corresponding to A and let α_0 be the negative of the highest root. Let $\alpha_0^{\vee}$ be

the coroot of α_0. Then $\tilde{A}$ is the $(\ell + 1) \times (\ell + 1)$ matrix of rank ℓ given by

$$\tilde{A}_{ij} = \langle \alpha_j, \alpha_i^{\vee} \rangle \qquad i, j \in \{0, 1, \dots \ell\}.$$

$\tilde{A}$ is a G.C.M. of affine type (although not every G.C.M. of affine type is

obtained in this way).

Let K be an algebraically closed field of characteristic 0 and $\tilde{G}(K)$ be

the simply connected Kac–Moody group over K with G.C.M. $\tilde{A}$. $\tilde{G}(K)$ is

generated by its elements

$$x_i(\lambda), \quad x_{-i}(\lambda) \qquad i \in \{0, 1, \dots \ell\} \quad \lambda \in K.$$

(Note that in this notation the elements $x_0(\lambda)$ and $x_{-0}(\lambda)$ are distinct !).

We shall now outline some work due to R.W. Carter and Y. Chen on the

automorphism group of $\tilde{G}(K)$.

In order to describe an automorphism of $\tilde{G}(K)$ it is sufficient to give its

effect on the generators $x_i(\lambda)$ and $x_{-i}(\lambda)$. We begin by describing some

specific automorphisms of $\tilde{G}(K)$.

(i)   Inner automorphisms. For each $g \in \tilde{G}(K)$ we have the

corresponding inner automorphism $\tau_g : x \to gxg^{-1}$.

(ii)   Diagonal automorphisms. For any set of elements $\xi = (\xi_0, \xi_1, \dots$

$\xi_\ell)$, $\xi_i \in K^*$, we may obtain an automorphism $d(\xi)$ of $\tilde{G}(K)$ such that

$$d(\xi) \quad : \quad x_i(\lambda) \to x_i(\xi_i \lambda)$$

$$x_{-i}(\lambda) \to x_{-i}(\xi_i^{-1} \lambda).$$

However some of these automorphisms $d(\xi)$ are inner automorphisms. Let

$$-\alpha_0 = m_1 \alpha_1 + \dots + m_\ell \alpha_\ell$$

be the expression of the highest root as a combination of simple roots. Then $d(\xi)$

can be shown to be an inner automorphism of $\tilde{G}(K)$ if and only if

$$\xi_0 \xi_1^{m_1} \dots \xi_\ell^{m_\ell} = 1.$$

In order to obtain automorphisms which are not inner it is therefore natural to consider the case when

$$\xi_1 = 1 \ \dots \ \xi_\ell = 1 \qquad \xi_0 \ \text{is arbitrary.}$$

Such automorphisms $d(\xi_0)$ for $\xi_0 \in K^*$ will be called diagonal automorphisms of $\tilde{G}(K)$. The diagonal automorphisms form a group isomorphic to K^*.

(iii)   Field automorphisms. If $f : K \to K$ is any automorphism of the field K we obtain a corresponding automorphism $a(f)$ of $\tilde{G}(K)$ given by

$$a(f) : \quad x_i(\lambda) \ \to \ x_i\big(f(\lambda)\big)$$
$$x_{-i}(\lambda) \to x_{-i}\big(f(\lambda)\big).$$

(iv)   Graph automorphisms. Let γ be a permutation of $\{0, 1, .. \ \ell\}$ such that $A_{ij} = A_{\gamma(i),\gamma(j)}$ for all i,j. γ is called an automorphism of the extended Cartan matrix $\tilde{A}$. It induces an automorphism $a(\gamma)$ of the group $\tilde{G}(K)$ given by

$$a(\gamma) : \quad x_i(\lambda) \ \to \ x_{\gamma(i)}(\varepsilon_i\lambda)$$
$$x_{-i}(\lambda) \to \ x_{-\gamma(i)}(\varepsilon_i\lambda) \qquad \text{for certain } \varepsilon_i \in \{1,-1\}.$$

Automorphisms $a(\gamma)$ of this kind are called graph automorphisms of $\tilde{G}(K)$.

(v)   The Cartan automorphism. The group $\tilde{G}(K)$ also has an automorphism ω, which we shall call the Cartan automorphism, such that

$$\omega : \ x_i(\lambda) \ \to \ x_{-i}(\lambda)$$
$$x_{-i}(\lambda) \to \ x_i(\lambda).$$

(Note that the corresponding type of automorphism for a Chevalley group is an inner automorphism induced by the longest element w_0 of the Weyl group when $w_0 = -1$. If $w_0 \neq -1$ the automorphism is obtained by combining the inner automorphism induced by w_0 with the opposition graph automorphism. However, in the Kac–Moody situation we are discussing the Weyl group W is an affine Weyl group which has no element of maximal length. In this case the Cartan automorphism cannot be described in terms of other types of automorphism previously discussed.)

We may now state our main theorem.

Theorem.   Let $\tilde{G}(K)$ be a simply connected Kac–Moody group over an algebraically closed field of characteristic 0 whose G.C.M. is an extended Cartan matrix. Then any automorphism of $\tilde{G}(K)$ can be expressed uniquely as inner × diagonal × field × graph automorphism

or as

inner × diagonal × field × graph × Cartan automorphism.

Note. There is a well known theorem of Steinberg which gives a corresponding result for automorphisms of Chevalley groups [S].

8.   The main technique used in the proof of the above theorem is a well known theorem of Borel and Tits on abstract homomorphisms between algebraic groups over fields. In order to apply this theorem we use the fact that the Kac–Moody group $\tilde{G}(K)$ is closely related to the Chevalley group $G(R)$ over the ring $R = K[t,t^{-1}]$ of Laurent polynomials over K . In fact there is a surjective homomorphism $\tilde{G}(K) \rightarrow G(R)$ whose kernel lies in the centre of $\tilde{G}(K)$ and is isomorphic to K^* . This kernel is a characteristic subgroup of $\tilde{G}(K)$ so any automorphism of $\tilde{G}(K)$ induces an automorphism of $G(R)$. Let $F = K(t)$ be the field of fractions of R and $\bar{F}$ be its algebraic closure. Then we have

$$G(K) \subset G(R) \subset G(F) \subset G(\bar{F}) .$$

Thus any automorphism of $G(R)$ induces a homomorphism from $G(K)$ into $G(F)$. The image of this homomorphism can be shown to be dense in $G(\bar{F})$. Using this density property, the fact that G is simply–connected, and the fact that $G(K)$ is a perfect group we may apply the Borel–Tits theorem to show that the given homomorphism from $G(K)$ into $G(F)$ is obtained by combining an isogeny of algebraic groups with a homomorphism of fields. This result enables us to apply a succession of simplifications to our original automorphism of $G(R)$, multiplying it by a field automorphism, a graph automorphism, an inner automorphism and a diagonal automorphism of $G(R)$ as appropriate. Eventually the automorphism can be reduced to the identity map in this way. The diagonal automorphisms of $G(R)$ are, by definition, those induced by conjugation by elements of $\hat{G}(R) = N_{G(\bar{F})}(G(R))$. In fact $\hat{G}(R)/G(R)$ turns out to be isomorphic to the fundamental group of G .

These various special kinds of automorphisms of $G(R)$ must be related to those of $\tilde{G}(K)$ in order to complete the proof of the theorem. The details are expected to appear in a forthcoming article in the Journal of Algebra.

References

[CC]   Carter R.W. and Chen Y. 'Automorphisms of affine Kac–Moody groups and related Chevalley groups over rings', Warwick University preprint. (1990).

[K1]   Kac V.G. 'Simple graded Lie algebras of finite growth', Func. Anal. Appl. 1 (1967) 328–329.

[K2]   Kac V.G. 'Infinite Dimensional Lie Algebras'. Birkhäuser (1983).

[K3]   Kac V.G. 'Constructing groups associated to infinite dimensional Lie algebras', Infinite Dimensional Groups with Applications (Ed. V. Kac), Springer (1985), 167–216.

[KP]   Kac V.G. and Peterson D.H. 'Defining relations of certain infinite dimensional groups'. Proc. of E. Cartan Conference, Lyon (1984).

[M]   Moody R.V. 'Lie algebras associated with generalized Cartan matrices', Bull. Amer. Math. Soc. 73 (1967), 217–221.

[S]   Steinberg R. 'Lectures on Chevalley groups'. Yale University (1967).

[T]   Tits J. 'Uniqueness and presentation of Kac–Moody groups over fields'. J. Algebra 105 (1987), 542–573.

Generalized Hexagons as Geometric Hyperplanes of Near Hexagons[1]

Ernest E. Shult

§1. Introduction.

The first theorem given here asserts that a geometric hyperplane H of a near hexagon, which intersects each quad at a star must be a generalized hexagon. The second theorem tells us that if a finite near hexagon with parameters possesses such a geometric hyperplane, then that near hexagon Γ must be the dual of a rank 3 polar space Δ. Moreover, there is a bijection $H \leftrightarrow$ quads of Γ, which induces an embedding of the hexagon H into the polar space Δ which is an epimorphism on points. Conceivably, there is a possibility that generalized hexagons might be represented as geometric hyperplanes of some of the "other" dual polar spaces, such as $\Omega(n, \mathbb{R})$ (with signature $(n - 3, 3)$), $Sp(6, k)$, $\Omega^-(8, k)$, $U(6, k)$ or $U(7, k)$. But the final theorem shows that if Γ is finite, such possibilities cannot happen; that in fact Γ (and Δ) are type $\Omega(7, q)$ (or $Sp(6, q)$ if q is even) and H is the hexagon of type $G_2(q)$ associated with the standard embedding of $G_2(q)$ (either as the stabilizer of an appropriate hyperplane in the 8-dimensional spin module for $\Omega(7, q)$ or as the stabilizer of a trilinear form in its natural 7-dimensional module – or the factor of this 7-space module by a 1-dimensional radical when q is even).

The author thanks Professor J. Tan for a valuable discussion, Queen Mary College, U. of London, and the Mathematisches Institute, Albert-Ludwigs Universität Freiburg for their kind hospitality during the writing of this work, and the Alexander von Humboldt Stiftung whose support made the research possible.

§2. Notation and Preliminary Results.
2.1. Basic Definitions.

A point-line incidence system is simply a rank 2 incidence system $\Gamma = (\mathcal{P}, \mathcal{L})$ with the elements of set $\mathcal{P}$ called **points**, those of set $\mathcal{L}$ called **lines**. The point shadow of a line L is the set of all points incident with L; we

---

[1] This work was partially supported by a grant from the National Science Foundation (U.S.) and a grant from the Science and Engineering Research Council (U.K.).

say L and M are **repeated** lines in Γ if they have the same point shadows. To avoid a lot of unnecessary awkwardness we assume for the purposes of this note, that **all point-line systems do not have repeated lines**. We may thus identify each line with its unique point shadow so that symbols for lines are operands for the various set-theoretic operators.

Any point-line incidence system $\Gamma = (\mathcal{P}, \mathcal{L})$ carries with it a natural metric: $d_\Gamma : \mathcal{P} \times \mathcal{P} \to \{0\} \cup \mathbb{Z}^+$ which is distance in the collinearity graph on $\mathcal{P}$. The usual notions "diameter", "geodesic" and "convex set" are defined relative to d_Γ. A subspace S of Γ is a subset of $\mathcal{P}$ such that for each line $L \in \mathcal{L}$, $|L \cap S| \geq 2$ implies $L \subseteq S$. The set $\mathcal{L}(S)$ of all lines of $\mathcal{L}$ contained in S are called the **lines of (subspace)** S. A subspace S is called a **geometric hyperplane** of Γ, if and only if $S \cap L \neq \emptyset$ for each line $L \in \mathcal{L}$. We write $x^\perp$ for all points (including x) in $\mathcal{P}$ which are collinear with x. Γ is said to have **thick lines** if every line L of $\mathcal{L}$ contains at least three points.

Γ is called a **near polygon** if for every point-line pair $(p, L) \in \mathcal{P} \times \mathcal{L}$, L contains a unique point nearest p ("nearest" means such that $d_\Gamma(p, x)$ is minimal.) A **generalized quadrangle** is a near polygon of diameter 2. It is **non-degenerate** if none of its points is collinear with all points; otherwise it consists of a point p and a collection of lines on p pairwise intersecting at p, a structure in this note called a **star**. A. Yanushka proved in [5] that the following two conditions were equivalent for a pair of points $(x, y) \in \mathcal{P} \times \mathcal{P}$ such that $d_\Gamma(x, y) = 2$ in a near polygon Γ.

(2.1) (i)   There are at least 2 distinct paths from x to y and one of these paths carries a thick line.

(ii) The convex closure $\langle x, y \rangle$ (intersection of all convex subspaces containing x and y) is a non-degenerate generalized quadrangle containing a thick line.

If either of the above two conditions holds for each pair of points at mutual distance 2, we say Γ is a **near polygon with quads**. The convex subspaces $\langle x, y \rangle$ which arise via (2.1)(ii) are called quads. If Γ is a near polygon with quads with diameter at least 3, the points, lines and the collection Q of all quads forms a rank 3 diagram geometry (in the sense of Buekenhout) with diagram:

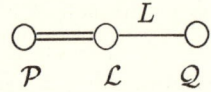

$$(\text{O}\!\!=\!\!=\!\!\text{O} = \text{ generalized quadrangle}; \text{O}\!\!\overset{L}{\rule{2cm}{0.4pt}}\!\!\text{O} = \text{ linear space}).$$

A **near hexagon** is a near polygon of diameter 3. A **generalized hexagon** is a near hexagon such each pair (x, y) of points at mutual distance 2 possesses a unique geodesic connecting them. A generalized hexagon is **non-degenerate** if every pair (x, y) of points at mutual distance 3 are connected by at least 2 distinct geodesics. The following is an elementary observation.

(2.2) Let Γ be a generalized quadrangle or generalized hexagon. Then Γ is non-degenerate if and only if every point lies on at least two lines.

2.2. Basic Properties of Near Hexagons with Quads.

In this subsection Γ is a near hexagon with quads. For each point p, let the set of lines on p and quads on p, be denoted $\mathcal{L}_p$ and Q_p respectively. Then the **point residual** Γ^p is the rank 2 incidence system $(\mathcal{L}_p, Q_p)$ where incidence is containment of subspaces. This incidence system is a **linear space** since any two lines A and B on p, if distinct, lie in a unique quad $\langle A, B \rangle$. We shall have need of this space in Section 3.

LEMMA 2.1.

(i)   *Let (p, Q) be a non-incident point-quad pair. Then either*

(a)   *(p, Q) is a **classical pair**, which means that $p^{\perp} \cap Q = \{q\}$ and $d_{\Gamma}(p, x) = 1 + d_{\Gamma}(q, x)$ as x ranges over Q.*

or

(b)   *(p, Q) is an **ovoid pair**, which means that the set of points of Q at distance 2 from p form an ovoid $O_p(Q)$ (a coclique meeting each line of Q at a point), the remaining points in $Q - O_p(Q)$ being distance three from q.*

(ii) *Every 5-circuit and 4-circuit in Γ lies in a quad.*

(iii) *For no point p is it true that $p^{\perp} \subseteq Q$ for some quad Q.*

PROOF:

(i) is developed fully in [5].

(ii) Clearly any 4-circuit lies in the convex closure of any of its non-collinear point-pairs. So assume (a, b, c, d, e) is a 5-circuit, and set $Q = \langle a, c \rangle$. Then if d is not in Q, (d, Q) is a classical pair so $d_{\Gamma}(d, a) = 2 + d_{\Gamma}(c, a) = 1 + 2 = 3$ against $d_{\Gamma}(d, a) = 2$. Thus $d \in Q$, whence also $e \in Q$ by convexity of Q.

(iii) Suppose $p^{\perp} \subseteq Q$. If $q \in p^{\perp} - \{p\}$, and $y \in q^{\perp} - Q$ then $\langle p, y \rangle$ is a quad

(iii) Suppose $p^\perp \subseteq Q$. If $q \in p^\perp - \{p\}$, and $y \in q^\perp - Q$ then $\langle p, y \rangle$ is a quad intersecting Q at the line pq, and so contains a line on p not in Q against our assumption. Thus also $q^\perp \subseteq Q$, and this argument can be repeated once more to yield $x^\perp \subseteq Q$ for all $x \in Q$. But then Q is a connected component of Γ whence $Q = \mathcal{P}$ by the near-polygon axiom. But then the diameter of Γ is only 2, a contradiction to Γ being a near hexagon. Thus $x^\perp$ cannot lie within a quad Q.

LEMMA 2.2.

(i) *Suppose* (p, Q) *is an ovoid-pair and* a *and* b *are distinct points of the ovoid* $O_p(Q)$ *of points of* Q *at distance 2 from* p. *Then the two quads* $\langle p, a \rangle$ *and* $\langle p, b \rangle$ *intersect only at the single point* $\{p\}$.

(ii) *Suppose* A *and* B *are two quads satisfying* $A \cap B = \{p\}$. *Then for any point* b *in* $B - p^\perp$, (b, A) *is an ovoid pair.*

PROOF:

(i) Suppose $\langle p, a \rangle \cap \langle p, b \rangle$ contained a line L on p. Then, $a^\perp \cap L = \{a_0\}$ and $b^\perp \cap L = \{b_0\}$ since $\langle p, a \rangle$ and $\langle p, b \rangle$ are generalized quadrangles. Suppose first that $a_0 \neq b_0$. Then for any $c \in a^\perp \cap b^\perp \subseteq Q$ we see that (a, c, b, b_0, a_0) is a 5-circuit and so, by Lemma 2.1(ii), lies in a quad R. But then $p \in a_0 b_0 \subseteq R = \langle a, b \rangle = Q$ which is absurd. Thus $a_0 = b_0$. But then $a_0^\perp \cap Q \supseteq \{a, b\}$, so $a_0 \in Q$ by convexity of Q. In that case (p, Q) is a classical pair, another contradiction. Thus $\langle p, a \rangle \cap \langle p, b \rangle$, contains p as an isolated point. If it contained any further point q, then $\langle p, q \rangle = \langle p, a \rangle = \langle p, b \rangle = \langle a, b \rangle = Q$ against our choice of q. Thus $\langle p, a \rangle \cap \langle p, b \rangle = b$.

(ii) If the conclusion here were false, (b, A) would be a classical pair, whence $b^\perp \cap A = \{a\}$. Clearly, $a \neq p$ by choice of b. If $d_\Gamma(a, p) = 1$, (p, c, b, a) is a 4-circuit for any $c \in p^\perp \cap b^\perp$ so $a \in \langle p, b \rangle = B$ against $A \cap B = \{p\}$. Also if $d_\Gamma(a, p) = 2$, (p, c, b, a, d) is a 5-circuit for $c \in p^\perp \cap b^\perp$ and $d \in a^\perp \cap p^\perp$. Again, by Lemma 2.1(ii), this 5-circuit lies in a quad R. But in that case $A = \langle a, p \rangle = R = \langle p, b \rangle = B$ contrary to hypothesis. The proof is complete.

2.3. Recognizing Hexagons in the Finite Case.

By an **embedding** of a point-line incidence system $\Gamma = (\mathcal{P}, \mathcal{L})$ in another $\Gamma' = (\mathcal{P}', \mathcal{L}')$ we mean an injection $f : \mathcal{P} \to \mathcal{P}'$ such that for each line $L \in \mathcal{L}$, $f(L)$ is a line of $\mathcal{L}'$. We shall have need to consider in the next section an embedding of a generalized hexagon into a rank 3 polar space $(\mathcal{P}, \mathcal{L})$ which in turn has a natural embedding into the ambient projective space $PG = (\mathcal{P}', \mathcal{L}')$ whose polarity defines it. The composition, then, is an embedding $e : H \to PG$ of the hexagon H into the projective space PG. We shall show that H has order (q, q). $PG \simeq PG(d, q)$, with the polar space $(\mathcal{P}, \mathcal{L})$ of type $Sp(6, q)$ or $\Omega(7, q)$. If $H_i(x) = \{y \in H \mid d_H(x, y) \le i\}$, we have

(2.3) (i) $e(H_1(x))$ is an isotropic plane of $(\mathcal{P}, \mathcal{L})$ for all $x \in H$.

(ii) $e(H_2(x)) = e(x)^\perp \cap e(H) = e(x)^\perp \cap \mathcal{P}$ where " $\perp$ " is the polarity of PG yielding $(\mathcal{P}, \mathcal{L})$.

In these circumstances the following criterion (which is based on theorems of Schellekens [4], Yanushka [7] and Ronan [2]) is useful.

PROPOSITION (THAS' CRITERION ([6])). *Let H be a finite generalized hexagon of order (q, q), $q > 1$. Suppose*

(Th) Given any three points x, y and $z \in H$, there exists a point u such that u is collinear with x and $d_H(u, y)$ and $d_H(u, z)$ are both at most 2.

Then H is a hexagon of type $G_2(q)$.

COROLLARY 2.3. *Let H be a generalized hexagon of order (q, q) and let $e : H \to PG(n-1, q)$ be an embedding such that $e(H)$ is the set of absolute points of $PG(n - 1, q)$ with respect to a non-degenerate polarity $\perp$ and (e, H) satisfies (2.3). Then H is type $G_2(q)$.*

PROOF: We need only verify Thas' criterion. Select three points x, y, z. $e(H_1(x))$ is an isotropic plane, and so there is an isotropic point $\overline{u}$ in the intersection of this plane with the two hyperplanes $e(y)^\perp$ and $e(z)^\perp$. By (3.2)(ii), all points of the isotropic plane $e(H_1(x))$ belong to $e(H)$ and so by (ii) applied to $e(H_2(y))$ and $e(H_2(z))$, we see $u = e^{-1}(\overline{u}) \in H_1(x) \cap H_2(y) \cap H_2(z)$, as required.

COROLLARY 2.4. *In the case of an embedding $H \to \Delta = (\mathcal{P}, \mathcal{L})$, of a generalized hexagon of order (q, q) into a polar space of type $Sp(6, q)$, satisfying (2.3), then q must be even.*

PROOF: All of the hypotheses of Theorems 3.1 and 3.2 of Cameron and Kantor ([1]) are in force. (Indeed the **conclusions** of Theorem 3.1 are supposed here also.) The assertion on q is contained in the conclusion of their Theorem 3.2.

Of course it should be remarked that there are two generalized hexagons "of type $G_2(q)$" when $q \not\equiv 0(3)$, and these are dual to each other. But it seems certain that only one of these has an embedding in $Sp(6, q)$, q even or $\Omega(7, q)$, q odd. For example when q is odd, the evidence runs as follows:

Since the distance-2 "perp" in H determines the associated polar space, and the embedding of the latter in V is universal when q is odd, (Ronan and Smith [3]) the automorphisms of H lift to orthogonal transformations of V so V becomes a module for Aut (H) and its subgroup $G \simeq G_2(q)$ as well. Then V is a factor of the 7- or 14-dimensional modules corresponding to the two fundamental weights of the Lie algebra of type G_2.

§3. Geometric Hyperplanes Which Are Generalized Hexagons.

THEOREM 1. *Let Γ be a near hexagon with quads and let H be a geometric hyperplane which intersects each quad at a star. Then H is a generalized hexagon.*

PROOF: We proceed by a series of short steps. For each quad Q, $Q \cap H$ is a "star" – i.e. $Q \cap H = p^{\perp} \cap Q$ for some point p. For convenience we call p the **center** of Q and write $p = z(Q)$.

Step 1. *The distance metric in Γ agrees with that on H. (As a consequence H is connected.)*

Let x and y be arbitrary points of H. If $d_{\Gamma}(x, y) = 0, 1$, clearly $d_H(x, y) = 0$ or 1 as H is a subspace. That $d_{\Gamma}(x, y) = 2$ implies $d_H(x, y) = 2$ follows from the fact that $H \cap \langle x, y \rangle$ is a star. Suppose $d_{\Gamma}(x, y) = 3$. Let Q be a quad on x. Then (y, Q) is either an ovoid pair or a classical pair. If (y, Q) is an ovoid pair, let L be a line of H on $z = z(Q)$ and x (this is unique if $x \neq z$). Then L contains a point v in the ovoid $O = \Delta_2(y) \cap Q$ and so $d_{\Gamma}(y, v) = d_H(y, v) = 2$ and as v is collinear with x via a line L of H, $d_H(y, x) = 3$. Otherwise, (y, Q) is classical and, choosing L and z as above, $d_{\Gamma}(x, y) = 3$ implies $\{p\} = y^{\perp} \cap Q$ is not contained in L. If now $\{v\} = p^{\perp} \cap L$, then $d_{\Gamma}(y, v) = d_H(y, v) = 2$ and $d_H(y, x) = 3$ as before.

Step 2. *For every point $p \in H$, there is a point $q \in H$ such that $d_H(p, q) = 3$. Thus $diam(H) = 3$.*

Suppose, by way of contradiction, that this fails for some point p. Then we see that for every quad Q not on p, $H \cap Q \subseteq (p^{\perp} \cup \Gamma_2(p))$ implies that (p, Q) is a classical pair with $p^{\perp} \cap Q = z(Q)$.

Now let R be any quad on p. Choose $v \in R - p^{\perp} - z(R)$. Then by Lemma 2.1(iii), v lies on a line vy not in R. We can choose $u \in v^{\perp} \cap p^{\perp} - z(R)$ and form $Q = \langle u, y \rangle$. Then as $p \in H$ and $u \notin z(R)$, the line vu cannot belong to H. But then $\{v\} = p^\perp \cap Q$ is not the center of Q against the conclusion of the previous paragraph.

Final Step. *H is a non-degenerate generalized hexagon.*

Since by Steps 1 and 2 and the hypothesis that $H \cap Q$ is always a star, H is a near hexagon without quads and so is a (possibly degenerate) generalized hexagon. It remains thus only to show that every pair of points p and q at distance 3 in H possess at least two disjoint paths of length 3 in H between them. So suppose $d_H(p,q) = 3$ and let $(pabq)$ be a path in the collinearity graph of H. Then by Lemma 2.1(iii), $q^\perp$ does not lie in the quad $R = \langle a, q \rangle$. Now $b = z(R)$ so q lies on a line N in R **not** belonging to H and a second line M not lying in R. The quad $T = \langle N, M \rangle$ contains a line L of H lying on q. Let $\{c\} = \Gamma_2(p) \cap L$. Then $L \neq R \cap T = M$. By Step 1, $d_H(p,c) = 2$ so there is a point $d \in p^\perp \cap c^\perp \cap H$ (in fact $d = z(\langle p, c \rangle)$). We now have the circuit $(pabqcd)$ lying in H. This proves H is non-degenerate and completes the proof of the Theorem.

COROLLARY 1. *For each point p in H, $p^\perp \not\subseteq H$.*

PROOF: Suppose $p^\perp \subseteq H$. By Theorem 1 there exists a point q in H with $d_H(p,q) = 3$. Let (p, a, b, q) be a path from p to q in H. Then $a = z(\langle p, b \rangle)$. But $p^\perp \subseteq H$ implies $p = z(\langle p, b \rangle)$ a contradiction as $p \neq a$.

§4. The Case Γ is a Near Hexagon With Parameters.

We suppose as before that H is a geometric hyperplane of a near hexagon Γ with quads such that H intersects each quad at a star. From Theorem 1 we know that the metric of Γ restricted to H is the metric on H and that H is a generalized hexagon. We say Γ **has parameters** (r, s, t) if and only if

(4.1)   (i)   Each point of Γ lies on r lines.

(ii) Each quad of Γ has order (s, t).

Then Γ has parameterized diagram

$$
\overset{\mathcal{P}}{\underset{s}{\circ}} = \overset{\mathcal{L}}{\underset{t}{\circ}} - \overset{\mathcal{Q}}{\underset{r}{\circ}}
$$

where parameter "x" below node i means each panel of cotype i lies in $x + 1$ chambers.

We shall show

THEOREM 2. *If a geometric hyperplane H of a finite near hexagon Γ with parameters, meets each quad at a star, then the near hexagon is a dual polar space.*

PROOF: Assume Γ has finite parameters (r, s, t) as in (4.1). Then $R_p = (\mathcal{L}_p, \mathcal{Q}_p)$ is a linear space with $r + 1$ "points" and "lines" of size $t + 1$. Then, since all "lines" on a "point" partition the remaining "points"

$$(4.2) \qquad \begin{aligned} r = m \cdot t \quad &\text{where } m = \text{number of "lines" of } R_p \text{ on a "point"} \\ &= \text{number of quads on a given line in } \Gamma. \end{aligned}$$

If p is a point of H, let $\mathcal{H}_p$ be the subset of $\mathcal{L}_p$ consisting of all lines on p lying within H. If A and B are two distinct lines on p within H, then p is the center of the quad $\langle A, B \rangle$. Moreover, if Q is any quad on p, Q contains a line of H on p, since Q has a center. Since $p^\perp$ is not contained in H (Lemma 2.1(iii)), $\mathcal{H}_p \neq \mathcal{L}_p$. Summarizing the three previous statments:

(4.4) $\mathcal{H}_p$ is a geometric hyperplane of the linear space $R_p = (\mathcal{L}_p, \mathcal{Q}_p)$.

Then, considering a "point" A of $R_p - \mathcal{H}_p$, we see that R_p a linear space, and $\mathcal{H}_p$ a geometric hyperplane on R_p, forces a bijection: "lines of R_p on A" $\Longleftrightarrow$ "points" of $\mathcal{H}_p$, whence

$$(4.5) \qquad\qquad\qquad |\mathcal{H}_p| = m.$$

But since $\mathcal{H}_p$ is itself a linear space with all "lines" of size $t + 1$, we obtain, analogous to (4.2), that

$$(4.6) \qquad\qquad\qquad m - 1 = m_H \cdot t$$

where m_H = number of "lines" of $\mathcal{H}_p$ on a "point" of $\mathcal{H}_p$. But m_H may be interpreted in Γ as being the number of quads on a line $L \in \mathcal{H}_p$ having p as center. Since no quad on L can have two centers, and the center of each quad on L lies on L, we see that the number of quads on L is

$$(4.8) \qquad\qquad\qquad m_H(1 + s) = m.$$

Then (4.6) and (4.8) imply whence $m_H = 1$. Then $r = 1 + t + t^2 = |R_p|$ so R_p is a projective plane for each point p. It follows that any two distinct quads having a non-empty intersection must meet at a line. Then by Lemma 2.2(i), all non-incident point-quad pairs are classical. This implies (by [5], Lemma 2.2.1) that Γ is a dual polar space. The proof is complete.

Open Problem: Determine whether the conclusion of Theorem 2 also holds when Γ does not have parameters, for example, when Γ is infinite and all quads are classical.

§5. The Finite Case.

THEOREM 3. *Assume H is a geometric hyperplane of a classical near hexagon (rank 3 dual polar space) Γ. Suppose $H \cap Q$ is a star for each quad Q. Then:*

(i)   *There is a bijection $Q : H \to$ Quads (Γ). Two points h_1 and h_2 of H are collinear if and only if the two quads $Q(h_1)$ and $Q(h_2)$ share a common line of H.*

(ii)   *Let $\Delta = (\mathcal{P}', \mathcal{L}')$ be the natural polar space of which Γ is a dual. Then there is an embedding $f : (H, \mathcal{L}(H)) \to (\mathcal{P}', \mathcal{L}')$ of the hexagon H into Δ so that the points of H are mapped bijectively into $\mathcal{P}'$ and lines of H are mapped into lines of $\mathcal{L}'$.*

(iii)   *In case Γ is finite, Δ is the polar space of type $\Omega(7, q)$ and H is a generalized hexagon of type $G_2(q)$.*

PROOF:

(i) If $p \in H$, then for each quad Q on p, there is exactly only line of H on p in Q, or else $p = z(Q)$ and all lines of Q on p belong the H. Thus the set $\mathcal{L}(H)_p$ is a subspace of $\Gamma^p = (\mathcal{L}_p, Q_p)$ which meets each "line" in one or all of its "points". Now Γ^p is a plane (since Γ is a dual polar space of rank 3) and so $\mathcal{L}(H)_p = \mathcal{L}_p$ or all lines on p within some quad of Q_p. The first alternative is impossible by Corollary 1. Thus we see:

(5.1) For each point p in H, there is a unique quad $Q(p)$ such that p is the center of $Q(p)$.

If p is collinear with another point q in H, then $q \in p^\perp \cap Q(p)$ and $p \in q^\perp \cap Q(q)$ so $Q(p) \cap Q(q)$ is the line pq. (The possibility that $Q(p) = Q(q)$ is out, since then $p = z(Q(p)) = q$.) Clearly, then Q is a bijection $H \to$ Quad (Γ) and collinearity of distinct points p and q in H implies that $Q(p) \cap Q(q) = pq$ is a line of H. One notes that the mapping $z :$ Quad $(\Gamma) \to H$ taking each quad to its center is a right and left inverse of the mapping $Q : H \to$ Quad (Γ).

(ii) The polar space $\Delta = (\mathcal{P}', \mathcal{L}')$ is simply the incidence system

$$(\text{Quad } (\Gamma), \mathcal{L}').$$

By (i) the mapping $Q : H \rightarrow$ Quad (Γ) is an embedding – since, as p ranges over a line L of H, the quads $Q(p)$ pairwise intersect at L and comprise all possible quads of Γ on L.

(iii) Now assume Γ and hence Δ is finite. Let P be the ambient projective space in which the polar space Δ is embedded. Then the points of the geometric hyperplane H of Γ are a family $\mathcal{H}$ of maximal singular subspaces (which are projective planes of P. The lines of H are the lines L of L such that every maximal singular subspace of P containing it lies in $\mathcal{H}$: this is so as H is a **subspace** of Γ. Now by (ii) the mapping $Q : H \rightarrow$ Quad (Γ) induces the bijection $Q^* : \mathcal{H} \rightarrow \mathcal{P}' =$ points of Δ, such that for each line L of H (which is already a line of Δ) Q^* bijectively maps the set $\mathcal{H}(L)$ of all maximal singular subspaces of Δ on L onto the point shadow of L in $\Delta : \mathcal{H}(L) \rightarrow L$. This means that the number $m := |\mathcal{H}(L)| = s := |L|$. But if Δ is type $\Omega^-(8,q)$, $U(6,q^2)$ or $U(7,q^2)$, $(m,s) = (q^2,q), (q^2,q)$ and (q^2,q^3) respectively against $m = s$. Thus Δ is type $Sp(6,q)$ or $\Omega(7,q)$ and $m = s = q$. Now if Π is a plane in $\mathcal{H}$, the lines of H incident with Π are precisely the lines of Π incident with $Q^*(\Pi)$. This means that in the embedding e of the generalized hexagon H in the projective space P obtained by composing the embedding of H in Δ given in (ii) with the natural embedding of Δ into its ambient projective space P, the lines of H through a point h of H are mapped to the set of lines on $e(h) = Q^*(\Pi)$ lying in the plane Π of Δ. Since there are $1 + q$ of these lines we see

(5.2) The embedding $e : H \rightarrow P = PG(d,q)$ (where P is the ambient space of $\Delta \simeq Sp(6,q)$ or $\Omega(7,q)$) is such that:

(i) H is a generalized hexagon of order (q,q).

(ii)   For each point $h \in H$, $\langle e(h^\perp) \rangle$ is a plane of P (Here " $\perp$ " is taken within H alone).

We shall show next, that in addition,

(iii) If h is a point of H, let $H_2(h) = \{k \in H \mid d_H(h,k) \leq 2\}$. Then $e(H_2(h))$ consists of all polar points of $\mathcal{P}'$ lying in the hyperplane $e(h)^\perp$ where " $\perp$ " is the polarity of P whose absolute points are $\mathcal{P}'$.

Since H is a generalized hexagon of order (q,q), $|H_2(h)| = 1 + q + q^2 + q^3 + q^4$, which is the cardinality of $e(h)^\perp \cap \mathcal{P}'$ as Δ is type $Sp(6,q)$ or $\Omega(7,q)$. Thus it suffices to show that $e(H_2(h)) \subseteq e(h)^\perp$.

Fix h and suppose q is a point of $H_2(h)$. We have seen that if $d_H(h,q) = 1$, $e(q)$ lies in the singular plane Π of Δ and so $e(q)$ lies in $e(h)^\perp$ in Δ. So suppose $d_H(h,q) = 2$. Now as points of Γ it follows that $z = z(\langle h,q \rangle)$ is a point of $h^\perp \cap q^\perp$ ("perps" here taken in Γ). Thus the quads Q_h and Q_q are distinct from $\langle h,q \rangle$ but intersect at the point z. Since Γ is a **classical** near hexagon Q_h and Q_q intersect at a line L, which does not belong to H. None the less, lines of Γ, the dual polar space, are precisely the lines of Δ, the corresponding polar space. But Q_h and Q_q are precisely the points $e(h)$ and $e(q)$ of P in $\mathcal{P}'$. Thus, as they are collinear by the line L of Δ, $e(q)$ lies in $e(h)^\perp \cap \mathcal{P}'$ as required.

Now from (3.2)(i), $|H_2(h)| = 1 + q + q^2 + q^3 + q^4$ as we have seen and (ii) and (iii) imply that (2.3)(i) and (ii) and all three conditions (e), (f) and (g) of the Cameron-Kantor Lemma hold. By Thas' criterion (the proposition of subsection 2.3) and its application here, it follows that H is type $G_2(q)$ (in both cases of the embeddings $e : H \rightarrow \Omega(7,q)$ or $H \rightarrow Sp(6,2^a)$).

REFERENCES

[1] Cameron, P. and Kantor, W. *2-transitive and antiflag transitive collineation groups of finite projective spaces*, J. Algebra **60** (1979) no. 2, 384-422.

[2] Ronan, M. *A geometric characterization of Moufang hexagons*, Inv. Math. **57** (1980), 227-262.

[3] Ronan, M. and Smith, S. *Sheaves on buildings and modular representations of Chevalley groups*, J. Alg. **96** (1985), 319-346.

[4] Schellekens, G.J. *On a hexagonic structure I, II,* (Nedrl. Akad. Wetensch. Proc. A65=) Indag. Math. **24** (1962), 201-234.

[5] Shult, E. and Yanushka, A. *Near n-gons and line systems*, Geom. Dedicata **9** (1980), 1-72.

[6] Thas, J. *A remark on the theorem of Yanushka and Ronan characterizing the "generalized hexagon" H(q) arising from the group $G_2(q)$*, J. Comb. Theory (A) **29**, (1980), 361-362.

[7] Yanushka, A. *Generalized hexagons of order (t,t)* Israel J. Math. **23** (1976), 309-324.

On simplicial complexes related to the Suzuki sequence graphs

Leonard H. Soicher

1 Introduction

In [9], describing the construction of his sporadic simple group Suz, Suzuki constructs a remarkable sequence of graphs Γ_n ($n = 1, \ldots, 6$), the *Suzuki sequence graphs*, starting with the empty graph on four vertices. The graphs $\Gamma_1, \ldots, \Gamma_6$ have respectively

$$4, \ 14, \ 36, \ 100, \ 416, \ 1782$$

vertices, and automorphism groups

$$S_4, \ L_3(2){:}\,2, \ U_3(3){:}\,2, \ J_2{:}\,2, \ G_2(4){:}\,2, \ Suz{:}\,2.$$

If $1 < n \leq 6$, then the following properties hold:

1. Γ_n is connected.

2. Γ_n is ordered-k-clique transitive ($k = 1, \ldots, n$).

3. Γ_n is locally Γ_{n-1} (i.e. the induced subgraph on the neighbours of a vertex of Γ_n is a copy of Γ_{n-1}).

One may ask to what extent properties (1)–(3) characterize Γ_n, and in this paper we attack a more general problem in the context of simplicial complexes (it turns out that only Γ_4, Γ_5 are characterized by (1)–(3)). Our investigations lead naturally to an ordered-k-clique transitive graph ($k = 1, \ldots, 7$), and a new flag-transitive rank 8 Buekenhout geometry, both having automorphism group $2 \times Co_1$. We also discover a new distance-transitive graph which is a triple cover of Γ_6.

2 Γ_n-complexes

We start with some definitions, in order to be able to describe our main objects of study, Γ_n-complexes, which are generalizations of the clique complexes of the Suzuki sequence graphs Γ_n ($n \geq 2$). Many of our notions come from the study of finite geometries (see [2]), and our notation for group structures and presentations is that of the ATLAS [3].

A collection K of subsets of a set Ω is a *simplicial complex* with *point set* Ω if all 1-subsets of Ω are in K and each subset of any element of K is again in K. Let K be a simplicial complex with point set Ω. Since Ω is recoverable from K we may call the elements of Ω the *points* of K. An element S of K is called a *simplex* of *rank* $\operatorname{rk} S = |S|$. The *rank* $\operatorname{rk} K$ of K is the supremum the ranks of its simplices. Two simplices are *incident* if one is contained in the other, and a set of pairwise incident simplices is called a *flag*. An automorphism of K is a permutation σ of the points Ω of K, such that $S \subseteq \Omega$ is a simplex exactly when $S\sigma$ is. We say that K is *flag-transitive* if given any flags F_1, F_2 of K such that $\{\operatorname{rk} S \mid S \in F_1\} = \{\operatorname{rk} S \mid S \in F_2\}$, there exists a $\sigma \in \operatorname{Aut} K$ such that $F_1\sigma = F_2$. Note that if K has finite rank then flag-transitivity is the same as the property that all maximal simplices are equivalent under $\operatorname{Aut} K$ and the stabilizer in $\operatorname{Aut} K$ of a maximal simplex M induces the full symmetric group on M.

Given a simplicial complex K, the *point graph* $\Gamma(K)$ is the graph whose vertices are the points of K and edges the rank 2 simplices of K. We call K *connected* if its point graph is. Conversely, given a graph Γ, we can form a simplicial complex $K(\Gamma)$, the *clique complex*, by taking the simplices to be the cliques of Γ. It is not always true that $K = K(\Gamma(K))$, but we always have $\Gamma = \Gamma(K(\Gamma))$. The *link* $\operatorname{lk} S = \operatorname{lk}_K S$ of a simplex S of K is the simplicial complex

$$\{R - S \mid R \in K,\ R \supseteq S\},$$

with the obvious point set.

We call K *link-connected* if $\operatorname{lk} S$ is connected for every $S \in K$ with $\operatorname{rk} S \leq \operatorname{rk} K - 2$. If $\operatorname{rk} K \geq 2$ then link-connected implies connected, as $\operatorname{lk} \{\} = K$.

It follows from properties (1)–(3) above that the clique complexes $K(\Gamma_n)$ of the the Suzuki sequence graphs Γ_n are flag-transitive and link-connected, and if $n \geq 2$ then

$$\operatorname{lk}_{K(\Gamma_n)}(\text{rank } n - 2 \text{ simplex}) \cong K(\Gamma_2).$$

Definition 1 A simplicial complex K is a Γ_n-*complex* if K is flag-transitive, link-connected, of rank $n \geq 2$, and $\operatorname{lk}_K(\text{rank } n - 2 \text{ simplex}) \cong K(\Gamma_2)$.

We would like to classify the Γ_n-complexes, and in this paper present a partial classification for n up to 8.

We pause to describe the graph Γ_2, which plays an important role in what follows. The vertices of Γ_2 can be taken to be the seven points and seven lines of the projective plane of order 2. A typical point (line) is joined just to those four lines (points) it is not on (does not contain). The graph Γ_2 can also be regarded as the incidence graph of the unique $2-(7,4,2)$ design.

The automorphism group $\operatorname{Aut}\Gamma_2$ of Γ_2 is obtained from the automorphism group $L_3(2)$ of the plane of order 2 by adjoining an automorphism which interchanges points and lines, and so $\operatorname{Aut}\Gamma_2 \cong L_3(2){:}2 \cong PGL_2(7)$. The graph Γ_2 is vertex, edge, and quadrangle transitive. The vertex stabilizer is S_4, acting naturally on the neighbourhood of a vertex, the edge stabilizer is $S_3 \times 2$, and the quadrangle stabilizer is D_{16}. Note that a quadrangle in Γ_2 is determined by any three of its vertices. Finally, we remark that the only subgroup of $\operatorname{Aut}\Gamma_2$ which acts transitively on the 56 incident (vertex,edge) pairs (flags) of Γ_2 is the whole group $L_3(2){:}2$.

We end this section with a useful result of G. Weetman. Note that we often write $\operatorname{lk} p$ for $\operatorname{lk}\{p\}$, when p is a point of a simplicial complex.

Theorem 1 (Weetman) *Let K be a connected simplicial complex of rank ≥ 3, $G \leq \operatorname{Aut}K$, and let $\overline{G_p}$ denote the restriction to $\operatorname{lk} p$ of the stabilizer G_p of a point p. Suppose that $(\overline{G_p})_q$ acts faithfully on $\operatorname{lk}_{\operatorname{lk} p}q$ for each rank 2 simplex $\{p,q\}$.*
Then G_p acts faithfully on $\operatorname{lk} p$ for each point p.

Proof Consider the point graph $\Gamma = \Gamma(K)$, and suppose that $H \leq G$ fixes each point at distance $\leq i$ from p, where i is fixed, $1 \leq i < \operatorname{diameter}(\Gamma)$ (if Γ has diameter 1 there is nothing to prove). We show that all points at distance $i+1$ are fixed by H, and so by induction, the stabilizer of p and its neighbours fixes all points.

Let x,y,z be a path in Γ such that x,y,z are at respective distances $i-1,i,i+1$ from p. In $\operatorname{lk} y$, H fixes x and each point of $\operatorname{lk}_{\operatorname{lk} y}x$, and so by assumption H is the identity on $\operatorname{lk} y$. Thus H fixes z.   □

Corollary 1 *Let K be a Γ_n-complex. Then the stabilizer in $\operatorname{Aut}K$ of a point p acts faithfully on $\operatorname{lk} p$.*

Proof In $\operatorname{Aut}K(\Gamma_2) \cong L_3(2){:}2$ the stabilizer S_4 of a point acts faithfully on the link of that point. For $n \geq 3$ apply Theorem 1 and induction.   □

3 Γ_n-maps

We now define the group U_n, and shall show that the automorphism group of any Γ_n-complex is a quotient of U_n. The group U_n is presented by a Coxeter graph together with a single additional relation.

Definition 2 The group U_n $(n \geq 2)$ is defined by the presentation

$$\left\langle \underset{u_n}{\circ}\text{—}\underset{u_{n-1}}{\circ}\cdots\underset{u_3}{\circ}\text{—}\underset{u_2}{\circ}\overset{8}{\text{———}}\underset{u_1}{\circ}\text{—}\underset{u_0}{\circ}\text{———}\underset{a}{\circ}\mid a = (u_1 u_2)^4\right\rangle \qquad (*)$$

Note that the Coxeter graph relations for $u_2, \ldots, u_n$ define the symmetric group S_n. We also remark that coset enumeration shows that $U_2 \cong L_3(2){:}2$.

Definition 3 Let $\phi : U_n \to G$ be an epimorphism, and denote $x\phi$ by $\bar{x}$. The simplicial complex $K(\phi)$ is defined as follows. The points of $K(\phi)$ are the right cosets in G of $H = \langle \bar{a}, \bar{u}_0, \ldots, \bar{u}_{n-1} \rangle$, and the simplices of $K(\phi)$ are the images under right multiplication by G of the subsets of

$$M = \{H, H\bar{u}_n, H\bar{u}_n\bar{u}_{n-1}, \ldots, H\bar{u}_n\bar{u}_{n-1}\ldots\bar{u}_2\}.$$

Keeping the notation of Definition 3 we have:

Theorem 2 *G acts flag-transitively on $K(\phi)$. Furthermore, $K(\phi)$ is connected, with M a maximal simplex, and if $H \neq G$ then $K(\phi)$ has rank n.*

Proof If $H = G$ then $K(\phi)$ has just one point and there is nothing to prove. Assume now that $H \neq G$, and let $S = \langle \bar{u}_2, \ldots, \bar{u}_n \rangle$. Now S is a quotient of the symmetric group S_n, and in fact $S \cong S_n$, for otherwise $\bar{u}_n$ would be identified with an element of H, and we would have $H = G$. Therefore $S \cap H = \langle \bar{u}_2, \ldots, \bar{u}_{n-1} \rangle \cong S_{n-1}$, and S acts naturally on M as S_n (if $p_i = H\bar{u}_n\bar{u}_{n-1}\ldots\bar{u}_{i+1}$, then $\bar{u}_i$ acts on M as $(p_{i-1}\ p_i)$ $(i = 2, \ldots, n)$). Flag-transitivity follows. Connectedness follows from the facts that $G = \langle H, \bar{u}_n \rangle$ and $\{H, H\bar{u}_n\}G$ is the edge set of $\Gamma(K(\phi))$. $\square$

Definition 4 A Γ_n-*map* is an epimorphism

$$U_n \to \operatorname{Aut} K, \ x \mapsto \bar{x}$$

such that K is a Γ_n-complex, and for some point p of K

1. $\operatorname{stab}_{\operatorname{Aut} K}(p) = \langle \bar{a}, \bar{u}_0, \ldots, \bar{u}_{n-1} \rangle$, and

2. the symmetric group $\langle \bar{u}_2, \ldots, \bar{u}_n \rangle$ acts naturally on a maximal simplex containing p.

Note that if $\phi : U_n \to \operatorname{Aut} K$ is a Γ_n-map then $K \cong K(\phi)$.

Theorem 3 *Let K be a Γ_n-complex, and $\{p_1, \ldots, p_n\}$ a maximal simplex of K. Then the following hold:*

1. *There exists a Γ_n-map $\phi : U_n \to \operatorname{Aut} K$, $x \mapsto \bar{x}$, such that $\langle \bar{a}, \bar{u}_0, \ldots, \bar{u}_i \rangle$ is the pointwise stabilizer in $\operatorname{Aut} K$ of $\{p_{i+1}, \ldots, p_n\}$ $(i = 0, \ldots, n)$, the symmetric group $\langle \bar{u}_2, \ldots, \bar{u}_n \rangle$ acts naturally on $\{p_1, \ldots, p_n\}$, and if $\psi : U_n \to \operatorname{Aut} K$ is any Γ_n-map then $\psi = \phi\tau$ for some inner automorphism τ of $\operatorname{Aut} K$.*

2. *No proper subgroup of $\operatorname{Aut} K$ acts flag-transitively on K.*

3. *If K_i are Γ_n-complexes, and $\phi_i : U_n \to \operatorname{Aut} K_i$ are Γ_n-maps $(i = 1, 2)$, then $\ker \phi_1 = \ker \phi_2$ if and only if $K_1 \cong K_2$.*

Proof Let $P_i = \{p_{i+1}, \ldots, p_n\}$ $(i = 0, \ldots, n)$, $G = \operatorname{Aut} K$, and denote by G_X and $G_{(X)}$ respectively the setwise and pointwise stabilizer in G of X.

We first find generators $\bar{x}$ of G which satisfy the defining relations of the generators x of U_n, subject to the requirements that $G_{p_n} = \langle \bar{a}, \bar{u}_0, \ldots, \bar{u}_{n-1} \rangle$, and $\langle \bar{u}_2, \ldots, \bar{u}_n \rangle$ acts naturally on $\{p_1, \ldots, p_n\}$. We shall find that our set of choices is then determined, up to applying an inner automorphism of $G_{\{p_1, \ldots, p_{n-1}\}}$.

The only flag-transitive subgroup of $\operatorname{Aut} K(\Gamma_2)$ is the whole group, and so $G_{(P_2)}$ must act as $L_3(2){:}2$ on $\operatorname{lk} P_2$. It follows from Corollary 1 that this action is faithful, and we deduce that $G_{(P_1)} \cong S_4$, $G_{(P_0)} \cong S_3$. Due to flag-transitivity, G_{P_0} must act as the symmetric group S_n on P_0, and so G_{P_0} is of shape $S_3.S_n$; but any group of this shape is isomorphic to $S_3 \times S_n$.

The requirements above on the map $x \mapsto \bar{x}$ imply that the images $\bar{a}, \bar{u}_0$, being in G_{p_n} and centralizing $\bar{u}_2, \ldots, \bar{u}_n$, must be in $G_{(P_0)} \cong S_3$. We thus take $\bar{a}, \bar{u}_0$ to be involutory generators of $G_{(P_0)}$.

Now $\bar{u}_2, \ldots, \bar{u}_n$ must be transpositions in the S_n-factor of $G_{P_0} \cong S_3 \times S_n$ (even if $n = 6$), and without loss of generality we may take $\bar{u}_2, \ldots, \bar{u}_n$ to be the transpositions acting as $(p_1\ p_2), (p_2\ p_3), \ldots, (p_{n-1}\ p_n)$ on P_0.

Now the image $\bar{u}_1$ is forced to lie in $G_{(P_1)}$, and so we take $\bar{u}_1$ to be the unique involution in $G_{(P_1)} \cong S_4$, such that

$$1 = (\bar{a}\bar{u}_1)^2 = (\bar{u}_0\bar{u}_1)^3.$$

Now $\bar{a}, \bar{u}_0, \bar{u}_1, \bar{u}_3, \bar{u}_4, \ldots, \bar{u}_n$ generate $G_{P_1} \cong S_4 \times S_{n-1}$, and since $\bar{u}_3, \ldots, \bar{u}_n$ generate the quotient S_{n-1} and centralize $G_{(P_0)} \cong S_3$, we have that $\bar{u}_3, \ldots, \bar{u}_n$ must centralize $G_{(P_1)} = \langle \bar{a}, \bar{u}_0, \bar{u}_1 \rangle$.

Now $\langle \bar{a}, \bar{u}_0, \bar{u}_1, \bar{u}_2 \rangle$ fixes P_2 pointwise, and so acts faithfully on $L = \operatorname{lk} P_2$. Let p be the unique point in $\operatorname{lk} P_1$ such that $p \neq p_1$ is fixed by $\bar{a}$. Now the points p_2, p_1, p are contained in a unique quadrangle Q of L. It is not hard to see that $\langle \bar{u}_1, \bar{u}_2 \rangle$ acts as D_8 on Q, which is fixed pointwise by $\bar{a}$, and so $\langle \bar{a}, \bar{u}_1, \bar{u}_2 \rangle$ of shape $2.D_8$ is the full quadrangle stabilizer D_{16}, and we have

$$\bar{a} = (\bar{u}_1\bar{u}_2)^4$$

(for otherwise D_{16} would be of shape $2 \times D_8$). We have thus finished showing that $\bar{a}, \bar{u}_0, \ldots, \bar{u}_n$ satisfy the defining relations of U_n, and that this set of images of the generators of U_n is determined up to an inner automorphism of G by the requirement that $x \mapsto \bar{x}$ defines a Γ_n-map.

We finish proving (1) by showing that $G_{(P_i)} = \langle \bar{a}, \bar{u}_0, \ldots, \bar{u}_i \rangle$. We do this by induction on i, and note that the assertion is certainly true for $i = 0, 1$. Now fix i, $1 \leq i \leq n - 1$, and assume that $G_{(P_i)} = H_i$, where $H_i = \langle \bar{a}, \bar{u}_0, \ldots, \bar{u}_i \rangle$. Therefore H_i is the stabilizer of p_{i+1} in the (transitive) action of $G_{(P_{i+1})}$ on the the points of lk P_{i+1}. Now $\bar{u}_{i+1}$ flips the edge $\{p_i, p_{i+1}\}$ in the connected point graph of lk P_{i+1}, and so H_{i+1} is transitive on the points of lk P_{i+1}. Therefore $G_{(P_{i+1})} = H_{i+1}$, and (1) is proved.

Now any flag-transitive subgroup T of G must contain $G_{(P_2)} \cong L_3(2){:}2$, and also a subgroup acting as the full symmetric group on P_0. It follows that $T = G$, and (2) is proved.

Assertion (1) shows that the isomorphism class of K uniquely determines the kernel of a Γ_n-map $U_n \rightarrow \text{Aut}\, K$. To finish the proof of (3), suppose that $\phi_i : U_n \rightarrow \text{Aut}\, K_i$ are Γ_n-maps ($i = 1, 2$), with $\ker \phi_1 = \ker \phi_2$. Then $K_1 \cong K(\phi_1) \cong K(\phi_2) \cong K_2$. $\qquad\square$

Theorem 4 *Let L be a Γ_{n-1}-complex $(n > 2)$, $\phi : U_n \rightarrow G$, $x \mapsto \bar{x}$ an epimorphism, and $H_i = \langle \bar{a}, \bar{u}_0, \ldots, \bar{u}_i \rangle$. Suppose that G acts faithfully on the right cosets of $H = H_{n-1}$, and let ψ be the restriction of ϕ to $\langle a, u_0, \ldots, u_{n-1} \rangle$.*

Then ϕ is a Γ_n-map with $\text{lk}_{K(\phi)}(H) \cong L$ if and only if ψ is a Γ_{n-1}-map with $K(\psi) \cong L$ and $H_{n-2} = H \cap H^{\bar{u}_n}$.

Proof Suppose ϕ is a Γ_n-map ($n > 2$) with $\text{lk}_{K(\phi)}(H) \cong L$. It follows from Theorem 3 that ψ is a Γ_{n-1}-map with $K(\psi) \cong L$, and that H_{n-2} is the pointwise stabilizer in G of the cosets $\{H, H\bar{u}_n\}$, and so $H_{n-2} = H \cap H^{\bar{u}_n}$.

Conversely, suppose that ψ is a Γ_{n-1}-map with $K(\psi) \cong L$, and that $H_{n-2} = H \cap H^{\bar{u}_n}$. Let $K = K(\phi)$, and $p_i = H\bar{u}_n \bar{u}_{n-1} \ldots \bar{u}_{i+1}$ ($i = 1, \ldots, n$). Now Theorem 2 tells us that K is connected and flag-transitive, and we shall show that K is a Γ_n-complex. It will then follow that ϕ is a Γ_n-map, since G is a flag-transitive subgroup of $\text{Aut}\, K$, and so by Theorem 3, $G = \text{Aut}\, K$.

Let T be the set of H-images of the subsets of $\{p_1, \ldots, p_{n-1}\}$. Since $H = G_{p_n}$, and G acts flag-transitively on K, we have that $T = \text{lk}_K(p_n)$. Since H_{n-2} is the stabilizer in H of p_{n-1} it is easy to see that T is isomorphic to $K(\psi) \cong L$, which is a Γ_{n-1}-complex. Thus K is a Γ_n-complex such that $\text{lk}_K(p_n) \cong L$, and we are done. $\qquad\square$

4   Classifying Γ_n-complexes

We now apply Theorems 3 and 4 to classify certain Γ_n-complexes.

Theorem 5 *There are precisely two Γ_3-complexes. They are the clique complexes $K(3\Gamma_3)$ and $K(\Gamma_3)$, where $3\Gamma_3$ is the proper triple cover of Γ_3 described in [4], p.317. The automorphism groups of $K(3\Gamma_3)$ and $K(\Gamma_3)$ are $(3 \times U_3(3)){:}2$ and $U_3(3){:}2$, respectively.*

Proof Coset enumeration shows that $\langle a, u_0, u_1, u_2 \rangle \cong L_3(2){:}2$ has index 108 in U_3. It follows that $U_3 \cong (3 \times U_3(3)){:}2$. There are thus just two essentially different Γ_3-maps, and so the Γ_3-complexes can only be $K(3\Gamma_3)$ and $K(\Gamma_3)$, with respective automorphism groups $(3 \times U_3(3)){:}2$ and $U_3(3){:}2$. $\square$

Unfortunately, the structure of U_4 is not known; U_4 may in fact be infinite. In U_3, the normal subgroup of order 3 is generated by $(u_0 u_1 u_2 u_3)^8$, and we define U_n^* to be U_n factored by the normal closure of $(u_0 u_1 u_2 u_3)^8$. In [6] and [7] it is shown that $U_4^*, \ldots, U_8^*$ are respectively isomorphic to

$$J_2{:}2, \quad G_2(4){:}2, \quad 3{\cdot}Suz{:}2, \quad Co_1 \times 2, \quad 2{\cdot}(Co_1 \wr 2).$$

Definition 5 A Γ_n^*-complex is a Γ_n-complex of rank ≥ 3, such that

$$\mathrm{lk}\,(\mathrm{rank}\ n - 3\ \mathrm{simplex}) \cong K(\Gamma_3).$$

It is straightforward to apply Theorems 3 and 4 to prove:

Theorem 6 *For $n =$*

$$4, 5, 6, 7, 8$$

there are exactly

$$1, 1, 2, 2, 2$$

non-isomorphic Γ_n^*-complexes whose respective automorphism groups are

$$J_2{:}2, \quad G_2(4){:}2, \quad 3{\cdot}Suz{:}2 \ \text{or} \ Suz{:}2, Co_1 \times 2 \ \text{or} \ Co_1, \quad 2{\cdot}(Co_1 \wr 2) \ \text{or} \ Co_1 \wr 2.$$

It seems worthwhile to describe the Γ_7^*-complex whose automorphism group is $2 \times Co_1$ (see also [8]). This complex is the clique complex of a graph Γ_7, whose vertices are the $3,091,200$ $3A$-elements of Co_1, with x joined to y precisely when xy is an involution. The central involution of the automorphism group simultaneously inverts all $3A$-elements. The link of a point of any Γ_7^*-complex is isomorphic to the clique complex of a proper triple cover $3\Gamma_6$ of the Suzuki graph Γ_6, and in [8] we show that $3\Gamma_6$ is distance-transitive, with intersection array (see [1])

$$\{416, 315, 64, 1; 1, 32, 315, 416\}.$$

The link of a point of any Γ_8^*-complex is isomorphic to $K(\Gamma_7)$.

Richard Parker has a recipe for constucting $GF(2)$-matrices which satisfy the relations of U_n^*. When $n = 8$ his matrices generate $Co_1 \wr 2$, and when $n = 9$ they generate $2^k.(Co_1 \wr 2)$ for some $k > 1$. Thus Γ_9^*-complexes exist.

5   Diagram geometries from Γ_n-complexes

In this section we assume the reader is familiar with the basic concepts of diagram geometries (see [2]).

Neumaier [4] shows, amongst more general results, that a Suzuki sequence graph Γ_n ($2 \leq n \leq 6$) gives rise to a connected flag-transitive geometry of rank $n + 1$, with diagram:

$$\circ\!\!-\!\!-\!\!-\!\!\circ \cdots \circ\!\!-\!\!-\!\!-\!\!\circ\!\!=\!\!\circ\overset{c}{-\!\!-\!\!-}\circ \qquad (**)$$

For $i = 0, \ldots, n - 1$ the i-elements of this geometry are the $(i + 1)$-cliques of Γ_n, and the n-elements are the complete multipartite subgraphs $K_{n \times 2}$ (having n parts of size 2) of Γ_n. (The node at distance i from the leftmost in the diagram corresponds to the i-elements.) Incidence is symmetrized inclusion.

In fact, any Γ_n-complex gives rise to a strongly-connected flag-transitive geometry with diagram $(**)$. Let $\phi : U_n \to G$, $x \mapsto \bar{x}$ be a Γ_n-map, $K = K(\phi)$, $H = \langle \bar{a}, \bar{u}_0, \ldots, \bar{u}_{n-1} \rangle$, and define the geometry $\mathcal{G}$ as follows. For $i = 0, \ldots, n - 1$ the i-elements of $\mathcal{G}$ are the rank $i + 1$ simplices of K. The n-elements of $\mathcal{G}$ are the G-images of

$$\bigcup_{i=1}^{n} \{H v_i, H w_i\}, \quad \text{where } v_i = \bar{u}_n \bar{u}_{n-1} \ldots \bar{u}_{i+1}, \ w_i = v_1 \bar{u}_1 \bar{u}_2 \ldots \bar{u}_i,$$

and incidence is defined by symmetrized inclusion. It is not difficult to show that $\mathcal{G}$ is a strongly-connected flag-transitive geometry with diagram $(**)$.

6   Concluding remarks

Neumaier [4] was interested in rectagraphs, and multiple extensions of rectagraphs, where a *rectagraph* is a triangle-free connected graph such that every path of length 2 is contained in a unique quadrangle. The graph Γ_2 is a rectagraph, and this paper can, in part, be seen as an answer to Problem 1 of [4]. In response to Problem 4 of [4], we are now in the process of using techniques similar to those in this paper to study extensions of other rectagraphs. For example, one graph which is locally the incidence graph of the (unique) $2-(11, 5, 2)$ design gives rise to a 144 point flag-transitive rank 4 geometry with diagram $(**)$, and automorphism group $M_{12} : 2$.

It would be nice to know when we could remove our assumption of flag-transitivity. It would also be desirable to have a non-computer proof of our results, as the results of [6] and [7] depend on computer coset enumeration. Simple-connectedness arguments like those of Pasini [5] should be useful here.

I thank Graham Weetman for many useful conversations.

References

[1] A.E. Brouwer, A.M. Cohen and A. Neumaier, *Distance-Regular Graphs*, Springer, 1989.

[2] F. Buekenhout, Diagram geometries for sporadic groups, in *Finite Groups - Coming of Age*, J.McKay ed., AMS Contemporary Mathematics **45**, 1985, pp. 1–32.

[3] J.H. Conway, R.T. Curtis, S.P. Norton, R.A. Parker and R.A. Wilson, *An ATLAS of Finite Groups*, Clarendon Press, Oxford, 1985.

[4] A. Neumaier, Rectagraphs, Diagrams, and Suzuki's sporadic simple group, in *Algebraic and Geometric Combinatorics*, (E. Mendelsohn, Ed.), Ann. Discrete Math. **15**, 1982, pp. 305–318.

[5] A. Pasini, A classification of a class of Buekenhout geometries exploiting amalgams and simple connectedness, preprint.

[6] L.H. Soicher, Presentations of some finite groups, Ph.D. Thesis, Cambridge, 1985.

[7] L.H. Soicher, Presentations for Conway's group Co_1, *Proc. Camb. Phil. Soc.* **102** (1987), 1–3.

[8] L.H. Soicher, Two new distance-transitive graphs related to Suzuki's sporadic group, preprint.

[9] M. Suzuki, A finite simple group of order 448,345,497,600, in *Symposium on Finite Groups* (Brauer and Sah, Eds.), Benjamin, New York, 1969, pp. 113–119.

Twin Buildings and Groups of Kac-Moody Type

Jacques Tits

To my wife

1 Introduction

The combinatorial objects called **buildings** were first introduced (cf. [11]) to provide a geometric approach to complex simple Lie groups – in particular the exceptional ones – and later on, more generally, to isotropic algebraic simple groups (cf. eg. [12], [17]). The buildings which do arise in that way are **spherical**, that is, have finite Weyl groups. This assertion has a partial converse, proved in [12]: *roughly speaking*, there is a one-to-one correspondence between the (isomorphism classes of) buildings of irreducible spherical type and rank $r \geq 3$ and the algebraic absolutely simple groups of relative rank r, where the notion of algebraic simple groups must be suitably extended so as to include, for instance, classical groups over division rings of infinite dimension over their centre. In order to have a similar statement in the rank 2 case, one must impose an extra condition, the **Moufang condition**, on the buildings under consideration [1].

The correspondence in question is established via the classification of buildings of irreducible, spherical types and rank ≥ 3. For non-spherical buildings, a full classification is known only for the affine types, in rank ≥ 4 (cf. [19]): buildings of such a type have a "spherical building at infinity", which is the essential tool for classification. A construction procedure for buildings of more general types given in [9] shows that there cannot be any hope for a complete classification of buildings of arbitrary types. On the other hand, there is another wide class of groups, the Kac-Moody groups, which give rise to buildings, usually of non-spherical and non-affine types; these are "concrete" objects, which one also wishes to characterize geometrically.

In establishing the classification of spherical buildings, the notion of **opposite chambers** plays a crucial role. Opposite chambers correspond to opposite minimal parabolic (eg. Borel) subgroups. The existence of such pairs of chambers in a spherical building reflects the existence of an element

---

[1] These results, concerning buildings of rank 2, have not been completely written up and still need verification.

of maximal length in a finite Coxeter group. Since no such element exists in infinite Coxeter groups, it appears at first that the field of application of the arguments used for the classification of spherical buildings has no chance to be extended to other cases. However, one knows that Kac-Moody groups do possess opposite Borel subgroups. The difference between the spherical and non-spherical case is that, in the latter, opposite Borel subgroups are no longer conjugate, and hence define *different buildings*. It was Mark Ronan's idea to try to study and possibly classify non-spherical buildings by attributing to the chambers of such a building opposite chambers situated in another building. This led us to the definition of **twin buildings** (or **twinned pairs** of buildings), a notion which generalizes that of spherical buildings because a spherical building turns out to be twinned with itself in a canonical way (see 2.2, Proposition 1). One wishes of course to extend the theory of spherical buildings and, in particular, the classification, to twin buildings, where algebraic simple groups would be replaced by Kac-Moody or similar groups.

This program has succeeded to a certain extent but not completely. Two key results of the theory of spherical buildings are the Theorems 4.1.1 and 4.1.2 of [12]. The first one extends easily to twin buildings (see 4.2, Theorem 1): it shows a certain rigidity of that notion and enables one, in concrete cases, to determine the full automorphism group of given twinned pairs; for instance, the automorphism group of the twinned pair associated to an adjoint Kac-Moody group is shown to be that same group extended by the automorphism group of the ground field (cf. 4.4, Proposition 9). The situation is less satisfactory for Theorem 4.1.2 of [12]. First of all, we cannot say anything when the Coxeter matrix describing the type of the buildings has coefficients equal to infinity; in the Kac-Moody situation, this means that only generalized Cartan matrices $(A_{ij})_{i,j\in I}$ such that $A_{ij}A_{ji} \le 3$ for all i,j can be taken into consideration, which is of course a severe restriction. Furthermore, the point of Theorem 4.1.2 of [12] is that it reduces the classification of spherical buildings to that of certain substructures, the **foundations** (cf. [9] and 6.1 below), which are finite "amalgams" of buildings of rank 2. For twin buildings, which are pairs of buildings (Δ_+, Δ_-) together with a certain relation $\vartheta \subset \Delta_+ \times \Delta_-$, the opposition relation, foundations can still be defined and the arguments of [12], §4, can be carried over, but they only prove (under the assumption that the coefficients of the Coxeter matrix are all finite) that a foundation of Δ_+ determines Δ_+ and Δ_- up to isomorphism, but not ϑ (cf. §5, Theorem 2). However, if one assumes that the twinning under consideration satisfies a certain, fairly natural condition, the **Moufang condition** (which is in fact fulfilled in all known cases of rank at least 3 and irreducible type), then it is often and perhaps always true (cf. the comments following the statement of Conjecture 1′ in 6.2, and 6.4) that the foundation does determine the whole twinning $(\Delta_+, \Delta_-, \vartheta)$. This gives a strong hold on the classification problem of all Moufang twinnings of given type (m_{ij}) for finite m_{ij}'s. As a result, we are led to conjecture the existence of a wide class of new groups which appear as some sort of "non-Galois twists" of Kac-Moody groups (non-Galois, since non-trivial such twists exist even over algebraically closed fields: cf. 6.5).

The present paper covers only part of the material presented at the Durham conference. I first recall some basic definitions and facts concerning buildings and twinnings, without adding much comments; further motivation beyond what has been said in this introduction can be found in [3], [4], [5], [8], [11], [12], [13], [17], [18], [21], [22]. I then present the theorems 1 and 2 and some consequences concerning, among others, automorphisms and the classification of Moufang twinnings. Detailed proofs are not given, but the references and indications of methods which are provided should in general enable the interested reader to reconstruct complete proofs. Actually, the proof of Theorem 1 is entirely omitted, as it is totally similar to that of Theorem 4.1.1 of [12], and is also described in some detail in [21]. On the other hand, although the proof of Theorem 2 runs parallel to that of Theorem 4.1.2 of [12], I give a new outline of it, because the long and technical proof of [12] becomes somewhat better motivated when spherical buildings are replaced by the more general twin buildings; thus §5 below may be of some help for the reader of §4 of [12]. Finally, §6 deals with some aspects of the classification problem and shows the kind of new groups one may expect to come out of the process.

As was already mentioned, part of the material presented here has been developed in collaboration with Mark Ronan; detailed proofs and further results will appear in joint papers in preparation.

2 Buildings and twin buildings

2.1 Buildings

Throughout this paper, the symbol I denotes a finite set and $M = (m_{ij})_{i,j \in I}$ is a **Coxeter matrix**, that is, a symmetric matrix with coefficients in $\mathbb{N} \cup \{\infty\}$ such that $m_{ii} = 1$ for all i and $m_{ij} \geq 2$ if $i \neq j$. If a subset J of I is such that $m_{ij} = 2$ whenever $i \in I - J$ and $j \in J$, the Coxeter matrix $M_J = (m_{ij})_{i,j \in J}$ is called a **direct factor** of M; if there exists no proper nonempty subset J of I with that property, M is said to be **irreducible**. We let $(W, (s_i)_{i \in I})$ be a Coxeter system of type M: this means that W is a group generated by the s_i's and defined by the relations $(s_i s_j)^{m_{ij}} = 1$ for $m_{ij} \neq \infty$. The **length** of an element $w \in W$ relative to the generating set $S = \{s_i | i \in I\}$ is denoted by $l(w)$.

A **building of type** M, or of type (W, S) (or, by abuse of language, of type W), is a set Δ, whose elements are called **chambers**, endowed with a "distance" $d : \Delta \times \Delta \to W$ satisfying the following axioms, where $x, y \in \Delta$ and $w = d(x, y)$:

(Bu 1) $w = 1$ *if and only if* $x = y$;

(Bu 2) *if* $z \in \Delta$ *is such that* $d(y, z) = s \in S$, *then* $d(x, z) = w$ *or* ws, *and if, furthermore,* $l(ws) = l(w) + 1$, *then* $d(x, z) = ws$;

(Bu 3) *if* $s \in S$, *there exists* $z \in \Delta$ *such that* $d(y, z) = s$ *and* $d(x, z) = ws$.

The function $l \circ d : \Delta \to \mathbf{N}$ is called the *l*-**distance**. The cardinality of I is the **rank** of M, of (W, S) and of the building Δ. For $J \subset I$, we denote by W_J the subgroup of W generated by $\{s_j | j \in J\}$, and if $c \in \Delta$, the set $\Sigma_J(c) = \{x \in \Delta | d(c, x) \in W_J\}$ is called the **sphere** of **radius** W_J, or simply of radius J, and **centre** c; it is a building of type M_J, therefore, we call it also a **sphere of rank** Card J. The **Weyl group** of a building of type M is, by definition, the group W. A building is said to be **thick** if its spheres of rank 1 have cardinality at least 3, and **spherical**, or **of spherical type**, if its Weyl group is finite. The group W itself, endowed with the distance function $(x, y) \mapsto x^{-1}y$, is a building and we give the name of **Coxeter building** of type M to all buildings isomorphic to it.

In the literature (cf. eg. the references given in the introduction), buildings are more commonly viewed as simplicial complexes. The relation with the definition adopted here is the following: the **vertices** of the building (Δ, d) are the spheres of **co-rank** 1 (ie. of rank Card $I - 1$) and a set of vertices is a **simplex** if the intersection of the spheres representing those vertices is not empty. Then that intersection is itself a sphere, the radius J of which is the intersection of the radii of the vertices; the simplex in question is said to be of **type** J.

2.2   Twin buildings

A **twinning**, or **twinned pair of buildings**, of type M is a pair $((\Delta_+, d_+), (\Delta_-, d_-))$ of buildings of that type endowed with a **codistance**

$$d^* : (\Delta_+ \times \Delta_-) \cup (\Delta_- \times \Delta_+) \to W$$

satisfying the following axioms, where $\epsilon \in \{+, -\}$, $x \in \Delta_\epsilon$, $y \in \Delta_{-\epsilon}$ and $w = d^*(x, y)$:

(Tw 1) $d^*(y, x) = w^{-1}$;

(Tw 2) *if* $z \in \Delta_{-\epsilon}$ *is such that* $d_{-\epsilon}(y, z) = s \in S$ *and* $l(ws) = l(w) - 1$, *then* $d^*(x, z) = ws$;

(Tw 3) *if* $s \in S$, *there exists* $z \in \Delta_{-\epsilon}$ *such that* $d_{-\epsilon}(y, z) = s$ *and* $d^*(x, z) = ws$.

Two chambers $x \in \Delta_+$ and $y \in \Delta_-$ are said to be **opposite** if their co-distance $d^*(x, y)$ is 1. It can be shown (cf. [21]) that the opposition relation between Δ_+ and Δ_- determines the twinning. The following propositions are easy.

Proposition 1 *Suppose W is finite, let w° be its longest element, and let (Δ, d) be a building of type M. Let Δ_ϵ, with $\epsilon = +$ or $-$, be two copies of Δ and let the functions $d_\epsilon : \Delta_\epsilon \times \Delta_\epsilon \to W$, $d^* : (\Delta_+ \times \Delta_-) \cup (\Delta_- \times \Delta_+) \to W$ be defined by $d_+ = d$, $d_- = w^\circ d w^\circ$, $d^* = w^\circ d$ on $\Delta_+ \times \Delta_-$ and $= d w^\circ$ on $\Delta_- \times \Delta_+$. Then $((\Delta_+, d_+), (\Delta_-, d_-), d^*)$ is a twinning and all twinnings of (spherical) type M are obtained in that way up to isomorphism.*

Thus the twinnings of spherical type M are "nothing else but" the buildings of type M.

Proposition 2 *Let (Δ, d) be a Coxeter building. Then, the function d taken as codistance defines a twinning of (Δ, d) with itself and any twinning of (Δ, d) with some building is isomorphic to that one.*

2.3   Apartments

An **apartment** in a building (Δ, d) of type M is a subset A of Δ which, together with the restriction of d to $A \times A$, is a Coxeter building of type M.

Let $((\Delta_\pm, d_\pm), d^*)$ be a twinning. If $c_+ \in \Delta_+$ and $c_- \in \Delta_-$ are opposite chambers, we set, for $\epsilon = +$ or $-$, $A(c_\epsilon, c_{-\epsilon}) = \{x \in \Delta_\epsilon | d_\epsilon(c_\epsilon, x) = d^*(c_{-\epsilon}, x)\}$; this is an apartment of Δ_ϵ.

Proposition 3   (I) *If A is an apartment of Δ_+, every chamber of Δ_- is opposite to at least one chamber of A.*

(II) *For a pair of apartments $A_+ \in \Delta_+$, $A_- \in \Delta_-$, the following properties are equivalent:*

   (i) *the restriction of d^* to $(A_+ \times A_-) \cup (A_- \times A_+)$ is a twinning of Coxeter buildings;*

   (ii) *A_+, A_- are the apartments $A(c_+, c_-)$, $A(c_-, c_+)$ for some opposite chambers c_+, c_-;*

   (iii) *a chamber of Δ_- belongs to A_- if and only if it is opposite to only one chamber of A_+.*

The proof is an easy exercise. Two apartments satisfying the conditions (i) to (iii) of (II) are said to be **opposite** to each other. An apartment of Δ_+ is called **admissible** (with respect to the given twinning) if it has an opposite, which is then unique by (iii). In a spherical building, every apartment is admissible (with respect to the " self-twinning" described in Proposition 1).

3   The group-theoretical approach

3.1   BN-pairs

Let (Δ, d) be a building of type (W, S) and let G be a group operating on Δ, preserving d and permuting the ordered pairs of chambers at given distance transitively. Choose a chamber $c_0 \in \Delta$ and let B denote its stabilizer in G. Then, clearly, the map $g \mapsto d(c_0, gc_0)$ of G in W factorizes through a bijection $\beta : B \backslash G / B \to W$. Conversely, let G be a group, B a subgroup and $\beta : B \backslash G / B \to W$ a bijection. Set $\Delta = G/B$ and let the function $d : \Delta \times \Delta \to W$ be defined by $d(x, y) = \beta(x^{-1}y)$. Then, for the pair (Δ, d), the axioms (Bu 1) to (Bu 3) take the following simple form, where we set $C(w) = \beta^{-1}(w)$:

(B1) *if* $w \in W$ *and* $s \in S$, *then* $C(ws) \subset C(w)C(s) \subset C(w) \cup C(ws)$, *and if, furthermore,* $l(ws) = l(w) + 1$, *then* $C(w)C(s) = C(ws)$.

Under that condition, (Δ, d) is a building, which is thick if and only if

(Th) $[C(s) : B] \geq 2$ *for all* $s \in S$.

In fact, when (Th) holds, the last assertion of (B1) is a consequence of the first one and (B1) can then be replaced by

(B) *for* $w \in W$ *and* $s \in S$, *one has* $C(ws) \subset C(w)C(s) \subset C(w) \cup C(ws)$.

Moreover, (Th) and (B) *imply* the injectivity of the map $C : W \to B \backslash G / B$ hence also, *assuming its surjectivity*, the existence of the bijection $\beta = C^{-1}$. Now, given a group G, a subgroup B, a *set* W and a bijection $C : W \to B \backslash G / B$ (in other words, a labeling of the double cosets of B in G), there are at most one group structure on W and one subset S of W such that (B) and (Th) hold, and (W, S) is then a Coxeter system. (For the proof of those results, cf. [2], IV, §2, exercise 3.)

A special case of the above situation is that of BN-pairs, or Tits systems (cf. [2],[12]). Here, the bijection C is given by means of a subgroup N of G together with a surjective homomorphism $\nu : N \to W$ whose kernel is contained in B: one sets $C(w) = B.\nu^{-1}(w).B$. The first inclusion of property (B) is then automatic, and (B) can be replaced by

(BN) *for* $w \in W$ *and* $s \in S$, *one has* $C(w)C(s) \subset C(w) \cup C(ws)$.

If W is finite, every system (G, B, β) satisfying (B) and (Th) comes from a BN-pair.

Examples.

(a) A variety of examples of BN-pairs can be found in [2], [4], [12], [20] and in other papers mentioned in the bibliography.

(b) Let us give an example of a system (G, B, β) which satisfies the axioms (B) and (Th) but does not come from a BN-pair. To that end, consider the group $\hat{G}$ of rational points of an almost simple simply connected isotropic algebraic group $\mathcal{G}$ over a locally compact local field $\hat{K}$. Following [3], [16], a building $(\Delta; d)$ of affine type is associated to this group. Then, any subgroup G of $\hat{G}$ which is dense (in the "natural", locally compact topology) is transitive on the set of pairs of chambers at any given distance, hence, by the above considerations, is part of a system (G, B, β) satisfying the axioms (B), (Th); in general, that system does not come from a BN-pair. One can, for instance, assume $\mathcal{G}$ defined over some global field K, dense in $\hat{K}$, and take $G = \mathcal{G}(K)$. The case where $\mathcal{G}$ is anisotropic over K is especially striking.

3.2   Twin BN-pairs

Let $((\Delta_\epsilon, d_\epsilon)_{\epsilon=+,-}, d^*)$ be a twinning of type (W, S) and let G be a group operating on Δ_+ and Δ_-, preserving the functions d_+, d_-, d^* and transitive on the set of pairs of opposite chambers. Choose two opposite chambers $c_+ \in \Delta_+$, $c_- \in \Delta_-$, set $A_\epsilon = A(c_\epsilon, c_{-\epsilon})$ (cf. 2.3), denote by B_ϵ the stabilizer of c_ϵ in G and by N the common stabilizer of A_+ and A_-. For $n \in N$, we have the equality $d_+(c_+, nc_+) = d_-(c_-, nc_-)$, the map $\nu : N \to W$ defined by $\nu(n) = d_\epsilon(c_\epsilon, nc_\epsilon)$ is a homomorphism and, if Δ_+, Δ_- are thick, the systems (G, B_ϵ, N, ν) are BN-pairs which, following 3.1, define the buildings $(\Delta_\epsilon, d_\epsilon)$. Furthermore, the map $g \mapsto d^*(c_+, gc_-)$ of G in N clearly factorizes through a bijection $B_+ \backslash G / B_- \mapsto W$, the inverse of which is easily seen to be the map $w \mapsto B_+ \nu^{-1}(w) B_-$.

Conversely, let (G, B_ϵ, N, ν) be two BN-pairs of type (W, S) with common groups G, N and homomorphism $\nu : N \to W$, such that the map $w \mapsto B_+ \nu^{-1}(w) B_-$ is a bijection of W onto $B_+ \backslash G / B_-$. For $w \in W$, set $C_\epsilon(w) = B_\epsilon \nu^{-1}(w) B_\epsilon$ and $C_\epsilon^*(w) = B_\epsilon \nu^{-1}(w) B_{-\epsilon}$. Set also $\Delta_\epsilon = G / B_\epsilon$, define $d_\epsilon : \Delta_\epsilon \times \Delta_\epsilon \to W$ by $C_\epsilon(d_\epsilon(x, y)) = x^{-1}y$ for $x, y \in \Delta_\epsilon$, and $d^* : (\Delta_+ \times \Delta_-) \cup (\Delta_- \times \Delta_+) \to W$ by $C_\epsilon^*(d^*(x, y)) = x^{-1}y$ for $(x, y) \in \Delta_\epsilon \times \Delta_{-\epsilon}$. The pairs $(\Delta_\epsilon, d_\epsilon)$ are thick buildings. For such $(\Delta_\epsilon, d_\epsilon)$ and d^*, the conditions (Tw 1) to (Tw 3) boil down to:

(TBN) *if $w \in W$ and $s \in S$ are such that $l(sw) = l(w) - 1$, then, for $\epsilon = +$ or $-$, $C_\epsilon(s) C_\epsilon^*(w) = C_\epsilon^*(sw)$.*

Consider now two arbitrary BN-pairs (G, B_ϵ, N, ν) with the same G, N and ν, and satisfying (TBN). Since, for all $s \in S$, $C_\epsilon(s) C_\epsilon(s) = B_\epsilon \cup C_\epsilon(s)$, the condition (TBN) implies

(TBN') *if $w \in W$ and $s \in S$ are such that $l(sw) = l(w) + 1$, then, for $\epsilon \in \{+, -\}$, $C_\epsilon(s) C_\epsilon^*(w) = C_\epsilon^*(w) \cup C_\epsilon^*(sw)$.*

From (TBN) and (TBN'), it follows that $B_\epsilon N B_{-\epsilon} = \bigcup_{w \in W} C_\epsilon^*(w) = G$. In other words, the map $C_\epsilon^* : W \to B_\epsilon \backslash G / B_{-\epsilon}$ is *surjective*. It is also injective if and only if

(TBN$_0$) *for all* $s \in S$, $B_+ . \nu^{-1}(s) \cap B_- = \emptyset$.

Indeed, assume that there exist $w', w'' \in W$ such that $w' \neq w''$ and $C_\epsilon^*(w') = C_\epsilon^*(w'')$. Induction on $\inf\{l(w'), l(w'')\}$ and (TBN) show that there exists $w \in W$ such that $w \neq 1$ and $C_\epsilon^*(w) = C_\epsilon^*(1)$. Let $s \in S$ be such that $l(sw) = l(w) - 1$; by (TBN) and (TBN'), we have

$$C_\epsilon^*(sw) = C_\epsilon(s) C_\epsilon^*(w) = C_\epsilon(s) C_\epsilon^*(1) = C_\epsilon^*(s) \cup C_\epsilon^*(1),$$

hence $C_\epsilon^*(s) = C_\epsilon^*(1)$, which contradicts the equality in (TBN$_0$). Conversely, the injectivity of C_ϵ^* clearly implies (TBN$_0$).

As a result of that discussion, we see that, if two BN-pairs (G, B_ϵ, N, ν), with the same G, N, ν, satisfy (TBN) and (TBN$_0$) – in which case we say that they are **twinned** or form a **twin set** of BN-pairs –, they determine a twinned pair of buildings on which G acts, transitively on pairs of opposite chambers. The condition that N be the full stabilizer in G of the two opposite apartments determined by the opposite chambers B_+, B_-, as was assumed at the beginning of this discussion, is expressed by the relation

(Sat) $Ker\ \nu = B_+ \cap B_-$,

which turns out to be equivalent to the condition that the two BN-pairs (or, still equivalently, one of them) be **saturated** in the sense of [12], 3.2.5.

Example. Take $G = \mathcal{G}(R)$, where $R = k[t, t^{-1}]$ is the ring of Laurent polynomials over a field k and $\mathcal{G}$ is an isotropic, simply connected almost simple k-group. Let $\mathcal{P}_+, \mathcal{P}_-$ be two opposite minimal k-parabolic subgroups of $\mathcal{G}$ and denote by $\mathcal{S}$ the (unique) maximal k-split torus in $\mathcal{P}_+ \cap \mathcal{P}_-$, by $\mathcal{N}$ the normalizer of $\mathcal{S}$ in $\mathcal{G}$, by N the group $\mathcal{N}(R)$ and by B_ϵ, for $\epsilon \in \{+, -\}$, the subgroup of G inverse image of $\mathcal{P}_\epsilon(k)$ in $\mathcal{G}(k[t^\epsilon]) \subset G$ (we set $t^+ = t$ and $t^- = t^{-1}$) under the reduction homomorphism $\mathcal{G}(k[t^\epsilon]) \to \mathcal{G}(k)$ induced by the k-homomorphism $k[t^\epsilon] \to k$ which annihilates t^ϵ. Then, it can be shown that there exist a Coxeter system (W, S) and a homomorphism $\nu : N \to W$, both unique by the discussion of 3.1, such that (G, B_ϵ, N, ν) are twin BN-pairs. In fact, the group W is the affine Weyl group of the relative root system of $\mathcal{G}$ with respect to $\mathcal{S}$ (here, a system of type BC_n behaves as C_n).

Remark. We have seen in 3.1 (cf. also [2], IV, §2, exercise 3) that one can formulate the main axioms of BN-pairs without introducing the group N, thus generalizing the notion somewhat (cf. 3.1, example (b)). There is no such generalization for twin BN-pairs, the reason being that a building twinned with another one has a canonical system of apartments, described in 2.3.

3.3   Roots and root groups

For any $s \in S$, let α_s denote the subset $\{w | l(sw) = l(w) + 1\}$ of W. A root of a Coxeter building of type (W, S) is defined as an image of some α_s by an isometry of W (considered as a Coxeter building: cf. 2.1) onto the building in question. We denote by $\Phi(W, S)$, or simply by Φ, the set of all roots of W, that is, the set of all subsets of the form $w\alpha_s$, for $w \in W$ and $s \in S$. The roots containing (resp. not containing) 1 are called **positive** (resp. **negative**); the set of those roots is denoted by Φ_+ (resp. Φ_-). The complement $\Sigma - \alpha$ of a root α in a Coxeter building Σ is simply denoted by $-\alpha$ if the choice of the ambient building Σ is made clear by the context. Thus, for instance, $\Phi_- = -\Phi_+$. A pair of roots $\{\alpha, \beta\}$ is said to be **prenilpotent** if $\alpha \cap \beta \neq \emptyset$ and $(-\alpha) \cap (-\beta) \neq \emptyset$, in which case, we set $[\alpha, \beta] = \{\gamma | \gamma \in \Phi, \alpha \cap \beta \subset \gamma, (-\alpha) \cap (-\beta) \subset (-\gamma)\}$ and $(\alpha, \beta) = [\alpha, \beta] - \{\alpha, \beta\}$. If $\alpha \in \Phi$, there is a unique element of W conjugate to some element of S and permuting (by left translation) α and $-\alpha$; we denote it by s_α: it is the **reflexion associated to the root** α. For instance, if $\alpha = \alpha_s$, then $s_\alpha = s$.

We shall be interested in systems $(G; (U_\alpha)_{\alpha \in \Phi})$ consisting of a group G and a family of subgroups indexed by the roots of W and satisfying the following axioms, where H and U_+ denote respectively the intersection of the normalizers of all U_α's and the subgroup of G generated by the U_α's for $\alpha > 0$:

(RGD 0) $U_\alpha \neq \{1\}$ *for all* $\alpha \in \Phi$;

(RGD 1) *if* $\{\alpha, \beta\}$ *is a prenilpotent pair of distinct roots, the commutator* (U_α, U_β) *is contained in the group generated by all* U_γ's *for* $\gamma \in (\alpha, \beta)$ *(with the notation introduced above)*;

(RGD 2) *if* $s \in S$ *and* $u \in U_{\alpha_s} - \{1\}$, *there exist elements* u', u'' *of* $U_{-\alpha_s} - \{1\}$ *such that, for all* $\beta \in \Phi$, *the product* $m(u) = u'uu''$ *conjugates* U_β *onto* $U_{s_\alpha(\beta)}$;

(RGD 3) *if* $s \in S$, *then* $U_{-\alpha_s}$ *is not contained in* U_+;

(RGD 4) *the group* G *is generated by* H *and the* U_α's.

It is well-known (cf. [14]) that, given u, the elements $u', u'', m(u)$ of (RGD 2) are unique. Set $U_- = \langle U_\alpha | \alpha < 0 \rangle$, $B_\epsilon = HU_\epsilon$ for $\epsilon \in \{+, -\}$ and $N = \langle H, m(u) | u \in U_{\alpha_s} - \{1\}, s \in S \rangle$. It is easily verified that there exists a unique homomorphism $\nu : N \to W$ such that $\nu(H) = \{1\}$ and $\nu(m(u)) = s$ for $u \in U_{\alpha_s} - \{1\}$.

Proposition 4 *The systems* (G, B_ϵ, N, ν) *for* $\epsilon = +, -$ *are twin BN-pairs of type* (W, S).

The proof, which is less easy than the familiar proof in the special case where W is finite, can be extracted from [20].

For lack of a better idea (or rather, because all appropriate names that I can think of seem to be already taken!), I shall call the systems $(G; (U_\alpha)_{\alpha \in \Phi})$ satisfying the above axioms RGD-**systems of type** M or of type (W, S) (the letters RGD stand for the *Root Groups Data*). I shall also say that the twin buildings deduced as in 3.2 from the twin BN-pairs of Proposition 4 is **associated** to the RGD-system $(G; (U_\alpha)_{\alpha \in \Phi})$, and that the opposite chambers of $\Delta_\epsilon = G/B_\epsilon$ ($\epsilon = +, -$) represented by B_+ and B_- are the **fundamental chambers** relative to that system.

Remarks.

(1) A necessary condition for the existence of an RGD-system of type M (and in fact also, as will be seen later, for the existence of a thick twinning of type M) is that all coefficients of the matrix M be equal to 1, 2, 3, 4, 6, 8, or ∞. This follows from [14],[23]. (N.B. There are further very strong restrictions on M when some coefficients are equal to 8).

(2) If M has no direct factor of rank 1 (cf. 2.1), the twin buildings which are associated to RGD-systems are characterized by a simple geometric property, the **Moufang condition**: see 4.3. For spherical buildings of irreducible type and rank ≥ 3, the Moufang condition always holds: cf. [12],[15]. The proof given in [15], which is uniform (i.e. without case distinctions), relies essentially on Theorem 4.1.2 of [12]. In some sense, Theorem 2 below generalizes that result of [12] to twin buildings of any type M with finite coefficients m_{ij}, however it turns out to be a bit too weak to yield a proof that any twinning of rank ≥ 3 and irreducible type with finite coefficients is Moufang. The question whether it is so remains open (cf. howver 6.2, after Conjecture 1, and 6.4).

(3) To approach that question, one might try to experiment on the affine type, in which case all buildings of rank ≥ 4 have been classified (cf. [19]): roughly speaking, they are the affine buildings of algebraic simple groups or classical groups over fields or skew fields endowed with a discrete, complete valuation. One can show that a necessary condition for such a building to be part of a Moufang twinning is that the local field or skew field involved have *equal characteristics*. It would be worthwhile trying to prove a similar result without the Moufang restriction.

We say that an RGD-system is **faithful** if G operates faithfully on the buildings of the associated twinning, and that it is **reduced** if it is faithful and if, moreover, G is generated by the U_α's. Let $(G; (U_\alpha)_{\alpha \in \Phi})$ be an arbitrary RGD-system. The actions of G on the two buildings of the associated twinning have the same kernel K; if U_α° denotes the canonical image of U_α in G/K and G° is the subgroup of G/K generated by all U_α°'s, then $(G^\circ, (U_\alpha^\circ)_{\alpha \in \Phi})$ is a reduced RGD-system having "the same" associated twinning as $(G; (U_\alpha)_{\alpha \in \Phi})$, and which we call its **reduction**. It can also be obtained as follows: G° is the quotient of the subgroup $\langle U_\alpha | \alpha \in \Phi \rangle$ of G by its centre and U_α° is the canonical image of U_α in G°. The canonical homomorphisms $U_\alpha \to U_\alpha^\circ$ are easily seen to be isomorphisms. Two RGD-systems are said to be **equivalent** if they have the same reduction.

Example. Let Λ and $\Lambda^\vee$ be two lattices (finitely generated free abelian groups), $\mathbf{Z}$-dual of each other, and let $(\alpha_i)_{i \in I}$, $(\alpha_i^\vee)_{i \in I}$ be two families of elements of Λ and $\Lambda^\vee$ respectively, indexed by I. Set $A_{ij} = \langle \alpha_i^\vee, \alpha_j \rangle$ and suppose that the matrix $(A_{ij})_{i,j \in I}$ is a **generalized Cartan matrix**, which means that $A_{ii} = 2$ for all i, $A_{ij} \in \mathbf{Z}$ for all i,j, $A_{ij} \leq 0$ whenever $i \neq j$ and $A_{ij} = 0$ implies $A_{ji} = 0$. Suppose M is the Coxeter matrix defined as follows: $m_{ii} = 1$ and, for $i \neq j$, $m_{ij} = 2, 3, 4, 6$ or ∞ according as $A_{ij}A_{ji}$ is equal to $0, 1, 2, 3$, or ≥ 4. Let Φ be as above. To the given system $D = (\Lambda, (\alpha_i)_{i \in I}, (\alpha_i^\vee)_{i \in I})$ and to any field k, [20], 3.6, associates an RGD-system $(\mathcal{G}_D(k); (U_\alpha(k))_{\alpha \in \Phi})$ (in *loc. cit.*, $\mathcal{G}_D$ is denoted by $\tilde{\mathcal{G}}_D$) which we shall call the **Kac-Moody RGD-system** of type D over k, the corresponding twinning being named accordingly (Kac-Moody twinning etc.). When Λ is generated by the α_i's, the Kac-Moody system is faithful; the converse is true if $k \not\cong \mathbf{F}_2$.

4   First theorem. Frames of reference. The Moufang condition. Automorphism groups

4.1   Notation

We keep the notation of the previous sections. In particular, $(\Delta_+, \Delta_-, d^*)$ denotes a twinning of type (W, S). If x is a chamber in some building Δ of type (W, S) and J a subset of I, we let $\Sigma_J(x)$ be the sphere of centre x and radius J (cf. 2.1) and, for any integer r, we set $E_r(x) = \bigcup \{\Sigma_J(x) | J \subset I, \text{ Card } J = r\}$; this can be viewed as the **neighbourhood of order** r of x in Δ. Later on, it will also be useful to regard $E_r(x)$, endowed with the distance function inherited from Δ, as an "amalgam" of buildings of rank r, this being justified by the following easy proposition:

Proposition 5 *The restriction of the distance function d of Δ to $E_r(x)$ is entirely determined by its restriction to the various $\Sigma_J(x)$ (Card $J = r$), that is, by the building structures of those spheres (given, of course, the inclusion maps $\Sigma_J(x) \to E_r(x)$).*

Proof. One shows that if $y \in \Sigma_J(x)$ and $y' \in \Sigma_{J'}(x)$, then $d(y,y')$ is the unique element of minimum length in the set $\{d(y,z).d(z,y') | z \in \Sigma_{J \cap J'}(x)\}$.

From now on, all buildings under consideration will be assumed to be thick, unless the contrary is stated explicitly or self-evident (e.g. when we deal with Coxeter buildings).

4.2   Theorem 1 *If $c_+ \in \Delta_+$, $c_- \in \Delta_-$ are opposite chambers, then the identity is the only automorphism of $(\Delta_+, \Delta_-, d^*)$ fixing (chamberwise) $E_1(c_+) \cup \{c_-\}$.*

In other words, $E_1(c_+) \cup \{c_-\}$ can serve as a "frame of reference" for the twinning. The proof of this theorem is similar to that of Theorem 4.1.1 of [12]. It is also described with some detail in [21].

4.3   The Moufang property

From now on, and until the end of the paper, we shall always assume, when the contrary is not specified, that M has no direct factor of rank 1.

Let us choose two opposite chambers $c_\epsilon \in \Delta_\epsilon$, $\epsilon \in \{+, -\}$, set $A_\epsilon = A(c_\epsilon, c_{-\epsilon})$ (cf. 2.3) and identify A_+ with W by $x \mapsto d(c_+, x)$. Thus, roots of (W, S) become subsets of A_+. Let $^\circ : A_+ \to A_-$ denote the opposition isomorphism (i.e., for $x \in A_+$, x° is the chamber of A_- opposite to x). As before, Φ will denote the set of all roots of W; thus, via the above identification, the elements of Φ are also subsets of A_+. For $\alpha \in \Phi$, we set $\overline{\alpha} = A_- - \alpha^\circ$.

Let x be a chamber of the root α. For $i \in I$, the intersection of $\Sigma_i(x)$ with α has cardinality 1 or 2. If for some i (resp. for no i) it has cardinality 1, we say that x lies **at the border** (resp. **in the interior**) of α. Because of the assumption made on M, one can show as in [15] that

4.3.1   *every root has at least one chamber in its interior.*

For $\alpha \in \Phi$, we denote by $\mathcal{A}_\alpha$ the set of all admissible apartments of Δ_+ which contain α and whose opposite contains $\overline{\alpha}$, and by U_α the group of all automorphisms of the twinning $(\Delta_+, \Delta_-, d^*)$ which fix (chamberwise) $\overline{\alpha}$ and all spheres $\Sigma \subset \Delta_+$ of l-radius 1 such that card $(\Sigma \cap \alpha) = 2$. Theorem 1 readily implies

Proposition 6 *If $\alpha (\subset A_+)$ is a root and x is a chamber lying in its interior, the group $U(\alpha, \overline{\alpha}; x)$ of all automorphisms of $(\Delta_+, \Delta_-, d^*)$ fixing chamberwise $\alpha, \overline{\alpha}$ and $E_1(x)$ operates freely on $\mathcal{A}_\alpha$.*

Indeed, if $A \in \mathcal{A}_\alpha$, if A' is the apartment of Δ_- opposite to A and if x' denotes the opposite of x in A', Theorem 1, in which one sets $c_+ = x$ and $c_- = x'$, implies that the only element of $U(\alpha, \overline{\alpha}; x)$ stabilizing A is the identity.

In view of 4.3.1, we have

Corollary 1 *For all roots* $\alpha \in \Phi$, *the group* U_α *operates freely on the set* $\mathcal{A}_\alpha$.

Hence also

Corollary 2 *If, for some root* α, U_α *operates transitively on* $\mathcal{A}_\alpha$, *one has* $U_\alpha = U(\alpha, \overline{\alpha}; x)$ *for every chamber* x *lying in the interior of* α.

We say that the twinning $(\Delta_+, \Delta_-, d^*)$ has the **Moufang property**, or "is Moufang", if

(Mo) *for every root* α, *the group* U_α *acts transitively (hence regularly, by Corollary 1) on the set* $\mathcal{A}_\alpha$.

At first sight, this condition may seem to depend on the initial choice of the chambers c_+ and c_-, but it is not so; indeed, using the (not quite trivial) Proposition 5 of [21], one can see that, when (Mo) holds, the group generated by all U_α's permutes transitively the pairs of opposite chambers.

Proposition 7 *If the twinning* $(\Delta_+, \Delta_-, d^*)$ *satisfies the condition (Mo) and* G° *denotes the group of automorphisms of that twinning generated by all* U_α *'s* $(\alpha \in \Phi)$, *the system* $(G^\circ; (U_\alpha)_{\alpha \in \Phi})$ *is a reduced RGD-system. Conversely, if* $(G; (U_\alpha)_{\alpha \in \Phi})$ *is any RGD-system of type* M, *the twinning associated to that system is Moufang and the automorphism groups of that twinning induced by the* U_α *'s coincide with the groups* U_α *defined above when one takes for* c_+ *and* c_- *the two fundamental chambers relative to the given RGD-system (cf. 3.3).*

For spherical types, this is known (cf. [12], Addenda, p. 278). The proof of the general case follows the same pattern but is more involved. An essential intermediate step is stated as part (i) of proposition 8 of [21], and one also has to use some results of [20], §5.

4.4 Moufang sets

Since this subsection deals with the rank 1 case of the Moufang condition, the general restriction imposed at the beginning of 4.3 ("no direct factor of rank 1") will be lifted for the time being. The following definition is motivated by Proposition 7.

A **Moufang set** is a system $(X; (U_x)_{x \in X})$ consisting of a set X and a family of groups of permutations of X indexed by X itself and satisfying the following conditions:

(MoS 1) U_x *fixes* x *and is simply transitive on* $X - \{x\}$;

(MoS 2) *in the full permutation group of* X, *each* U_x *normalizes the set of subgroups* $\{U_y | y \in X\}$.

This notion appears in the literature under various denominations; for instance, it is essentially equivalent to the notion of **split BN-pair of rank 1**, except that in most papers dealing with such pairs, extra conditions (e.g. nilpotency) are imposed on the groups U_x (cf. [7], [10]). For an approach of 1-dimensional projective geometry in a similar spirit, see [6].

Clearly, if $(X; (U_x)_{x \in X})$ is a Moufang set, the group G generated by all U_x is doubly transitive on X. If furthermore Card $X \geq 3$ and $0, \infty$ are two points of X, the system $(G; (U_0, U_\infty))$ is a reduced RGD-system of type $M = (1)$. Conversely, if $(G; (U_0, U_\infty))$ is an arbitrary RGD-system of that same type and if H denotes the intersection of the normalizers of U_0 and U_∞ in G, then $X = G/(HU_\infty)$ has a natural structure of Moufang set, stable by G, namely the structure consisting of the permutation groups of X induced by the conjugates of U_0 in G.

Examples.

(i) If k is a division ring, the projective line $X = \mathbf{P}_1(k) = k \cup \{\infty\}$ has a structure of Moufang set in which the groups U_x are the conjugates of the group of translations $x \mapsto x + t$, $\infty \mapsto \infty$, in the group of all projective transformations $x \mapsto (ax + b)(cx + d)^{-1}$ (with a, c not both zero, b, d not both zero and $ac^{-1} \neq bd^{-1}$). By a well-known theorem of Hua (cf. e.g. [12], 8.12.3),

the Moufang set $\mathbf{P}_1(k)$ determines the division ring k up to isomorphism or antiisomorphism.

(ii) Consider a nondegenerate quadratic or, more generally, pseudo-quadratic form q of Witt index 1 (cf. [12], 8.2). Then, the corresponding "oval hyperquadric", whose points are the singular lines of q, has a natural structure of Moufang set in which the group U_x is the group of quadratic or unitary transvections with centre x.

If $(\Delta_+, \Delta_-, d^*)$ is a Moufang twinning, the spheres of rank 1 of Δ_+ and Δ_- have natural Moufang structures obtained as follow. Let X be a sphere of radius $i \in I$ in, say, Δ_+. Choose the chambers c_+, c_- of 4.3 so that $c_+ \in X$. Set $\alpha = \alpha_{s_i}$. Then, the groups of permutations of X induced by the conjugates of U_α in the group generated by U_α and $U_{-\alpha}$ constitute a Moufang structure which is easily seen to be independent of the choice of c_+, element of X, and of c_-, opposite to c_+.

We shall see in 5.6 (ii) that, when all m_{ij} are finite, the spheres of rank 1 of *any* twinning whose type has no direct factor of rank ≥ 2, have "induced" Moufang structures.

4.5   Automorphisms

After the parenthesis of 4.4, we assume again that M has no direct factor of rank 1. Let $(G^\circ; (U_\alpha)_{\alpha \in \Phi})$ be a reduced RGD-system and let H° be the intersection of the normalizers of all U_α's in G°. Between the full automorphism group $\overline{G}$ of the twinning $(\Delta_+, \Delta_-, d^*)$ associated to the given RGD-system (group of all pairs consisting of a permutation of Δ_+ and a permutation of Δ_-, preserving the distances and the codistance) and the full automorphism group $\overline{H}$ of the RGD-system (group of all automorphisms of G° normalizing each U_α), one has the following relations:

> $\overline{G}$ is the quotient of the semi-direct product $\overline{H} \ltimes G^\circ$, relative to the natural action of $\overline{H}$ on G°, by the subgroup $\{(\operatorname{int} h, h^{-1}) | h \in H^\circ\}$ of that product;

> G°, and hence the U_α's, being identified with the corresponding subgroups of $\overline{G}$, $\overline{H}$ is the intersection of the normalizers of the U_α's in $\overline{G}$.

There is an obvious one-to-one correspondence between the subgroups of $\overline{G}$ containing G° and the subgroups of $\overline{H}$ containing H°, hence also between the isomorphism classes of faithful RGD-systems equivalent to $(G^\circ; (U_\alpha)_{\alpha \in \Phi})$ and the conjugacy classes of subgroups of $\overline{H}$ containing H°.

Thus, in order to compute the full automorphism group of the Moufang twinning associated to a given RGD-system $(G^\circ; (U_\alpha)_{\alpha \in \Phi})$, say reduced, and to classify all faithful equivalent systems, one needs only to determine the groups $\overline{H}$ and H°. We shall see that the above theorem and results of the forthcoming sections reduce that problem to its special case in rank 1 and 2, hence involving groups which are often well-known and easy to handle.

For $J \subset I$, let W_J be the Coxeter group generated by $\{s_j | j \in J\}$, let Φ_J be the set of roots of that Coxeter group and let $G^\circ_J, H^\circ_J, \overline{H}_J$ denote respectively the subgroup $\langle U_\alpha | \alpha \in \Phi_J \rangle$ of G°, the intersection of the normalizers of the U_α ($\alpha \in \Phi_J$) in G°_J and the full automorphism group of the system of groups $(G^\circ_J; (U_\alpha)_{\alpha \in \Phi_J})$. If $i, j \in I$, we also write $H^\circ_i, \overline{H}_i, H^\circ_{ij} \dots$ for $H^\circ_{\{i\}}, \overline{H}_i, H^\circ_{\{i,j\}}, \dots$. For $J \subset J' \subset I$, there are obvious canonical homomorphisms $\overline{H}_{J'} \to \overline{H}_J$ (the "restriction"), $H^\circ_J \to H^\circ_{J'}$, (the inclusion) and $H^\circ_J \to \overline{H}_J$. The following proposition is an easy consequence of Theorem 1.

Proposition 8 *The homomorphism* $\overline{H} \to \prod_{i \in I} \overline{H}_i$, *direct product of the canonical homomorphisms* $\overline{H} \to \overline{H}_i$, *is injective.*

Thus, $\overline{H}$ can be identified with a subgroup of $\prod_{i \in I} \overline{H}_i$; this product being, in some sense, a "first approximation" of $\overline{H}$, we shall denote it by $\overline{H}^{(1)}$. Identifying similarly $\overline{H}_J$, for $J \subset I$, with its image in the product $\prod_{j \in J} \overline{H}_j$, we set $\overline{H}^{(2)} = \bigcap_{i,j \in I} \operatorname{pr}_{ij}^{-1}(\overline{H}_{ij})$ $(\subset \overline{H}^{(1)})$, where $\operatorname{pr}_{ij} : \overline{H}^{(1)} \to \overline{H}_i \times \overline{H}_j$ is the (i,j)-th projection. Clearly $\overline{H}$, as a subgroup of $\overline{H}^{(1)}$, is contained in $\overline{H}^{(2)}$. In fact, it is often, and probably always, equal to $\overline{H}^{(2)}$ (cf. 6.2, 6.4). Note that, in order to prove this equality in any given case, it suffices to exhibit an RGD-system $(G^{(2)}; (U_\alpha)_{\alpha\in\Phi})$ with $G^{(2)} \supset G^\circ$ (the U_α's being unchanged), such that the intersection of the normalizers of the U_α's in $G_\alpha^{(2)}$ is precisely $H^{(2)}$; that often is a simple matter.

Let us illustrate this by the example of Kac-Moody twinnings (cf. the end of §3). Let k be a field, let $D = (\Lambda, (\alpha_i)_{i\in I}, (\alpha_i^\vee)_{i\in I})$ be as in 3.3, suppose that the graph of the generalized Cartan matrix $(A_{ij}) = (\langle \alpha_i^\vee, \alpha_j \rangle)$ is connected (i.e. that the corresponding Coxeter matrix M – see 3.3 – is irreducible) and that Λ is freely generated by α_i's (this can be called the "adjoint case"), let $(G = \mathcal{G}_D(k); (U_\alpha)_{\alpha\in\Phi})$ be the Kac-Moody RGD-system corresponding to those data (cf. 3.3), set $H_i = \mathrm{Hom}\,(\mathbf{Z}\alpha_i, k^\times)$ and $H = \mathrm{Hom}\,(\Lambda, k) = \bigcap_{\alpha\in\Phi} \mathcal{N}(U_\alpha)$, where $\mathcal{N}$ means "normalizer". Since $\Lambda = \coprod_{i\in I} \mathbf{Z}\alpha_i$, the group H can and will be identified with $\prod_{i\in I} H_i$. Let now G° be the subgroup of G generated by the U_α's. For this choice of the RGD-system $(G^\circ; (U_\alpha)_{\alpha\in\Phi})$, the group $\overline{H}_i$ is an extension of Aut k by H_i: there is a *canonical epimorphism* $\kappa_i : \overline{H}_i \to \mathrm{Aut}\,k$ with kernel H_i. An elementary computation shows that, for $m_{ij} \geq 3$, $\overline{H}_{ij}$ consists of all $(h_i, h_j) \in \overline{H}_i \times \overline{H}_j$ such that $\kappa_i(h_i) = \kappa_j(h_j)$. Consequently, $\overline{H}^{(2)}$ is the group of all $(h_i)_{i\in I} \in \overline{H}^{(1)}$ such that $\kappa_i(h_i)$ is independent of i; thus, it is the semi-direct product of Aut k by H relative to the obvious action of the former on the latter. As a result of the functorial nature of $\mathcal{G}_D$, there is a natural action of Aut k on $G = \mathcal{G}_D(k)$. We let $G^{(2)}$ be the semi-direct product of Aut k by G relative to that action. The system $(G^{(2)}; (U_\alpha)_{\alpha\in\Phi})$ clearly has the properties required above to ensure that $\overline{H} = \overline{H}^{(2)}$. Thus, we can state:

Proposition 9 *If the Kac-Moody data $D = (\Lambda, (\alpha_i)_{i\in I}, (\alpha_i^\vee)_{i\in I})$ are such that the lattice Λ is freely generated by the α_i's, the full automorphism group of the Kac-Moody twinning associated to the system $(\mathcal{G}_D(k); (U_\alpha)_{\alpha\in\Phi})$ is the (natural) semi-direct product of Aut k by $\mathcal{G}_D(k)$.*

For H° there is no such simple description, but Proposition 10 below still gives a construction involving subsystems of rank 1 and 2, hence, most of the time, well manageable. For $i, j \in I$ and $u, u' \in U_{\alpha_i} - \{1\}$, let $\eta_j(u, u') \in \overline{H}_j$ be the automorphism of $(G_j^\circ; U_{\alpha_j}, U_{-\alpha_j})$ which is the restriction to G_j° of the conjugation by $m(u)m(u')^{-1}$ (for the notation m, see 3.3, axiom (RGD 2)), set $\eta(u, u') = \prod_{j\in I} \eta_j(u, u') \in \overline{H}^{(1)}$ (notation introduced above) and let ρ_u be the automorphism of $\overline{H}^{(1)}$ defined as follows: if $h = \prod_{l\in I} h_l \in \overline{H}^{(1)}$, with $h_l \in \overline{H}_l$, then $\rho_u(h) = \prod_{l\in I} h_l'$ with $h_l' = h_l$ for $l \neq i$ and $h_i' = \eta_i(u, {}^{h_i}u).h_i$. Note that, via the canonical inclusions $H^\circ \to \overline{H} \to \overline{H}^{(1)}$ (cf. Proposition 8), $\eta(u, u')$ is nothing else but $m(u).m(u')^{-1}$ (element of H_i°), and that, if $h \in \overline{H}$, then $\rho_u(h)$ is the conjugate of h by $m(u)$, as is shown by the following computation:

$$m(u).h.m(u)^{-1} = m(u).h.m(u)^{-1}.h^{-1}.h = m(u).m({}^h u)^{-1}.h$$
$$= \eta_i(u, {}^h u).h.$$

Proposition 10 *The group $H°$, identified with its canonical image in $\overline{H}^{(1)}$ (via the monomorphism of Proposition 8), is the smallest subgroup of $\overline{H}^{(1)}$ containing all $\eta(u, u')$ for $u, u' \in U_i - \{1\}$, $i \in I$, and stable by all ρ_v for $v \in U_i - \{1\}$, $i \in I$.*

The proof is left to the reader.

5   Second theorem. Characterization of the two halves of a twinning by local data

From now on, the coefficients m_{ij} of the type M are always assumed to be finite.

Theorem 4.1.2 of [12] suggests that, possibly, a twinning $(\Delta_+, \Delta_-, d^*)$ might be determined up to isomorphism by $E_2(c_+)$, for $c_+ \in \Delta_+$. I do not know whether it is so, but this section presents results pointing in that direction.

5.1   The canonical isomorphism between opposite spheres

If $(\Delta_+, \Delta_-, d^*)$ and $(\Delta'_+, \Delta'_-, d'^*)$ are twinnings of the same type M, a bijection of a subset X of $\Delta_+ \cup \Delta_-$ onto a subset X' of $\Delta'_+ \cup \Delta'_-$ will be called an **isometry** if it maps $X \cap \Delta_+$ onto $X' \cap \Delta'_+$ and if it preserves distances and codistances (this terminology is introduced here for later use). As before (cf. 4.1), if x is a chamber in a building of type M and J is a subset of I, the sphere of centre x and radius J in the building is denoted by $\Sigma_J(x)$.

Let $x_+ \in \Delta_+$, $x_- \in \Delta_-$ be opposite chambers and let J be a **spherical subset** of I, i.e. a subset such that the Coxeter group $W_J = \langle s_j | j \in J \rangle$ is finite. Then, for $y \in \Sigma_J(x_+)$, there is a unique $y' \in \Sigma_J(x_-)$ "closest to y", i.e. such that $d^*(y, y')$ is the longest element of W_J; the map $y \mapsto y'$ is an isometry of $\Sigma_J(x_+)$ onto $\Sigma_J(x_-)$ which we denote by $\mathrm{op}_J(x_+, x_-)$. If $x_+, x'_+ \in \Delta_+$ are both opposite to $x_- \in \Delta_-$, then, by composing $\mathrm{op}_J(x_+, x_-)$ and $\mathrm{op}_J(x_-, x'_+)$, one gets an isometry of $\Sigma_J(x_+)$ onto $\Sigma_J(x'_+)$, which we denote by $\iota_J(x_+, x'_+; x_-)$.

5.2   Adjacent chambers. Galleries

The following terminology is more or less standard.

Two chambers of a building are said to be **adjacent** if their l-distance is 1. A sequence $(a_0, \ldots, a_n)$ of $n+1$ chambers such that any two consecutive terms are adjacent or identical is called a **gallery** of **length** n, and the gallery is said to be **geodesic** if n is equal to the l-distance of a_0 and a_n. An important property of buildings is that a geodesic gallery is contained in any apartment containing its origin and its extremity (cf. [12], 3.4). A **closed gallery** is a gallery whose origin and extremity coincide.

If $\Gamma = (a_0, \ldots, a_n)$ is a gallery, we represent by Γ^{-1} the same gallery in reverse order, that is, $(a_n, \ldots, a_0)$. Two galleries are said to be **equivalent** if one can can deduce the second one from the first by a succession of operations consisting, either in replacing a closed subgallery of the form $\Gamma\Gamma^{-1}$ by its origin (this includes the cancellation of a term equal to its predecessor) or, conversely, in replacing a term by a closed gallery of that form having the term in question as its origin and extremity (e.g. doubling a term). Clearly, the equivalence classes of closed galleries with given origin form a group. A closed gallery will be called an **elementary cycle** if it consists of the elements of an apartment of a sphere of rank 2, all occurring exactly once, except for the extremities (which are equal, of course). We define an **elementary loop** as a closed gallery of the shape $\Gamma\Gamma_e\Gamma^{-1}$, where Γ_e is an elementary cycle.

Lemma 1 *In any building, the group of equivalence classes of closed galleries with a given origin is generated by the classes of elementary loops it contains.*

We must show that a closed gallery $(a_0, \ldots, a_n = a_0)$ is equivalent to a product of elementary loops with origin a_0. For $i = 1, 2, \ldots, n-1$, let Γ_i be a geodesic gallery joining a_i and a_0. Since the gallery $\Gamma_i\Gamma_i^{-1}$ is equivalent to (a_i), the given gallery is equivalent to $a_0\Gamma_1^{-1}\Gamma_2\Gamma_2^{-1}\ldots\Gamma_{n-1}$. Thus, all we have to show is that, for $2 \leq i \leq n-1$, the gallery $\Gamma_{i-1}^{-1}\Gamma_i$ is equivalent to a product of elementary loops. If $a_{i-1} \neq a_i$ and if the l-distance of a_0 and a_{i-1} is smaller (resp. bigger) than the l-distance of a_0 and a_i, the gallery $\Gamma_{i-1}^{-1}a_i$ (resp. $a_{i-1}\Gamma_i$) is geodesic, therefore $\Gamma_{i-1}^{-1}\Gamma_i$ is contained in any apartment containing a_0 and a_i (resp. a_0 and a_{i-1}), and that is of course also true if $a_i = a_{i-1}$. But apartments are Coxeter buildings and, in a Coxeter building, our assertion readily follows from the standard presentation of Coxeter groups.

5.3 Proposition 11 *Let Δ_+, Δ'_+ be two buildings of type M, let $A \subset \Delta_+$, $A' \subset \Delta'_+$ be two apartments, let $c \in A, c' \in A'$ be two chambers and let $E_2(A)$ (resp. $E_2(A')$) denote the union of all $E_2(x)$ for $x \in A$ (resp. $x \in A'$). Suppose that there exist twinnings $(\Delta_+, \Delta_-, d^*), (\Delta'_+, \Delta'_-, d'^*)$ with respect to which the apartments A, A' are admissible. Then:*

(i) *Every isometry $\phi : E_2(c) \to E_2(c')$ mapping $E_2(c) \cap A$ onto $E_2(c') \cap A'$ extends in a unique way to an isometry $\tilde{\phi} : E_2(A) \to E_2(A')$ mapping A onto A'.*

(ii) *Suppose M has no direct factor of rank 2. Let α be a root of A containing c in its interior. Suppose that $\Delta_+ = \Delta'_+$, $c = c'$ and $\alpha \subset A'_+$, and that ϕ fixes $E_1(c)$ (chamberwise). Then, for every chamber x lying in the interior of α, $E_1(x)$ is also fixed by ϕ.*

The next two sections will give an outline of a proof of this proposition under a mild additional hypothesis (cf. 5.4(b)), always satisfied when the spheres of rank 1 in Δ_+ are "not too small" (cf. 5.7).

5.4 Proof of (i)

Let $\omega : A \to A'$ be the isometry mapping c onto c', choose twinnings $(\Delta_+, \Delta_-, d^*), (\Delta'_+, \Delta'_-, d'^*)$ as in the statement of the proposition, let A_- and A_+ be the apartments of Δ_- and Δ'_- respectively opposite to A and A' and let $^\circ$ denote the two opposition isomorphisms $A \to A_-$ and $A' \to A'_-$. We first show

(a) If $x, y \in A$ are two adjacent chambers and $\xi : E_2(x) \to E_2(\omega(x))$ is an isometry mapping $E_2(x) \cap A$ onto $E_2(\omega(x)) \cap A'$, then, there exists a unique isometry $\eta : E_2(y) \to E_2(\omega(y))$ which coincides with ξ on $E_2(x) \cap E_2(y)$ and maps $E_2(y) \cap A$ onto $E_2(\omega(y)) \cap A'$.

Set $d(x, y) = s_i$. We have only to show that, for any $J \subset I$ of cardinality 2, there exists a unique isometry $\eta_J : \Sigma_J(y) \to \Sigma_J(\omega(y))$ which coincides with ξ on $E_2(x) \cap \Sigma_J(y)$ and maps $\Sigma_J(y) \cap A$ onto $\Sigma_J(\omega(y)) \cap A'$; indeed, those Σ_J will then agree on the spheres $\Sigma_j(y)$ of rank 1, since $E_1(y)$ is contained in $E_2(x)$. If $i \in J$, the existence and uniqueness of η_J is clear since $\eta_J(y)$ is then contained in $E_2(x)$. Suppose therefore that i does not belong to J. The uniqueness of η_J follows from Theorem 1 (or Theorem 4.1.1 of [12]), again because $E_1(y) \subset E_2(x)$. To prove the existence, let us choose an element z of $\Sigma_i(x) = \Sigma_i(y)$, distinct from x and y, set $z_- = \mathrm{op}_J(x, x^\circ)(z)$, $z' = \omega(z)$ and $z'_- = \mathrm{op}_J(\omega(x), \omega(x)^\circ)(z')$. Then, one can verify that $\eta_J = \iota_J(\omega(x), \omega(y); z'_-) \circ \xi \circ \iota_J(y, x : z_-)$ has the desired properties.

With the notation of (a), let us set $\eta = T_{xy}.\xi$, and proceed to the second step of the proof.

(b) If $x_0, x_1, \ldots, x_m = x_0$ is a closed gallery in A, then

$$T_{x_{m-1}, x_m}.T_{x_{m-2}, x_{m-1}}. \ \cdots \ .T_{x_0, x_1}$$

is the identity.

Since T_{xy} and T_{yx} are obviously inverse of each other, it suffices, by Lemma 1, to prove (b) when $x_0, \ldots, x_m$ are the elements of a sphere of rank 2 in A. Let us therefore assume that $\{x_0, \ldots, x_{m-1}\} = \Sigma_J(x_0) \cap A$, for some $J \subset I$ of cardinality 2, let $\xi_0 : E_2(x_0) \to E_2(\omega(x_0))$ be an isometry mapping $E_2(x_0) \cap A$ onto $E_2(\omega(x_0)) \cap A'$ and, for $r = 1, 2, \ldots, m$, set $\xi_r = T_{x_{r-1}, x_r}.T_{x_{r-2}, x_{r-1}}. \ \cdots \ .T_{x_0, x_1}.\xi_0$. We must show that $\xi_m = \xi_0$. In $\Sigma_J(x_0)$ let us choose a chamber z opposite to all x_r (in the sense of [12]: remember that $\Sigma_J(x_0)$ is a building of *spherical type*). The existence of such a z is the restrictive hypothesis announced at the end of 5.3. The remarks in 5.7 will show that it is indeed a mild restriction. Set $z' = \xi_0(z)$. The chambers $\mathrm{op}_J(x_r, x_r^\circ)(z)$ and $\mathrm{op}_J(\omega(x_r), \omega(x_r)^\circ)(z')$ do not depend on r and will be called z_- and z'_- respectively. For $r = 0, 1, \ldots, m$, one verifies that the isometries

$$\iota_{J'}(\omega(x_0), \omega(x_r); z'_-) \circ \xi_0 \circ \iota_{J'}(x_r, x_0; z_-) \ : \ \Sigma_{J'}(x_r) \to \Sigma_{J'}(\omega(x_r)),$$

where J' runs over the set of all subsets of cardinality 2 of I, agree pairwise on the intersections of their domains of definition, that is, are the restrictions of an isometry $E_2(x_r) \to E_2(\omega(x_r))$ which we denote by ξ'_r. Moreover, for all $r \geq 1$, ξ'_r maps $E_2(x_r) \cap A$ onto $E_2(\omega(x_r)) \cap A'$ and coincides with ξ'_{r-1} on $E_2(x_r) \cap E_2(x_{r-1})$. Therefore, $\xi'_r = T_{x_{r-1},x_r}.\xi'_{r-1}$. Since $\xi'_0 = \xi_0$, this implies that $\xi'_r = \xi_r$ for all r. In particular, $\xi_m = \xi'_m = \xi_0$, hence (b).

From (b), it readily follows that, if we consider all possible (nonnecessarily closed) galleries $c = x_0, x_1, \ldots, x_m$ in A, starting at c, the corresponding maps

$$T_{x_{m-1},x_m}.T_{x_{m-2},x_{m-1}}. \; \cdots \; .T_{x_0,x_1}.\phi$$

"fit together", that is, are the restrictions to the sets $E_2(x_m)$ of a single bijection $\tilde{\phi} : E_2(A) \to E_2(A')$, mapping A onto A' and whose restriction to each $E_2(x)$, for $x \in A$, is an isometry onto $E_2(\omega(x))$. There remains to prove that $\tilde{\phi}$ is an isometry. This is a special case of a property of buildings which has nothing to do with twinnings and which will be stated in 6.3, where brief indications of proof will be given.

5.5   Proof of (ii)

We take up the hypotheses and notation of (ii). Let $\alpha^{\circ\circ}$ denote the set of all chambers $x \in \alpha$ such that $E_2(x) \cap A \subset \alpha$. We first prove that

5.5.1   *if $x \in \alpha^{\circ\circ}$, then $E_2(x)$ is fixed by $\tilde{\phi}$.*

By [15], 2.5, there exists a gallery $c = x_0, x_1, \ldots, x_m = x$ joining c and x, and all elements of which, except possibly the first one, are contained in $\alpha^{\circ\circ}$. By induction on m, we may – and shall – assume that x is adjacent to c. Thus, $d(c, x) = s_i$ for some $i \in I$. In order to prove 5.5.1, it suffices, by Theorem 1 (or, more particularly, by Theorem 4.1.1 of [12]) to show that $\tilde{\phi}$ fixes $\Sigma_j(x)$ for all $j \in I$. Since it fixes $\Sigma_i(x) = \Sigma_i(c)$, we may take $j \in I - \{i\}$. Let us set $J = \{i, j\}$. Since $\tilde{\phi}$ fixes $\Sigma_i(c)$, $\Sigma_j(c)$ and $\Sigma_J(c) \cap A = \Sigma_J(x) \cap A$ (because $x \in \alpha^{\circ\circ}$), Theorem 1, again, implies that it fixes $\Sigma_J(c) = \Sigma_J(x)$, hence, in particular, $\Sigma_j(x)$; the assertion 5.5.1 is thereby proved.

Now, any chamber x which lies in the interior of α is adjacent to a chamber $y \in \alpha^{\circ\circ}$ (by [15], 2.6 and 2.5), therefore, (ii) follows from 5.5.1 applied to y, since $E_1(x) \subset E_2(y)$.

5.6   Corollary 3 *Let Δ_+ be a building of type M without direct factor of rank ≤ 2, which is part of a twinning. We recall that the coefficients of M are supposed to be finite.*

   (i) *If $i, j \in I$ are such that $m_{ij} \geq 3$, then all spheres of radius $\{i, j\}$ are Moufang buildings.*

(ii) *If $i, j, k \in I$ are pairwise distinct and such that $m_{ij} \neq 2$ and $m_{jk} \neq 2$, then, for $c \in \Delta_+$, the Moufang buildings $\Sigma_{ij}(c)$ and $\Sigma_{jk}(c)$ induce the same structure of Moufang set on $\Sigma_j(c)$ (cf. 4.4).*

The proof of these assertions as consequences of Proposition 11 is left as an exercise to the reader, who may find some inspiration in [15].

5.7 Remarks on the extra-hypothesis used in the proof of Proposition 11 (cf. 5.4 (b))

We take up again the notation of 5.4 (b).

(a) The following lemma shows that the hypothesis in question can be false only if the spheres of rank 1 in $\Sigma_J(x_0)$ are finite and, indeed, very small.

Lemma 2 *If a (thick) building of rank 2 and type $\begin{pmatrix} 1 & m \\ m & 1 \end{pmatrix}$ has the property that all spheres of rank 1 have cardinality at least $m + 1$, then, for any apartment, there is a chamber in the building which is opposite to all chambers of the apartment.*

In this proof, we view the buildings of rank 2 as graphs, the edges and vertices of which are respectively the chambers and the spheres of rank 1. An apartment is a cycle $p_{-m}, p_{-m+1}, \ldots, p_{m-1}, p_m = p_{-m}$. To prove the lemma, one shows by induction on an integer $m' \in \{0, 1, \ldots, m\}$, that

$(*_{m'})$ *there exists a vertex $q_{m'}$ opposite to $p_{-m'}, p_{-m'+2}, \ldots, p_{m'-2}, p_{m'}$.*

The induction starts, obviously. Assuming $(*_{m'})$, there exist, in the graph, unique neighbours $r_{-m'-1}, r_{-m'+1}, \ldots, r_{m'-1}, r_{m'+1}$ of $q_{m'}$ such that, for $h \in \{-m'-1, -m'+1, \ldots, m'+1\}$, the vertex r_h is not opposite to p_h. Then, any neighbour of $q_{m'}$, different from the r_h's satisfies $(*_{m'+1})$. Finally, the lemma follows from $(*_m)$ by a similar argument.

(b) By Lemma 2, if the extra hypothesis of 5.4 (b) fails to be satisfied, some sphere of rank 1 in $\Sigma_J(x_0)$ must be finite. If m_J is odd, this implies that all spheres of rank 1 in $\Sigma_J(x_0)$ and hence $\Sigma_J(x_0)$ itself are finite. By a well-known theorem of W. Feit and G. Higman, this can happen only if $m_J = 3$. I ignore whether, for m even $\neq 2$, the finiteness of one sphere of rank 1 in a thick building of type $\begin{pmatrix} 1 & m \\ m & 1 \end{pmatrix}$ implies the finiteness of all of them, but if so, the theorem of Feit-Higman then implies $m_J = 2, 4, 6, 8$. Now, we know already by Corollary 5.6 (i) and [14],[23], that all coefficients of the type M of a twinning are necessarily equal to $2, 3, 4, 6, 8$. (It might be mentioned here, that Corollary 5.6 can be given a simple direct proof, independent of Proposition 11).

(c) Replacing the proof of 5.4 (b) by the similar but somewhat more elaborate argument of [12], 4.15, one can replace the extra hypothesis of 5.4 (b) by the following, weaker one: there exists a sphere of rank 1 in $\Sigma_J(x_0)$ such that any two x_j's are opposite simultaneously to some element of that sphere. By [12], 3.36, this is automatically satisfied if $m_J \leq 5$. Thus, altogether, only $m_J = 6$ and 8 may still cause trouble.

5.8 Theorem 2 *Let* $(\Delta_+, \Delta_-, d^*)$ *and* $(\Delta'_+, \Delta'_-, d'^*)$ *be two twinnings of type* M *and let* $(c_+, c_-) \in \Delta_+ \times \Delta_-$, $(c'_+, c'_-) \in \Delta'_+ \times \Delta'_-$ *be two pairs of opposite chambers. Then, any isometry* ϕ *of* $E_2(c_+) \cup \{c_-\}$ *onto* $E_2(c'_+) \cup \{c'_-\}$ *extends to an isometry of* $\Delta_+ \cup \{c_-\}$ *onto* $\Delta'_+ \cup \{c'_-\}$.

We first observe that

5.8.1 *the isometry* ϕ *extends uniquely to an isometry*

$$\overline{\phi}: \ E_2(c_+) \cup E_2(c_-) \to E_2(c'_+) \cup E_2(c'_-) \ .$$

Indeed, if $\overline{\phi}$ exists, its restriction to $\Sigma_J(c_-)$, for some $J \subset I$ of cardinality 2, cannot but be the map

$$\mathrm{op}_J(c'_+, c'_-) \circ \phi \circ \mathrm{op}_J(c_+, c_-)^{-1} \ ,$$

and one verifies easily that all those maps fit together, constituting with ϕ itself the desired isometry $\overline{\phi}$.

Let us set $A_- = A(c_-, c_+)$, $A'_- = A(c'_-, c'_+)$ (cf. 2.3) and denote by $\tilde{\phi}_-$ the isometry of $E_2(A_-) = \bigcup\{E_2(x_-) | x_- \in A_-\}$ onto $E_2(A'_-) = \bigcup\{E_2(x'_-) | x'_- \in A'_-\}$ which coincides with $\overline{\phi}$ on $E_2(c_-)$ (cf. Proposition 11). For $z \in A_+$ (resp. A'_+) let $\pi(z)$ denote the element of A_- (resp. A'_-) defined by $d(c_-, \pi(z)) = d^*(c_-, z)$ (resp. $d(c'_-, \pi(z)) = d'^*(c'_-, z)$); it is opposite to z. For any pair $(x, x') \in \Delta_+ \times \Delta'_+$ and any chamber $y \in \Delta_+$ adjacent to x, i.e. such that $d(x, y) = s_i$ for some $i \in I$, we define a chamber $\tau(x, x'; y)$ of Δ'_+ by

$$\tau(x, x'; y) = (\mathrm{op}_i(\pi(x'), x') \circ \tilde{\phi}_- \circ \mathrm{op}_i(x, \pi(x)))(y) \ .$$

The function τ enables us to "copy" galleries of Δ_+ into Δ'_+: if (x, x') is as above and if $\Gamma = (x_0 = x, x_1, \ldots, x_m)$ is a gallery in Δ_+, we let $\tau(x, x')(\Gamma)$ denote the gallery $(x'_0 = x', x'_1, \ldots, x'_n)$ defined inductively by $x'_r = \tau(x_{r-1}, x'_{r-1}; x_r)$ for $r = 1, 2, \ldots, m-1$. The function $\tau(x, x')$ is "multiplicative" and "compatible with the inverse" in the following sense: if Γ and Γ' are two galleries such that the last term of Γ coincides with the first term of Γ', we now denote by $\Gamma\Gamma'$ the succession of those galleries with fusion of the two terms in question, and, as before, by Γ^{-1}, the gallery Γ in the reverse order; then

5.8.2 *if* z, z' *are the last terms of* Γ *and* $\tau(x, x')(\Gamma)$, *one has*

(1) $\tau(x, x')(\Gamma\Gamma') = \tau(x, x')(\Gamma).\tau(z, z')(\Gamma')$,

(2) $\tau(x, x')(\Gamma)^{-1} = \tau(z, z')(\Gamma^{-1})$.

Now, the crux of the proof of Theorem 2 is the following fact:

5.8.3 *if the gallery* Γ *is closed, then also* $\tau(x, x')(\Gamma)$ *is closed.*

To prove this, it suffices, by Lemma 1 and 5.8.2 (2), to consider the case where Γ consists of the elements of an apartment of a sphere of rank 2 in Δ_+; in that case, verifying 5.8.3 is straightforward.

From 5.8.2 and 5.8.3, it follows that, if two galleries Γ_1 and Γ_2 of Δ_+ with common origin x also have the same end z, then also $\tau(x, x')(\Gamma_1)$ and $\tau(x, x')(\Gamma_2)$ have the same end, which we denote by $\tau(x, x')(z)$. (N.B. With this convention, the chamber noted $\tau(x, x'; y)$ above coincides with $\tau(x, x')(y)$.)

Finally, it is easy to check that the map $\tau(c_+, c'_+) : \Delta_+ \to \Delta'_+$ coincides with ϕ on $E_2(c_+)$ and that, extended by $c_- \mapsto c'_-$, it is an isometry. This completes our sketch of proof of Theorem 2.

5.9 Remarks

(a) The above argument does not establish the uniqueness of the isometry $\Delta_+ \cup \{c_-\} \to \Delta'_+ \cup \{c'_-\}$ extending ϕ, a uniqueness which was asserted, perhaps hurriedly, in [22], 5.1; however, the isometry constructed in the proof is certainly "natural", and as good as unique for all practical purposes, as is illustrated for instance by the next remark.

(b) Denote by τ_+ the function τ of the above proof. Exchanging the roles of Δ_+ and Δ_-, Δ'_+ and Δ'_-, c_+ and c_-, ..., ϕ and the restriction of $\bar\phi$ on $E_2(c_-) \cup \{c_+\}$ (cf. 5.8.1), one similarly defines an isometry $\tau_-(c_-, c'_-) : \Delta_- \to \Delta'_-$. From the procedure followed to obtain τ_+, it readily follows that

5.9.1 *if there exists an isometry* $\Delta_+ \cup \Delta_- \to \Delta'_+ \cup \Delta'_-$ *extending* ϕ, *it coincides with* $\tau_+(c_+, c'_+)$ *on* Δ_+ *and with* $\tau_-(c_-, c'_-)$ *on* Δ_-.

(We already knew by Theorem 1 that if such an isometry exists, it is unique.)

(c) It would be interesting to know whether, always under the same general conditions as above (in particular, assuming that all m_{ij} are finite and M has no direct factor of rank 1), it is automatically true that

$$(*) \quad \tau_+(c_+, c'_+) \cup \tau_-(c_-, c'_-) : \Delta_+ \cup \Delta_- \to \Delta'_+ \cup \Delta'_- \text{ is an isometry.}$$

If $(\Delta'_+, \Delta'_-, d'^*) = (\Delta_+, \Delta_-, d^*)$ and if M has no direct factor of rank ≤ 2, the validity of $(*)$ for all choices of $c_+, c_-, c'_+, c'_-, \phi$ is equivalent with the Moufang condition: the proof is similar to that of [15], 3.5. Whether $(*)$ holds or not, Theorem 2 and Proposition 11 readily imply the following weaker form of the Moufang condition:

As before, we assume that all m_{ij} are finite and that M has no direct factor of rank ≤ 2. Let $(\Delta_+, \Delta_-, d^*)$ be a twinning and let A, A' be two admissible apartments of Δ_+, the intersection of which is a root α of A. Then, there exists an automorphism of Δ_+ mapping A onto A', fixing α (chamberwise) and fixing all spheres of rank 1 whose centre lies in the interior of α.

(d) We take up the notation of 5.8. For an isometry $\phi_+ : E_2(c_+) \to E_2(c'_+)$, the condition that it extend to an isometry of $E_2(c_+) \cup \{c_-\}$ onto $E_2(c'_+) \cup \{c'_-\}$ is equivalent to the condition that ϕ_+ map $E_2(c_+) \cap A(c_+, c_-)$ onto $E_2(c'_+) \cap A(c'_+, c'_-)$. With the terminology introduced in 6.1, those two intersections are "apartments of the foundations $E_2(c_+)$ and $E_2(c'_+)$".

(e) We keep the hypotheses and notation of 5.8 and assume further that $\Delta_+ = \Delta'_+$, $\Delta_- = \Delta'_-$, $c_+ = c'_+$, $c_- = c'_-$ and $\phi =$ id. Then, the automorphisms $\tau(x, x')$ of Δ_+, for $x, x' \in \Delta_+$ opposite to c_-, clearly form a group U simply transitive on the set of all chambers opposite to c_-. If the elements of U can be extended (as isometries) to $\Delta_+ \cup \Delta_-$, the group U, thus extended, is the group of all automorphisms of the twinning $(\Delta_+, \Delta_-, d^*)$ which fix $E_1(c_-)$ (chamberwise). It may seem surprising, considering the data we started with, that only $E_1(c_-)$ – and not $E_2(c_-)$ – comes into play here but, in fact, one should observe that only the restriction of $\tilde{\phi}_1$ to $E_1(A_-) = \bigcup \{E_1(x) | x \in A_-\}$ plays a role in the definition of τ_+.

(f) When we defined twin buildings, M. Ronan and I conjectured that the arguments leading to Theorem 4.1.2 of [12] can be carried over with only minor adjustments to the framework of twinnings. This was indeed checked by Ronan (unpublished). The proof given here of the resulting theorem is somewhat different from his, in that it displays the main idea of the argument more explicitly in the form of the handy Proposition 11, which has no equivalent in [12].

6 Foundations. The classification problem. An example

In this section, we assume that, unless the contrary is specified, M has no direct factor of rank ≤ 2. Recall also that the m_{ij}'s are supposed to be finite.

6.1 Foundations

A **foundation of type M** is a pointed set $(E, *)$ endowed with a **distance function** $d : E \times E \to W = W(M)$, submitted to the following three axioms, where $\Sigma_J(x)$ denotes, as always, the sphere of radius W_J and center x, we set $E_J = \Sigma_J(*)$ (hence $E_\emptyset = \{*\}$), and write W_{ij}, E_{ij}, etc. for $E_{\{i,j\}}$, $W_{\{i,j\}}$, etc.:

(F1) *the set E is the union of all E_{ij} ($i, j \in I$);*

(F2) *with the induced metric, E_{ij} is a building of type* $\begin{pmatrix} 1 & m_{ij} \\ m_{ij} & 1 \end{pmatrix}$, *if* $i \neq j$;

(F3) *if $J, J' \subset I$ are sets of cardinality ≤ 2 and $x \in E_J$, $y \in E_{J'}$, then $d(x, x')$ is the unique element of smallest length in the set $d(x, y).d(y, x')$, where y runs over $E_{J \cap J'}$.*

Thus, the metric d in E is determined by its restriction to the E_{ij}'s, that is, by the building structures given on those sets; with the same heuristic language as in 4.1, we may say that a foundation is an amalgam of buildings of rank 2 (observe that $E_J \cap E_{J'} = E_{J \cap J'}$). Foundations were introduced in [9] with a definition formally different from, but essentially equivalent to the above one.

If E is a foundation and E_J is defined as above, an **apartment** of E is a subset A of E such that $* \in A$, Card $(A \cap E_i) = 2$ for all i and $A \cap E_{ij}$ is an apartment of the building E_{ij} for any $i, j \in I$, $i \neq j$. A **Moufang foundation** is a foundation E such that the building (generalized polygon) E_{ij} satisfies the Moufang condition whenever $m_{ij} \geq 3$ and such that, for $i \in I$, the structure of Moufang set induced on M_i by M_{ij} (cf. 4.4) for $j \in I$ and $m_{ij} \geq 3$, is independent of j.

Lemma 3 *The automorphism group of a Moufang foundation permutes transitively the apartments of the foundation.*

Indeed, let A, A' be two apartments of a Moufang foundation E of type M. For each $i \in I$, the structure of Moufang set on E_i provides a distinguished permutation u_i of E_i fixing $*$ and mapping $A \cap E_i$ onto $A' \cap E_i$. Now, it is easy to see that, for $i, j \in I$, $i \neq j$, the maps u_i and u_j can be extended simultaneously to an automorphism u_{ij} of E_{ij} mapping $A \cap E_{ij}$ onto $A' \cap E_{ij}$. The collection of all u_{ij} fit together and provide an automorphism of the foundation E which maps A onto A'.

Now, let $(\Delta_+, \Delta_-, d^*)$ be a twinning. For $c \in \Delta_+$, $E_2(c)$ is a foundation *which does not, up to isomorphism, depend on the choice of c.* To prove this, it suffices to show that if two chambers $c, c' \in \Delta_+$ are adjacent, then $E_2(c), E_2(c')$ are isomorphic. Let A be an admissible apartment of Δ_+ containing c and c', let A_- be the apartment of Δ_- opposite to A, let $^\circ : A \to A_-$ be the opposition isomorphism and, in the sphere of rank 1 of Δ_- containing c° and c'°, let us choose an element z distinct from c° and c'° (hence opposite to c and c'). When J runs over all subsets of cardinality 2 of I, the isomorphisms $\iota_J(c, c'; z) : \Sigma_J(c) \to \Sigma_J(c')$ (cf. 5.1) fit together and compose an isomorphism $E_2(c) \to E_2(c')$, as announced. (Similar arguments have been used in 5.4).

Abusing language, we call $E_2(c)$ *the* foundation of the twinning $(\Delta_+, \Delta_-, d^*)$ (relative to the given ordering of the pair Δ_+, Δ_-). By Corollary 3 (cf. 5.6), *it is a Moufang foundation.* If $c_- \in \Delta_-$ is a chamber opposite to c, $A(c, c_-) \cap E_2(c)$ is an apartment of that foundation. It has already been mentioned in different terms, in 5.9 (d), that this apartment determines the restriction of d^* to $E_2(c) \cup \{c_-\}$. More precisely, one has he following easy

Lemma 4 *Let* $i, j \in I$, $i \neq j$, *and* $x \in \Sigma_{ij}(c)$. *If* w_{ij} *denotes the longest element of* W_{ij}, *and* c_1 *the element of* $E_2(c) \cap A(c, c_-)$ *at distance* w_{ij} *of* c, *then* $d^*(x, c_-) = d(x, c_1).w_{ij}$.

Collecting the information we have just gathered, we see that the isometry class of the foundation $E_2(c)$ (i.e. the foundation of the twinning under consideration) determines the isometry class of $(E_2(c), E_2(c) \cap A(c, c_-))$ (Lemma 3), hence the isometry class of $E_2(c) \cup \{c_-\}$ (Lemma 4), hence the isomorphism class of the building Δ_+ (Theorem 2) as well as that of Δ_- (Remark 5.9 (b)).

6.2   On the classification problem. Three "conjectures".

Mainly for the sake of convenience, for later references, I take the risk of stating as a conjecture the general validity of the assertion $(*)$ of 5.9 (c), which means, in the present set-up:

Conjecture 1 *A twinning is determined up to isomorphism by its foundation.*

Here, we shall briefly discuss the folllowing special case of that conjecture, which can be approached with simpler, purely group-theoretical methods, and appears to be within closer reach, although substantial progress towards a proof of Conjecture 1 itself has recently been reported by Mark Ronan (private communication).

Conjecture 1$'$ *A Moufang twinning is determined up to isomorphism by its foundation.*

In fact, what appears to be a complete proof of that assertion does exist, but it is long and involved, and has not been written up in detail so far; thus, to state that result as a theorem would be at least premature. On the other hand, in important special cases, Conjecture 1$'$ can indeed be established at much lesser cost, as will be seen in 6.4.

In view of Proposition 7, Conjecture 1$'$ can be restated in group-theoretical terms as follows. Let $(G, (U_\alpha)_{\alpha \in \Phi})$ be a reduced RGD-system of type M (cf. 3.3). For $i, j \in I$, let $G_{ij} = G_{ji}$ denote the subgroup of G generated by U_{α_i}, $U_{-\alpha_i}$, U_{α_j}, $U_{-\alpha_j}$ (where $\alpha_i = \alpha_{s_i}$: cf. 3.3). Then, we have:

Conjecture 1′ (group-theoretical version) *The reduced RGD-system $(G; (U_\alpha)_{\alpha \in \Phi})$ is determined up to unique isomorphism by the "amalgam" consisting of the groups G_{ij} ($i, j \in I$, $G_{ij} = G_{ji}$), U_{α_i}, $U_{-\alpha_i}$ ($i \in I$), and all canonical inclusions $U_{\alpha_i} \to G_{ij}$ and $U_{-\alpha_i} \to G_{ij}$.*

More precisely, if $(G', (U'_\alpha)_{\alpha \in \Phi})$ is another RGD-system of type M and if the subgroups G'_{ij} of G' are defined in the same way as the G_{ij}'s, then, any system of isomorphisms $G_{ij} \to G'_{ij}$, $U_{\alpha_i} \to U'_{\alpha_i}$, $U_{-\alpha_i} \to U'_{-\alpha_i}$ compatible with the inclusions extends uniquely to an isomorphism $G \to G'$ mapping the U_α's onto the U'_α's.

Remarks.

(i) In the above statement, one can replace the G_{ij}'s by the quotients $G_{ij}/Z(G_{ij})$ and the inclusions $U_{\pm\alpha_i} \hookrightarrow G_{ij}$ by the corresponding monomorphisms. The statement one obtains in that way looks less natural than the above one but it is a closer translation of Conjecture 1′.

(ii) In the "amalgam", one can of course omit the G_{ij}'s for i, j such that $m_{ij} = 2$, because for such i, j's, the commutativity of $U_{\pm\alpha_i}$ with $U_{\pm\alpha_j}$ is included in the definition of RGD-systems of type M.

(iii) The above group-theoretical version of Conjecture 1′ "looks like" a presentation of G by amalgamation. In fact, when M is spherical (i.e. W is finite), the sum of the G_{ij}'s amalgamated along the $U_{\pm\alpha_i}$ is indeed a mere central extension of G (cf. [12], 13.32 for a closely related result; here, the G_{ij}'s for $m_{ij} = 2$ can of course *not* be omitted!). This is definitely no longer true in general: the conjecture only asserts that G is *determined* by the given "amalgam" *taking into account the axioms RGD.*

We now come to the classification. Assuming the validity of Conjecture 1 (resp. Conjecture 1′), the problem of classifying all twinnings (resp. all Moufang twinnings) of type M can be decomposed as follows:

(A) classify all Moufang foundations of type M;

(B) find necessary and sufficient conditions for a given Moufang foundation to be the foundation of a twinning (resp. a Moufang twinning).

In principle, problem (A) is straightforward, once all Moufang buildings of rank 2 are known: one may just, for complicated M, have trouble with bookkeeping and/or formulation of the result. A simple example will be treated in 6.5. As for (B), it might be taken care of by the following, perhaps too optimistic

Conjecture 2 *A Moufang foundation E of type M is the foundation of a Moufang twinning as soon as*

(so) *for any spherical triple $J \subset I$ (i.e. Card $J = 3$ and W_J is finite) the subfoundation of type $(m_{ij})_{i,j \in J}$ supported by the set $\bigcup \{E_{ij} | i, j \in J\}$ is the foundation of a spherical building.*

That this might be true is suggested by the main result of [9]. We shall say that a foundation is **sound** if it satisfies (so). That condition is of course necessary for the existence of a twinning with the given foundation.

Observe that all conjectures stated here are true for spherical types M.

6.3 Thickenings of Coxeter buildings

Let A be a Coxeter building of type M. We define a **thickening** of A as a set Θ containing A, endowed with a distance function $d : \Theta \times \Theta \to W$ which extends the distance function of A and which satisfies the axioms (Bu1) and (Bu2) of 2.1 and the following two conditions:

(Θ 1) *for $J \subset I$ of cardinality 2 and $a \in A$, the "sphere" $\Sigma_J(a) = \{x \in \Theta | d(a, x) \in W_J\}$, with the distance function induced by d, is a building of type $M_J = \begin{pmatrix} 1 & m_J \\ m_J & 1 \end{pmatrix}$;*

(Θ 2) $\Theta = \bigcup \{\Sigma_J(a) | a \in A, J \subset I, \text{ Card } J = 2\}.$

(N.B. It would perhaps be more natural to call this a 2-thickening and to define r-thickenings for an arbitrary integer $r \leq \text{Card } I$ in a similar way, replacing 2 by r in (Θ 1), (Θ 2), but we shall have no use here for that more general notion.)

If Δ is a building of type M and $A \subset \Delta$ is an apartment of Δ, then $E_2(A) = \bigcup_{a \in A} E_2(a)$ is a thickening of A. We then say that the building Δ **extends** the thickening $E_2(A)$, and that the latter is **extensible**. Proposition 13 below shows that an extensible thickening of A is essentially nothing more than an "amalgam" of the buildings of rank two $\Sigma_J(a)$, $a \in A$, Card $J = 2$.

Let α be a root of A and let Θ be any thickening of A. We denote by $\mathcal{B}(\alpha)$ the set of all $x \in \alpha$ lying at the boundary of α. Thus, for $a \in \mathcal{B}(\alpha)$, there is a unique sphere of rank 1 in Θ whose intersection with α is $\{a\}$; we denote it by $\Sigma(a, \alpha)$. We call **fringe** of α any map $\theta : \mathcal{B}(\alpha) \to \bigcup \{\Sigma(a, \alpha) | a \in \mathcal{B}(\alpha)\}$ preserving distances and such that $\theta(a) \in \Sigma(a, \alpha) - \{a\}$ for all $a \in \mathcal{B}(\alpha)$ or, alternatively, the image of such a map.

Lemma 5 *For $a \in \mathcal{B}(\alpha)$ and $x \in \Sigma(a, \alpha) - \{a\}$, there is at most one fringe of α containing x. If it exists, it is entirely determined by the data consisting of the spheres of rank 2 centered at elements of $\mathcal{B}(\alpha)$, their inclusion in Θ and their structures as buildings.*

Indeed, if $a' \in \mathcal{B}(\alpha)$, there exists a sequence of elements of $\mathcal{B}(\alpha)$, $a = a_0, a_1, \ldots, a_r = a'$, such that, for $i = 0, \ldots, r - 1$, a_{i+1} belongs to a sphere of rank 2 centered at a_i. But, working inside that sphere, one sees that $\theta(a_{i+1})$ is determined by $\theta(a_i)$. Hence the claim.

When it exists, we denote by $\theta(\alpha, x)$ the fringe of α containing x, as above. If it exists for all α, $a \in \mathcal{B}(\alpha)$ and $x \in \Sigma(a, \alpha) - \{a\}$, we say that the thickening Θ is **coherent**. We shall denote by $U(\alpha)$ the set of all fringes of α, and by $\overline{U}(\alpha)$ the set $U(\alpha) \cup \{\mathcal{B}(\alpha)\}$. If $a \in \mathcal{B}(\alpha)$, each element u of $\overline{U}(\alpha)$, viewed as a subset of Θ, intersects $\Sigma(a, \alpha)$ in exactly one element, hence a map $\overline{U}(\alpha) \to \Sigma(a, \alpha)$, which we denote by $\pi(a, \alpha)$; if Θ is coherent, then $\pi(a, \alpha)$ is a bijection for all α and $a \in \mathcal{B}(\alpha)$.

An extensible thickening Θ is coherent: to see that, one considers an apartment A' containing α and x in an extension Δ of Θ (such an apartment exists, by [18], 3.7.4 (7)) and takes for $\theta(a, \alpha)$ the restriction of the reflection of A' permuting α and $A' - \alpha$ to $\mathcal{B}(\alpha)$.

Let Δ be a building which, for the time being, we do not assume to be thick, and let X be a sphere in Δ. We recall (cf. [12], §3) that there exists a retraction $\mathrm{pr}_X : \Delta \to X$, the **projection** of Δ onto X (the notation and terminology in [12] are slightly different), characterized by the identity

(pr 1) *for $y \in \Delta$, $l(d(y, \mathrm{pr}_X y)) = \inf \{l(d(y, x)) | x \in X\}$.*

Furthermore,

(pr 2) *if $y \in \Delta$ and $x \in X$, then $d(y, x) = d(y, \mathrm{pr}_X y) . d(\mathrm{pr}_X y, x)$.*

If Δ is thick, (pr 2) also characterizes pr_X.

Two spheres X and Y are said to be **parallel** if the maps $\mathrm{pr}_X|_Y : Y \to X$ and $\mathrm{pr}_Y|_X : X \to Y$ are bijective.

Proposition 12 *Let X, Y be two spheres in the building Δ, and let A be an apartment having a nonempty intersection with X and Y.*

(i) *The spheres X and Y are parallel if and only if $X \cap A$ and $Y \cap A$ are parallel (as spheres of the Coxeter building A).*

(ii) *Suppose X and Y are parallel. Then, the bijections $\mathrm{pr}_X|_Y$ and $\mathrm{pr}_Y|_X$ are inverse of each other. Furthermore, for $x \in X$, the element $d(x, \mathrm{pr}_Y x)$ of W depends only on X and Y, and not on x. We denote it by $d(X, Y)$ and call it the distance between X and Y.*

(iii) *In all cases, the set $\mathrm{pr}_X Y$ is a sphere which meets A. In fact, the restriction of pr_X to $Y \cap A$ is equal to the projection $\mathrm{pr}_{X \cap A}|_{Y \cap A}$ in the Coxeter building A; in particular, $\mathrm{pr}_X(Y \cap A) = \mathrm{pr}_{X \cap A}(Y \cap A)$.*

(iv) *The spheres $\mathrm{pr}_X Y$ and $\mathrm{pr}_Y X$ are parallel.*

(v) *If $X' = pr_X Y$, $Y' = pr_Y X$, $x \in X$, $y \in Y$, $x' = pr_{X'} x$, $y' = pr_{Y'} y$ and $y'' = pr_{Y'} x'$, then, one has the decomposition*

$$d(x, y) = d(x, x').d(X', Y').d(y'', y').d(y', y),$$

with additivity of the length l.

The proof is easy. The assertions (i) to (iv) are established by the techniques used in [12], §§3 and 12; in fact, some of them are just reformulations of statements in *loc. cit.* The relation (v) follows from successive applications of (pr 2). Let us just add one word concerning (iv). To prove it, it suffices, by (i) and (iii), to consider the case where Δ is a Coxeter building. Then, one can use the representation of the building Δ as a simplicial decomposition (in fact, a partition in *simplicial cones*) of a certain convex cone in a real vector space, namely the space of the contragredient representation of W (cf. e.g. [2], Chap. 5, 4.6). The spheres of Δ are in one-to-one correspondence with the simplicial cones in question. Two spheres are parallel if and only if the corresponding cones span the same linear space, and the span of the cone corresponding to $pr_X Y$ is the sum of the spans of the cones corresponding to X and Y, hence (iv).

Proposition 13 *Let Θ be an extensible thickening of a Coxeter building A. Then, the distance function $d : \Theta \times \Theta \to W$ is entirely determined by the data $\mathcal{D}$ consisting of (the set underlying) Θ, the inclusion $A \to \Theta$, the family of all subsets of Θ of the form $\Sigma_J(a)$ ($a \in A$, $J \subset I$, Card $J = 2$) and the structures of buildings on A and on those subsets (i.e. the restrictions of d to $A \times A$ and to $X \times X$ for $X = \Sigma_J(a)$ as above).*

We shall only sketch the proof of this.

Let $\mathcal{S}$ denote the set of all spheres of the form $\Sigma_J(a)$, with $a \in A$ and Card $J \leq 2$. This set is determined by the data $\mathcal{D}$ since, in any building, the spheres of rank 1 are also the spheres of rank 1 in the spheres of rank 2 considered as buildings. Since Θ is extensible, we can talk about the projection $pr_X : \Theta \to X$ of Θ onto a sphere $X \in \mathcal{S}$, and about parallel elements of $\mathcal{S}$: because of (pr 1), those notions are independent of the choice of an extension Δ of Θ.

Whether two spheres $X, Y \in \mathcal{S}$ are parallel or not is determined by $\mathcal{D}$, as follows from Proposition 12 (i). If they are parallel, their distance $d(X, Y)$ is also determined by $\mathcal{D}$ since, by Proposition 12 (ii) and (iii), it is equal to the distance $d(X \cap A, Y \cap A)$ inside A. In all cases, $pr_X Y$ is determined by $\mathcal{D}$ (and X, Y) because of Proposition 12 (iii). Finally, by Proposition 12 (iv) and (v), our assertion is reduced to showing that

if $X, Y \in \mathcal{S}$ are parallel, then the bijection $pr_Y|_X : X \to Y$ is determined by $\mathcal{D}$.

If rk X = rk Y = 1, this follows from Lemma 5. Indeed, suppose that $X = \Sigma_h(a)$ and $Y = \Sigma_i(b)$, with $h, i \in I$ and $a, b \in A$, are parallel; then a and b lie at the boundary of a root α of A, $\mathrm{pr}_Y|_X$ maps any element x of X on the unique element of Y which belongs to the same fringe of α as x, and Lemma 5 asserts that this fringe is determined by $\mathcal{D}$.

Finally, let $X = \Sigma_J(a)$, $Y = \Sigma_K(b)$ with $a, b \in A$ and Card J = Card K = 2, and suppose again that X and Y are parallel. Then, the bijection $\mathrm{pr}_Y|_X$ of the building X of type W_J onto the building Y of type W_K is a "semi-isomorphism": it transforms the distances by a fixed isomorphism $W_J \to W_K$ mapping $\{s_j | j \in J\}$ onto $\{s_k | k \in K\}$. By Proposition 12 (iii), the restriction of pr_Y to $X \cap A$ is determined by $\mathcal{D}$. For $a \in X \cap A$ and $j \in J$, the restriction of pr_X to $\Sigma_j(a)$ is also determined by $\mathcal{D}$, as follows from what we have just seen on the rank 1 situation. Now, Theorem 4.1.1 of [12] implies that $\mathcal{D}$ determines the whole projection $\mathrm{pr}_X|_Y$. This completes our outline of the proof of Proposition 13.

We say that a thickening Θ of a Coxeter building A of type M has the **Moufang property**, or **is Moufang**, if it satisfies the following conditions, where we use the notation $\mathcal{B}(\alpha)$, $\Sigma(a, \alpha)$, $\pi(a, \alpha)$, $\overline{U}(\alpha)$ introduced at the beginning of 6.3:

(i) for $a \in A$ and $J \subset I$ of cardinality 2 such that $m_J \geq 3$, the sphere $\Sigma_J(a)$ is a Moufang building (or generalized polygon);

(ii) if $a \in A$ and $h, i, j \in I$ are such that $m_{hi} \geq 3$ and $m_{ij} \geq 3$, then $\Sigma_{hi}(a)$ and $\Sigma_{ij}(a)$ induce the same structure of Moufang set on $\Sigma_i(a)$ (cf. 4.4); thus, all spheres of rank 1 centered on A (that is, meeting A) have a canonical structure of Moufang set, inherited from Θ;

(iii) for any root α of A and any two elements a, a' of $\mathcal{B}(\alpha)$, the images of the Moufang structures of $\Sigma(a, \alpha)$ and $\Sigma(a', \alpha)$ by $\pi(a, \alpha)^{-1}$ and $\pi(a', \alpha)^{-1}$ are the same Moufang structure on $\overline{U}(\alpha)$; thus $\overline{U}(\alpha)$ has a canonical structure of Moufang set, also inherited from Θ.

I do not know whether (iii) is a consequence of (i) and (ii).

If Θ is a thickening of a Coxeter building A whose automorphism group is W, the group Aut Θ acts on A via a homomorphism $\nu : \mathrm{Aut}\ \Theta \to W$. By abuse of language, we shall say that Θ is **homogenous** if ν is an epimorphism.

Suppose Θ has the Moufang property. To every root $\alpha \in \Phi$ is then naturally associated a group which we denote by $U_\alpha(\Theta)$, namely the group simply transitive on $\overline{U}(\alpha) - \{\mathcal{B}(\alpha)\}$ (notation as above) which the structure of Moufang set on $\overline{U}(\alpha)$ associates to the point $\mathcal{B}(\alpha) \in \overline{U}(\alpha)$. Any element g of Aut Θ clearly induces an isomorphism of $U_\alpha(\Theta)$ onto $U_{\nu(\alpha)}(\Theta)$. If Θ is the thickening of the apartment $A(c_+, c_-)$ of the building Δ_+ associated to an RGD-system $(G; (U_\alpha)_{\alpha \in \Phi})$ as in 3.3, the system of groups $U_\alpha(\Theta)$ is of course nothing else but the system of groups $(U_\alpha)_{\alpha \in \Phi}$ divested from the embracing group G (while retaining, however, part of the structure inherited from G: cf. 6.4).

The following assertion is a slightly improved version of Corollary 3 of 5.6; it is proved in a similar way, as an easy corollary of Proposition 11, but one must, in addition, use some argument taken from the proof of that proposition:

In a building of type M which is part of a twinning, the thickening of any apartment is homogenous and Moufang.

6.4   Proof of Conjecture 1' for "sufficiently thick" twinnings.

We keep the notation of the previous sections (see especially 3.3, 4.1); in particular, $(G; (U_\alpha)_{\alpha \in \Phi})$ denotes an RGD-system of type M, $(\Delta_+, \Delta_-, d^*)$ the associated twinning, c_+, c_- the fundamental chambers relative to the given RGD-system, N the normalizer in G of the set of groups $\{U_\alpha | \alpha \in \Phi\}$, H the intersection of the normalizers of the U_α's and U_+ the subgroup of G generated by the U_α's for $\alpha > 0$. We set $E = E_2(c_+)$. Conjecture 1' asserts that *if the RGD-system is reduced, it can be recovered from the foundation E* (plus a distinguished apartment, if one wants the construction to be canonical). We shall, more generally, discuss the following problem, *assuming only that the RGD-system under consideration is faithful:*

Problem 1 *Reconstruct the system $(G; (U_\alpha)_{\alpha \in \Phi})$, knowing the foundation E, the group H and the action of H on E.*

A first reduction of that problem is provided by the following, purely group-theoretical observation.

Proposition 14 *The faithful system $(G; (U_\alpha)_{\alpha \in \Phi})$ is entirely determined by the following data:*

(i) *the groups U_α ($\alpha \in \Phi$) and N, and the action by conjugation of N on the U_α's;*

(ii) *the group U_+ and the inclusion maps $U_\alpha \hookrightarrow U_+$ for $\alpha > 0$.*

This is an easy consequence of [20], Theorem 2 (cf. also [22], Propositions 2 and 4) which provides a presentation of G in terms of the subgroups HU_α. In the version of [22], Proposition 4, the main relations are the commutation relations of the form

$$[u_\alpha, u_\beta] = \Pi\{u_\gamma | \gamma \in (\alpha, \beta)\}, \tag{1}$$

where $\{\alpha, \beta\}$ is a prenilpotent pair of roots (cf. 3.3), (α, β) is defined as in 3.3 and $u_\xi \in U_\xi$. All such relations can be deduced from the given data:

indeed, by definition of prenilpotent pairs, $\alpha \cap \beta$ contains a chamber and, upon conjugating by a suitable element of N, we may assume that $\alpha \cap \beta$ contains the fundamental chamber 1, in which case, (1) is a relation inside U_+. The presentation also involves the groups $\langle U_{\alpha_i}, U_{-\alpha_i}, H\rangle$; the structure of those groups can be deduced by standard arguments à la Steinberg from the data (i), the relations (1) and the hypothesis that $(G; (U_\alpha)_{\alpha \in \Phi})$ is a faithful RGD-system.

Coming back to Problem 1, we suppose given E (with a distinguished apartment, in order to rigidify the situation) and the group H acting on E. Proposition 11 of 5.3 implies that, from E, one can reconstruct the thickening Θ of the apartment $A = A(c_+, c_-)$ in Δ_+, hence the system of groups $(U_\alpha(\Theta))_{\alpha \in \Phi}$ (cf. 6.3). There are canonical isomorphisms $U_\alpha \to U_\alpha(\Theta)$, but they are not given to us among the data of the problem. The given action of H on E induces an action of H on Θ, hence also on the groups $U_\alpha(\Theta)$. To every $i \in I$ and $u \in U_{\alpha_i} - \{1\}$, there corresponds an element $m(u)$ of N (cf. 3.3, (RGD 2)). The action of $m(u)$ on Θ is determined by E and the canonical image of u in $U_{\alpha_i}(\Theta)$. Indeed, working inside $\Sigma_{ij}(c_+)$, for $j \in I$, one sees that those data determine the action of $m(u)$ on $\Sigma_j(c_+)$, hence also, varying j, its action on $E_1(c_+)$; but, as in the proof of Proposition 11, repeated use of Theorem 4.1.1 of [12] shows that the action of $m(u)$ on $E_1(c_+)$ determines its action on Θ. Since H and the elements $m(u)$ of the above type generate N, we conclude that the action of N on Θ, hence the data (i) of Proposition 14, can be reconstructed, knowing only E, H and the action of H on E.

On the other hand, Theorem 2 and its proof enable us, from the same knowledge, to reconstruct Δ_+ and U_+. Consequently, in order to solve Problem 1, there remains only, by Proposition 14, to recover also the canonical injections $U_\alpha(\Theta) \to U_+$ (via the so far unknown canonical isomorphisms $U_\alpha(\Theta) \to U_\alpha$) for $\alpha > 0$.

The next step consists in verifying (details will be omitted) that the arguments used in the proof of Theorem 2 and a little extra work enable one, from the knowledge of E and H, to deduce not only Δ_+ and U_+, but also:

(a) the action of U_+ on Δ_+;

(b) the embedding of Θ (and in particular A) in Δ_+;

(c) for any given $w \in W$, the group $U_w \subset U_+$ generated by all U_α, for $\alpha > 0$ such that $w\alpha < 0$, and the action of H on U_w by conjugation.

The second part of (c) is clear since everything here is H-equivariant.

For $\alpha \in \Phi$, $\alpha > 0$, let $\tilde{U}_\alpha$ denote the group of all elements of U_+ fixing α (chamberwise). From the definitions of $U_\alpha(\Theta)$ and $\tilde{U}_\alpha$, one readily deduces a natural homomorphism $\tilde{U}_\alpha \to U_\alpha(\Theta)$ whose restriction to U_α is the canonical isomorphism $U_\alpha \to U_\alpha(\Theta)$. Furthermore, it can be shown that those homomorphisms $\tilde{U}_\alpha \to U_\alpha(\Theta)$ may be reconstructed from the knowledge of E. The same is therefore true of the homomorphisms $U_\alpha(\Theta) \to U_+$, once the system of subgroups $(U_\alpha)_{\alpha \in \Phi}$ of U_+ is known.

Let us summarize the results of the above discussion:

From the data of Problem 1, i.e. the foundation of E, the group H and the action of H on E, it is possible to reconstruct the group U_+ generated by all U_α ($\alpha > 0$); to solve the problem completely, it suffices to characterize the groups U_α ($\alpha > 0$) themselves inside U_+.

I know of no easy way to do that in general, but there is a simple solution under the following, rather mild restriction: one supposes that the groups U_w are nilpotent (this is probably no restriction at all) and that

(For) *if $\alpha, \alpha' \in \Phi$ are distinct positive roots, the H-groups (groups with operators indexed by H) are foreign to each other, that is, have no isomorphic nontrivial (H-)subquotients.*

When this holds, U_α can be characterized, for any given $\alpha > 0$, as a subgroup of U_+ which is contained in some U_w (any w such that $w\alpha < 0$ will do), is the image of an H-equivariant monomorphism $U_\alpha(\Theta) \to U_+$ and is maximal with those properties. This is an immediate consequence of Lemma 3.3 in [1], p.72.

We already indicated that the condition (For) is often satisfied. An important example is that of Kac-Moody systems *over infinite fields* when the α_i's are linearly independent in Λ (notation of 3.3). (This condition imposed on the α_i's may be unnecessary.)

6.5   An example: classification of the sound Moufang foundations of type $M = (m_{ij})_{i,j \in I}$ when all m_{ij} are ≤ 3

For any graph Γ (nonoriented, without loop) with set of vertices I, let us denote by $\mathcal{E}(\Gamma)$ the set of its edges and by $M(\Gamma)$ the Coxeter matrix $(m_{ij})_{i,j \in I}$ defined as follows

$$m_{ij} = \begin{cases} 1 & \text{if } i = j, \\ 3 & \text{if } \{i,j\} \in \mathcal{E}(\Gamma), \\ 2 & \text{otherwise.} \end{cases}$$

We wish to classify all *sound* Moufang foundations of type $M(\Gamma)$ (cf. 6.2). For the sake of simplicity, we shall make the following hypothesis:

(CNC) *the graph Γ is connected and is not the complete graph on I;*

the general case is not more difficult to treat but the statement of the final result is more complicated.

Let $(E, *)$ be a sound, Moufang foundation of type $M(\Gamma)$ and let E_i $(i \in I)$ and E_{ij} $(i, j \in I)$ be defined as in 6.1. The sets E_i are Moufang sets and, when $\{i, j\} \in \mathcal{E}(\Gamma)$, E_{ij} is the flag building of two Moufang projective planes, dual of each other, which we denote by Π_{ij} and $\Pi_{ji} = \Pi'_{ij}$, choosing notation in such a way that E_i and E_j are respectively lines in Π_{ij} and Π_{ji}. Simple computations and classical projective geometry tell us the following facts.

(a) Suppose $i, j \in I$, $\{i, j\} \in \mathcal{E}(\Gamma)$ and Π_{ij} is desargesian. Then, the affine plane $\Pi_{ij} - E_i$ is the affine plane over a division ring k_{ij} *which is canonically defined*: indeed, it is the ring of homothetic transformations of the affine plane in question in which a point zero is arbitrarily chosen; up to canonical isomorphism, the ring is independent of that choice. There is a canonical isomorphism between the additive group of k_{ij} and the simply transitive group which the structure of Moufang set on E_i associates to $(*)$; thus, *the additive group of k_{ij} and, in particular, its underlying set, depends only on i.* We recall that the structure of Moufang set of E_i also determines the pair of opposite division rings k_{ij} and k_{ij}° (where, as usual, $^{\circ}$ stands for "opposite"). The planes Π_{ij} and Π_{ji} being dual of each other, the rings k_{ij} and k_{ji} are antiisomorphic ; more precisely, the inclusions of E_i and E_j in Π_{ij} and $\Pi_{ji} = \Pi'_{ij}$ *determine* a class ϕ_{ij} of isomorphisms $k_{ji} \to k_{ij}^{\circ}$ modulo inner automorphisms of k_{ji}; one has $\phi_{ji} = \phi_{ij}^{-1}$.

(b) If $\{i, j\} \in \mathcal{E}(\Gamma)$ and if the Moufang set E_i is isomorphic to a projective line over a division ring, then Π_{ij} is desarguesian.

(c) If $\overset{h}{\bullet}\!\!\rule[0.5ex]{1.2em}{0.4pt}\!\!\overset{i}{\bullet}\!\!\rule[0.5ex]{1.2em}{0.4pt}\!\!\overset{j}{\bullet}$ is a full subgraph of type A_3 in Γ, the planes Π_{ih} and Π_{ij} are desarguesian and $k_{ij} = k_{ih}^{\circ}$; this expresses the soundness of the foundation.

From (a), (b), (c) and (CNC), it follows that all Π_{ij} are indeed desarguesian. Furthermore, it is easily verified that the foundation is entirely determined by the k_{ij}'s and the ϕ_{ij}'s and that all the conditions necessary for a system of k_{ij}'s and ϕ_{ij}'s to determine a foundation have been stated on the way. Therefore, we can state:

Proposition 15 *The sound, Moufang foundations of type $M(\Gamma)$ (where Γ satisfies (CMC)) are classified by the systems $\mathcal{K} = (k_{ij}, \phi_{ij}|i, j \in I; \{i, j\} \in \mathcal{E}(\Gamma))$, where*

k_{ij} *is a division ring, the additive group (and, in particular the underlying set) of which depends only on i;*

given $i \in I$, the rings k_{ij} for $\{i,j\} \in \mathcal{E}(\Gamma)$ are pairwise identical or opposite; if $h,i,j \in I$ are such that $\{h,i\} \in \mathcal{E}(\Gamma)$, $\{i,j\} \in \mathcal{E}(\Gamma)$ and $\{h,j\} \notin \mathcal{E}(\Gamma)$, then $k_{ij} = k_{ih}^{\circ}$;

ϕ_{ij} is a class of isomorphisms $k_{ji} \to k_{ij}^{\circ}$ modulo inner automorphisms of k_{ji}; one has $\phi_{ij} = \phi_{ji}^{-1}$.

The system $\mathcal{K}$ can be viewed as some sort of "sheaf of division rings" over Γ.

By (CNC), all k_{ij}'s are pairwise isomorphic or antiisomorphic. In particular, if some k_{ij} is commutative, all of them are. Observe that this is the case when Γ has a full subgraph of type

(D4)

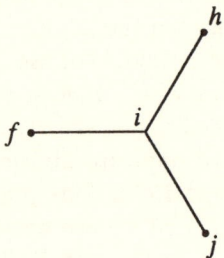

Indeed, we then have, by (c), $k_{if} = k_{ih}^{\circ} = k_{ij} = k_{if}^{\circ}$. (Compare [12], Theorem 6.12.)

Suppose the k_{ij}'s are commutative and choose a vertex $0 \in I$ of Γ. All fields k_{0i}, for $\{0,i\} \in \mathcal{E}(\Gamma)$, are now the same; let us denote them by k. Giving a system $\mathcal{K}$ as above is then clearly equivalent to giving the field k and a homomorphism $\phi : \pi_0(\Gamma,0) \to \mathrm{Aut}\ k$.

If the conjectures 1' and 2 are true, to any system $\mathcal{K}$ as above, there corresponds a twinning $(\Delta_+(\mathcal{K}), \Delta_-(\mathcal{K}), d^*)$ and also an RGD-system $(G(\mathcal{K}); (U_\alpha(\mathcal{K}))_{\alpha \in \Phi})$. Here, one could take for $G(\mathcal{K})$ the group of all automorphisms of the twinning, but it is perhaps more natural to take only the automorphisms which induce inner automorphisms of all k_{ij}'s; we shall do that here. It would be interesting, and perhaps not too difficult, to give a direct construction (and existence proof) of the group $G(\mathcal{K})$ in general. There is a case where this group is well-known, namely when the k_{ij}'s are commutative and the homomorphism ϕ introduced above is trivial; then $G(\mathcal{K})$ is nothing else but the Kac-Moody group of adjoint type over k corresponding to the generalized Cartan matrix (A_{ij}) defined by $A_{ii} = 2$ and, for $i \neq j$, $A_{ij} = 0$ or -1 according as $\{i,j\} \notin$ or $\in \mathcal{E}(\Gamma)$. In general, we see that $G(\mathcal{K})$ appears as some sort of Kac-Moody group over a sheaf of division rings.

There is another case where $G(\mathcal{K})$ has a simple interpretation in classical terms, namely, when the k_{ij}'s are commutative and Γ is a single cycle, say of length n. Then the homomorphism ϕ is determined by a single automorphism

τ of k (once an orientation of Γ is chosen) and if R denotes a non-commutative extension $k[t, t^{-1}]$, where $txt^{-1} = \tau(x)$ for $x \in k$, the group $G(\mathcal{K})$ can be identified with a subgroup of $PGL_n(R)$ of index n.

References

[1] A. Borel et J. Tits, Groupes réductifs, Publ. Math. I.H.E.S. 27 (1965), 55–150.

[2] N. Bourbaki, Groupes et algèbres de Lie, Chap. 4, 5 et 6, Actu. Scient. et Ind. 1337, Hermann, Paris, 1968.

[3] K.S. Brown, Buildings, Springer-Verlag, 1989.

[4] F. Bruhat et J. Tits, Groupes réductifs sur un corps local. I. Données radicielles valuées, Publ. Math. I.H.E.S. 41 (1972), 5–251

[5] F. Bruhat et J. Tits, Groupes réductifs sur un corps local. II. Schémas en groupes. Existence d'une donnée radicielle valuée, ibid. 60 (1984), 5–184.

[6] F. Buekenhout, Foundations of one dimensional projective geometry based on perspectivities, Abhandlungen Math. Sem. Univ. Hamburg 43 (1975), 21–29.

[7] C. Hering, W.M. Kantor and G.M. Seitz, Finite groups having a split BN-pair of rank 1, J. Algebra 20 (1972), 435–475.

[8] M. Ronan, Lectures on buildings, Perspectives in Math., vol. 7, Academic Press, 1989.

[9] M. Ronan and J. Tits, Building buildings, Math. Annalen 278 (1987), 291–306.

[10] M. Suzuki, Group theory I., Grundlehren der Math. Wiss. Bd. 247, Springer Verlag, 1982.

[11] J. Tits, Sur certaines classes d'espaces homogènes de groupes de Lie, Mém. Acad. Roy. Belg. 29 (3) (1955).

[12] J. Tits, Buildings of spherical type and finite BN-pairs, Springer Lecture Notes in Math. 386 (1974) (2nd augm. ed. 1986).

[13] J. Tits, On buildings and their applications, Proc. Intern. Congress Math. Vancouver 1974 (1975), vol. 1, 209–220.

[14] J. Tits, Non-existence de certains polygones généralisés, I. Inv. Math. 36 (1976), 275–284; II. Ibid. 51 (1979), 267–269.

[15] J. Tits, Endliche Spiegelungsgruppen, die als Weylgruppen auftreten, Inv. Math. 43 (1977), 283–295.

[16] J. Tits, Reductive groups over local fields, Proc. Symp. Pure Math. 33 (Summer Inst. on Group representations and automorphic forms, Corvalis, 1977) (1979), vol. 1, 26–69.

[17] J. Tits, Buildings and Buekenhout geometries, in Finite Simple Groups II, Proc. Lond. Math. Soc. Res. Symposium, Durham, 1978, ed. M.J. Collins, Academic Press, 1980, 309–320.

[18] J. Tits, A local approach to buildings, in The geometric vein, the Coxeter Festschrift, Springer-Verlag 1981, 519–547.

[19] J. Tits, Immeubles de type affine, in Buildings and the geometry of diagrams, Como 1984, Springer Lecture Notes in Math. 1181 (1976), 159–190.

[20] J. Tits, Uniqueness and presentation of Kac-Moody groups over fields, J. Algebra 105 (1987), 542–573.

[21] J. Tits, Résumé de cours, Annuaire du Collège de France, 89e année, 1988–1989, 81–95.

[22] J. Tits, Résumé de cours, Annuaire du Collège de France, 90e année, 1989–1990, 87–103.

[23] R. Weiss, The nonexistence of certain Moufang polygons, Inv. Math. 51 (1979), 261–266.

Some Remarks on the Structure of Finite Subgroups of Simple Algebraic Groups

This paper is a survey of some results by the author in the study of the subgroup structure of the finite simple groups of Lie type. Throughout the paper G is a simple algebraic group; if G is defined over a finite field F_q then σ is some Steinberg endomorphism of G. We shall omit index G in notations like $N_G(X)$, $C_G(X)$.

1 Subgroups of simple groups of exceptional type.

The first result of the paper is a reduction theorem for the maximal subgroups of finite exceptional groups similar to the well-known result of M. Aschbacher [1] for finite classical groups. In the case of classical groups Theorem 1 doesn't give any new information, but it may be useful in the study of simple groups of exceptional type. Another and more explicit version of a reduction theorem was obtained recently by M. W. Liebeck and G. M. Seitz [2].

Theorem 1 *[3]. Let G be defined over the finite field F_q, $G_\sigma \cong G(F_q)$ and $G_0 = G_\sigma'$. Let $G_0 \leq G_1 \leq \mathrm{Aut}\, G_0$ and let M be a subgroup in G_1. Then one of the following statements is valid:*

(a) $M \leq N_{G_1}(H_\sigma \cap G_0) \leq N_G(H) \cap G_1$ for some proper connected nontrivial σ-invariant subgroup $H < G$;

(b) M is an almost simple group, i.e. $S \leq M \leq \mathrm{Aut}\, S$ for some simple group S;

(c) $M \leq N_G(J)$ for some Jordan subgroup J in G ;

(d) G is of type E_8, $\mathrm{char} F_q = p > 5$ and $M \leq N_{G_1}(X)$, where $X \cong Alt_5 \times Alt_6$.

Here, a Jordan subgroup [4] is an elementary abelian r-subgroup J satisfying the following properties:

(a) $N_G(J)$ is finite;

(b) J is a minimal normal subgroup in $N_G(J)$;

(c) if $A \geq J$ is an elementary abelian r-subgroup and $N_G(J) \leq N_G(A)$ then $N_G(J) = N_G(A)$.

Jordan subgroups have been classified in [4], [5], [6]. We shall discuss them later.

The following theorem gives us the explicit method of constructing the subgroup X mentioned in Theorem 1.

Theorem 2 *[7]. Let G be a simple algebraic group of type E_8 defined over the finite field $F_p, p > 5$ is a prime number, $F_q = F_p[\sqrt{5}]$, σ is a Steinberg endomorphism of G with the group of fixed points $G_\sigma \cong G(F_q)$. The the following statements are valid:*

(a) *the group G_σ contatins a unipotent element u such that $C(u) = UY$, where $U = R_u(C(u)), Y \cong Sym_5$ and $Y \leq G_\sigma$;*

(b) *$C(Y) \cong PSL_2$;*

(c) *$C(Y)_\sigma \cong PGL_2(F_p)$ or $PGL_2(F_{p^2})$ and contains a subgroup A which is isomorphic to Alt_5;*

(d) *$C(A) = C(A)_\sigma \cong Sym_6$;*

(e) *if we denote $B = C(A)'$, then $B \cong Alt_5$, $B > Y$ and $C(B) = A$;*

(f) *if $X = A \times B$, then $N(X) = N(X)_\sigma$ and $N(X)/X$ is a four-group.*

Certainly the statement (a) of the theorem follows from the result of Mizuno [8] (see also A. V. Alekseevski [9]). The element u belongs to the only unipotent conjugate class of E_8 with a nonsolvable factorgroup $C(u)/C(u)^0$ (see [8], [9]).

The subgroup $B \cong Alt_6$ in the group $G = E_8(\overline{F})$ has a strange property: there is a subgroup $B_1 \cong Alt_6$ in G, which is not conjugate to B, but whose elements belong to the same conjugacy classes of G . So restrictions of characters of the group G to subgroups B and B_1 don't allow us to distinguish them. This observation permits us to pose the following (maybe well-known) problem:

Problem 1 *Describe the ring of invariants of the action of the simple algebraic group G by simultaneous conjugation on $G^n = G \times \ldots \times G$.*

Theorem 3 *[7]. Under the conditions of the Theorem 2 the group $N = N(X)$ is maximal in $G_\sigma \cong E_8(q)$. The group G_σ has only one conjugacy class of maximal subgroups isomorphic to N.*

The proof of Theorem 3 depends on the Classification of the Finite Simple Groups (it is the only result in the paper depending on this Classification).

2   Centralizers in simple algebraic groups.

The main tool of the proof of Theorem 1 is the following result which is a generalization of a well-known result by R. Steinberg [10].

Theorem 4 *Let* $charF_p > 5$ *and* H *is a closed subgroup of* G. *If* $C = C(H)$ *and* $\overline{C} = C/C^0$ *then either* $\overline{C}$ *is solvable or* G *is of type* E_8 *and*

$$C(H) = C(H)^0 * Z$$

(semidirect product) where $Z \cong Alt_5, Sym_5$ *or* Sym_6 *is conjugate in* G *to a subgroup of* $N(X)$.

Problem 2 *Prove an analog of Theorem 4 for the case of characteristic 2, 3 or 5.*

3   Jordan subgroups and orthogonal decompositions.

We use in the following table standard notations [11].

G	J	$N(J)$
$A_{p^n-1}, n \geq 1$	p^{2n}	$p^{2n}.Sp(2n,p)$
$B_n, n \geq 2$	2^{2n}	$2^{2n}.Sym_{2n+1}$
$C_{2^{n-1}}, n \geq 3$	2^{2n}	$2^{2n}.O^-(2n,2)$
$D_{2^{n-1}}, n \geq 3$	2^{2n}	$2^{2n}.O^+(2n,2)$
$D_{n+1}, n \geq 3$	2^{2n}	$2^{2n}.Sym_{2n+2}$
E_6	3^3	$3^3.3^3.SL(3,3)$
E_8	5^3	$5^3.SL(3,5)$
E_8	2^5	$2^5.2^{10}.SL(5,2)$
F_4	3^3	$3^3.SL(3,3)$
G_2	2^3	$2^3.SL(3,2)$

Table 1: Jordan subgroups of simple
algebraic groups [4], [5]

In each case G contains only one Aut$G$-class of Jordan subgroups [4], [5].

The notion of Jordan subgroup is closely related to the notion of *orthogonal decomposition of a simple Lie algebra*, that is of the decomposition $L = H_0 + H_1 + \ldots + H_r$ of a simple Lie algebra L over the complex numbers into the direct sum of Cartan subalgebras $H_i, 0 \leq i \leq r$, pairwise orthogonal with respect to the Killing form. Orthogonal decompositions were introduced by A. I. Kostrikin, I. A. Kostrikin and V. A. Ufnarovski [12] (see also W. H. Hesselink [13]). The initial point of their considerations was Thompson's construction of his sporadic simple group Th via weight decomposition of Lie algebra E_8 related to the Jordan subgroup $J \cong 2^5$ in the corresponding

Lie group $G = E_8(C)$ [14]; this is a reason for the sporadic simple group Th to inherit Dempwolff subgroup $2^5.SL(5,2)$ (non-split extension) from $N_G(J) \cong 2^5.2^{10}.SL(5,2)$. It's interesting that the analogous construction beginning with Lie algebra A_4 gives us Leech lattice and Conway's group Co_1 [15]. The most important for various applications are two types of orthogonal decomposition: *irreducible* (their automorphism groups act irreducibly on L), and *transitive* (their automorphism groups are transitive on the set $\{H_i\}$).

The following result gives a uniform construction of all known irreducible and transitive orthogonal decompositions. There is strong evidence that the list is complete [16].

Let L be a simple complex Lie algebra, $G = \mathrm{Aut}(L)^0$ and J be a Jordan subgroup in G. Let us call a subgroup $A < J$ *toric* if $C_G(A)^0$ is a maximal torus in G and A is maximal among subgroups in J with this property.

Theorem 5 *[17]. In this notation J contains toric subgroups and all toric subgroups in J are conjugate in $N_G(J)$. If $\{A_0, \ldots, A_r\}$ is a set of toric subgroups having maximal size with respect to the property $A_i A_j = J, i \neq j$ then the subalgebras $H_i = C_L(A_i)$ form an orthogonal decomposition*

$$L = H_0 + \ldots + H_r$$

of the algebra L.

Moreover, one can choose the set $\{A_i\}$ in such a way that some subgroup $M \leq N_G(J)$ acts transitively on sets $\{A_i\}$ and $\{H_i\}$, and irreducibly on the algebra L.

References

[1] Aschbacher M. – *On the maximal subgroups of the finite classical groups* – Invent. Math., 1984, v.76, No.3, p.469-514.

[2] Liebeck M. W., Seitz G. M. – *Maximal subgroups of exceptional groups of Lie type, finite and algebraic* – Geom. Ded. 35 (1984), p.353-387.

[3] Borovik A. V. – *On the structure of finite subgroups of the simple algebraic groups* – Algebra and Logic, 1989, v.28, No.3 , p.249-279 (in Russian).

[4] Alekseevski A. V. – *On Jordan finite commutative subgroups of the simple classical Lie groups* – Funct. analysis and appl., 1974, v.8, No.4, p.1-4 (in Russian).

[5] Borovik A. V. – *The Jordan subgroups of the simple algebraic groups* – Algebra and logic, 1989, v.28, No.2, p.144-159 (in Russian).

[6] Liebeck M. W. – *The local maximal subgroups of the finite simple groups* – Proc. Symp. Pure Math., 1987, v.47, p.455-461.

[7] Borovik A. V. – *The maximal subgroup in the simple finite group $E_8(q)$* – Preprint.

[8] Mizuno K. – *The conjugate classes of elements of the Chevalley groups E_7 and E_8* – Tokyo J. Math., 1980, v.3, No.2, p393-460.

[9] Alekseevski A. V. – *The groups of the components of the centralizers of unipotent elements of the simple algebraic groups* – Trudy Tbilisskogo matem. Instituta, 1979, v.62, p.5-97 (in Russian).

[10] Steinberg R. – *Torsion in reductive groups* – Adv. Math., 1979, v.15, No.1, p63-92.

[11] Conway I. H., Curtis R. T., Norton S. P. e.a. – *Atlas of finite groups* – Oxford: Clarendon Press, 1985.

[12] Kostrikin I. A., Kostrikin, Ufnarovski V. A. – *Orthogonal decompositions of simple Lie algebras* – Soviet Math. Dokl., 1981, v.24, No.2, p.292-296

[13] Hesselink W. H. – *Special and Pure gradings of Lie algebras.* – Math. Z., 1982, B.179, No.1, S.135-149.

[14] Thompson J. – *A conjugacy theorem for E_8* – J. Algebra, 1976, v.38, No.2, p.525-530.

[15] Bondal A. I., Kostrikin A. I., Pham Huu Tiep – *Invariant lattices, Leech lattice and its even unimodular analogs in Lie algebras A_{p^n-1}.* – Matem. Sb., 1987, v.58, No.2, p.435-465 (in Russian).

[16] Pham Huu Tiep – *On the irreducible orthogonal decompositions of simple Lie algebras.* – Int. conf. dedic. to A. I. Mal'tsev (1909-1967)– Novosibirsk, 1989, p.204.

[17] Borovik A. V. – *Jordan subgroups and orthogonal decompositions* – Algebra and Logic, 1989, v.28, No.4, p.372-382.

SOME (ALMOST) MULTIPLICITY-FREE COSET ACTIONS

R. Lawther

In this paper we shall be concerned with the decomposition of the permutation characters of coset actions of a certain type. Specifically, we let G be a connected reductive algebraic group over an algebraically closed field K of characteristic $p > 0$, and σ be a Frobenius map of G: for convenience we set $F = \sigma^2$, and write G^σ and G^F for the groups of fixed points. The action with which this paper deals is then that of G^F on cosets of G^σ: this covers actions such as $G(q^2)$ on $G(q)$ and (where appropriate) $G(q^2)$ on $^2G(q^2)$. We find that actions of this sort are either almost or actually multiplicity-free (an action is said to be multiplicity-free if no irreducible constituent of the permutation character occurs more than once): moreover some which are not multiplicity-free can be made so by the application of outer automorphisms. We shall classify precisely which almost simple actions of this type are multiplicity-free in Theorems 4, 10 and 11: these results are used in [3]. The material that follows is taken from the author's Ph.D. thesis [10]: we also mention a recent preprint of Kawanaka [8] which covers very similar ground.

Preliminaries

If H is a finite group, we shall denote by $\hat{H}$ the set of (complex) irreducible characters of H. We shall mainly use the standard notation for finite groups of Lie type as given in [1]. However, in a few instances we employ a slight modification, necessitated by our use of more than one Frobenius map. We let T_0 be a fixed maximally split F-stable maximal torus of G: then we say that the F-stable maximal torus T of G is F-twisted by $w \in W$ if there exists $g \in G$ with $T = {}^gT_0$ such that the element $g^{-1}F(g) \in N_G(T_0)$ corresponds to w under the natural projection $\pi : N_G(T_0) \to N_G(T_0)/T_0 = W$. For such a T, the common value taken by all Deligne-Lusztig generalized characters $R_{T,\theta}$ for $\theta \in \hat{T}^F$ at the unipotent element $u \in G^F$ is written $Q_T^{G,F}(u)$: thus $Q_T^{G,F}$ is a Green function on the set $\mathcal{U}(G^F)$ of unipotent elements of G^F.

Since σ is a Frobenius map of G, there is an injection $i : G \to GL_n(K)$ and an integer m such that $i(\sigma^m(g)) = F_{q'}(i(g))$ for all $g \in G$ for some $q' = p^e$ (where $F_{q'} : GL_n(K) \to GL_n(K)$ is defined by $F_{q'}((a_{ij})) = (a_{ij}{}^{q'})$). We let $q = (q')^{\frac{1}{m}} \in \mathbf{R}^+$ and call q the modulus of σ.

The permutation character $1_{G^\sigma}{}^{G^F}$

The first stage in the decomposition of the permutation character $1_{G^\sigma}{}^{G^F}$ is to calculate its scalar product with Deligne-Lusztig generalized characters $R_{T,\theta}$. We have proved the following:

Theorem 1. *Let G be a connected reductive algebraic group with Frobenius map σ of modulus q, and let $F = \sigma^2$: let T be an F-stable maximal torus of G which is F-twisted by w_1, and θ be an element of $\hat{T}^F$. Take $g_0 \in G$ such that $T = {}^{g_0}T_0$ and $\pi(g_0{}^{-1}F(g_0)) = w_1$, and define θ_0 by $\theta = {}^{g_0}\theta_0$. There is a constant q_1, depending only on G, such that if $q \geq q_1$, then*

$$(1_{G^\sigma}{}^{G^F}, R_{T,\theta}) = |\{w \in W : w\sigma(w) = w_1 \text{ and } \theta_0(t) = 1 \text{ for all } t \in T_w{}^\sigma\}|$$

and

$$\frac{1}{|G^\sigma|} \sum_{u \in \mathcal{U}(G^\sigma)} Q_T^{G,F}(u) = \sum_{\substack{w \in W \\ w\sigma(w) = w_1}} \frac{1}{|T_w{}^\sigma|},$$

where $T_w{}^\sigma = \{t \in T_0 : {}^w\sigma(t) = t\}$.

We note the similarity of the second formula to a result of Deligne and Lusztig in [5, 7.11.4], which states that if T is σ-stable then

$$\frac{1}{|G^\sigma|} \sum_{u \in \mathcal{U}(G^\sigma)} Q_T^{G,\sigma}(u) = \frac{1}{|T^\sigma|}.$$

This result holds without restriction on the size of q, and so expresses the Green function sum as a polynomial in q. If we assume that the Green function sum in the Theorem is also a polynomial in q, then the restriction on q can be dropped: we shall make this assumption in all that follows (it will be shown later that this is justified). This will enable the calculation of the scalar product $(1_{G^\sigma}{}^{G^F}, \chi)$ for any χ which is a uniform function, i.e. a linear combination of Deligne-Lusztig generalized characters: in fact we shall find that the same will be true of many non-uniform irreducible χ.

We then turn to the consideration of unipotent characters for simple G: here we assume that F acts trivially on W, as the only case which this fails to cover is that of the action of ${}^3D_4(q^6)$ on the cosets of ${}^3D_4(q^3)$, which is easily dealt with. The unipotent characters are formed in two stages from the Deligne-Lusztig generalized characters $R_{T,1}$: we begin by defining the almost characters R_ϕ for $\phi \in \hat{W}$ by

$$R_\phi = \frac{1}{|W|} \sum_{w \in W} \phi(w) R_{T_w,1},$$

where T_w is an F-stable maximal torus F-twisted by w. From Theorem 1 it is simple to obtain the following:

Lemma 2. *For all $\phi \in \hat{W}$,*

$$(1_{G^\sigma}{}^{G^F}, R_\phi) = \frac{1}{|W|} \sum_{w \in W} \phi(w\sigma(w)) = \begin{cases} 1 & \text{if } \sigma \text{ fixes } \phi; \\ 0 & \text{otherwise.} \end{cases}$$

Thus if σ acts on W either trivially or as conjugation by w_0, then σ fixes all $\phi \in \hat{W}$ and so the scalar products $(1_{G^\sigma}{}^{G^F}, R_\phi)$ are all 1. In addition, we produce the following result, which is of use later:

Corollary 3. *If G is simple and F acts trivially on W,*

$$\sum_{\substack{\phi \in \hat{W} \\ \sigma(\phi)=\phi}} \phi(1) = |\{w \in W : w\sigma(w) = 1\}|.$$

For the second step we require the concept of families: the unipotent characters χ and χ' are said to be in the same family if there are chains $\chi = \chi_1, \chi_2, \ldots, \chi_n = \chi'$ of unipotent characters and $\phi_1, \ldots, \phi_{n-1}$ of irreducible characters of W such that

$$(\chi_i, R_{\phi_i}) \neq 0 \neq (\chi_{i+1}, R_{\phi_i}) \quad \text{for } i = 1, \ldots, n-1.$$

The irreducible characters of W are divided into families in similar fashion.

Lusztig has shown in [12] that to each such family $\mathcal{F}$ of unipotent characters there may be associated a finite group Γ chosen from the list

$$1, \quad \underbrace{C_2 \times \cdots \times C_2}_{e \text{ factors}}, \quad S_3, \quad S_4, \quad S_5.$$

Given such a group Γ, the set $M(\Gamma)$ is defined to have as elements pairs (x, α), where x runs through a set of conjugacy class representatives and α is an irreducible character of $C_\Gamma(x)$. The size of the set $M(\Gamma)$ is thus

Γ	1	$C_2 \times \cdots \times C_2$	S_3	S_4	S_5
$|M(\Gamma)|$	1	2^{2e}	8	21	39

There is then a bijection between the set of unipotent characters in the family $\mathcal{F}$ and the set $M(\Gamma)$, where Γ is the group associated to $\mathcal{F}$. Moreover, given any two pairs (x, α) and (y, β) in $M(\Gamma)$, we may define the complex number $\{(x, \alpha), (y, \beta)\}$ by

$$\{(x, \alpha), (y, \beta)\} = \frac{1}{|C_\Gamma(x)|} \frac{1}{|C_\Gamma(y)|} \sum_{\substack{g \in \Gamma \\ x.gyg^{-1}=gyg^{-1}.x}} \alpha(gyg^{-1})\overline{\beta(g^{-1}xg)}.$$

This gives a $|M(\Gamma)| \times |M(\Gamma)|$ matrix, which Lusztig calls a non-abelian Fourier transform matrix. All entries are in fact real for the groups Γ concerned, and it is symmetric, orthogonal and involutory: moreover the bijection between $\mathcal{F}$ and $M(\Gamma)$ may be arranged such that, writing $\chi^{\mathcal{F}}_{(x, \alpha)}$ for the unipotent character in $\mathcal{F}$ corresponding to $(x, \alpha) \in M(\Gamma)$, and χ_ϕ for the principal series character of G corresponding to $\phi \in \hat{W}$, we have

$$(\chi^{\mathcal{F}}_{(x, \alpha)}, R_\phi) = \begin{cases} \Delta(\chi^{\mathcal{F}}_{(x, \alpha)})\{(x, \alpha), (y, \beta)\} & \text{if } \chi_\phi = \chi^{\mathcal{F}}_{(y, \beta)}; \\ 0 & \text{if } \chi_\phi \notin \mathcal{F}. \end{cases}$$

Here $\Delta(\chi^{\mathcal{F}}_{(x,\alpha)}) = \pm 1$: in fact there are only three families, one when G is of type E_7 and two when G is of type E_8, where it is not identically equal to 1. Thus each R_ϕ with $\chi_\phi \in \mathcal{F}$ is the class function created by taking the linear combination of unipotent characters in $\mathcal{F}$ given by one of the rows of the matrix (with signs in the three exceptional cases). However, not all of the class functions thus created are of the form R_ϕ for some $\phi \in \hat{W}$: those that are not take the value zero on all semisimple classes of G^F. As in [2] we shall call these class functions $Y_1, Y_2, \ldots, Y_m$. It then follows that each $\chi \in \mathcal{F}$ is the linear combination of the R_ϕ and Y_i given by the corresponding column of the matrix (again, with signs in the exceptional cases). The row or column of the matrix labelled by the pair $(1,1)$ clearly has all its entries positive: the corresponding class function has the form R_ϕ rather than Y_i, and the character ϕ concerned is said to be special.

Since by Lemma 2 we know the scalar product of the permutation character with each R_ϕ, it would be a simple matter to use the Fourier transform matrix to calculate the multiplicity in $1_{G^\sigma}{}^{G^F}$ of each unipotent irreducible character if it were not for the Y_i, whose scalar products with $1_{G^\sigma}{}^{G^F}$ are not known. In fact in the majority of cases this is not a problem, since we know that the multiplicities must be non-negative integers: using this it is possible to deduce the scalar products $(1_{G^\sigma}{}^{G^F}, Y_i)$ and hence obtain the multiplicities $(1_{G^\sigma}{}^{G^F}, \chi)$ for unipotent $\chi \in \hat{G}^F$. However, there are a few cases for which this type of reasoning leads to a small number of possibilities, to decide between which requires further information.

We now survey the possibilities for G^F and G^σ. In the case where G is of type A_ℓ, each family is of size 1 and the Fourier transform matrix is the identity matrix: thus the almost characters are in fact irreducible, and as the action of σ on W here is either trivial (for the subfield case) or as conjugation by w_0 (for the twisted case), each unipotent character appears in $1_{G^\sigma}{}^{G^F}$ with multiplicity 1. This agrees with the result of [7] that these actions are multiplicity-free.

In order to give the results for the other classical groups, we must know a little of how the unipotent characters are grouped into families. We begin with the case where G is of type B_ℓ or C_ℓ. The unipotent characters of G^F are here parametrized by symbols Λ of the form

$$\begin{pmatrix} \lambda_1 & \lambda_2 & \ldots & \lambda_a \\ \mu_1 & \ldots & \mu_b \end{pmatrix}$$

satisfying $0 \leq \lambda_1 < \lambda_2 < \ldots < \lambda_a$, $0 \leq \mu_1 < \ldots < \mu_b$, with λ_1 and μ_1 not both zero, and $d = a - b$ odd and positive: these give pairs of partitions $\alpha = (\lambda_1, \lambda_2 - 1, \ldots, \lambda_a - a + 1)$ and $\beta = (\mu_1, \mu_2 - 1, \ldots, \mu_b - b + 1)$. The rank of such a symbol is defined as

$$\sum \lambda_i + \sum \mu_j - \left\lfloor \left(\frac{a+b-1}{2}\right)^2 \right\rfloor$$

(where $\lfloor z \rfloor$ is the greatest integer $\leq z$), and symbols of rank ℓ, giving pairs (α, β) with $|\alpha| + |\beta| = \ell$, correspond to unipotent characters of $B_\ell(q^2)$ or $C_\ell(q^2)$. The number d is called the defect of the symbol: symbols of defect 1 correspond to characters in the principal series. Two characters lie in the same family if and only if their symbols contain the same entries with the same multiplicities. Each family contains a unique special symbol Λ_0, satisfying $d = 1$ and

$$\lambda_1 \leq \mu_1 \leq \lambda_2 \leq \mu_2 \leq \ldots \leq \mu_b \leq \lambda_{b+1}.$$

Let the number of entries appearing only once in the symbol be $2k+1$: then the size of the family can be seen to be 2^{2k}. Assuming the action not to be that of $B_2(q^2)$ on $^2B_2(q^2)$ (which will be mentioned below), we find that if Λ is any symbol in the family and χ_Λ is the corresponding unipotent character then

$$(1_{G^\sigma}{}^{G^F}, \chi_\Lambda) = \begin{cases} 2^k & \text{if } \Lambda = \Lambda_0; \\ 0 & \text{otherwise.} \end{cases}$$

The case of D_ℓ is similar, but this time the symbols

$$\begin{pmatrix} \lambda_1 & \lambda_2 & \ldots & \lambda_a \\ \mu_1 & \mu_2 & \ldots & \mu_b \end{pmatrix} \quad \text{and} \quad \begin{pmatrix} \mu_1 & \mu_2 & \ldots & \mu_b \\ \lambda_1 & \lambda_2 & \ldots & \lambda_a \end{pmatrix}$$

are regarded as the same, and the defect $d = |a - b|$ must be divisible by 4: symbols of defect 0 correspond to characters in the principal series. Let the number of entries appearing only once in the symbol be $2k$. If $k \geq 1$ the size of the family is $2^{2(k-1)}$: we find that

$$(1_{G^\sigma}{}^{G^F}, \chi_\Lambda) = \begin{cases} 2^{k-1} & \text{if } \Lambda = \Lambda_0; \\ 0 & \text{otherwise.} \end{cases}$$

However, if $k = 0$ the symbol gives rise to two unipotent characters, each of which comprises a family of size one: these appear with multiplicity 1 in the permutation character of the subfield action, but do not appear if the action is twisted as they are then interchanged by σ.

This leaves the exceptional groups, together with the twisted B_2 action and the $^3D_4$ action (which is dealt with in [10]). It is here that we find the families for which the type of reasoning described above fails to be conclusive, but yields a small number of possibilities. Three of these families are those for which the signs $\Delta(\chi^{\mathcal{F}}_{(x,\alpha)})$ are not identically one: they contain the exceptional characters of $E_7(q^2)$ and $E_8(q^2)$ corresponding to characters of the Weyl group of degree 512 and 4096 respectively.

These characters are known to have properties slightly different from those of other unipotent characters in the principal series, as is shown by the necessity for the signs $\Delta(\chi^{\mathcal{F}}_{(x,\alpha)})$: for example, the corresponding characters of the Weyl group are the only ones whose fake degrees fail to have palindromic coefficients, and they are the only ones to correspond to irreducible representations of the Hecke algebra of W which are not rational. In view of the close similarity between the two principal series characters in each instance, we conjecture that their multiplicities in the permutation character will be equal.

The remaining cases where there is ambiguity of this kind are the twisted B_2 action, the subfield and twisted G_2 actions, and the subfield and twisted F_4 actions. The first has been treated in [11], and the second and third in [9], where the complete decomposition of the permutation characters is given: the twisted B_2 action is in fact found to be multiplicity-free. For the F_4 actions it is possible to use the character table of $F_4(2)$ given in [4] to calculate some of the class functions Y_i: if we then conjecture that the values taken by these class functions are polynomials in q, this gives sufficient information to decide between the possibilities.

For convenience we list here the unipotent constituents in each of the exceptional actions. We use the notation of [1], except for the action of $B_2(q^2)$ on $^2B_2(q^2)$ for which we employ that of [6]. Where we have been unable to decide between a number of possibilities, they are bracketed together: that which is conjectured is indicated with an obelus.

$B_2(q^2)$ on $^2B_2(q^2)$: $\theta_0 + \theta_2 + \theta_3 + \theta_4$

$G_2(q^2)$ on $G_2(q)$: $\phi_{1,0} + \phi_{1,3}{}' + \phi_{1,3}{}'' + \phi_{1,6} + 2\phi_{2,1}$

$G_2(q^2)$ on $^2G_2(q^2)$: $\phi_{1,0} + \phi_{1,6} + 2\phi_{2,2}$

$^3D_4(q^6)$ on $^3D_4(q^3)$: $\phi_{1,0} + \phi_{1,3}{}' + \phi_{1,3}{}'' + \phi_{1,6} + 2\phi_{2,1}$

$F_4(q^2)$ on $F_4(q)$: $\phi_{1,0} + \phi_{9,2} + \phi_{8,3}{}' + \phi_{8,3}{}'' + \phi_{8,9}{}' + \phi_{8,9}{}'' + \phi_{9,10} + \phi_{1,24}$
$$+ 2\phi_{4,1} + 2\phi_{4,13}$$
$$+ \left\{ \begin{array}{l} 4\phi_{12,4} + 2\phi_{6,6}{}'' + \phi_{4,7}{}' + \phi_{4,7}{}'' + \phi_{4,8} + F_4{}^{II}[1] \\ 5\phi_{12,4} + 2\phi_{6,6}{}'' + B_2, \epsilon' + B_2, \epsilon'' + F_4{}^{I}[1] \\ 3\phi_{12,4} + 2\phi_{6,6}{}'' + \phi_{9,6}{}' + \phi_{9,6}{}'' + \phi_{6,6}{}' + F_4{}^{I}[1]^\dagger \\ 4\phi_{12,4} + 3\phi_{6,6}{}'' + \phi_{6,6}{}' + B_2, r \end{array} \right\}$$

$F_4(q^2)$ on ${}^2F_4(q^2)$: $\phi_{1,0} + \phi_{9,2} + \phi_{9,10} + \phi_{1,24}$

$$+ \left\{ \begin{array}{c} \phi_{4,1} + B_2, 1 \\ \phi_{2,4}{}' + \phi_{2,4}{}''{}^\dagger \end{array} \right\} + \left\{ \begin{array}{c} \phi_{4,13} + B_2, \epsilon \\ \phi_{2,16}{}' + \phi_{2,16}{}''{}^\dagger \end{array} \right\}$$

$$+ \left\{ \begin{array}{l} 2\phi_{12,4} + 3\phi_{4,8} + \phi_{4,7}{}' + \phi_{4,7}{}'' + F_4{}^{II}[1] \\ 3\phi_{12,4} + 2\phi_{4,8} + B_2, \epsilon' + B_2, \epsilon'' + F_4{}^{I}[1] \\ \phi_{12,4} + 2\phi_{4,8} + \phi_{9,6}{}' + \phi_{9,6}{}'' + \phi_{6,6}{}' + F_4{}^{I}[1]^\dagger \\ 2\phi_{12,4} + 2\phi_{4,8} + \phi_{6,6}{}' + \phi_{6,6}{}'' + B_2, r \end{array} \right\}$$

$E_6(q^2)$ on $E_6(q)$: $\phi_{1,0} + \phi_{6,1} + \phi_{20,2} + \phi_{64,4} + \phi_{60,5} + \phi_{81,6} + \phi_{24,6}$

$$+ \phi_{81,10} + \phi_{60,11} + \phi_{24,12} + \phi_{64,13} + \phi_{20,20} + \phi_{6,25} + \phi_{1,36}$$

$$+ 2\phi_{30,3} + 2\phi_{30,15} + 2\phi_{80,7} + \phi_{90,8} + \phi_{10,9}$$

$E_6(q^2)$ on ${}^2E_6(q^2)$: $\phi_{1,0} + \phi_{6,1} + \phi_{20,2} + \phi_{64,4} + \phi_{60,5} + \phi_{81,6} + \phi_{24,6}$

$$+ \phi_{81,10} + \phi_{60,11} + \phi_{24,12} + \phi_{64,13} + \phi_{20,20} + \phi_{6,25} + \phi_{1,36}$$

$$+ 2\phi_{30,3} + 2\phi_{30,15} + 2\phi_{80,7} + \phi_{90,8} + \phi_{10,9}$$

$E_7(q^2)$ on $E_7(q)$: $\phi_{1,0} + \phi_{7,1} + \phi_{27,2} + \phi_{21,3} + \phi_{189,5} + \phi_{210,6} + \phi_{105,6}$

$$+ \phi_{168,6} + \phi_{189,7} + \phi_{378,9} + \phi_{210,10} + \phi_{105,12} + \phi_{210,13}$$

$$+ \phi_{378,14} + \phi_{105,15} + \phi_{189,20} + \phi_{210,21} + \phi_{105,21}$$

$$+ \phi_{168,21} + \phi_{189,22} + \phi_{21,36} + \phi_{27,37} + \phi_{7,46} + \phi_{1,63}$$

$$+ 2\phi_{56,3} + 2\phi_{120,4} + 2\phi_{405,8} + 2\phi_{420,10} + 2\phi_{420,13}$$

$$+ \left\{ \begin{array}{l} 2\phi_{512,11} \\ \phi_{512,11} + \phi_{512,12}{}^\dagger \\ 2\phi_{512,12} \end{array} \right\} + 2\phi_{405,15} + 2\phi_{120,25} + 2\phi_{56,30}$$

$$+ 2\phi_{315,7} + \phi_{280,9} + \phi_{70,9} + 2\phi_{315,16} + \phi_{280,18} + \phi_{70,18}$$

$E_8(q^2)$ on $E_8(q)$: $\phi_{1,0} + \phi_{8,1} + \phi_{35,2} + \phi_{560,5} + \phi_{567,6} + \phi_{3240,9} + \phi_{525,12}$

$$+ \phi_{4536,13} + \phi_{2835,14} + \phi_{6075,14} + \phi_{4200,15} + \phi_{2100,20}$$

$$+ \phi_{4200,21} + \phi_{2835,22} + \phi_{6075,22} + \phi_{4536,23} + \phi_{3240,31}$$

$$+ \phi_{525,36} + \phi_{567,46} + \phi_{560,47} + \phi_{35,74} + \phi_{8,91} + \phi_{1,120}$$

$$+ 2\phi_{112,3} + 2\phi_{210,4} + 2\phi_{700,6} + 2\phi_{2268,10} + 2\phi_{2240,10} + 2\phi_{4200,12}$$

$$+ 2\phi_{2800,13} + 2\phi_{5600,15} + 2\phi_{5600,21} + 2\phi_{4200,24} + 2\phi_{2800,25}$$

$$+ 2\phi_{2240,28} + 2\phi_{2268,30} + 2\phi_{700,42} + 2\phi_{210,52} + 2\phi_{112,63}$$

$$+ \begin{Bmatrix} 2\phi_{4096,11} \\ \phi_{4096,11} + \phi_{4096,12}{}^\dagger \\ 2\phi_{4096,12} \end{Bmatrix} + \begin{Bmatrix} 2\phi_{4096,26} \\ \phi_{4096,26} + \phi_{4096,27}{}^\dagger \\ 2\phi_{4096,27} \end{Bmatrix}$$

$$+ 2\phi_{1400,7} + \phi_{1008,9} + \phi_{448,9} + 2\phi_{1400,8} + \phi_{1575,10} + \phi_{175,12}$$

$$+ 2\phi_{1400,32} + \phi_{1575,34} + \phi_{175,36} + 2\phi_{1400,37} + \phi_{1008,39} + \phi_{448,39}$$

$$+ 3\phi_{4480,16} + 2\phi_{3150,18} + \phi_{4200,18} + 2\phi_{4536,18}$$

$$+ 2\phi_{5670,18} + \phi_{420,20} + \phi_{1400,20} + E_8{}^I[1]$$

By putting together the results of [7] and [11] with those above we have thus shown:

Theorem 4. *The only multiplicity-free actions of G^F on G^σ, for G a simple algebraic group of adjoint type, are those of $A_\ell(q^2)$ on $A_\ell(q)$ or $^2A_\ell(q^2)$ where q is any prime power, and $B_2(q^2)$ on $^2B_2(q^2)$ where q^2 is an odd power of 2.*

We now consider the multiplicities of general irreducible characters in $1_{G^\sigma}{}^{G^F}$: here we drop the requirement that G be simple. We begin by showing that the condition

$$\theta_0(t) = 1 \text{ for all } t \in T_w{}^\sigma$$

in Theorem 1 is equivalent to

$$\sigma(^{g_1}\theta_0) = (^{g_1}\theta_0)^{-1},$$

where $g_1 \in G$ satisfies $\pi(g_1{}^{-1}\sigma(g_1)) = w$. It follows that if $(1_{G^\sigma}{}^{G^F}, R_{T,\theta}) \neq 0$ then $\sigma(R_{T,\theta}) = R_{T,\theta^{-1}}$. We now introduce the opposition graph automorphism τ, defined by

$$\tau(x_\alpha(\lambda)) = x_{-w_0(\alpha)}(\lambda) \qquad \text{and} \qquad \tau(t) = (t^{w_0})^{-1} \text{ for all } t \in T_0.$$

We set $\sigma' = \sigma\tau$: since τ is an involution and commutes with σ, we have $\sigma'^2 = \sigma^2 = F$. It is straightforward to see that $\tau(R_{T,\theta}) = R_{T,\theta^{-1}}$ for all pairs (T, θ), and so if $(1_{G^\sigma}{}^{G^F}, R_{T,\theta}) \neq 0$ then σ' fixes $R_{T,\theta}$. The pairs (T, θ) and (T', θ') are said to lie in the same geometric conjugacy class if $T' = {}^gT$ and $\theta' = {}^g\theta$ for some $g \in G^F$: by extension, the geometric conjugacy class containing (T, θ) is said to contain $R_{T,\theta}$ and any irreducible constituent of $R_{T,\theta}$. By consideration of the regular representation of G^F (which is known to be a uniform function, i.e. a linear combination of Deligne-Lusztig generalized characters), we thus deduce the following:

Theorem 5. *If $(1_{G^\sigma}{}^{G^F}, \chi) \neq 0$ for some irreducible character χ, then the geometric conjugacy class κ containing χ is σ'-stable.*

Thus we need only consider geometric conjugacy classes κ which are fixed by σ'. We now impose the condition that the centre of G be connected, as this is required for Lusztig's Jordan decomposition of characters. According to this, the irreducible characters in κ correspond to unipotent characters of the centralizer in the dual group G^* of a semisimple element whose conjugacy class corresponds to κ: and if κ is σ'-stable then the corresponding class is $\sigma^{*\prime}$-stable (where $\sigma^{*\prime} = \sigma'^*$ is the Frobenius map of G^* dual to σ'). We therefore choose a class representative s^* which is fixed by $\sigma^{*\prime}$, and let $S^* = C_{G^*}(s^*)$: as $Z(G)$ is connected, so is S^*, and $\sigma^{*\prime}$ is a Frobenius map of S^*. Write $F^* = \sigma^{*\prime 2}$ so that F^* is dual to F; let (T, θ) be a pair lying in κ and (s^*, T^*) be a corresponding pair in G^*, so that T^* is a $\sigma^{*\prime}$-stable maximal torus of G^* in duality with T. We then show:

Theorem 6. *With the notation established,*

$$(1_{G^\sigma}{}^{G^F}, R_{T,\theta}) = (1_{S^{*\sigma^{*\prime}}}{}^{S^{*F^*}}, R_{T^*,1}).$$

In fact Theorem 1 shows that each side is equal to the size of a certain subset of the appropriate Weyl group: we establish a bijection between the two subsets. However, Lusztig's Jordan decomposition of characters shows that there is a bijection $\chi \to \chi_u$ between irreducible characters of G^F lying in κ and unipotent characters χ_u of S^{*F^*}, such that

$$(\chi, \epsilon_G \epsilon_T R_{T,\theta}) = (\chi_u, \epsilon_{S^*} \epsilon_{T^*} R_{T^*,1})$$

for all pairs $(T, \theta) \in \kappa$ (where $\epsilon_G = (-1)^{\text{rel.rank} G}$): it is easy to show that if κ is $\sigma^{*\prime}$-stable then the signs can be dropped. Thus we have shown:

Theorem 7. *Let G be a connected reductive algebraic group with connected centre. If $\chi \in \hat{G}^F$ is a uniform function and lies in a σ'-stable geometric conjugacy class κ, let s^* be a $\sigma^{*\prime}$-stable element of the semisimple conjugacy class of G^* corresponding to κ, and $S^* = C_{G^*}(s^*)$: let χ_u be the unipotent character of S^{*F^*} corresponding to χ. Then*

$$(1_{G^\sigma}{}^{G^F}, \chi) = (1_{S^{*\sigma^{*\prime}}}{}^{S^{*F^*}}, \chi_u).$$

We have therefore achieved a reduction to the unipotent case for uniform χ. The group S^* need not be simple of adjoint type, but it is not difficult to reduce further to this case: in doing so, some instances of diagonal action may also occur, as can be seen by considering

$$G = G_1 \times G_1, \qquad \sigma : G \to G \text{ given by } \sigma(g, g') = (\sigma_1(g'), \sigma_1(g)),$$

for σ_1 a Frobenius map of the connected reductive algebraic group G_1. Since unipotent characters of simple groups of adjoint type which are uniform functions comprise families on their own, and the actions of σ and σ' on such characters are the same because τ fixes each $R_{T,1}$, we have shown:

Theorem 8. *If $Z(G)$ is connected and $\chi \in \hat{G}^F$ is a uniform function, then*

$$(1_{G^\sigma}{}^{G^F}, \chi) = \begin{cases} 1 & \text{if } \sigma'(\chi) = \chi; \\ 0 & \text{otherwise.} \end{cases}$$

In fact, even if χ is not uniform, its multiplicity in $1_{G^\sigma}{}^{G^F}$ may often be obtained by this type of reduction, as it was seen that the scalar products $(1_{G^\sigma}{}^{G^F}, R_\phi)$ were sufficient to determine the multiplicities in the permutation character of the majority of unipotent characters. We conjecture that this procedure will in fact give the multiplicities of all irreducible χ.

At a very late stage in the above (in fact during the Durham conference), the author learned that N. Kawanaka was working independently on the same problem, and subsequently received a preprint of his paper [8]. Kawanaka defines the twisted Frobenius-Schur indicator of a character of G^F, and the twisting operator on the space of class functions of G^F: using them he proves statements corresponding to Theorems 5 and 8. He then produces an analogue of Theorem 1, under the condition (as here) that q be large: however he quotes a preprint of Lusztig [13] which deals with a more general situation. Lusztig's results reduce to those of [8] and Theorem 1 in the situation considered here, and require no condition on the size of q: they are proved only for the case of odd characteristic, but Kawanaka comments that this assumption is not needed in the cases he (and the present author) treats. (Thus the assumption made after Theorem 1 concerning Green function sums being polynomials in q is shown to be correct, and all the results given subsequently do in fact hold for all values of q.) After producing results corresponding to Lemma 2 and Corollary 3, Kawanaka turns to explicit calculation of multiplicities of unipotent constituents in $1_{G^\sigma}{}^{G^F}$. He is able to determine all such multiplicities with no such ambiguity as that seen above, and his results confirm the present author's conjectures concerning the subfield F_4 action and the E_7 and E_8 actions: however his methods require good characteristic for exceptional groups, and so he does not treat the twisted G_2 and F_4 actions (or the subfield actions in bad characteristic). Finally, he uses Lusztig's Jordan decomposition of characters to reduce the consideration of non-unipotent characters in a classical group with connected centre to the unipotent case, and comments that it is extremely likely that the reduction holds for any connected reductive algebraic group with connected centre.

Outer automorphisms

As is shown in [11] and [9], although the subfield B_2 action and the two G_2 actions just fail to be multiplicity-free, if the characteristic is such that there is a non-trivial graph automorphism then its application succeeds in removing the single multiplicity. It is this possibility of applying outer automorphisms to make actions multiplicity-free which we now consider. As we are working in a simple algebraic group of adjoint type, the only outer automorphisms to be considered are field and graph automorphisms. We shall show that the field automorphisms never have any effect on multiplicities of unipotent constituents of $1_{G^\sigma}{}^{G^F}$, and then treat the various instances of graph automorphisms one by one.

Let f be a field automorphism of G^F: then our method of dealing with the effect of f on $1_{G^\sigma}{}^{G^F}$ is as follows. We consider the principal series character $1_{B^F}{}^{G^F}$ as a permutation character: since G^F acts transitively on the cosets of G^σ, we have

$$(1_{G^\sigma}{}^{G^F}, 1_{B^F}{}^{G^F}) = \text{ the number of } B^F, G^\sigma\text{-double cosets in } G^F.$$

Now $(1_{B^F}{}^{G^F}, 1_{B^F}{}^{G^F}) = |W|$, and the elements of the Weyl group provide a set of B^F, B^F-double coset representatives: as f acts trivially on W, this shows that all multiplicities in $1_{B^F}{}^{G^F}$ must be preserved when f is applied. We shall show that each B^F, G^σ-double coset is also preserved by f: this will then show that

$$(1_{G^\sigma}{}^{G^F}, 1_{B^F}{}^{G^F}) = (1_{G^\sigma\langle f\rangle}{}^{G^F\langle f\rangle}, 1_{B^F\langle f\rangle}{}^{G^F\langle f\rangle}),$$

and hence that no multiplicities of unipotent constituents of $1_{G^\sigma}{}^{G^F}$ lying in the principal series are broken down. As the only unipotent constituents of $1_{G^\sigma}{}^{G^F}$ lying outside the principal series have multiplicity one, this will prove that f has no effect on multiplicities of unipotent constituents of $1_{G^\sigma}{}^{G^F}$.

Now given a double coset $B^F g G^\sigma$, consider the element $g\sigma(g)^{-1}$. This is clearly unaffected if we postmultiply g by an element of G^σ: and if we write $g\sigma(g)^{-1} = uh\dot{w}v$ using the Bruhat decomposition, we see that premultiplying g by an element of B^F has no effect on the Weyl group element $w = \pi(\dot{w})$. Thus each double coset $B^F g G^\sigma$ gives rise to a unique Weyl group element: and as f commutes with σ and acts trivially on W, $f(g)\sigma(f(g))^{-1} = f(g\sigma(g)^{-1})$ and we see that $B^F g G^\sigma$ and $B^F f(g) G^\sigma$ give rise to the same Weyl group element. We must show that distinct double cosets give rise to distinct Weyl group elements: this will imply that $B^F g G^\sigma = B^F f(g) G^\sigma$ for all $g \in G^F$, so that f preserves all double cosets.

We do this simply by counting. First note that $g\sigma(g)^{-1}$ is inverted by σ, and so

$$\sigma(u)\sigma(h)\sigma(\dot{w})\sigma(v) = v^{-1}\dot{w}^{-1}h^{-1}u^{-1} :$$

thus $\sigma(w) = w^{-1}$ and so $w\sigma(w) = 1$. Assuming that F acts trivially on W, the number of such Weyl group elements is

$$\sum_{\substack{\phi \in \dot{W} \\ \sigma(\phi)=\phi}} \phi(1)$$

by Corollary 3. We shall show two things: firstly, that any Weyl group element inverted by σ arises from some B^F, G^σ-double coset, and secondly that

$$(1_{G^\sigma}{}^{G^F}, 1_{BF}{}^{G^F}) = \sum_{\substack{\phi \in \hat{W} \\ \sigma(\phi) = \phi}} \phi(1).$$

This will then imply that distinct double cosets give rise to distinct elements of W inverted by σ as required.

For the first, let $w \in W$ be inverted by σ and take $x \in N_G(T_0)$ with $\pi(x) = w$. Then as $w\sigma(w) = 1$, we have $x\sigma(x) \in T_0$: let $x\sigma(x) = t'$. Define $\psi : T_0 \to T_0$ by $\psi(t) = \left(\sigma(t^{-1})\right)^w$ (note that this is an endomorphism as T_0 is abelian), and by Lang's theorem take $t \in T_0$ satisfying $t\psi(t)^{-1} = (t'^{-1})^w$. Then $tx^{-1}\sigma(t)x = x^{-1}t'^{-1}x$, and so

$$1 = xtx^{-1}\sigma(t)t' = xtx^{-1}t'\sigma(t) = xt\sigma(x)\sigma(t) = xt\sigma(xt),$$

and xt is inverted by σ. Thus by [10, Lemma 3.1.3] we have $xt = g\sigma(g)^{-1}$ for some $g \in G^F$ and $B^F g G^\sigma$ gives rise to w.

For the second, we need some notation. Let $\mathcal{F}_1, \ldots, \mathcal{F}_r$ be the families of unipotent characters of G^F, with Fourier transform matrices $M_1, \ldots, M_r$: let $|\mathcal{F}_j| = d_j$ so that M_j is a $d_j \times d_j$ matrix. Set $d = d_1 + \cdots + d_r$, and let M be the $d \times d$ matrix with blocks $M_1, \ldots, M_r$ down the diagonal and zeroes elsewhere: then M is real, symmetric, orthogonal and involutory. Let the unipotent characters in $\mathcal{F}_j$ be $\chi_{d_{j-1}+1}, \ldots, \chi_{d_j}$, so that $\chi_{d_{j-1}+k}$ corresponds to the kth column of M_j if $1 \leq k \leq d_j$: then χ_k corresponds to the kth column of M. Let R_k be the class function which is the linear combination of $\chi_1, \ldots, \chi_d$ given by the kth row of M. Let $I = \{1, \ldots, d\}$, and J be the subset of I consisting of those j with χ_j in the principal series: thus if $\phi \in \hat{W}$, then $\chi_\phi = \chi_j$ for some $j \in J$. Then if $j \in J$ and $\chi_\phi = \chi_j$ we have $R_j = R_\phi$, unless χ_ϕ belongs to one of the families (say $\mathcal{F}_k$) of $E_7(q^2)$ or $E_8(q^2)$ mentioned earlier. In these cases, however, the signs $\Delta(\chi_{(x,\sigma)}^{\mathcal{F}_k})$ have the effect of merely interchanging the first and second, and third and fourth rows of M_k, so that $R_j = R_{i(\phi)}$ where i is the involution on $\hat{W}$ defined in [1, p.373] as interchanging these ϕ and fixing all others.

Let $V = \mathbf{R}^d$, so that M acts on V on the right: let $\mathbf{m}, \mathbf{n}, \mathbf{v} \in V$ be given by

$$m_j = (1_{G^\sigma}{}^{G^F}, R_j), \quad n_j = (1_{G^\sigma}{}^{G^F}, \chi_j), \quad v_j = (1_{BF}{}^{G^F}, \chi_j) = \begin{cases} \phi(1) & \text{if } \chi_j = \chi_\phi; \\ 0 & \text{if } j \notin J. \end{cases}$$

Now as $M^2 = 1$, the only eigenvalues of M are ± 1: let V_1, V_{-1} be the corresponding eigenspaces of V, then $V = V_1 \perp V_{-1}$. Inverting the equations for R_ϕ in terms of R_w gives

$$R_w = \sum_{\phi \in \hat{W}} \phi(w) R_\phi;$$

since $R_w = 1_{BF}{}^{G^F}$ if $w = 1$, we have

$$v_j = (\chi_j, 1_{BF}{}^{G^F}) = (\chi_j, \sum_{\phi \in \dot{W}} \phi(1) R_\phi) = \sum_{\phi \in \dot{W}} \phi(1)(\chi_j, R_\phi) = (\mathbf{v}M)_j$$

(since $\phi(1) = i(\phi)(1)$). Thus we have shown:

Lemma 9. $\mathbf{v} \in V_1$.

Now as $\mathbf{m}M = \mathbf{n}$, we have $\mathbf{m} - \mathbf{n} \in V_{-1}$: thus $\mathbf{v} \perp (\mathbf{m} - \mathbf{n})$, i.e. $\mathbf{v m}^T = \mathbf{v n}^T$. But if $j \in J$, then $m_j = 1$ or 0 according as ϕ is or is not fixed by σ, where $\chi_j = \chi_\phi$: thus

$$\mathbf{v m}^T = \sum_{\substack{\phi \in \dot{W} \\ \sigma(\phi) = \phi}} \phi(1),$$

while

$$\mathbf{v n}^T = \sum_{\phi \in \dot{W}} \phi(1)(1_{G^\sigma}{}^{G^F}, \chi_\phi) = \sum_{\phi \in \dot{W}} (1_{G^\sigma}{}^{G^F}, \chi_\phi)(1_{BF}{}^{G^F}, \chi_\phi) = (1_{G^\sigma}{}^{G^F}, 1_{BF}{}^{G^F}),$$

so that, as required, we have

$$(1_{G^\sigma}{}^{G^F}, 1_{BF}{}^{G^F}) = \sum_{\substack{\phi \in \dot{W} \\ \sigma(\phi) = \phi}} \phi(1).$$

We assumed earlier that F acts trivially on W: again, the only case which this fails to cover is the $^3D_4$ action. However, as σ^3 acts trivially on W here, the only elements of W inverted by σ are σ-stable involutions (and the identity): as $W^\sigma \cong D_{12}$, there are eight such elements. Since $(1_{G^\sigma}{}^{G^F}, 1_{BF}{}^{G^F}) = 8$ here, the result still holds. Hence we have shown:

Theorem 10. *The application of field automorphisms has no effect on the multiplicities of unipotent constituents of $1_{G^\sigma}{}^{G^F}$ if G is simple of adjoint type.*

We now consider the action of graph automorphisms γ. The only instances of nontrivial γ occur in the two A_ℓ actions, the two D_ℓ actions, the two E_6 actions, the $^3D_4$ action, the two B_2 actions in characteristic 2, the two F_4 actions in characteristic 2, and the two G_2 actions in characteristic 3. The two A_ℓ actions, and that of $B_2(q^2)$ on $^2B_2(q^2)$ where q^2 is an odd power of 2, are already multiplicity-free: and calculation of pairing of suborbits in [10] shows that the application of γ removes the multiplicity from the permutation character in the cases of $B_2(q^2)$ on $B_2(q)$ in characteristic 2 and the two G_2 actions in characteristic 3. In the F_4 subfield action the multiplicity of $\phi_{12,4}$ is too large for the application of γ to make it multiplicity-free. This leaves the two D_ℓ actions, the two E_6 actions, the $^3D_4$ action and the twisted F_4 action to deal with.

Our approach is to choose a parabolic subgroup P of G which is fixed by both σ and γ, and has the property that $1_{PF}{}^{G^F}$ contains a constituent which occurs in $1_{G^\sigma}{}^{G^F}$ with multiplicity two. Again, we consider P^F, G^σ-double cosets: if $P^F g G^\sigma$ is one such, once more the element $g\sigma(g)^{-1} = u h \dot{w} v$ is unaffected if g is postmultiplied by an element of G^σ. This time, however, premultiplication of g by an element of P^F may alter the Weyl group element $w = \pi(\dot{w})$: but if we write $W_P < W$ for the Weyl group of P then the W_P, W_P-double coset containing w is uniquely defined. In each instance we shall give W_P, W_P-double coset representatives which are inverted by σ and consider the action of γ upon them. (As γ must commute with σ to be an automorphism of G^σ, as before $\gamma(g)\sigma(\gamma(g))^{-1} = \gamma(g\sigma(g)^{-1})$ and the action of γ on the labelling Weyl group elements will determine its action on the P^F, G^σ-double cosets.)

We begin with the two D_ℓ actions where $\ell \geq 5$, which we take together. Here we let P be of type $A_1 \times D_{\ell-2}$: thus taking the simple roots of G to be

$$e_1 - e_2, \quad e_2 - e_3, \quad \ldots, \quad e_{\ell-2} - e_{\ell-1}, \quad e_{\ell-1} - e_\ell, \quad e_{\ell-1} + e_\ell,$$

we see that W_P is obtained from W by omitting the second distinguished generator. As stated above, the irreducible characters of W are indexed by unordered pairs of partitions: using [14] we calculate

$$1_{W_P}{}^W = \phi_{(\ell,-)} + \phi_{(\ell-1\ 1,-)} + \phi_{(\ell-2\ 2,-)} + \phi_{(\ell-1,1)} + \phi_{(\ell-2\ 1,1)} + \phi_{(\ell-2,2)}.$$

Thus, using the same notation for the principal series characters of G^F, we have

$$1_{PF}{}^{G^F} = \phi_{(\ell,-)} + \phi_{(\ell-1\ 1,-)} + \phi_{(\ell-2\ 2,-)} + \phi_{(\ell-1,1)} + \phi_{(\ell-2\ 1,1)} + \phi_{(\ell-2,2)}.$$

The symbols for these characters are

$$\begin{pmatrix} \ell \\ 0 \end{pmatrix}, \quad \begin{pmatrix} 1 & \ell \\ 0 & 1 \end{pmatrix}, \quad \begin{pmatrix} 2 & \ell-1 \\ 0 & 1 \end{pmatrix}, \quad \begin{pmatrix} \ell-1 \\ 1 \end{pmatrix}, \quad \begin{pmatrix} 1 & \ell-1 \\ 0 & 2 \end{pmatrix}, \quad \begin{pmatrix} \ell-2 \\ 2 \end{pmatrix},$$

and hence, by the results of the previous section, their multiplicities in $1_{G^\sigma}{}^{G^F}$ are 1, 1, 0, 1, 2 and 1 respectively. Thus $(1_{G^\sigma}{}^{G^F}, 1_{PF}{}^{G^F}) = 6$ and so there are six P^F, G^σ-double cosets. Now the six Weyl group elements

$$1, \quad w_{2-3}, \quad w_0, \quad w_{2-3}w_{1-2}w_{3-4}w_{2-3},$$

$$w_{2-3}w_{3-4}\cdots w_{(\ell-2)-(\ell-1)}w_{(\ell-1)-\ell}w_{(\ell-1)+\ell}w_{(\ell-2)-(\ell-1)}\cdots w_{3-4}w_{2-3},$$

$$w_{2-3}w_{1-2}w_{3-4}\cdots w_{(\ell-2)-(\ell-1)}w_{(\ell-1)-\ell}w_{(\ell-1)+\ell}w_{(\ell-2)-(\ell-1)}\cdots w_{3-4}w_{2-3}$$

are all clearly inverted by σ (whether it acts on W trivially or as conjugation by the matrix $\mathrm{diag}(1,1,\ldots,1,-1)$, where we are representing W naturally as $\ell \times \ell$ matrices). To show that they lie in different W_P, W_P-double cosets, we note that pre- or postmultiplying a matrix by an element of W_P can only permute the four elements lying in the first two rows and columns, since the elements of $W(D_{\ell-2})$ fix this 2×2 submatrix, while w_{1-2} interchanges its rows or columns. The six elements listed have as this submatrix

$$\begin{pmatrix} 1 & 0 \\ 0 & 1 \end{pmatrix}, \begin{pmatrix} 1 & 0 \\ 0 & 0 \end{pmatrix}, \begin{pmatrix} -1 & 0 \\ 0 & -1 \end{pmatrix}, \begin{pmatrix} 0 & 0 \\ 0 & 0 \end{pmatrix}, \begin{pmatrix} 1 & 0 \\ 0 & -1 \end{pmatrix}, \begin{pmatrix} 0 & 0 \\ 0 & -1 \end{pmatrix}.$$

Thus they do indeed lie in different W_P, W_P-double cosets, and so label the six P^F, G^σ-double cosets. Since each is fixed by γ, there are still six $P^F\langle\gamma\rangle, G^\sigma\langle\gamma\rangle$-double cosets and so $(1_{G^\sigma\langle\gamma\rangle}{}^{G^F\langle\gamma\rangle}, 1_{P^F\langle\gamma\rangle}{}^{G^F\langle\gamma\rangle}) = 6$: so the multiplicity of $\phi_{(\ell-2\ 1,1)}$ remains when γ is applied.

We next consider G of type D_4: the behaviour is different here, and depends on σ. Let γ_1 be the graph automorphism which acts on W as conjugation by $\mathrm{diag}(1,1,1,-1)$, and γ_2 be the triality graph automorphism which sends w_{3-4} to w_{3+4}, w_{3+4} to w_{1-2} and w_{1-2} to w_{3-4}. This time we take P of type $A_1{}^3$: and here $1_{W_P}{}^W$ has a slightly different decomposition from that given above, because the character $\phi_{(2,2)}$ of $W(B_4)$ splits into two characters $\phi_{(2,2)}'$ and $\phi_{(2,2)}''$ of $W(D_4)$. Thus we have

$$1_{PF}{}^{GF} = \phi_{(4,-)} + \phi_{(3\ 1,-)} + \phi_{(2\ 2,-)} + \phi_{(3,1)} + \phi_{(2\ 1,1)} + \phi_{(2,2)}' + \phi_{(2,2)}''.$$

We treat first the subfield case: the multiplicities in $1_{G^\sigma}{}^{G^F}$ of these constituents are then 1, 1, 0, 1, 2, 1 and 1 respectively. Thus $(1_{G^\sigma}{}^{G^F}, 1_{PF}{}^{GF}) = 7$ and there are seven P^F, G^σ-double cosets. This time we take the seven Weyl group elements

$$1, \quad w_{2-3}, \quad w_0, \quad w_{2-3}w_{1-2}w_{3-4}w_{3+4}w_{2-3},$$

$$w_{2-3}w_{1-2}w_{3-4}w_{2-3}, \quad w_{2-3}w_{1-2}w_{3+4}w_{2-3}, \quad w_{2-3}w_{3-4}w_{3+4}w_{2-3},$$

which are all involutions and so inverted by σ. The same reasoning as above shows that all belong to different W_P, W_P-double cosets, with the possible exception of the fifth and sixth: these have matrices

$$\begin{pmatrix} 0 & 0 & 1 & 0 \\ 0 & 0 & 0 & 1 \\ 1 & 0 & 0 & 0 \\ 0 & 1 & 0 & 0 \end{pmatrix} \quad \text{and} \quad \begin{pmatrix} 0 & 0 & 1 & 0 \\ 0 & 0 & 0 & -1 \\ 1 & 0 & 0 & 0 \\ 0 & -1 & 0 & 0 \end{pmatrix}$$

respectively, and as there is no way of negating just one of the last two rows with elements of W_P, they do in fact belong to different W_P, W_P-double cosets. Thus these elements label the 7 P^F, G^σ-double cosets.

If we now apply γ_1, the two matrices just given are interchanged and the other five elements are fixed; it follows that there are only six $P^F\langle\gamma_1\rangle, G^\sigma\langle\gamma_1\rangle$-double cosets, and so $(1_{G^\sigma\langle\gamma_1\rangle}{}^{G^F\langle\gamma_1\rangle}, 1_{PF\langle\gamma_1\rangle}{}^{G^F\langle\gamma_1\rangle}) = 6$. However, γ_1 also fuses $\phi_{(2,2)}{}'$ and $\phi_{(2,2)}{}''$, so that

$$1_{PF\langle\gamma_1\rangle}{}^{G^F\langle\gamma_1\rangle} = \phi_{(4,-)} + \phi_{(3\ 1,-)} + \phi_{(2\ 2,-)} + \phi_{(3,1)} + \phi_{(2\ 1,1)} + \phi_{(2,2)}$$

(where the pairs of partitions are now ordered): thus the multiplicity of $\phi_{(2\ 1,1)}$ in the permutation character remains. If instead we apply γ_2, this fixes the first four Weyl group elements listed and permutes the other three cyclically: hence there are just five $P^F\langle\gamma_2\rangle, G^\sigma\langle\gamma_2\rangle$-double cosets, and $(1_{G^\sigma\langle\gamma_2\rangle}{}^{G^F\langle\gamma_2\rangle}, 1_{PF\langle\gamma_2\rangle}{}^{G^F\langle\gamma_2\rangle}) = 5$. This time γ_2 fuses $\phi_{(2,2)}{}'$, $\phi_{(2,2)}{}''$ and $\phi_{(3,1)}$, and again we see that the multiplicity of $\phi_{(2\ 1,1)}$ must remain. Since any graph automorphism of G is a conjugate of either γ_1 or γ_2, we have shown that the action of $D_4(q^2)$ on $D_4(q)$ does not become multiplicity-free under the application of any graph automorphism.

In the action of $D_4(q^2)$ on ${}^2D_4(q^2)$, the details are slightly different: to begin with, γ_2 may not be applied since it does not commute with σ. Here $\phi_{(2,2)}{}'$ and $\phi_{(2,2)}{}''$ do not appear in $1_{G^\sigma}{}^{G^F}$, although the other multiplicities are the same as in the subfield case, and so $(1_{G^\sigma}{}^{G^F}, 1_{PF}{}^{G^F}) = 5$. Thus there are only five P^F, G^σ-double cosets, labelled by the first, second, third, fourth and seventh Weyl group elements listed above, each of which is inverted by σ. Since γ_1 fixes each of these elements, and the constituents of $1_{PF}{}^{G^F}$ which do appear in $1_{G^\sigma}{}^{G^F}$, the multiplicity of $\phi_{(2\ 1,1)}$ once more remains.

Finally, let $G^F = {}^3D_4(q^6)$, $G^\sigma = {}^3D_4(q^3)$. This time γ_1 does not commute with σ, and so cannot be applied. Instead of taking P as above, we simply note that the action of γ_2 on W is the same as that of σ, and so it fixes the eight B^F, G^σ-double cosets mentioned above under the treatment of field automorphisms: hence the multiplicity of $\phi_{2,1}$ remains. This completes the treatment when G is of type D_4.

We next turn to the two E_6 actions, which we treat together: here we take P of type A_5. By considering $1_{W_P}{}^W$, we find

$$1_{PF}{}^{G^F} = \phi_{1,0} + \phi_{6,1} + \phi_{20,2} + \phi_{30,3} + \phi_{15,4}.$$

The previous section shows that the multiplicities in $1_{G^\sigma}{}^{G^F}$ are 1, 1, 1, 2 and 0 respectively: thus we have $(1_{G^\sigma}{}^{G^F}, 1_{PF}{}^{G^F}) = 5$ and there are five P^F, G^σ-double cosets. It proves to be convenient to use a non-standard notation for the simple roots of G: we choose them to be

$$e_2-e_3, \quad e_3-e_4, \quad e_4-e_5, \quad e_5-e_6, \quad e_6-e_7, \quad \tfrac{1}{2}(-e_1-e_2-e_3-e_4+e_5+e_6+e_7+e_8),$$

and write the Weyl group elements as 8×8 matrices in the obvious way. For brevity, write w_a for the reflection in the last simple root, and w_0' for the element of maximal length in $W_P = \langle w_{2-3}, w_{3-4}, w_{4-5}, w_{5-6}, w_{6-7} \rangle$ (while w_0 is the element of maximal length in W): then the five elements

$$1, \quad w_a, \quad w_a w_{4-5} w_{3-4} w_{5-6} w_{4-5} w_a, \quad w_a w_0' w_a, \quad w_0$$

are visibly inverted by σ, whether its action on W is trivial or as conjugation by w_0. The effect of W_P on matrices is merely to permute the central six rows and columns, so that the 2×2 submatrix comprising the four corner entries is unchanged: the five elements here have as this submatrix

$$\begin{pmatrix} 1 & 0 \\ 0 & 1 \end{pmatrix}, \quad \begin{pmatrix} \frac{3}{4} & \frac{1}{4} \\ \frac{1}{4} & \frac{3}{4} \end{pmatrix}, \quad \begin{pmatrix} \frac{1}{2} & \frac{1}{2} \\ \frac{1}{2} & \frac{1}{2} \end{pmatrix}, \quad \begin{pmatrix} \frac{1}{4} & \frac{3}{4} \\ \frac{3}{4} & \frac{1}{4} \end{pmatrix}, \quad \begin{pmatrix} 0 & 1 \\ 1 & 0 \end{pmatrix},$$

and so lie in different W_P, W_P-double cosets. Since each is fixed by γ, the multiplicity of $\phi_{30,3}$ remains.

Finally we turn to the twisted F_4 action: here there is no maximal parabolic subgroup fixed by γ, which complicates matters a little. In addition, we had various possibilities above for the unipotent constituents of $1_{G^\sigma}{}^{G^F}$: however, we need not consider those involving the first or second row of the third bracket, as those each have a multiplicity of 3, which is too large to be broken down by γ as in the subfield action. We take the simple roots of G to be

$$a = e_2 - e_3, \quad b = e_3 - e_4, \quad c = e_4, \quad d = \tfrac{1}{2}(e_1 - e_2 - e_3 - e_4),$$

and write the Weyl group elements as 4×4 matrices. For P we take the parabolic subgroup of type B_2, with $W_P = \langle w_b, w_c \rangle$. Then, by considering $1_{W_P}{}^W$, we find

$$1_{PF}{}^{G^F} = \phi_{1,0} + 3\phi_{9,2} + 2\phi_{8,3}{}' + 2\phi_{8,3}{}'' + 2\phi_{4,1} + \phi_{2,4}{}' + \phi_{2,4}{}''$$
$$+ \phi_{12,4} + \phi_{4,8} + \phi_{9,6}{}' + \phi_{9,6}{}'' + \phi_{6,6}{}'' + 2\phi_{16,5}.$$

We find that, whichever of the possibilities given above holds, $(1_{G^\sigma}{}^{G^F}, 1_{PF}{}^{G^F}) = 11$, and so we require eleven Weyl group elements. We take

$$1, \quad w_a w_d, \quad w_a w_b w_c w_d, \quad w_d w_c w_b w_a, \quad w_0, \quad w_a w_d w_b w_c w_b w_c w_a w_d,$$

$$w_a w_b w_c w_b w_a w_d w_c w_b w_c w_d, \quad w_d w_c w_b w_c w_d w_a w_b w_c w_b w_a,$$

$$w_a w_d w_b w_c w_b w_c w_a w_d w_b w_c w_b w_c w_a w_d,$$

$$w_a w_d w_b w_c w_b w_a w_d w_c w_b w_c w_a w_d, \quad w_a w_d w_c w_b w_c w_a w_d w_b w_c w_b w_a w_d,$$

each of which is clearly inverted by σ (as $\sigma(w_a) = w_d$, $\sigma(w_b) = w_c$). To show that these represent different W_P, W_P-double cosets, we note that this time W_P acts on matrices only by permuting and possibly negating the last two rows and columns, and so leaves alone the 2×2 submatrix of entries in the first two rows and columns. The eleven elements listed have as this submatrix

$$\begin{pmatrix} 1 & 0 \\ 0 & 1 \end{pmatrix}, \begin{pmatrix} \frac{1}{2} & \frac{1}{2} \\ \frac{1}{2} & -\frac{1}{2} \end{pmatrix}, \begin{pmatrix} \frac{1}{2} & \frac{1}{2} \\ -\frac{1}{2} & \frac{1}{2} \end{pmatrix}, \begin{pmatrix} \frac{1}{2} & -\frac{1}{2} \\ \frac{1}{2} & \frac{1}{2} \end{pmatrix}, \begin{pmatrix} -1 & 0 \\ 0 & -1 \end{pmatrix}, \begin{pmatrix} 0 & 0 \\ 0 & 0 \end{pmatrix},$$

$$\begin{pmatrix} 0 & 1 \\ -1 & 0 \end{pmatrix}, \begin{pmatrix} 0 & -1 \\ 1 & 0 \end{pmatrix}, \begin{pmatrix} -\frac{1}{2} & -\frac{1}{2} \\ -\frac{1}{2} & \frac{1}{2} \end{pmatrix}, \begin{pmatrix} -\frac{1}{2} & \frac{1}{2} \\ -\frac{1}{2} & -\frac{1}{2} \end{pmatrix}, \begin{pmatrix} -\frac{1}{2} & -\frac{1}{2} \\ \frac{1}{2} & -\frac{1}{2} \end{pmatrix},$$

and so lie in different W_P, W_P-double cosets: thus they do label the eleven P^F, G^σ-double cosets. Now γ acts on W as σ does, and so fixes the first, second, fifth, sixth and ninth elements, which are involutions, while it interchanges the third and fourth, seventh and eighth, and tenth and eleventh. Hence there are eight $P^F \langle \gamma \rangle, G^\sigma \langle \gamma \rangle$-double cosets, and $(1_{G^\sigma\langle\gamma\rangle}^{G^F\langle\gamma\rangle}, 1_{PF\langle\gamma\rangle}^{G^F\langle\gamma\rangle}) = 8$. However, calculation in $W\langle\gamma\rangle$ shows that the application of γ to $1_{W_P}^W$ not only fuses $\phi_{2,4}{}'$ and $\phi_{2,4}{}''$, $\phi_{9,6}{}'$ and $\phi_{9,6}{}''$, and $2\phi_{8,3}{}'$ and $2\phi_{8,3}{}''$, but also breaks up the $3\phi_{9,2}$ into $2\phi_{9,2}{}^0 + \phi_{9,2}{}^1$ (where $\phi_{9,2}{}^0$ takes the value 1 on the class of $W\langle\gamma\rangle$ containing γ), $2\phi_{4,1}$ into $\phi_{4,1}{}^0 + \phi_{4,1}{}^1$, and $2\phi_{16,5}$ into $\phi_{16,5}{}^0 + \phi_{16,5}{}^1$: thus

$$1_{PF\langle\gamma\rangle}^{G^F\langle\gamma\rangle} = \phi_{1,0}{}^0 + 2\phi_{9,2}{}^0 + \phi_{9,2}{}^1 + 2\phi_{8,3} + \phi_{4,1}{}^0 + \phi_{4,1}{}^1 + \phi_{2,4}$$
$$+ \phi_{12,4}{}^0 + \phi_{4,8}{}^0 + \phi_{9,6} + (\phi_{6,6}{}'')^0 + \phi_{16,5}{}^0 + \phi_{16,5}{}^1.$$

It follows that for the scalar product to be 8, γ must lift both constituents $\phi_{4,8}$ of $1_{G^\sigma}^{G^F}$ to $\phi_{4,8}{}^0$ (which agrees with the result for $q^2 = 2$ obtained by direct calculation from [4]). Thus once more γ does not make the action multiplicity-free.

Putting all these results together, we have proved:

Theorem 11. *The only actions of G^F on G^σ for G simple of adjoint type which are made multiplicity-free by the application of graph automorphisms are those of $B_2(q^2)$ on $B_2(q)$ where q is a power of 2, $G_2(q^2)$ on $G_2(q)$ where q is a power of 3, and $G_2(q^2)$ on ${}^2G_2(q^2)$ where q^2 is an odd power of 3.*

Thus the only almost simple multiplicity-free actions of the type considered here are those listed in Theorems 4 and 11.

References

1. Carter, R.W., *Finite groups of Lie type*, John Wiley, Chichester (1985).

2. Chang, B. and Ree, R., "The characters of $G_2(q)$", Symp. Math. XIII, Academic Press, London (1974), 395-413.

3. Cohen, A., Liebeck, M. and Saxl, J., "Distance-transitive graphs with automorphism group exceptional of Lie type", to appear.

4. Conway, J.H., Curtis, R.T., Norton, S.P., Parker, R.A. and Wilson, R.A., *Atlas of finite groups*, Oxford University Press (1985).

5. Deligne, P. and Lusztig, G., "Representations of reductive groups over finite fields", Ann. of Math. **103** (1976), 103-161.

6. Enomoto, H., "The characters of the finite symplectic group $Sp(4, q)$, $q = 2^f$", Osaka J. Math. **9** (1972), 75-94.

7. Gow, R., "Two multiplicity-free permutation representations of the general linear group $GL(n, q^2)$", Math. Z. **188** (1984), 45-54.

8. Kawanaka, N., "On subfield symmetric spaces over a finite field", to appear in Osaka J. Math.

9. Lawther, R., "Some coset actions in $G_2(q)$", Proc. London Math. Soc. **61** (1990), 1-17.

10. Lawther, R., *On certain coset actions in finite groups of Lie type*, Ph.D. Thesis, Cambridge (1990).

11. Lawther, R. and Saxl, J., "On the actions of finite groups of Lie type on the cosets of subfield subgroups and their twisted analogues", Bull. London Math. Soc. **21** (1989), 449-455.

12. Lusztig, G., *Characters of reductive groups over a finite field*, Princeton University Press (1984).

13. Lusztig, G., "Symmetric spaces over a finite field", The Grothendieck Festschrift, vol. III, Birkhäuser (1990), 57-81.

14. Mayer, S.J., "On the characters of the Weyl group of type C", J. Alg. **33** (1975), 59-67.

Acknowledgement

The work covered in this paper was done while the author was in receipt of a Science and Engineering Research Council grant.

Orbits in Internal Chevalley Modules

Gerhard Röhrle

§1 Introduction

Let G be a simple, connected algebraic group over an algebraically closed field F of characteristic $p \geq 0$. Let $P = LQ$ be a parabolic subgroup of G, where Q is the unipotent radical of P and L is a Levi subgroup of P. Here L acts on Q via conjugation. This induces an L-action on consecutive subquotients of the lower central series of Q. Provided with a suitable F-vector space structure these quotients can be regarded as L-modules. They are called internal Chevalley modules for L.

There exists a unique parabolic subgroup P^- of G such that $P \cap P^- = L$. We refer to P^- as the opposite of P and $Q^- = R_u(P^-)$ is called the opposite unipotent radical of P. The internal Chevalley modules that occur in Q^- are dual to the ones in Q.

In this note we describe some results from [6] regarding the structure of orbits for the action of L on these modules and give some information on the associated stabilizers for arbitrary characteristic.

(1.1) A motivation for this is a result of R. Richardson [4] asserting that L has only finitely many orbits on each of its internal Chevalley modules. We show that there is a close connection between the L-orbits on $Q^-/(Q^-)'$ and (P,P)-cosets of G. For details and further information we refer to [7].

We say that p is a 'very bad' prime for G, if p occurs as a structure constant in Chevalley's commutator relations for G. In this situation there are degeneracies in these relations affecting the structure of orbits. We assume throughout these notes that p is not very bad for G. That means we exclude the cases when (G,p) is one of $(B_n, 2)$, $(C_n, 2)$, $(F_4, 2)$, or $(G_2, 2 \text{ or } 3)$. It is known that if p is not very bad for G, then the internal Chevalley modules for any Levi factor are completely reducible. The number of simple summands and their highest weights depend only on the type of G and can be described in terms of the root system of G [1]. This is used in [7] to obtain the L-orbits in $Q^-/(Q^-)'$ when $P = LQ$ is a maximal standard parabolic subgroup of the classical group G. In particular formulas for the number of orbits are obtained.

§2 Orbits and Double Cosets

Let W denote the Weyl group of G.

(2.1) LEMMA. *Let $P = LQ$ be a standard parabolic subgroup of G. Then:*

(i) *Q^-P is open dense in G,*

(ii) *for any $w \in W$ we have $PwP \cap Q^-$ is nonempty.*

Proof. Observe that Q^-P is an orbit under the connected group $Q^- \times P$ in G; the orbit map being $Q^- \times P \to G$, $(q,p) \mapsto q \cdot 1 \cdot p^{-1}$. So Q^-P is open in its closure which is G, since $\dim Q^-P = \dim G$ and G is connected. This shows (i). Likewise we see that PQ^- is open dense in G. Therefore $PQ^-P = PQ^- \cdot Q^-P = G$ by [2] 1.3(a). This implies (ii). $\qquad\square$

Remark: It is possible to refine (2.1)(ii) as follows. One can give a procedure for a canonical choice of an orbit representative in $(PwP \cap Q^-)(Q^-)'/(Q^-)'$ in terms of a reduced expression for the distingushed double coset representative w. In particular each double coset PwP gives rise to at least one L-orbit in $Q^-/(Q^-)'$ [7].

(2.2) PROPOSITION. *Let $P = LQ$ be a standard parabolic subgroup of G, $w \in W$. Then:*

(i) *$PwP \cap Q^-$ is an irreducible subvariety of Q^-.*

(ii) *$\dim PwP \cap Q^- = \dim PwP - \dim P$.*

Proof. It follows from (2.1) that $PwP \cap Q^-P$ is open dense in PwP. Now PwP is an irreducible variety, since it is an orbit under the connected group $P \times P$. Thus $PwP \cap Q^-P$ is irreducible.

Consider the morphism $G \to G/P$. Then, as $PwP \cap Q^-P$ is irreducible, so is its image $(PwP \cap Q^-)P/P$. Since $Q^- \cong Q^-P/P$ as varieties, we have $PwP \cap Q^- \cong (PwP \cap Q^-)P/P$ as varieties and (i) follows. Finally

$$\dim PwP = \dim PwP \cap Q^-P = \dim(PwP \cap Q^-)P = \dim PwP \cap Q^- + \dim P.$$

This proves (ii). $\qquad\square$

(2.3) COROLLARY. *Let $P = LQ$ be a maximal standard parabolic subgroup of G with Q abelian. Then:*

(i) *The number of L-orbits on Q^- is finite and*
$$|\{L\text{-orbits on } Q^-\}| \geq |P\backslash G/P|.$$

(ii) *For $w \in W$, $PwP \cap Q^-$ contains a unique dense L-orbit.*

(iii) *Let x^L be the unique dense L-orbit in $PwP \cap Q^-$, $w \in W$, then*

$$\dim x^L = \dim PwP - \dim P.$$

Proof. Richardson's theorem (1.1) implies that $PwP \cap Q^-$ is a finite union of L-classes, since $PwP \cap Q^-$ is nonempty and L-invariant. This implies (i). Clearly, (ii) and (iii) follow from (2.2). $\square$

Remark: Evidently, $\dim PwP - \dim P$ only depends on the distinguished double coset representative of PwP and can be expressed in terms of lenghts of Weyl group elements.

(2.4) THEOREM. *Assume that p is not very bad for G. Let $P = LQ$ be a maximal standard parabolic subgroup of G with Q abelian. Then:*

(i) *For $w \in W$, we have $PwP \cap Q^- = x^L$ for some $x \in Q^-$. So there is a bijective correspondence*

$$\{L\text{-orbits on } Q^-\} \longleftrightarrow P\backslash G/P.$$

(ii) *For $w \in W$ and $PwP \cap Q^- = x^L$ we have*

$$\dim x^L = \dim PwP \cap Q^- = \dim PwP - \dim P.$$

Remarks: (2.4)(i) was first proved in [6] in a case-by-case analysis. Now there exists a general geometric argument for good characteristic [5].

Note that the assumption on p in (2.4) is necessary. For instance if G is of type B_n and $P = LQ$ is the maximal standard parabolic corresponding to the simple root α_1, then Q is abelian. The number of L-orbits on Q^- is 4 if char $F = 2$ and it is 3 otherwise, while the number of (P, P)-cosets in G is 3 in all characteristics.

Here is an interesting consequence of (2.4) relating G-conjugacy to L-conjugacy in Q^-.

(2.5) COROLLARY. *Assume that p is not very bad for G. Let $P = LQ$ be a maximal standard parabolic subgroup of G with Q abelian. Let $x \in Q^-$, then*

$$x^G \cap Q^- = x^L.$$

In particular the number of unipotent classes of G that meet Q^- is the number of (P, P)-cosets in G.

Proof. Since Q is abelian $x^{Q^- P} = x^P \subseteq PxP$. Thus by continuity of the orbit map $G \to x^G$ we have

$$x^G = x^{\overline{Q^- P}} \subseteq \overline{x^{Q^- P}} = \overline{x^P} \subseteq \overline{PxP}.$$

Hence

$$x^G \cap Q^- \subseteq \overline{PxP} \cap Q^- = PxP \cap Q^- \ \dot\cup \ \bigcup_{i=1}^{m}(Px_iP \cap Q^-)$$

for some $x_i \in Q^-$ and $\dim Px_iP < \dim PxP$ for $1 \leq i \leq m$.

Suppose there exists a $y = x^g \in Px_iP \cap Q^-$ for some i. Then by the same argument as above

$$x^G = y^G \subseteq \overline{PyP} = \overline{Px_iP}.$$

But as $x^P \subseteq x^G$ we get

$$x^P \subseteq PxP \cap \overline{Px_iP} = \emptyset$$

because of dimension. But this is absurd. Hence $x^G \cap Q^- \subseteq PxP \cap Q^-$. Now $PxP \cap Q^-$ is a single L-orbit by (2.4). So $x^G \cap Q^- = x^L$. $\square$

Remark: In fact the following holds. Let $P = LQ$ be any parabolic subgroup of G and $x \in Z(Q^-)$, the center of Q^-, then $x^G \cap Z(Q^-) = x^L$. A proof due to G. Seitz which is idependent of (2.4) can be found in [7].

(2.6) PROPOSITION. *Let G be classical and assume p is not very bad for G. Let $P = LQ$ be the opposite of a maximal standard parabolic subgroup of G. Then*

$$|\{L\text{-orbits on } Q^-/(Q^-)'\}| \leq |P \backslash G/P|.$$

Proof. This is proved in [6] in a case-by-case study. $\square$

Many non-classical examples suggest the following.

(2.7) CONJECTURE. *The result (2.6) also holds for G exceptional.*

§3 SOME RESULTS ON STABILIZERS

Again assume that p is not very bad for G. For G classical representatives for all L-orbits in Q/Q' are described in [6], where $P = LQ$ is a maximal standard parabolic subgroup of G. From that it is easy to work out the structure of the stabilizers in terms of matrices. As a consequence one obtains the following.

(3.1) PROPOSITION. *Let G be classical and let $P = LQ$ be a maximal parabolic subgroup of G. If $0 \neq x \in Q/Q'$ not in the dense orbit, then $\mathrm{Stab}_L(x)$ is not reductive.*

It is known that already in characteristic zero (3.1) may fail for exceptional G. However, there is a weaker version of (3.1) that holds for all simple G.

(3.2) THEOREM. *Let $P = LQ$ be a maximal standard parabolic subgroup of G. If $0 \neq x \in Q/Q'$ is not in the dense orbit, then $\mathrm{Stab}_L(x)^0$ is contained in a proper parabolic subgroup of L.*

Proof. For the case of G classical this follows from (3.1), as then $\mathrm{Stab}_L(x)^0$ is not reductive and therefore contained in a proper parabolic subgroup of L by a result of Borel and Tits [3]. The case for G exceptional is handled in [8]. □

BIBLIOGRAPHY

1. H. Azad, M. Barry, and G. Seitz, *On the structure of parabolic subgroups*, Com. in Algebra **18** (1990), 551–562.
2. A. Borel, "Linear Algebraic Groups," W. A. Benjamin, New York, 1969.
3. A. Borel and J. Tits, *Élements unipotents et sous-groupes paraboliques de groupes réductifs, I*, Inv. Math. **12** (1971), 95–104.
4. R. Richardson, *Finiteness Theorems for Orbits of Algebraic Groups*, Indag. Math. **88** (1985), 337–344.
5. R. Richardson and G. Röhrle, *Parabolic subgroups with abelian unipotent radical*, (to appear).
6. G. Röhrle, *Orbits in Internal Chevalley Modules*, Ph.D. Thesis, University of Oregon, 1990.
7. G. Röhrle, *On the Structure of Parabolic Subgroups in Algebraic Groups*, (to appear).
8. G. Röhrle, *On certain Stabilizers in Algebraic Groups*, (to appear).

Subgroups of finite and algebraic groups

Gary M. Seitz

§1. Introduction

The purpose of this article is to survey some recent results on the subgroup structure of finite groups of Lie type and related algebraic groups. A great deal of progress has been made, although there remain challenging problems. We emphasize certain of these problems.

Throughout we let $G = G(q)$ denote a quasisimple group of Lie type, where $q = p^a$ and p is prime. Then G arises from a simple algebraic group $\overline{G}$ defined over the algebraic closure F of F_p. Indeed, we have $G = 0^{p'}(\overline{G}_\sigma)$, where σ is a Frobenius morphism of $\overline{G}$. There is a strong connection between the subgroup structure of of G and that of $\overline{G}$. One result illustrating this connection is the following well-known result of Steinberg.

Theorem 1.1. (Steinberg, [22]) Suppose $G = SL(n,q)$ and let $\phi : X(p^b) \to G$ be an absolutely irreducible representation. Then ϕ can be extended to a rational representation $\overline{\phi} : \overline{X} \to \overline{G}$, where $\overline{X}$ is a simple algebraic group corresponding to $X(p^b)$ and where $\overline{G} = SL(n,F)$.

From the point of view of group embeddings this shows that an important class of subgroups of $SL(n,q)$ lift to closed, connected, subgroups of $SL(n,F)$. Steinberg's theorem has recently been extended as follows

Theorem 1.2. ([21]) Let $\phi : X(p^b) \to G(p^a)$ be a homomorphism of finite groups of Lie type. Assume that the image of $X(p^b)$ is not contained in a proper parabolic subgroup of $G(p^a)$. If p is sufficiently large, then there exists an extension $\overline{\phi} : \overline{X} \to \overline{G}$. No restriction on p is required if G is a classical group.

If G is an exceptional group, the prime restriction in (1.2) depends on both X and G and is stated precisely in Theorem 2 of [21]. We will be content to point out that $p > 13$ suffices unless $\overline{X} = A_1$, A_2, or B_2. Possibly a much weaker restriction will do, but we have been unable to establish this. Such an improvement would be an important step in the study of maximal subgroups

In view of these results our strategy is to first obtain results on the subgroup structure of simple algebraic groups and then aim for similar results for the finite groups. The analysis differs for the classical and exceptional groups.

To avoid technicalities we will state results only for the quasisimple groups, although most of these results allow for automorphisms. The reader is referred to the literature for the complete results.

§2. Classical algebraic groups.

In this section we let $G = SL(V)$, $Sp(V)$ or $SO(V)$, where V is a finite dimensional vector space over the algebraically closed field K. To avoid trivialities we assume $\dim(V) > 1$. The first result here is an easy reduction theorem

Theorem 2.1. ([16], Thm.3) Let X be a maximal among closed connected subgroups of G. Then one of the following holds.

(i). X leaves invariant a proper subspace of V.

(ii). X preserves a tensor decomposition of V.

(iii). $(X, G) = (Sp(V), SL(V))$, $(SO(V), SL(V))$, or $(SO(V), Sp(V))$ with $p = 2$.

(iv). X is a simple group with $V \mid X$ irreducible and tensor indecomposable.

Maximal subgroups of types (i)-(iii) are easily understood, so this leaves groups of type (iv). The key result here is the following abbreviated version of a theorem which appears in [16]. The theorem uses results of Testerman [25] and extends famous work of Dynkin to fields of arbitrary characteristic.

Theorem 2.2. ([16], Thm.1) Let X be a closed connected subgroup of $SL(V)$ such that $V \mid X$ is irreducible and tensor indecomposable. Then with specified exceptions X equals or is maximal in one of the groups $SL(V)$, $Sp(V)$, or $SO(V)$.

Given a group X satisfying the hypotheses of (2.2), Table 1 of [16] gives a precise description of all triples (X, Y, V) where Y is a closed connected overgroup of X in $SL(V)$ with $Y \neq SL(V)$, $Sp(V)$, or $SO(V)$. Such triples are relatively rare, so (2.2) really shows that X is usually maximal in one of the classical groups.

But which classical group? To answer this we must first be able to tell whether or not $V \mid X$ is self-dual, and if the answer is yes we need to determine whether or not X fixes a quadratic form.

Write $V = V_X(\lambda)$ to indicate that V is an irreducible high weight module for X, corresponding to the dominant weight λ. A standard result (e.g. see 31.6 of [9]) shows that $V^* = V_X(-\lambda w_o)$, where w_o denotes the long word in the Weyl group of X. Hence is it is easy to determine whether or not X fixes a nondegenerate bilinear form on V. For example, if the Dynkin diagram of X does not admit a graph automorphism, then all irreducible modules for X are self-dual.

So now assume X fixes a form on V. If K does not have characteristic 2, there is very nice way to determine which form is fixed. Fix a parameterization of the elements of X and set

$$z = \prod h_\alpha(-1),$$

where the product is taken over all positive roots α in the root system associated with X.

Proposition 2.3. ([23], Lemma 79).

(i). $z \in Z(X)$.

(ii). $z^2 = 1$.

(iii). X preserves a quadratic form on $V = V_X(\lambda)$ if and only if $\lambda(z) = 1$.

In particular, if X does not have a central involution, then X necessarily preserves a quadratic form on V. For example, all irreducible modules for E_8 are quadratic.

When K has characteristic 2, the situation is much more subtle and remains open, although there exist partial results. Here we have $SO(V) < Sp(V)$.

Recall we are assuming V | X is tensor indecomposable, so we may assume λ is a restricted weight. One approach is to start with characteristic 0 representations, where we already have complete information. So let X_C denote a simply connected group of the same type as X but over the complex field and $V_{X_C}(\lambda)$ the corresponding irreducible rational module of high weight λ.

If V is not a 2n-dimensional symplectic module for $X = C_n$, then it is possible to produce a nontrivial quadratic form on the corresponding Weyl module, $W_X(\lambda)$, for which the radical of the underlying bilinear form is M, the unique maximal submodule of $W_X(\lambda)$. We obtained one proof of this and another was communicated to us by J. Jantzen.

If M is totally singular under the the quadratic form, then there is an induced quadratic form on $V_X(\lambda)$. Moreover, any nonzero quadratic form on $V_X(\lambda)$ must arise in this way. On the other hand, if M is not totally singular, then the set of singular vectors in M is a submodule, M_o, of codimension 1 in M. Hence, $W_X(\lambda)/M_o$ is a nontrivial extension of $V_X(\lambda)$ by the trivial module.

These ideas can be utilized to analyze special cases. Many such were settled in unpublished joint work of the author, P. Kleidman, and D. Testerman. The following general result was also obtained. It is an easy consequence of (2.3) and the above remarks.

Theorem 2.4. Suppose λ is 2-restricted and X_C preserves a nondegenerate symplectic form on $V_{X_C}(\lambda)$. Then either X preserves a nondegenerate quadratic form on V or $X = C_n$ and V is a 2n-dimensional symplectic module for X.

§3.   Finite classical groups

In this section we let $G = G(p^a) = G(V)$ denote a finite classical group, where V is a finite dimensional vector space over a field F of order p^a (p^{2a} in the unitary case). As in §2 we assume $\dim(V) > 1$.

The first result is a version of (2.1) for finite groups, due to Aschbacher. In [1] Aschbacher produces a family C of natural subgroups of G. Included in the family are stabilizers of subspaces and sets of subspaces, classical groups over subfields and extension fields, stabilizers of tensor product decompositions, and certain local subgroups. He then proves

Theorem 3.1. ([1]) If R $\leq$ G, then either R is contained in a member of C or X = F^*(R) is quasisimple, V $\mid$ X is absolutely irreducible, and V $\mid$ X is not realizable over a proper subfield of F.

Therefore we must study absolutely irreducible, quasisimple subgroups X of G, for which N_G(X) is not contained in a member of C. Given such a group X, the question is whether or not N_G(X) is maximal in G. If not, then using the Schreier conjecture (now a theorem) and (3.1) we obtain a quasisimple group Y with X < Y < G. So just as in §2 it is necessary to study triples (X, Y, V).

However, here the problem is considerably more difficult and much remains open. Part of the difficulty stems from the many possibilities for X and Y. But there are serious difficulties even when X and Y are explicitly given, due to a lack of information from representation theory.

We organize the discussion according to the isomorphism type of X. We have progressively less information in the various cases.

Generic Case. X = $X(p^b)$ is of Lie type in characteristic p.

The first problem here is to determine the possibilities for Y. This was accomplished in [10] where the main result of shows that one of the following holds: Y is of Lie type in characteristic p; (X,Y,V) is explicitly known; Y is a sporadic group. The latter two possibilities yield relatively few possibilities, so typically Y is of Lie type in characteristic p. Assume this is the case, so that both groups and the module are defined over fields of characteristic p.

If Y is not an exceptional group with p small, we can apply (1.2) to obtain $\overline{X} \leq \overline{Y} < \overline{G}$, an embedding of algebraic groups. At this point (2.2) can be applied to yield detailed information regarding (X, Y, V). So the generic case is in relatively good shape.

Cross characteristic case. X = $X(r^b)$ is of Lie type in characteristic r $\neq$ p.

Here there is currently no analog of the result in [10]. Most likely, Y is of Lie type in characteristic r or p. As a test case we considered the former case with Y a classical group.

Theorem 3.2. ([17]) Assume X = $X(r^b)$ is contained in no member of C and r^b > 3. Further assume that Y = $Y(r^c)$ is of classical type and minimal among quasisimple overgroups of X. Then the pair (X/Z(X),Y/Z(Y)) is one of the following

$$(PSp(2n, q), PSL^\varepsilon(2n, q)) \quad (P\Omega(n-1, q), P\Omega(n, q))$$
$$(PSp(2n, q^d), PSp(2nd, q) \quad (G_2(q), P\Omega(7, q)).$$

For most of these embeddings there does exist an appropriate module V, at least for certain primes p. Only the third possibility permits long chains of embeddings of irreducible subgroups. Indeed, for q odd the smallest nontrivial irreducible irreducible representation of Sp(2n, q) over a field of characteristic r has degree $\frac{1}{2}(q^n - 1)$. Such a representation for Sp(2nd,q) necessarily restricts to an irreducible representation of Sp(2n,q^d). In this way one can produce arbitrarily long chains of subgroups. For example, Sp(2n, q^8) < Sp(4n, q^4) < Sp(8n, q^2) < Sp(16n, q) all act irreducibly on a module of dimension $\frac{1}{2}(q^{8n} - 1)$.

The proof of the theorem makes use of an upper bound for degrees of irreducible representations of groups of Lie type. For most types this upper bound is just the index of a Borel subgroup. Unfortunately the only proof we know of this theorem is deep in that it uses Lusztig's Jordan decomposition of characters.

A final comment here is that (3.2) only gives the group embedding. It does not determine all modules V which can occur. This remains open even for the embedding Sp(2n,q) < SL(2n,q).

Alternating case. $X = Alt_m$.

This appears to be a difficult case. There does not yet exist a result restricting the possibilities for Y. Even the the case where $Y = Alt_n$ seems to be difficult. Since it is easier to work with representations of symmetric groups, we consider $Sym_m < Sym_n$. If the field has characteristic 0, this has been completely settled by Saxl [15]. In general, for both groups to be irreducible on V either m = n -1 or the subgroup acts transitively on n points.

Consider then the problem of trying to determine those modules V for Sym_n which are irreducible upon restriction to Sym_{n-1}. Write $V = V(\lambda)$, where λ is a partition of n. For representations over fields of characteristic 0 or p > n, there is an easy answer resulting from the Branching theorem (see (9.1) of [6]). Namely, this occurs if and only if λ is a partition with equal parts. However, when p ≤ n, the problem is much more subtle.

In recent work we established the following result which produces a fairly large family of such modules. Let $\lambda = (\lambda_1^{a_1}, ..., \lambda_k^{a_k})$ be a p-regular partition of n with $\lambda_1 > ... > \lambda_k$. We assume $\lambda \neq (n)$, as otherwise V is the trivial module.

Theorem 3.3. ([18], Thm. 3.2) Let M be the irreducible $FSym_n$-module corresponding to $\lambda \neq (n)$. Assume $\lambda_i - \lambda_{i-1} + a_i + a_{i-1} \equiv 0 \pmod{p}$ for i = 2, ... , k. Then M restricts to an irreducible representation of Sym_{n-1}.

We conjecture that the modules described in (3.3) are the only nontrivial $FSym_n$-modules which restrict to irreducible modules for Sym_{n-1}, but we have been unable to prove this.

Theorem (3.3) arose naturally from the correspondence between representations of GL_n and representations of Sym_n based on the Schur functor. As part of our work we determine the fibres of the Schur functor.

Sporadic case. X is a sporadic group.

This case is completely open, although it should be possible to obtain results for many of the sporadic groups given current information on modular representations.

§4. Exceptional algebraic groups

In this section G will denote an exceptional group over an algebraically closed field K of characteristic p. Theorems (4.1) and (4.2) to follow require mild prime restrictions which are indicated in the following table

	E_8	E_7	E_6	F_4	G_2
A_1	7	7	5	3	3
A_2	5	5	3	3	
B_2, G_2	5	3	3	2^*	
B_3	2	2	2		
A_3, C_3, B_4	2				

A comment is required concerning the last entry in the F_4 column. The asterisk indicates that in Theorem 4.1, no prime restriction is needed, whereas for Theorem 4.2, the restriction is char(K) ≠ 2.

As was the case for classical groups we first determine the maximal closed connected subgroups of G. The following result extends the work of Dynkin on exceptional Lie algebras over C.

Theorem 4.1. ([19]) Let G be a simple algebraic group of exceptional type and let X be maximal among proper closed connected subgroups of G. If X is simple, assume char(K) = 0 or char(K) is strictly greater than the integer specified in the above table. Then either X contains a maximal torus of G or X is semisimple and the pair (G, X) is given below. Moreover, maximal subgroups of each type exist and are unique up to conjugacy in Aut(G).

	X simple	X not simple
G_2	A_1 (p = 0 or p ≥ 7)	
F_4	A_1 (p = 0 or p ≥ 13), G_2 (p = 7)	$A_1 G_2$ (p ≠ 2)
E_6	A_2, G_2, F_4, C_4 (p ≠ 2)	$A_2 G_2$
E_7	A_1 (2 classes, p = 0 or p ≥ 17, 19, resp.), A_2	$A_1 A_1$ (p ≠ 2,3), $A_1 G_2$ (p ≠ 2), $A_1 F_4$, $G_2 C_3$
E_8	A_1 (3 classes, p = 0 or p ≥ 23, 29, 31, resp.), B_2	$A_1 A_2$ (p ≠ 2,3), $G_2 F_4$

We first note that the maximal subgroups containing a maximal torus are easy to determine. Either they are maximal parabolic subgroups or they arise from maximal subsystems of the root system of G. The prime restrictions following entries in the table are required for the existence of the various subgroups. Existence of certain of the groups, in particular the groups of type A_1, is due to Testerman [26], [27].

An outline of the proof of (4.1) is given in [20]. We merely point out that the proof makes use of a particular 1-dimensional torus T of X. Namely, T consists of elements T(c), where for $0 \neq c \in$ K we set $T(c) = \prod h_\alpha(c)$, the product taken over all positive roots in the root system of X. Notice that T(-1) is just the element z mentioned prior to (2.3). It turns out that if X is maximal and does not contain a maximal torus of G, then T determines a labelling of the Dynkin diagram of G by by 0's and 2's and this labelling yields detailed information on the composition factors of X on L(G), the Lie algebra of G. Our proof of (4.1) is based on the analysis of L(G) | X.

Special cases of (4.1) have been obtained by other authors using different methods. In particular, Aschbacher and Maagard settled the cases of groups of type E_6 and F_4, respectively (see [2] and [13]).

Theorem 4.1 tells us what to expect for the finite groups, but it turns out that an extension of (4.1) is required to obtain results at the level of finite groups. What is needed is a classification of subgroups maximal with respect to having positive dimension and being σ-invariant, where σ is a Frobenius morphism.

This is resolved in the following result. For this theorem assume that K = $\overline{F}_p$ or that K is an algebraically closed field of characteristic 0. In the former case let σ denote a Frobenius morphism of G.

Theorem 4.2. ([12]) Let S < G be maximal among closed subgroups of positive dimension or closed σ-invariant subgroups of positive dimension. Set X = S^o. If X is simple with $C_G(X) = 1$, assume p = 0 or p is greater than the integer in the table above. Then one of the following holds

(i) X contains a maximal torus of G.

(ii) X is maximal among closed connected subgroups of G.

(iii) $G = E_8$ and $X = A_1 G_2 G_2$ ($\leq F_4 G_2$).

(iv) $G = E_7$, char(K) $\neq$ 2, and $S = (Z_2 \times Z_2 \times D_4)Sym_3$.

(v) $G = E_8$, char(K) $\neq$ 2, 3, or 5, and $S = A_1 \times Sym_5$.

If (i) holds then as above either S is a maximal parabolic subgroup of G or S is reductive and corresponds to the subsystem of the root system of G. If (ii) holds, then S is the normalizer of one of the groups given in (4.1). Apart from these, there are just three other possibilities.

We shall sketch the proof of (4.2). In particular, we indicate some of the ideas leading to the especially interesting group in (v). Assume X does not contain a maximal torus, as otherwise (i) holds. Then X must be semisimple and we write $X = X_1 \circ \cdots \circ X_k$, where each X_i is simple.

The full force of maximality was not required in the proof of (4.1). Indeed, in Theorem 2 of [19] the possibilities for X were determined assuming that the following conditions hold:

(i) X acts on L(G) as an adjoint group. That is, all weights of L(G) I X are integral combinations of roots in $\Sigma(X)$.

(ii) $C_{L(G)}(X) \leq Z(L(G))$.

(iii) $N_G(X_i)^0 = X$ for i = 1, ... , k.

(iv) $C_{L(G)}(L(X)) = 0$ if $X = A_1$.

First assume $C_G(X) = 1$. If X is simple, then S is just X together with a group of graph automorphisms and we are fairly close to the hypotheses of (4.1). Here we verify the conditions (i) - (iv) above and obtain the result. Suppose X is not simple. If S is intransitive on $\{X_1,...,X_k\}$, then the result follows from Theorem 3 of [19]. The most difficult case is when S acts transitively. This requires a number of tricky arguments. There is no restriction here on the characteristic of K and certain subtleties arise for small characteristic.

Now suppose $R = C_G(X) \neq 1$. The maximality of S forces R to be a finite group. Since $X \leq C_G(R)$, we have $C_G C_G(R) \leq C_G(X) = R$, and so $C_G C_G(R)$ is finite. This double centralizer condition has nice consequences as indicated in the following result

Proposition (4.3). ((1.2) of [12]) Let $R \leq G$ and assume $C_G C_G(R)$ is finite. Then R is finite and I R I is only divisible by bad primes. If Fit(R) = 1, then $G = E_8$ and $F^*(R) = Alt_5$ or Alt_6.

In our situation consider Fit(R). If Fit(R) $\neq$ 1, then S is a local subgroup and hence Theorem 2 of [5] shows that (iv) holds. Therefore, assume Fit(R) = 1. Then by (4.3) $G = E_8$ and $F^*(R) = Alt_5$ or Alt_6. Technical arguments show that K does not have characteristic 2, 3, or 5.

Let $f \in R$ be an element of order 5. From the containment $C_G C_G(f) \leq C_G C_G(R)$ we conclude $C_G C_G(f)$ is finite and hence $Z(C_G(f))$ must be finite. Information on centralizers of semisimple elements of G forces $C_G(f) = A_4 \circ A_4$. Of course X is contained in this group and additional arguments show that $X = A_1$ and that X projects to an irreducible subgroup of each A_4. In particular, this determines the conjugacy class of the unipotent elements of X. If $1 \neq u \in X$ is unipotent, it follows that $C_G(u) = V \cdot Sym_5$, where V is a unipotent group. As $R = C_G(X) \leq C_G(u)$, this immediately gives $F^*(R) = Alt_5$. Moreover, additional work shows that actually $R = Sym_5$. In this way we obtain the group $A_1 \times Sym_5$ indicated in (v).

§5. Finite exceptional groups

In this section we indicate how results of the previous section yield information about the finite exceptional groups. The main result here is a reduction theorem for finite exceptional groups similar to (3.1) and based on ideas of A. Borovik [4].

Let L be a finite simple exceptional group. Write $L = O^{p'}(G_\sigma)$, where G is a simple algebraic group and σ is a Frobenius morphism of G.

Suppose we are given a maximal subgroup, say M, of L. If Fit(M) $\neq$ 1, then M is a local subgroup of G, and the possibilities for such groups are determined in Theorem 1 of [5]. Among the groups that arise here are groups such as $(Z_5 \times Z_5 \times Z_5)\cdot SL(3,5)$ in $E_8(q)$ and local subgroups containing a maximal torus of L.

Now suppose Fit(M) = 1 and write $F^*(M) = X_1 \circ \cdots \circ X_k$, a direct product of simple groups. If k>1, then from first principles we have $[C_G C_G(X_i),$ $C_G C_G(X_j)] = 1$ for $i \neq j$ and hence M normalizes $D = \prod C_G C_G(X_i)$. Now D is either finite or of positive dimension. In the former case (4.3) shows that for each i we have $X_i = Alt_5$ or Alt_6. In the latter case we embed D in a maximal subgroup of G invariant under M and σ, and apply (4.2) (no prime restriction is required for this application of (4.2)).

Arguments along these lines eventually lead to the following reduction theorem, which reduces the study of maximal subgroups of finite exceptional groups to the almost simple case. The theorem is based on (4.2), but only those parts for which there is no prime restriction. Consequently, the following theorem holds with no restriction on the characteristic.

Theorem 5.1. ([12], Thm. 2) Let M be a maximal subgroup of L. Either $F^*(M)$ is simple, or one of the following holds:

(a) $M = N_L(D_\sigma)$, where $D = D^0 = D^\sigma$ is either parabolic or reductive of maximal rank.

(b) $M = N_L(E)$, where E is an elementary abelian group.

(c) M is the centralizer of a graph, field, or graph-field automorphism of L of prime order.

(d) $G = E_8$, $p > 5$ and $M = Sym_5 \times PGL_2(2, q)$ or $(Alt_5 \times Sym_6).2$.

(e) $F^*(M)$ is as in the table below and $F^*(M) = O^{p'}(X_\sigma)$, where $X = X^\sigma$ is as in the second column of the table in (4.1) or as in (iii) of Theorem (4.2).

In (d) and (e), $F^*(M)$ is unique up to G-conjugacy.

L	$F^*(M)$
$F_4(q)$	$L_2(q) \times G_2(q)$ $(p > 2, q > 3)$
$E_6^\varepsilon(q)$	$L_3(q) \times G_2(q)$, $U_3(q) \times G_2(q)$ $(q > 2)$
$E_7(q)$	$L_2(q) \times L_2(q)$ $(p > 3)$, $L_2(q) \times G_2(q)$ $(p > 2, q > 3)$
	$L_2(q) \times F_4(q)$ $(q > 3)$, $G_2(q) \times PSp_6(q)$
$E_8(q)$	$L_2(q) \times L_3^\varepsilon(q))$ $(p > 3)$, $G_2(q) \times F_4(q)$
	$L_2(q) \times G_2(q) \times G_2(q)$ $(p > 2, q > 3)$, $L_2(q) \times G_2(q^2)$ $(p > 2, q > 3)$

Maximal subgroups of type (a) are determined in [11] and those of type (b) in [5]. The first of the groups in (d) is obtained from the group in (4.2)(v) by taking fixed points. The second is obtained from the first in the following way. Choose a group $Alt_5 < L_2(q)$ (this requires $q^2 \equiv 1 \pmod 5$). This Alt_5 is clearly centralized by the Sym_5 centralizing $L_2(q)$. The full centralizer is in fact Sym_6.

The final result we mention is an application of (1.2) to the almost simple case arising from (5.1).

Theorem 5.2. ([12], Thm. 3) Suppose M is maximal in L and $F^*(M)$ is simple of Lie type over a field of characteristic p. If p is sufficiently large, then one of the following holds.

(i) $F^*(M) = 0^{p'}(G_\delta)$, where δ is a field or graph-field automorphism of G.

(ii) $F^*(M) = 0^{p'}(X_\sigma)$, where X is a closed connected σ-stable subgroup of G containing a maximal torus of G.

(iii) $F^*(M) = 0^{p'}(X_\sigma)$, where X is as in the first column of the table in (4.1).

Since the groups of type (i) - (iii) are completely determined (use [11] for (ii)) this settles the generic case, at least for large primes p.

The maximal subgroups of certain of the exceptional groups are known. In particular there are nearly complete results for $Sz(q)$, $G_2(q)$, $^2G_2(q)$, $^3D_4(q)$, $^2F_4(q)$, $F_4(q)$, and $E_6(q)$ (see [24], [3], [7], [8], [13], [14], and [2]). However, the difficult cases of $E_7(q)$ and $E_8(q)$ remain open.

References

1. Aschbacher, M., On the maximal subgroups of the finite classical groups, Inven. Math., 76, (1984), 469-514.

2. Aschbacher, M., The 27-dimensional module for E_6, I-IV, Invent.Math 89, (1987), 159-195; J. London Math. Soc., 37, (1988), 275-293; (to appear Trans. AMS); (to appear J. Alg.).

3. Aschbacher, M., Chevalley groups of type G_2 as the group of a trilinear form, J. Alg., 109, (1987), 193-259.

4. Borovik, A., The structure of finite subgroups of simple algebraic groups, Algebra and Logic, 28, (1989), 249-279, (in Russian).

5. Cohen, A., Liebeck, M., Saxl, J., and Seitz, G., The local maximal subgroups of the exceptional groups of Lie type, Proc. LMS, to appear.

6. James, G., The representation theory of the symmetric groups, Springer Lecture Notes 682, New York/Berlin, 1978.

7. Kleidman, P., The maximal subgroups of the Chevalley groups $G_2(q)$ with q odd, of the Ree groups $^2G_2(q)$, and of their automorphism groups, J. Alg., 117, (1988), 30-71.

8. Kleidman, P., The maximal subgroups of the Steinberg triality groups $^3D_4(q)$ and of their automorphism groups, J. Alg., 115, (1988), 182-199.

9. Humphreys, J., "Linear Algebraic Groups", Springer-Verlag, New York, Berlin, 1975.

10. Liebeck, M., Saxl, J, and Seitz, G., On the overgroups of irreducible subgroups of the finite classical groups, Proc. London Math. Soc., 55, (1987), 507-537.

11. Liebeck, M., Saxl, J, and Seitz, G., Subgroups of maximal rank in finite exceptional groups, Proc. LMS, to appear.

12. Liebeck, M. and Seitz, G., Maximal subgroups of exceptional groups of Lie type, finite and algebraic, Geom. Ded., 35, (1990), 353-387.

13. Maagard, K., Maximal subgroups of groups of type F_4, Ph.D. thesis Cal. Tech.

14. Malle, G., The maximal subgroups of the Ree groups $^2F_4(q)$, to appear J. Alg.

15. Saxl, J., The complex characters of the symmetric groups that remain irreducible in subgroups, J. Alg., 111, (1987), 210-219.

16. Seitz, G., The maximal subgroups of classical algebraic groups, Memoirs Amer. Math. Soc., 365, (1987), 1-286.

17. Seitz, G., Cross-characteristic embeddings of finite groups of Lie type, Proc. London Math. Soc., 60, (1990), 166-200.

18. Seitz, G., On the representation theory of the symmetric groups, (preprint).

19. Seitz, G., The maximal subgroups of exceptional algebraic groups, Memoirs Amer. Math. Soc., 90, (1991), 1-197.

20. Seitz, G., Maximal subgroups of exceptional groups, Contemporary Math, 82, (1989), 143-157.

21. Seitz, G. and Testerman, D., Extending morphisms from finite to algebraic groups, J. Alg., 131, (1990), 559-574.

22. Steinberg, R., Representations of algebraic groups, Nagoya Math. J., 22, (1963), 33-56.

23. Steinberg, R., Lectures on Chevalley groups, Yale University lecture notes, 1968.

24. Suzuki, M., On a class of doubly transitive group, Annals of Math., 75. (1962), 105-145.

25. Testerman, D., Irreducible subgroups of exceptional algebraic groups, Mems Amer. Math. Soc., 390, (1988), 1-188.

26. Testerman, D., A construction of certain maximal subgroups of the algebraic groups E_6 and F_4, J. Alg., 122, (1989), 299-322.

27. Testerman, D., The construction of maximal A_1's in the exceptional algebraic groups, (to appear).

Irreducible Representations of Finite Chevalley Groups Containing a Matrix with a Simple Spectrum

I.D. Suprunenko and A.E. Zalesskii

For a matrix M the set of its eigenvalues, counting multiplicities, is called the spectrum of M. If the multiplicity of every eigenvalue of M is equal to 1, we say that M has a simple spectrum. In certain situations it would be desirable to determine finite linear groups containing a matrix with a simple spectrum. We shall discuss some aspects of this problem.

Matrices with simple spectra often occur as matrices of linear transformations acting irreducibly on the underlying space. For finite fields the following problem has been studed by Hering [2, 3] for determining doubly transitive permutation groups.

Problem. Describe all linear groups $H \subset GL_n(q)$ which contain irreducible cyclic subgroup.

Observe that if a matrix $c \in GL_n(q)$ generates an irreducible subgroup, then the spectrum of c is simple. Note that Hering has considered the case where c is of prime order. Kantor [4] has described subgroups of $GL_n(q)$ which contain a cyclic subgroup of order $q^n - 1$. In a more general investigation [5] Seitz has described (under certain restrictions) all subgroups of finite Chevalley groups containing a maximal torus.

If one tries to determine all finite linear groups containing a matrix with a simple spectrum, then one is forced to use the classification of finite simple groups and the representation theory since no other approach is available. This leads to a rather large program because it is necessary to examine representations of many classes of groups in various characteristics. We shall discuss in detail the following particular case which is an essential ingredient of the general problem.

Problem 1. Determine the irreducible representations of finite quasisimple Chevalley groups over an algebraically closed field of defining characteristic which contain a matrix with a simple spectrum.

We have obtained an almost complete solution of this problem. Our approach is based on the earlier description of the representations of algebraic groups which contain a matrix with a simple spectrum. We begin by recalling these results on algebraic groups.

Let G be a simple algebraic linear group over an algebraically closed field P of characteristic $p > 0$. Let $\omega_1, \ldots, \omega_n$ be the set of fundamental weights of G, labelled as in Bourbaki [1]. Define a set $\Omega(G)$ of weights of G by the following table. The weights ω_0 and ω_{n+1} for $A_n(P)$ which appear in the table for certain a, j are interpreted as 0. We shall write $\mathrm{Irr}G$ for the set of classes of equivalent irreducible rational representations of G.

Table

G	$\Omega(G)$
$A_1(P)$	$a\omega_1 \; (0 \le a < p)$
$A_n(P) \; (n > 1)$	$\omega_2, \ldots, \omega_{n-1}, \; (p - 1 - a)\omega_j + a\omega_{j+1}$ $(0 \le a < p, \; 0 \le j \le n)$
$B_n(P) \; (n > 2, p > 2)$	$0, \; \omega_1, \; \omega_n$
$C_n(P)(n > 1, p > 2)$	$0, \omega_n \; (n = 2, 3), \; \omega_1, \omega_{n-1} + \frac{(p-3)}{2}\omega_n, \; \frac{(p-1)}{2}\omega_n$
$C_n(P) \; (n > 1, \; p = 2)$	$0, \; \omega_1, \; \omega_n$
$D_n(P) \; (n > 3)$	$0, \; \omega_1, \; \omega_{n-1}, \; \omega_n$
$E_6(P)$	$0, \; \omega_1, \; \omega_6$
$E_7(P)$	$0, \; \omega_7$
$E_8(P)$	0
$F_4(P)$	$0, \; \omega_1$ for $p = 3$, $\; 0$ for $p \ne 3$
$G_2(P)$	$0, \; \omega_1, \; \omega_2$ for $p = 3$, $\; 0, \omega_1$ for $p \ne 3$

We shall say that ϕ is an SS-representation if its image contains a matrix with a simple spectrum.

Theorem 1 *(Seitz [6, 6.1]) Let G be an universal simple algebraic group over P and let $\phi \in \mathrm{Irr}G$ be infinitesimally irreducible. Suppose that ϕ is a tensor indecomposable SS-representation. Then the highest weight of ϕ is contained in $\Omega(G)$.*

Remark. Theorem 6.1 of [6] only gives necessary conditions for ϕ to be SS-representation. We have omitted a few C_n-representations where weight spaces are not actually all 1-dimensional.

Denote by $\mathrm{Irr}_1 G$ the set of $\phi \in \mathrm{Irr}\, G$ whose highest weights $\lambda = \lambda_o + \lambda_1 p + \ldots + \lambda_k p^k$ satisfy the conditions: $\lambda_i \in \Omega(G) \; (i = 1, \ldots, k)$ and $\lambda_{i+1} \ne \omega_1 (i =$

$0, \ldots, k-1$) if either $G = C_n(P)$, $n > 1$, $p = 2$, $\lambda_i = \omega_n$; or $G = G_2(P)$, $p = 2$, $\lambda_i = \omega_1$; or $G = G_2(P)$, $p = 3$, $\lambda_i = \omega_2$.

Theorem 2 *([8]) Let G be as in Theorem 1. Let $\phi \in IrrG$ and λ be the highest weight of ϕ. Then ϕ is an SS-representation if and only $\lambda \in Irr_1 G$.*

The following lemma and proposition are well-known:

Lemma 1 *Every semisimple element $g \in G$ is contained in a maximal torus T of G, so the weight spaces of T are contained in the eigenspaces of g.*

Proposition 1 *Let G be a connected algebraic group and $\phi \in IrrG$. The following conditions are equivalent:*
(1) ϕ is an SS-representation;
(2) all the weight multiplicities of ϕ are equal to 1.

Let $q = p^r$ be a field parameter and let $G(q)$ be the finite group of the respective type (including twisted groups for which this parameter is defined as in [7, §11]). Let $\Phi(G(q))$ be the set of all $\phi \in IrrG$ whose highest weights $a_1\omega_1 + \ldots + a_n\omega_n$ satisfy the condition: $0 \leq a_1, \ldots, a_n < q$ except groups ${}^2F_4(q)$ for $i = 3,4$ and ${}^2C_2(q)$, ${}^2G_2(q)$ for $i = 1$ where $0 \leq a_i < pq$. Every irreducible representation of $G(q)$ is a restriction of a representation from $\Phi(G(q))$, see [7, §13]. If $\phi(G(q))$ is an irreducible SS-representation, then by Proposition 1 $\phi \in Irr_1 G$. We put $\Phi_1 = \Phi(G(q)) \cap Irr_1 G$. Therefore our problem is reduced to the following one:

Problem 2. Let $\phi \in \Phi_1(G(q))$. Let $G(q) \subset G$ be a finite group (of normal or twisted type). Does there exist $g \in G(q)$ such that the matrix $\phi(g)$ has a simple spectrum? More precisely, for every $G(q)$ determine $\phi \in \Phi_1(G(q))$ such that $\phi \mid G(q)$ is an SS-representation.

One would like to think that such $g \in G(q)$ always exists. However this is not the case even for $SL_2(q)$ with q odd. Nevertheless for all $G(q)$ which we have completely analyzed a slightly weaker form of this conjecture turns out to be true. In particular, for $p > 2$ this is true for groups different from $A_n(q)$, ${}^2A_n(q)$ and ${}^2G_2(q)$. However, for $G(q) \in \{A_n(q), {}^2A_n(q)\}$ such an element g can be found inside the extended groups $GL_{n+1}(q) \supset SL_{n+1}(q) \cong A_n(q)$ and $U_{n+1}(q) \supset SU_{n+1}(q) \cong {}^2A_n(q)$. Observe that usually (not always) we can choose g as an element of maximal order.

Theorem 3 *Let $G(q) = SL_l(q)$ (resp. $G(q) = SU_l(q)$) and $\phi \in \Phi_1(G(q))$. Then ϕ can be continued to $GL_l(q)$ (resp. to $U_l(q)$) so that $\phi(g)$ has a simple spectrum for some element $g \in G(q)$ of order $q^l - 1$ (resp. $q^l + 1$, if l odd, and $q^l - 1$ or $q^{l-1} + 1$, if l is even).*

Exact determination of SS-representations of $SL_l(q)$ and $SU_l(q)$ seems to be very difficult. Observe that $\phi \in \Phi_1(G(q))$ is an SS-representation provided the highest weight of ϕ is not very "large":

Theorem 4 *Let $q = p^r, r > 1, G = SL_l(q)$ or $SU_l(q)$ and $\phi \in \mathrm{Irr} G$. Suppose that the highest weight of ϕ has the form $\lambda = \lambda_o + \lambda_1 p + \ldots + \lambda_{r-2} p^{r-2}$ with $\lambda_o, \ldots, \lambda_{r-2} \in \Omega(G)$. Then $\phi(g)$ has a simple spectrum for some element $g \in SL_l(q)$ of order $q^{l-1} + \ldots + q + 1$ and some element $g \in SU_l(q)$ of order $(q^l + (-1)^{l+1})/(q+1)$.*

Theorem 5 *Let $G(q)$ be different from $C_n(q), {}^2C_2(q), E_6(q), {}^2E_6(q), {}^3D_4(q), G_2(q)$ with $p = 2$ for all these cases, ${}^2G_2(q)$ and $G_2(q)$ with $p = 3$, $A_n(q), {}^2A_n(q)$. Let $\phi \in \Phi_1(G(q))$. Then $\phi(g)$ is an SS-representation.*

We have not completely analyzed the groups $E_6(q), {}^2E_6(q), {}^3D_4(q)$ with q even. The groups

$$(*) \quad C_n(q), \ {}^2C_2(q), \ G_2(q) \text{ with } p = 2, \ {}^2G_2(q) \text{ and } G_2(q) \text{ with } p = 3$$

have been excluded for a more essential reason. The analysis for $C_n(P)$ with $p = 2$ and $G_2(P)$ with $p = 2, 3$ in Theorem 2 requires some refinements for finite groups of these types. Let Fr denote the Frobenius morphism of algebraic groups. Let $\phi, \varphi \in \mathrm{Irr}\ G$. We say that ϕ, φ are weakly equivalent with respect to $G(q)$ if $\phi \mid G(q) \cong Fr^i \varphi \mid G(q)$ for some $i \in \mathbf{N}$. It sometimes happens that $\phi \in \Phi_1(G(q))$ and $\varphi \notin \Phi_1(G(q))$ for weakly equivalent ϕ, φ. The simplest example is produced for $q = 4$ by $\phi, \varphi \in \mathrm{Irr} Sp_{2n}(P)$ whose weights are $\omega_1 + 2\omega_n$ and $\omega_n + 2\omega_1$, respectively. It is clear that $\phi(G(q))$ contains no matrix with a simple spectrum. Therefore we must remove from $\Phi_1(G(q))$ the representations that are weakly equivalent to a representation outside $\Phi_1(G(q))$. To be precise, define for $q = p^r$ and the groups of the list $(*)$ the set $R(G(q)) \subseteq \Phi_1(G(q))$ as follows. For nontwisted groups $R(G(q))$ is the set of all $\phi \in \Phi_1(G(q))$ whose highest weights $\lambda = \lambda_o + \lambda_1 p + \ldots + \lambda_{r-1} p^{r-1}$ for $r > 1$ satisfy the condition: $(\lambda_{r-1}, \lambda_0) \neq (\omega_2, \omega_1)$ for $G(q) = G_2(q)$ with $p = 3$; $(\lambda_{r-1}, \lambda_0) \neq (\omega_1, \omega_1)$ for $G(q) = G_2(q)$ with $p = 2$; $(\lambda_{r-1}, \lambda_0) \neq (\omega_n, \omega_1)$ for $G(q) = C_n(q)$ with $p = 2, n > 1$; for $G(q) = {}^2C_2(q)$ and ${}^2G_2(q)$ we must take $\lambda = \lambda_o + \lambda_1 p + \ldots + \lambda_r p^r$ with $\lambda_r \neq \omega_2$ and our condition is $(\lambda_r, \lambda_0) \neq (\omega_1, \omega_1)$.

Theorem 6 *Let $G(q) \in \{C_n(q)(p = 2), {}^2C_2(q), {}^2G_2(q), G_2(q)$ with $p = 2, 3\}$ and $\phi \in R(G(q))$. Then $\phi \mid G(q)$ is an SS-representation.*

It is natural to consider a more precise version of Problem 2:

Problem 3. Given $\phi \in \Phi_1(G(q))$ describe the set $\{g \in G(q) \mid \phi(g)$ has simple spectrum$\}$.

This problem is much more difficult than Problem 2. At least we do not see any good way to solve it.

Perhaps, another problem which seems to be connected with Problem 2 merits attention.

Problem 4. Let $\phi \in \mathrm{Irr} G$ and $\phi \mid G(q)$ be irreducible. Does $g \in G(q)$ exist such that the eigenspaces of $\phi(g)$ are just the weight spaces of ϕ?

The following can be viewed as an ingredient of Problem 4:

Problem 5. Let G be a simple $\mathbf{F}_q$-defined algebraic linear group and $\phi \in \Phi_1(G(q))$. Does there exist an $\mathbf{F}_q$-defined 1-dimensional torus T such that the weight spaces of $\phi \mid T$ coincide with the weight spaces of ϕ?

Problem 6. Describe reducible SS-representations of $G(q)$.

We have obtained some results on Problems 4,5,6 in the course of our analysis of Problem 2. One of them is

Proposition 2 *Let $G = SL_n(P), p > 2$. Let ϕ be a direct sum of the irreducible representations of G with the highest weights $0, \omega_1, \ldots, \omega_{n-1}$ and let $\varphi = \phi \otimes Fr\phi \otimes \ldots \otimes Fr^{r-1}\phi$. Then there exists an element $g \in SL_n(q)$ of order $q^{n-1} + \ldots + q + 1$ such that $\varphi(g)$ has a simple spectrum.*

For groups of certain types our proof of Theorem 5 is straightforward: we fix a maximal torus T of G, choose an appropriate element $g \in G(q)$ and find its conjugate g_1 in T; this gives an expression for g_1 as a product of semisimple root elements of T. Then we explicitly find the weights of ϕ and determine the eigenvalues of g by straightforward calculation. Of course, these eigenvalues are described by more or less complicated formulas because q is an arbitrary prime power; for classical groups the parameter n is also arbitrary. In order to prove that the spectrum of $\phi(g)$ is simple we are forced to solve certain rather complicated congruences modulo the order of g. The cases $p = 2, 3$ require a more refined analysis, especially for $p = 2$. Direct calculations are not very effective for groups of types $A_n(q), ^2 A_n(q)$ and $C_n(q)$ since these groups have more weights in Ω and have more complicated combinatorics of eigenvalues than the groups of other types. In particular, we have encountered major difficulties when trying to determine the SS-representations of $A_n(q)$ and $^2 A_n(q)$. Fortunately, we can replace these groups by the extended group $GL_{n+1}(q)$ and $U_{n+1}(q)$. For them there exists another way of determining the eigenvalue multiplicities based on some results on Problem 3 which, in turn, are based on a calculation of spectra for certain complex representations of the group $GL_{n+1}(q)$ and $U_{n+1}(q)$ followed by a reduction to characteristic p.

For the groups $E_6(q), ^2 E_6(q), E_7(q)$ with $p > 2$ we consider restrictions of their representations to appropriate subgroups of classical type and then use some results on Problem 6. Sometimes this leads to rather long calculations.

References.

[1] Bourbaki N. Groupes et algebres de Lie, chap. 4,5,6. Hermann, Paris, 1968.

[2] Hering Ch. Transitive linear groups and linear groups which contain irreducible subgroups of prime order. Geometriae Dedicata (2)1974, 425 - 460.

[3] Hering Ch. Transitive linear groups and linear groups which contain irreducible subgroups of prime order, II. J. Algebra (33)1985, 151 - - 164.

[4] Kantor W.M. Linear groups containing a Singer cycle. J. Algebra (62)1980, 232 - 234.

[5] Seitz G. The root subgroups of maximal tori in finite groups of Lie type. Pacific J. Math. 106(1983), 153 - 244.

[6] Seitz G. The maximal subgroups of classical algebraic groups. Memoirs Amer. Math. Soc. (67)1987, no.365, 286 pp.

[7] Steinberg R. Lectures on Chevalley groups. Yale Univ. 1967.

[8] Suprunenko I.D. and Zalesskii A.E. The representations of dimensions $(p^n \pm 1)/2$ of the symplectic group of degree $2n$ over a field of characteristic p (in Russian). Vesti Acad. Navuk BSSR, ser. fiz.-mat. navuk, 1987, no.6, 9 - 15.

Overgroups of unipotent elements in simple algebraic groups

Donna M. Testerman

In this note, we announce recent progress on the determination of the over-groups of unipotent elements in simple algebraic groups. The proofs of the results stated here will appear in [17]. Our main result is the following

Theorem 1 Let G be a simple algebraic group defined over an algebraically closed field of characteristic p. Assume p is a good prime for G. Let $u \in G$ such that $u^p = 1$. Then u is contained in a subgroup of type A_1 in G.

We point out that Theorem 1 is the group analogue of

Theorem (Jacobson-Morozov [6], Pommerening [11]) Let G be a simple algebraic group defined over an algebraically closed field of characteristic p. Assume p is a good prime for G. Let $\mathcal{G}$ denote the Lie algebra of G and let e be a nilpotent element in $\mathcal{G}$. Then e is contained in an sl_2 subalgebra of $\mathcal{G}$.

Except for the assumption that p is a good prime, Theorem 1 is the best possible group theoretic version of the Jacobson-Morozov theorem; that is, a unipotent element lies in an A_1 exactly when its order allows it to. After introducing some notation, we will state in Theorem 2 a technical "formula", depending upon the class of the element, for the order of a unipotent element in a simple algebraic group. A useful consequence of Theorems 1 and 2 is the following

Corollary Let G and p be as in Theorem 1. Assume in addition that $p > h - 1$, where h is the Coxeter number of G. Then every unipotent element of G lies in a subgroup of type A_1.

In addition to the intrinsic interest of Theorem 1, it sheds light on the prime restrictions in the following theorem of Seitz which appears in his work on the classification of the maximal subgroups of the exceptional algebraic groups.

Theorem (4.2 of [12]): Let G be a simple algebraic group of exceptional type, defined over an algebraically closed field of characteristic p. Assume that $p > 3, 3, 5, 7, 7$ for G of type G_2, F_4, E_6, E_7, E_8, respectively. Suppose $X \leq G$ is a simple closed connected subgroup of type A_1 which is maximal among closed connected subgroups of G. Then X is determined up to conjugacy in $Aut(G)$ by the class of its unipotent elements. The classes which can give rise to such an X are described below; the class is either regular or corresponds to a labelled Dynkin diagram as in the Bala-Carter classification of unipotent conjugacy classes. In each case such an X can exist only under the indicated restriction on p.

(a) $p > 5$, $G = G_2$, regular;

(b) $p > 11$, $G = F_4$, regular;

(c) $p > 17$, $G = E_7$, regular;

(d) $p > 13$, $G = E_7$, $\quad 2 \quad 2 \quad \begin{smallmatrix} 0 \\ 2 \end{smallmatrix} \quad 2 \quad 2 \quad 2$;

(e) $p > 29$, $G = E_8$, regular;

(f) $p > 23$, $G = E_8$, $\quad 2 \quad 2 \quad \begin{smallmatrix} 0 \\ 2 \end{smallmatrix} \quad 2 \quad 2 \quad 2 \quad 2$;

(g) $p > 19$, $G = E_8$, $\quad 2 \quad 2 \quad \begin{smallmatrix} 0 \\ 2 \end{smallmatrix} \quad 2 \quad 0 \quad 2 \quad 2$.

Seitz did not address the existence of A_1's satisfying the conditions of his theorem. However, in [18], we established the existence of a maximal A_1 for each of the unipotent classes indicated in Seitz's theorem, under the restriction on p given in this theorem. While Seitz had arrived at the restrictions on p by studying the extension theory for A_1's and the possible actions of a maximal A_1 on $\mathcal{G}$, the Lie algebra of G, our work led us to conjecture that the restriction on p was precisely the restriction necessary to insure that the unipotent elements in the class corresponding to the A_1 have order p. The truth of this conjecture is a consequence of Theorem 2. We will discuss other possible applications of Theorems 1 and 2 at the end of the paper. We now introduce the notation to be used throughout.

Notation Let G be a simple algebraic group defined over an algebraically closed field of characteristic p. Let T be a maximal torus of G and let Σ be the corresponding root system with base Π and associated positive roots Σ^+. Let α_0 be the highest root in Σ, with respect to Π. For $J \subseteq \Pi$ and for $\beta = \sum_{\alpha \in \Pi} c_\alpha \alpha$, define $\mathrm{ht}_J(\beta) = \sum_{\alpha \notin J} c_\alpha$. Let B be a Borel subgroup corresponding to Π. For a parabolic $P \supseteq B$ corresponding to the subset $J \subseteq \Pi$, define $\mathrm{ht}(P) = \mathrm{ht}_J(\alpha_0)$. Finally, write $\Sigma(J)$ for $(\sum_{\alpha \in J} \mathbf{Z}\alpha) \cap \Sigma$ and $\Sigma^+(J)$ for $\Sigma(J) \cap \Sigma^+$.

We must recall some facts about the classification of the unipotent conjugacy classes in G. (See [4] for more details.) A unipotent element u is said to be distinguished if $C_G(u)^\circ$ contains no nonidentity semisimple element. A parabolic P with Levi decomposition LQ is distinguished if $\dim L = \dim Q/Q'$. The unipotent conjugacy classes in G are described in the following:

Theorem (Bala-Carter [2] and [3], Pommerening [10] and [11]): Let G be a simple algebraic group defined over an algebraically closed field of good characteristic p.

(i) There exists a bijective map between the conjugacy classes of unipotent elements of G and G-classes of pairs $(L, P_{L'})$, where L is a Levi subgroup of G and $P_{L'}$ is a distinguished parabolic of L'. The class corresponding to $(L, P_{L'})$ contains the dense orbit of $P_{L'}$ on $R_u(P_{L'})$.

(ii) There exists a bijective map between conjugacy classes of distinguished unipotent elements of G and conjugacy classes of distinguished parabolic subgroups of G. The class corresponding to a given distinguished parabolic P contains the dense orbit of P on $R_u(P)$.

In view of this classification, it will suffice to establish the existence of A_1's containing distinguished unipotent elements of order p. Moreover, we need only give a "formula" for the order of all distinguished unipotent elements in all simple groups. This follows in

Theorem 2 Let G be a simple algebraic group defined over an algebraically closed field of good characteristic p. Let P be a distinguished parabolic of G. Let $u \in P$ lie in the dense orbit of P on $R_u(P)$. Then

$$o(u) = \min\{p^a | p^a > \mathrm{ht}(P)\}$$

For example, the order of the regular unipotent element is the smallest power of p greater than the height of the highest root.

The proof of Theorem 2 is straightforward, but tedious. For the exceptional groups, we have explicit information about representatives of the distinguished unipotent conjugacy classes. In particular, one of three situations occurs. Either the unipotent element lies in a maximal rank subgroup and we may work in the smaller group in order to calculate the order of the element, or the element is regular and is conjugate to the product, in any order, of $|\Pi|$ nonidentity root elements, one from each of the root groups corresponding to a simple root, or the group has type E_n and and the element is conjugate to a known one of the elements explicitly described in [9] as a product of root elements. In the last two cases, we calculate the action of these root elements on $\mathcal{G}$ and thereby associate to each root element a $\dim \mathcal{G} \times \dim \mathcal{G}$ matrix with integer entries. We then use a computer to form the appropriate product of root elements and to determine the order of this matrix, modulo p, for all primes p less than $\dim \mathcal{G}$. (If $p \geq \dim \mathcal{G}$, then every unipotent element in $SL(\mathcal{G})$ has order p.) This gives the result for the exceptional groups. For the classical groups, we use the connection between the Jordan form of a distinguished unipotent element u and the corresponding parabolic P containing u in the dense orbit of P on $R_u(P)$, as given in [14], for example. Of course, given the Jordan form of a unipotent element, it is easy to determine its order (and to see that it lies in an A_1 when it has order p). Thus, we will restrict our attention now to the proof of Theorem 1 for distinguished elements in exceptional groups.

In proving Theorem 1, we actually construct, under suitable restrictions on p, a $(P)SL_2(F)$ in $G(F)$, the adjoint Chevalley group of type Σ over F, for F an arbitrary field of characteristic p. Our proof closely resembles the construction of the adjoint type Chevalley groups. In particular, we start with $\mathcal{L}$, a semisimple Lie algebra over $\mathbf{C}$ with root system of type Σ and Chevalley basis $\mathcal{B} = \{e_\alpha, f_\alpha, h_\gamma | \alpha \in \Sigma^+, \gamma \in \Pi\}$. Here e_α and f_α denote the root vectors corresponding to the roots α and $-\alpha$, respectively, and $h_\gamma = [e_\gamma, f_\gamma]$. Let $\mathcal{L}(\mathbf{Z})$ be the $\mathbf{Z}$ span of the Chevalley basis and let $\mathcal{L}(\mathbf{Z}_{(p)}) = \mathcal{L}(\mathbf{Z}) \otimes_\mathbf{Z} \mathbf{Z}_{(p)}$, a Lie algebra over $\mathbf{Z}_{(p)}$ (the localization of $\mathbf{Z}$ at the prime ideal $p\mathbf{Z}$. Now we form the F Lie algebra, $\mathcal{L}(F) = \mathcal{L}(\mathbf{Z}_{(p)}) \otimes_{\mathbf{Z}_{(p)}} F$, with basis $\mathcal{B}' = \{e_\alpha \otimes 1, f_\alpha \otimes 1, h_\gamma \otimes 1 | \alpha \in \Sigma^+, \gamma \in \Pi\}$.

Let $e \in \mathcal{L}(\mathbf{Z})$ such that $(\operatorname{ad} e)^k/k!$ preserves $\mathcal{L}(\mathbf{Z}_{(p)})$ for all $k \geq 0$. We will associate to e and to each $t \in F$ an automorphism of $\mathcal{L}(F)$. For $\lambda \in \mathbf{C}$, let $A(\lambda)$ be the matrix representing $\exp(\operatorname{ad} \lambda e)$ with respect to the basis $\mathcal{B}$; then the entries of $A(\lambda)$ lie in $\mathbf{Z}_{(p)}[\lambda]$. Now for $t \in F$, form the matrix $\bar{A}(t)$ by applying the natural ring homomorphism $\mathbf{Z}_{(p)}[x] \to F[x]$ to the entries of $A(x)$ and then evaluating at t. Since $\exp(\operatorname{ad} \lambda e)$ is an automorphism of $\mathcal{L}(\mathbf{Z}_{(p)})$, $\bar{A}(t)$ represents an automorphism of $\mathcal{L}(F)$ with respect to the basis $\mathcal{B}'$. Call this automorphism $x(t)$. Also, if we start with $e, f \in \mathcal{L}(\mathbf{Z})$ such that $\langle e, f \rangle$ is isomorphic to an $sl_2(\mathbf{C})$ subalgebra of $\mathcal{L}$ and such that $(\operatorname{ad} e)^k/k!$ and $(\operatorname{ad} f)^k/k!$ preserve $\mathcal{L}(\mathbf{Z}_{(p)})$ for all $k \geq 0$, then forming $x(t)$, respectively $y(t)$, corresponding to e, respectively f, we obtain $\langle x(t), y(t) | t \in F \rangle \simeq (P)SL_2(F) \leq Aut(\mathcal{L}(F))$. Moreover, p good implies that $x(t), y(t) \in G(F) \subseteq Aut(\mathcal{L}(F))$.

Up to this point, we have introduced no new ideas; indeed, we are simply following the construction of adjoint type Chevalley groups (as in [5] or [15], for example), replacing the $\mathbf{Z}$ Lie algebra $\mathcal{L}(\mathbf{Z})$ with the $\mathbf{Z}_{(p)}$ Lie algebra $\mathcal{L}(\mathbf{Z}_{(p)})$. The question we must address is: how can we insure that $(\operatorname{ad} e)^k/k!$ preserves $\mathcal{L}(\mathbf{Z}_{(p)})$ for primes p for which $(\operatorname{ad} e)^p$, viewed as a linear transformation of $\mathcal{L}(\mathbf{Z}_{(p)})$, is nonzero? For this, we use the following technical result.

Proposition 1 (i) Let $J \subseteq \Pi$ such that $e = \sum_{\alpha \in \Sigma^+ - \Sigma(J)} c_\alpha e_\alpha, c_\alpha \in \mathbf{Z}$ and such that $p > \operatorname{ht}_J(\alpha_o)$, Assume as well that $e = x + y$, $x = \sum_{\alpha \in \Sigma^+ - \Sigma(J')} c_\alpha e_\alpha$ and $y = \sum_{\alpha \in \Sigma^+ - \Sigma(J'')} c_\alpha e_\alpha$, for some $J', J'' \subseteq \Pi$ with

$p > 2\,\mathrm{ht}_{J'}(\alpha_o)$ and $p > 2\,\mathrm{ht}_{J''}(\alpha_o)$. Then $(\mathrm{ad}\,e)^k/k!$ preserves $\mathcal{L}(\mathbf{Z}_{(p)})$ for all $k \geq 0$.

(ii) In particular, if $n = \sum_{\alpha \in \Sigma^+} b_\alpha e_\alpha, b_\alpha \in \mathbf{Z}$ and $p > \mathrm{ht}(\alpha_o)$ then $(\mathrm{ad}\,n)^k/k!$ preserves $\mathcal{L}(\mathbf{Z}_{(p)})$ for all $k \geq 0$.

Heuristics of proof: In the universal enveloping algebra of a restricted Lie algebra over a field of characteristic p, we have the following identity:

$$(*) \quad (a+b)^p = a^p + b^p + \sum_{i=1}^{p-1} s_i(a,b),$$

where $is_i(a,b)$ is the coefficient of λ^{i-1} in $(\mathrm{ad}(\lambda a + b))^{p-1}(a)$, for an indeterminate λ. (See Chapter V of [7].) Now if $a, b \in \langle e_\alpha \,|\, \mathrm{ht}_J(\alpha) \geq 1\rangle$ and if $p > \mathrm{ht}_J(\alpha_o)$, then $(\mathrm{ad}(\lambda a + b))^{p-1}(a) = 0$. So

$$(**) \quad \text{if} \quad p > \mathrm{ht}\,_J(\alpha_o), \quad \text{then} \quad (a+b)^p = a^p + b^p.$$

Modification of the proofs of identities $(*)$ and $(**)$ produce similar identities which hold in a $\mathbf{Z}_{(p)}$ lattice in the universal enveloping algebra of $\mathcal{L}$ (a characteristic 0 Lie algebra). Careful use of these identities yields the Proposition.

At this point, we need only construct A_1's for regular unipotent elements and for specific distinguished unipotent elements listed in [9]. Each of those elements corresponds to some subset J of Π, via the Bala-Carter classification. Proposition 1 is applied to exponentiate ad-nilpotent elements in $\mathcal{L}(\mathbf{Z})$ to obtain unipotent elements in $Aut(\mathcal{L}(F))$. In each case, the condition $p > \mathrm{ht}_J(\alpha_o)$ implies the existence of subsets J' and J'' satisfying the conditions of the proposition. Moreover, if $F = \bar{F}$, an algebraically closed field of characteristic p, we may use the following results to predetermine the class of the unipotent element by exponentiating a particular ad-nilpotent element. Let $\{x_\alpha(t) \,|\, t \in \bar{F}\}$ denote the T- root group of G corresponding to the root α.

Proposition 2: Let $J \subseteq \Pi$, and let $e = \sum b_\alpha e_\alpha, b_\alpha \in \mathbf{Z}$. Assume $(\mathrm{ad}\,e)^k/k!$ preserves $\mathcal{L}(\mathbf{Z}_{(p)})$ for all $k \geq 0$. Form $x(1) = \exp(\mathrm{ad}\,e) \in Aut(\mathcal{L}(\bar{F}))$ as described above. Then $x(1) = \prod_{\alpha \in \Sigma^+ - \Sigma(J)} x_\alpha(d_\alpha)$ and $d_\alpha = b_\alpha$ for all α such that $\mathrm{ht}_J(\alpha) = 1$.

Lemma (5.8.5 of [4]): Let $P \supseteq B$ be a distinguished parabolic of G corresponding to $J \subseteq \Pi$ and let $x = \prod_{\alpha \in \Sigma^+ - \Sigma(J)} x_\alpha(d_\alpha)$ lie in the dense orbit of P on $R_u(P)$. Then the $R_u(P)$ orbit containing x consists of $\{\prod_{\alpha \in \Sigma^+ - \Sigma(J)} x_\alpha(a_\alpha) \,|\, a_\alpha = d_\alpha$ for all α with $\mathrm{ht}_J(\alpha) = 1\}$.

Given a distinguished unipotent element u we may choose $e \in \mathcal{L}(\mathbf{Z})$ so that $\exp(\mathrm{ad}\, e)$ is conjugate to u. In each case, we can find $f \in \mathcal{L}(\mathbf{Z})$ such that $\langle e, f \rangle \simeq sl_2(\mathbf{C})$ and such that $\mathrm{ad}\, f$ can also be exponentiated. In this way, we obtain the desired A_1.

Concluding remarks: It is natural to phrase the original question for finite groups of Lie type as well as for the algebraic groups. It is clear from our discussion of the proof of Theorem 1 that $(P)SL_2$'s corresponding to certain unipotent elements can be constructed in the non-twisted groups. It remains to be determined if this information will yield a complete finite group version of Theorem 1.

It should also be possible to settle the question of what happens if p is a bad prime. Certainly, we can use the explicit information about the conjugacy classes in the exceptional groups (given in [9], [13] and [16]) to compute the orders of the unipotent elements. (There are a few additional classes of elements in certain bad characteristics.) For the special case of characteristic 2, one might hope to extend the arguments used in [1] to establish the existence of A_1's containing involutions.

Other related questions seem answerable in view of the results obtained thus far. For example, what are the semisimple overgroups of unipotent elements, in general? In particular, what happens when a unipotent element does not have order p? Some information is readily available: the regular unipotent element in E_6 lies in F_4 for all primes p; the regular unipotent element in $A_{2\ell}$ lies in B_ℓ for all p. A more interesting example is the following: the regular unipotent element of A_5 (which has order the smallest power of p greater than 5) lies in a subgroup of type G_2 only when $p = 2$. A complete classification of the overgroups of the regular unipotent elements will probably follow easily from Seitz's work on the maximal subgroups and from knowing the orders of the elements.

Finally, one would hope to obtain some information about the conjugacy classes of A_1's containing a given unipotent element. In particular, can we characterize in some way the A_1's that we are obtaining, perhaps via their Lie algebras? Until some result along these lines is available, the usefulness of Theorem 1 will be significantly more restricted than that of the Jacobson-Morozov theorem.

REFERENCES

1. M. Aschbacher and G.M. Seitz, Involutions in Chevalley groups over fields of even order, *Nagoya Math. J.*, **63** (1976), 1–91.

2. P. Bala and R.W. Carter, Classes of unipotent elements in simple algebraic groups I, *Math. Proc. Cambridge Phil. Soc.*, **79** (1976), 401–425.

3. P. Bala and R.W. Carter, Classes of unipotent elements in simple algebraic groups II, *Math. Proc. Cambridge Phil. Soc.*, **80** (1976), 1–17.

4. R.W. Carter, *Finite Groups of Lie Type: Conjugacy Classes and Complex Characters*, John Wiley and Sons, Chichester (1985).

5. R.W. Carter, *Simple Groups of Lie Type*, John Wiley and Sons, London (1972).

6. N. Jacobson, A note on three dimensional simple Lie algebras, *J. Math. Mech.*, **7** (1958), 823–831.

7. N. Jacobson, *Lie Algebras*, Interscience Publishers, New York (1962).

8. K. Mizuno, The conjugate classes of Chevalley groups of type E_6, *J. Fac. Sci. Univ. Tokyo*, **24** (1977), 525–563.

9. K. Mizuno, The conjugate classes of the Chevalley groups of type E_7 and E_8, *Tokyo J. Math.*, **3** (1980), 391–461.

10. K. Pommerening, Über die unipotenten Klassen reduktiver Gruppen, *J. Algebra*, **49** (1977), 525–536.

11. K. Pommerening, Über die unipotenten Klassen reduktiver Gruppen II, *J. Algebra*, **65** (1980), 373–398.

12. G.M. Seitz, Maximal subgroups of exceptional algebraic groups, *Memoirs Amer. Math. Soc.*, to appear.

13. K. Shinoda, The conjugacy classes of Chevalley groups of type (F_4) over finite fields of characteristic 2, *J. Fac. Sci. Univ. Tokyo*, **21** (1974), 133–159.

14. T. Springer and R. Steinberg, Conjugacy classes, *Lecture Notes in Mathematics*, **131**, Springer-Verlag, Berlin/New York (1970).

15. R. Steinberg, *Lectures on Chevalley Groups*, notes by J. Faulkner and R. Wilson, Yale University (1968).

16. U. Stuhler, Unipotente und nilpotente Klassen in einfachen Gruppen und Lie algebren von Typ G_2, *Indag. Math.*, **33** (1971), 365–378.

17. D.M. Testerman, Overgroups of unipotent elements in simple groups of Lie type, finite and algebraic, in preparation.

18. D.M. Testerman, The construction of the maximal A_1's in the exceptional algebraic groups, submitted to *Proc. Amer. Math. Soc.*

Some open problems on permutation groups

PETER J. CAMERON

In the last decade, very many problems about permutation groups have been resolved using the classification of finite simple groups. In this article, I describe some problems which are still open, and some fields of research which are still relatively untilled. The problems concern transitivity, order, minimal degree and base size, fixed-point-free elements, reconstruction, etc.

1. INTRODUCTION

Even before the classification of finite simple groups was completed in 1980, it was clear that its impact on permutation group theory would be very dramatic. By an accident of fate, I happened to be considering these matters at the time. (My thoughts then are contained in my survey paper (1981).) The succeeding decade saw this impact working through the subject, with many expected and unexpected conclusions.

Among the consequences are (in no particular order):

(i) the determination of the 2-transitive groups, and of the rank 3 groups;

(ii) the determination of primitive groups of odd degree (except for affine groups, see later);

(iii) primitive groups of degree n are small (order at most $n^{c \log \log n}$) with "known" exceptions;

(iv) a transitive permutation group of degree greater than 1 contains a fixed-point-free element of prime-power order.

In connection with the last fact, the existence of a fixed-point-free element is trivial, and was known to Frobenius; but that it can be taken to have prime-power order requires the full force of the classification. The result has an application in number theory (Fein, Kantor and Schacher 1981). However, a closely related problem, raised much earlier by Isbell (1960) in the context of game theory, is still unsolved, and shows the limits of our knowledge. If the 2-part of n is sufficiently large compared to the odd part, can the element be taken to have 2-power order?

In Section 2, I will give a brief sketch of the O'Nan-Scott theorem, which enables the classification to be brought to bear on questions about primitive groups. The discussion highlights the rôle of recent advances in our knowledge of subgroup structure and modular representations of groups which are close to being simple, especially those of Lie type. In subsequent sections, I give a number of problems which are still open. For some of these, our knowledge of simple groups doesn't seem to help very much!

2. THE O'NAN-SCOTT METHODOLOGY

The O'Nan-Scott theorem (Scott 1980) provides a reduction method for proving theorems about primitive permutation groups. It is analogous to the familiar reductions from arbitrary permutation groups to transitive ones, and from transitive groups to primitive ones.

A primitive permutation group which can be "reduced" is embedded, as a "large" subgroup, in a wreath product $H \operatorname{Wr} K$ with product action. (This is the action on the set of functions from Δ to Γ, where H and K act on Γ and Δ respectively.) The bottom group H is necessarily primitive, and can be taken to be "irreducible" in this sense, i.e., no further reduction can be applied to it. The top group K is only required to be transitive, so that its analysis is more complicated: in general, we have to find its primitive components and apply the O'Nan-Scott reduction to each of them. Fortunately, it turns out in practice that many questions can be decided on the basis of detailed information about H alone.

The O'Nan-Scott theorem describes groups which cannot be "reduced" in the above sense. They are of four types:
  (i) groups of diagonal type;
  (ii) twisted wreath products;
  (iii) almost simple groups;
  (iv) affine groups.
The first two types are very specific, and usually are fairly easy to handle. (See one of the standard expositions, e.g., Liebeck, Praeger & Saxl (1988), for precise definitions.) G is *almost simple* if $N \trianglelefteq G \leq \operatorname{Aut}(N)$ for some simple group N; all we know is that the stabiliser G_α is a maximal subgroup of G. An *affine* group G has a regular normal subgroup which is the translation group of a finite vector space V; the stabiliser G_0 of the zero vector is an irreducible linear group (indeed, a primitive linear group, if G is not reducible in the sense described above). In this case, Aschbacher's reduction theorem (1984) (an analogue for classical groups of the O'Nan-Scott theorem for the symmetric group) shows that either G_0 (and hence G) is known, or $F^*(G_0)$ is quasisimple (and absolutely irreducible).

So we are led to the study of the maximal subgroups and the irreducible modular representations of groups which are close to simple groups. Of course, by the classification, we know that groups of Lie type make up the majority of these. Both topics have been covered extensively at the Symposium (for example, in the lecture series by Roger Carter and Gary Seitz). The developments in these areas over the past decade have contributed much to advancing our knowledge of permutation groups. It is no coincidence that the organisers of the Symposium have been prominent in the study of both permutation groups and subgroup structure of groups of Lie type!

A source of difficulty, which we will encounter later on, is that a question may be formulated for transitive groups; we may have the tools to settle it for primitive groups, but be unable to reduce to the primitive case. An extreme example of this concerns questions about p-groups, where the only primitive groups are cyclic of prime order.

I conclude with a problem concerned with the O'Nan-Scott reduction.

Problem 2.1. How difficult is it to decide, for a given primitive group G,

(i) whether $G \leq H \operatorname{Wr} K$ with product action (and, if so, to construct H and K), and

(ii) if not, to which of the types in the theorem G belongs?

(For example, can these questions be decided in polynomial time? By comparison, it is known that we can test a transitive group for primitivity, and if imprimitive, find a minimal block, in polynomial time (Atkinson 1975).)

3. ORDER, MINIMAL DEGREE AND BASE SIZE

Apart from the classification of multiply transitive groups, one of the first applications of the classification was to show that primitive groups are very small, with "known" exceptions. Order is closely connected with base size and minimal degree, though these have been less intensively studied.

A *base* for a permutation group G is a sequence of points whose pointwise stabiliser is the identity. It is *irredundant* if no point is fixed by the pointwise stabiliser of its predecessors. Bases arose in the algoritms for computing with permutation groups developed by Charles Sims. Since an arbitrary group element is completely determined by its effect on the points of a base, it can be stored as a b-tuple (where b is the base size) instead of an n-tuple.

Proposition 3.1.   If G has degree n and has an irredundant base of size b, then $\log |G| / \log n \leq b \leq \log |G| / \log 2$.

(Indeed, the lower bound holds for any base at all.) As with any inequality, this raises the question: which groups have bases at or near one of the bounds? (Clearly, the lower bound is met only by regular permutation groups, and the upper only by groups with all orbits of length 2.)

The *minimal degree* of G is the smallest number of points moved by a non-identity element of G. It is a classical invariant, whose origins are suggested by the simple observation that a primitive group containing a transposition or a 3-cycle must be the symmetric or alternating group. The fact that the minimal degree of a primitive group of degree n other than S_n or A_n tends to infinity with n was proved last century by Jordan. Recent developments, and applications (for example, to the study of primitive permutation groups which are Galois groups of extensions of $\mathbb{Q}(t)$ of given genus), have been discussed at the Symposium, for example in Bob Guralnick's talk.

It is straightforward to show the following, using the fact that any base meets the support of any non-identity element:

Proposition 3.2.   A transitive group of degree n with minimal degree m and minimal base size b satisfies $bm \geq n$.

Again, we can ask when equality holds. Known bounds for the order of primitive groups yield upper bounds for base size and lower bounds for minimal degree; these bounds have been improved in some cases.

It is of some interest in computational permutation group theory to find quickly a small base for a group. The *greedy algorithm* chooses each base point in an orbit of maximum size of the stabiliser of its predecessors. Blaha (to appear) showed the following:

Theorem 3.3.   In a permutation group of degree n having minimum base size b,

(i) any irredundant base has size at most $b \log n$;
(ii) any base found by the greedy algorithm has size at most $2b \log \log n$;
(iii) these bounds are essentially best possible in general.

In spite of (iii) above, there is some evidence that more is true for primitive groups. For a start, almost simple groups acting primitively should have base size bounded by an absolute constant, with "known" exceptions. Specifically, I conjecture the following:

Conjecture 3.4. There is a constant c such that, if G is a primitive group and $N \trianglelefteq G \leq \mathrm{Aut}(N)$ for some simple group N, then one of the following holds:
    (i) G has base size at most c;
    (ii) G is a symmetric or alternating group, acting on an orbit of subsets or of partitions of the natural G-set;
    (iii) G is a classical group, acting on an orbit of subspaces of its natural module.

Also, the following problem on primitive groups is open:

Conjecture 3.5. There is an absolute constant c such that, if a primitive group has minimal base size b, then any base found by the greedy algorithm has size at most cb.

The extremal case here should be the symmetric group S_n acting on 2-sets. A base is a graph on n vertices for which no non-trivial permutation fixes all the edges. The smallest base is made up of disjoint paths of length 2, and has $\frac{2}{3}n$ edges (in the case where 3 divides n; there are small "end effects" otherwise). It is instructive to track the greedy algorithm. It first covers all but at most seven points by disjoint edges. Then the remaining pairs contained in the union of these edges all lie in orbits of length 4; the greedy algorithm goes back and chooses extra edges of this form so that all but a few points lie in disjoint paths of length 3. The final base has size about $\frac{3}{4}n$.

If this really is the extremal case, then the constant would be $\frac{9}{8}$.

One could also ask:

Problem 3.6. Is there a general class of groups in which the greedy algorithm guarantees to find a base of smallest size?

Another interesting question is: which groups permute their irredundant bases transitively? If this holds, then the irredundant bases are the bases of a matroid (and G is a Jordan group of automorphisms of this matroid). Hence, if G is primitive, then it is doubly transitive. Groups satisfying this condition can be determined, using a (pre-classification) theorem of Kantor (1975). Another special case which can be handled by "elementary" means is that where the rank of the matroid (the number of elements in a basis) is at least 7 (Zil'ber, personal communication). However, the complete determination of such groups of rank greater than 1 (due to Tracey Maund (1989)) does use the classification.

This suggests a further problem:

Problem 3.7. For which groups do the irredundant bases (or perhaps just those of minimal size) form the bases of a matroid?

Another question which has received some attention is: what is the largest permutation group with given degree and base size? Let us consider just the smallest non-trivial case, namely base size 2. It is reasonable to restrict attention to transitive groups. (If an orbit contains a base, then the action of the group on that orbit is faithful. The alternative is that the two base points are chosen from different orbits. No example is known of a value of n for which the largest group with degree n and base size 2 is of this kind, though no good reason why this can not happen has been found.)

If G is transitive, then the existence of a base of size 2 is equivalent to that of a non-trivial regular (and faithful) orbit of a point stabiliser. Let $f(n)$ be the largest size of such a suborbit in a transitive group of degree n. The function f has been studied by Babai, Cameron, Deza & Singhi (1981). In all known cases, $f(n) \geq \frac{1}{2}n$, although it is thought that this is not always the case. However, the investigation has concentrated on imprimitive groups. The answer to the following question is not known:

Problem 3.8. Let G be primitive of degree n, and suppose that G_α acts regularly on a suborbit of length at least $\frac{1}{2}n$. Show that either G is sharply 2-transitive, or G is A_5 (degree 10).

4. TRANSITIVITY

Since the classification, we know that the only 6-transitive groups are the symmetric and alternating groups. More generally, we have a complete list of 2-transitive groups, and a list of primitive groups of rank 3. Rather than just extend these results to groups of higher and higher rank, we can pose a general problem:

Problem 4.1. Show that there is a function f, tending to infinity with n, such that primitive groups of degree n having rank less than $f(n)$ is explicitly known.

The rank of a transitive group is just the number of orbits on ordered pairs. Kantor, Liebeck and Macpherson (1989) showed a similar result for the number of orbits on 5-tuples, and remarked that 5 can certainly be improved by their techniques. Their result was proved for an application to model theory.

Another property which includes both 2-transitivity and rank 3 is that of having multiplicity-free permutation character. (The automorphism group of a distance-transitive graph also has this property.) Recent progress on this was reported at the Symposium by Baddeley and Lawther.

Cheryl Praeger raised the question whether, with known exceptions, primitive groups are k-closed for all sufficiently large k. (A permutation group G on Ω is k-*closed* if any permutation of Ω which preserves every G-orbit on Ω^k belongs to G.) This is related to an earlier problem, since a group with base size b is k-closed for all $k \geq b$.

In all these problems, the O'Nan-Scott methodology provides a siege-engine to break down the initial defences. Saying this is not to belittle the ingenuity required to solve the problems, but to contrast them especially with those in the next section.

5. FIXED-POINT-FREE ELEMENTS

It is a triviality that a transitive permutation group (of degree greater than 1) contains a fixed-point-free permutation. In abstract terms, a proper subgroup of a finite group is disjoint from some conjugacy class. This fact was known to Frobenius, as a consequence of the fact that the average number of fixed points of elements in a transitive group is 1.

In connection with a question on relative Brauer groups of global field extensions, Fein, Kantor and Schacher (1981) showed that there is a fixed-point-free element of prime power order. The proof is much less elementary. There is a straightforward reduction to the case where the group is primitive and simple; then the classification of simple groups is invoked, and the groups treated on a case-by-case basis. Surely this result deserves a more elementary proof!

What control do we have about which prime power occurs? Much earlier, Isbell (1960) had made the following conjecture:

Conjecture 5.1. There is a function f such that, if $n = 2^a.b$ with b odd and $a \geq f(b)$, then a transitive group of degree n must contain a fixed-point-free 2-element.

(The proof he gave in a later paper is not correct, and the problem is still open.) Isbell formulated the problem in terms of game theory (in the sense of von Neumann and Morgenstern); there is a purely combinatorial form, as follows. There is a family of pairwise intersecting subsets of an n-set, having maximum size (i.e. 2^{n-1}) and admitting the group G, if and only if G contains no fixed-point-free 2-element. Isbell's conjecture is that this cannot occur for transitive G if the 2-part of n is sufficiently large compared to the odd part.

The truth of Isbell's conjecture would follow from a stronger conjecture:

Conjecture 5.2. There is a function g such that, if a 2-group of permutations has b orbits, each of size at least $2^{g(b)}$, then it contains a fixed-point-free element.

A transitive 2-group contains a fixed-point-free (fpf) involution. Also, strictly more than half of the elements of a transitive 2-group are fpf; so a 2-group with two non-trivial orbits contains a fpf element. Other than this, all that is known is that a 2-group with 3 orbits each of size at least 4 contains a fpf element (Cameron, Kovács, Newman & Praeger (1985)). (This is best possible: consider a dihedral group, acting on vertices, edges, and orientations of a polygon.) So Isbell's function is defined for $b = 3$, with $f(3) = 2$.

The reason that this problem becomes difficult (and that Isbell's approach fails) is that there are transitive 2-groups of arbitrarily large degree in which fewer than two-thirds of the elements are fpf (Cameron *et al*). The obvious strategy for the problem is induction. The proportion of fixed-point-free elements does not increase when we replace the given action of G by the action on a system of blocks of imprimitivity (for example, the orbits of a central involution), so the inductive step is trivial. But starting the induction is not so easy!

Failing this, we can revert to the more traditional approach. Liam Halpenny (to appear) has worked on the primitive case. But even having a proof for primitive groups, we would have little idea about how to deal with arbitrary transitive groups.

To return to the combinatorial problem:

Problem 5.3. Suppose that every transitive group of degree n contains a fixed-point-free 2-element. By how much must the size of a transitive intersecting family of subsets of an n-set fall short of 2^{n-1}?

This question is settled for n a power of 2 and for $n = 12$ by Cameron, Frankl & Kantor (1989). This paper also contains constructions of transitive groups in which all 2-elements have fixed points, and includes a proof that groups with this property exist for almost all degrees.

Of course, both Conjectures 5.1 and 5.2 extend immediately to primes other than 2. Both are open for all primes; less is known in general.

Another problem on fpf elements in p-groups was raised by Bob Odoni (personal communication), again in a number-theoretic context.

Problem 5.4. Given two transitive actions of a p-group such that the same elements are fpf in both actions, must the permutation characters be equal? In other words, is the number of fixed points of any element the same in both actions?

This is false for general groups, or for intransitive actions.

Bob Guralnick raised a related question.

Problem 5.5. Let H, K be subgroups of the p-group P; let π_H and π_K be the permutation characters of P on the cosets of H and K respectively. Suppose that π_H is a summand of π_K. Is it true that K is contained in a subgroup K^* with $\pi_{K^*} = \pi_H$? Dually, does H contain a subgroup H^* with $\pi_{H^*} = \pi_K$?

I conclude this section with a problem which grew from the above analysis, but has some independent interest. Let G be a group, $\mathbb{F}$ a field, and V an $\mathbb{F}G$-module. The *covering depth* of V is the maximum codimension of an $\mathbb{F}$-subspace W of V for which the union of the images of W under G cover V. The *covering depth* of $\mathbb{F}G$ is just the maximum covering depth of any $\mathbb{F}G$-module. There is also a relativised version, where a subgroup H of G is given and W is required to be a $\mathbb{F}H$-submodule.

The relevance to what went before is simply that, if $\mathbb{F} = \mathrm{GF}(2)$ and G has odd order, then the split extension $V.G$, acting on the cosets of W (or $W.H$ in the relativised case), contains no fixed-point-free 2-elements.

Problem 5.6. Investigate covering depth!

An interesting extension would replace the module V by an arbitrary group (not necessarily abelian) on which G acts.

6. RECONSTRUCTION

The celebrated "reconstruction conjecture" for graphs comes in two forms, for vertices and for edges. The remarks in this section refer to the edge-reconstruction conjecture. This asserts that a finite graph with more than three edges is determined up to isomorphism by the multiset of isomorphism classes of edge-deleted subgraphs. That is, from a deck of cards on which pictures of the edge-deleted subgraphs have been drawn, it is possible to "reconstruct" the original graph. (A more formal statement, in a more general context, will be given shortly.)

Graphs with at most three edges are not necessarily edge-reconstructible (see Fig. 1).

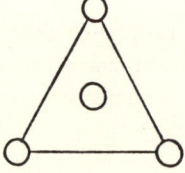 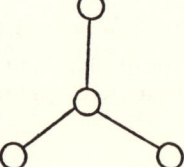

Fig. 1.

The best result on the edge-reconstruction conjecture for graphs is a theorem of Müller, extending an idea of Lovász: a graph on n vertices with more than $n \log_2 n$ edges can be reconstructed from the list of its edge-deleted subgraphs. (For an account of this, see the chapter on reconstruction in Beineke and Wilson (1979).) Lovász pointed out to me that the entire argument can be formulated in terms of permutation groups, as follows.

Let G be a permutation group on a set Ω. An *isomorphism* between finite subsets Δ_1, Δ_2 of Ω is a bijection which is induced by an element of G; two subsets are *isomorphic* if there is an isomorphism between them. A *hypomorphism* between Δ_1 and Δ_2 is a bijection h such that, for all $\alpha \in \Delta_1$, $\Delta_1 \setminus \{\alpha\}$ and $\Delta_2 \setminus \{\alpha h\}$ are isomorphic; two subsets are *hypomorphic* if there is a hypomorphism between them. (It is *not* required that the restriction of h to $\Delta_1 \setminus \{\alpha\}$ is an isomorphism.) A set Δ is *reconstructible* if every set hypomorphic to it is isomorphic to it; the *reconstruction number* of G is the least d such that every set of size d or greater is reconstructible.

For example, let G be the symmetric group S_n, and Ω the set of 2-element subsets of $\{1, \ldots, n\}$. A subset of Ω is the edge set of a graph on n vertices; isomorphism in the sense defined agrees with the usual concept (isomorphisms between graphs are induced by permutations of the vertex set), and hypomorphism agrees with the concept used in work on the reconstruction conjecture. For example, the two graphs of Fig. 1 are hypomorphic but not isomorphic; the edge-reconstruction conjecture asserts that, for $n \geq 4$, the reconstruction number of S_n (acting on 2-sets) is 4.

It is not obvious that a permutation group has a reconstruction number. But Lovász's observation is that the arguments of Lovász and Müller immediately give the following theorem:

Theorem 6.1 The reconstruction number of a permutation group G does not exceed $\log_2 |G|$.

Problem 6.2. Investigate reconstruction number. How does it relate to more familiar concepts of permutation group theory?

As a preparatory exercise for anyone considering this problem, I recommend proving Theorem 6.1 using the arguments of Lóvasz and Müller, following the reference cited.

It is not true that there is an absolute bound for the reconstruction number of any permutation group. For example, let $G = \mathrm{PSL}(n, 2)$, acting on the points of the projective space. By Theorem 6.1, the reconstruction number of G is at most n^2. But a set of n independent points, and a set of n points whose sum is zero but which satisfy no other linear relation, are hypomorphic; so the reconstruction number is greater than n. What is the true value?

REFERENCES

Atkinson, M. D. (1975), An algorithm for finding the blocks of a permutation group, *Math. Comput.* **29**, 911–913.

Babai, L., Cameron, P. J., Deza, M., & Singhi, N. M. (1981), On sharply edge-transitive permutation groups, *J. Algebra* **73**, 573–585.

Beineke, L. W. & Wilson, R. J. (1979), *Selected Topics in Graph Theory*, Academic Press, London.

Blaha, K. (to appear).

Cameron, P. J. (1981), Finite permutation groups and finite simple groups, *Bull. London Math. Soc.* **13**, 1–22.

Cameron, P. J., Frankl, P. & Kantor, W. M. (1989), Intersecting families of finite sets and fixed-point-free 2-elements, *Europ. J. Combinatorics* **10**, 149–160.

Cameron, P. J., Kovács, L. G., Newman, M. F. & Praeger, C. E. (1985), Fixed-point-free permutations in transitive permutation groups of prime-power order, *Quart. J. Math. Oxford* (2) **36**, 273–278.

Fein, B., Kantor, W. M. & Schacher, M. (1981), Relative Brauer groups II, *J. Reine Angew. Math.* **328**, 39–57.

Halpenny, L. (to appear).

Isbell, J. (1960), Homogeneous games II, *Proc. Amer. Math. Soc.* **11**, 159–161.

Kantor, W. M. (1975), 2-transitive designs, pp. 365–418 in *Combinatorics* (ed. Hall, M. Jr & van Lint, J. H.), D. Reidel, Dordrecht.

Kantor, W. M., Liebeck, M. W. & Macpherson, H. D. (1989), $\aleph_0$-categorical structures smoothly approximated by finite substructures, *Proc. London Math. Soc.* (3) **59**, 439–463.

Liebeck, M. W., Praeger, C. E., & Saxl, J. (1988), On the O'Nan–Scott theorem for finite primitive permutation groups, *J. Austral. Math. Soc. (Ser. A)* **44**, 389–396.

Maund, T. (1989), Thesis, Oxford University.

Scott, L. L. (1980), Representations in characteristic p, pp. 319–331 in *The Santa Cruz Conference on Finite Groups* (ed. B. Cooperstein & G. Mason), *Proc. Symp. Pure Math.* **37**, Amer. Math. Soc., Providence, R. I.

The Genus of a Permutation Group

Robert M. Guralnick*

1. Introduction.

If ϕ is a nonconstant analytic function between compact connected Riemann surfaces X and Y, we denote by $Mon(\phi) = Mon(X, Y, \phi)$ the monodromy group of the corresponding cover. Note that $Mon(\phi)$ is the Galois group of $\mathbf{C}(Z)/\mathbf{C}(Y)$ where Z is the Galois closure of the cover. Set

$$S(X) = \cup_{\phi, Y} cf(Mon(X, Y, \phi))$$

where $cf(G)$ denotes the set of composition factors of a finite group G. It is well known that for all primes p, C_p (the cyclic group of order p) and for all $n \geq 5$, A_n (the alternating group of degree n) are in $S(X)$ for all X. Thus we define $S^*(X)$ to be those nonabelian groups in $S(X)$ except for alternating groups. If $\mathcal{M}$ is a family of compact connected Riemann surfaces, let $S^*(\mathcal{M})$ denote the union of $S^*(X)$ over X in $\mathcal{M}$. If g is a nonnegative integer, let $\mathcal{M}_g$ denote the moduli space of all genus g surfaces. Then set $\mathcal{E}(g) = \cup_{i=0}^{g} S^*(\mathcal{M}_i)$. It was conjectured in [GT] that $\mathcal{E}(g)$ is finite for each g.

This situation can be translated into group theory as follows. Let $g(X)$ be the genus of the surface X. Let n be the degree of the cover ϕ. Then $B = \{y \in Y : |\phi^{-1}(y)| < n\}$ is a finite set of cardinality r. Choose a base point $p \in Y - B$. Then $\pi_1 = \pi_1(Y - B, p)$ is generated by elements $\gamma_1, ..., \gamma_{2m}, \delta_1, ..., \delta_r$ (where $m = g(Y)$ is the genus of Y) subject to the one relation $[\gamma_1, \gamma_2]...[\gamma_{2m-1}, \gamma_{2m}]\delta_1...\delta_r = 1$. Since ϕ is a covering map of degree n from $X - \phi^{-1}(B)$ to $Y - B$, we obtain a homomorphism $T_\phi : \pi_1 \to S_n$. Since X is connected, $G = T_\phi(\pi_1) = Mon(X, Y, \phi)$ is a transitive subgroup of S_n. Moreover, by the Riemann-Hurwitz formula we have

$$2(g(X) - 1) = 2n(g(Y) - 1) + \sum_{i=1}^{r} ind(x_i),$$

$$G = \langle y_1, .., y_{2m}, x_1, ...x_r \rangle, \text{ and} \tag{1}$$

$$[y_1, y_2]...[y_{2m-1}, y_{2m}]x_1...x_r = 1.$$

Here y_i is the image of γ_i and x_i is the image of δ_i. Recall that for $x \in S_n$, $ind(x) = n - orb(x)$, where $orb(x)$ is the number of orbits of x on $\{1, ..., n\}$.

* the author was partially supported by the NSF

Conversely, if G is a transitive subgroup of S_n satisfying (1) and Y is given, then since G is a homomorphic image of π_1, there is some covering map ϕ with $T_\phi(\pi_1) = G$. Riemann's existence theorem (see [F1]) asserts the existence of a Riemann surface X with $\phi : X \to Y$ realizing G as $Mon(\phi)$.

One consequence of the remarks above is that if $x_1, \ldots, x_r$ are permutations on a set of size n whose product is one and which generate a transtive group, then $\Sigma_{i=1}^r \geq 2n - 2$. This was first observed by Ree in this geometrical setting. This was finally proved via elementary group theory in [FLS]. Scott [S] proved an analagous result for linear transformations which includes the permutation result.

We illustrate the power of (1) by examining the case of Galois covers of spheres. Note that in this case, the permutation representation is the regular representation. Thus for $x \in G = Mon(\phi)$, we have $ind(x) = ((d - 1)/d)n$, where d is the order of x and $n = |G|$ is the degree of the cover. Choose notation as above so that x_i has order d_i with $1 < d_1 < \ldots < d_r$. The Riemann-Hurwitz formula then yields

$$n \sum_{i=1}^{r}((d_i - 1)/d_i) = 2(n + g - 1).$$

This yields the classical result:

Theorem 0. *Let $\phi : X \to Y$ be a Galois cover of degree n with X of genus g and Y of genus 0. Let $G = Mon(\phi)$. Then one of the following holds:*
(i) *$n \leq 84(g - 1)$,*
(ii) *$g = 1, (d_1, \ldots, d_r) = (2,2,2,2), (2,3,6), (2,4,4),$ or $(3,3,3)$ and $G'' = 1$,*
(iii) *$g = 0$ and G is cyclic, dihedral, A_4, S_4 or A_5 with $(d_1, \ldots, d_r) = (d,d),$ $(2,2,d), (2,3,3), (2,3,4),$ or $(2,3,5),$ respectively.*

The proofs in the non-Galois situation follow this general pattern. The difficulty is that elements have fixed points in representations other than the regular representation. However, most primitive groups do not contain nontrivial elements with very many fixed points.

We can factorize ϕ via

$$X = X_0 \to^{\phi_1} X_1 \to^{\phi_2} X_2 \to \ldots \to^{\phi_w} X_w = Y$$

where $\mathbf{C}(X_i)$ is a maximal subfield of $\mathbf{C}(X_{i-1})$ for $i = 1, 2, \ldots, w$. This corresponds to $G_i = Mon(\phi_i, X_{i-1}, X_i)$ being a primitive group of permutations on the subgroup corresponding to X_i. By the Riemann Hurwitz formula, it follows that $g(X_i) \geq g(X_{i-1})$. Moreover in [GT] it was shown that $cf(G) \subseteq \cup_{i=1}^w cf(G_i)$. Thus in trying to verify that $\mathcal{E}(g)$ is finite, we can restrict our attention to primitive permutation groups. Moreover, by

(1), if $g(Y) \geq 2$, then $n \leq (g(X) - 1)/(g(Y) - 1) \leq g(X) - 1$. Thus it suffices to assume that $g(Y)$ is at most 1. We will show in section 2 that when $g(Y) = 1$ and the cover is primitive and ramified that there is a similar bound unless G is A_n or S_n. Thus the critical situation to study is that of primitive covers of spheres (i.e. $g(Y) = 0$). The advantage to studying primitive covers or equivalently primitive groups is that their structure is known. In particular, the following holds (see [AS]):

Suppose G is a finite group and H is a maximal subgroup of G such that $ker_G(H) = \cap_{g \in G} H^g = 1$ (this is equivalent to saying that G is a primitive faithful group of permutations on the set of cosets of H). Let Q be a minimal normal subgroup of G, L a minimal normal subgroup of Q, and let $\Delta = \{L = L_1, \ldots, L_t\}$ be the set of G conjugates of L. Then $G = HQ$ and precisely one of the following holds:

(A) L is of prime order p, Q is an elementary abelian p-group, and $H \cap Q = 1$.
(B) $F^*(G) = Q \times R$, where $Q \cong R$ and $H \cap Q = 1$.
(C1) $F^*(G) = Q$ is nonabelian, $H \cap Q = 1$, and $t > 1$.
(C2) $F^*(G) = Q$ is nonabelian, $H \cap Q \neq 1 = H \cap L$.
(C3) $F^*(G) = Q$ is nonabelian, $H \cap L_i = H_i \neq 1$, and $H \cap Q = H_1 \times \ldots \times H_t$.

Recall that $F^*(G)$ is the generalized Fitting subgroup . In all cases except (A), L will be a nonabelian simple group and is called a component of G.

Let G be a transitive faithful group of permutations on the finite set Ω. Let $G(r)$ be the set of r-tuples of nontrivial elements of G which generate the group and whose product is 1. Let $E = (x_1, \ldots, x_r)$ be an element of $G(r)$. In view of (1), we define the genus of (G, Ω, E) to be g where $2(n + g - 1) = \sum_{i=1}^r ind(x_i)$. Let the genus of (G, Ω) be the minimum value of the genus of (G, Ω, E) as E ranges over all possibilities. We say G is a (primitive) group of genus g if G acts faithfully and transitively (primitively) on a set Ω with the genus of (G, Ω) at most g. Let $\mathcal{E}^*(g)$ be the set of components of primitive genus g groups other than alternating groups. By the main result of [N1] and [GT, Section 9] it follows that $\mathcal{E}^*(g)$ is cofinite in $\mathcal{E}(g)$. It has been shown that in cases (A), (C1), and (C2) that either $G'' = 1$ and $g = 0$ or the degree is bounded by a linear function in the genus (see [N],[GT],and [A]). Moreover, it has been shown in case (B) that there are no genus zero groups (see [S1]), and one can show easily by the same methods that the degree is bounded by a linear function of the genus. Indeed by [GT] it follows easily that in case (B), the degree grows linearly in the genus unless the component L has a centralizer of index less than 85 (of course, this eliminates all but finitely many possibilities for L). Thus case (C3) is the crucial (and most difficult) case.

Let G act transitively and faithfully on a finite set Ω. If $x \in G$, define $\theta(x, \Omega) = |\Omega(x)|/|\Omega|$, where $\Omega(x)$ is the set of fixed points of x on Ω. Set $\theta(G, \Omega) = max\{\theta(x, \Omega) : 1 \neq x \in G\}$ and $\theta(G) = max\,\theta(G, \Omega)$. If L is a nonabelian simple group, it is convenient to define $\epsilon(L) = max\,\theta(M)$, where M ranges over all subgroups of Aut L containing L. A minor modification of the proof of [GT, Theorem E] shows:

Theorem 1. *Let G act primitively and faithfully on Ω. Assume that (G, Ω) has genus g. Let L be a component of G. Fix a $0 < \delta < (1/85)$. If $\epsilon(L) \leq \delta$, or $cf(G/F^*(G))$ contains any noncyclic group other than A_5, then $|\Omega|$ is bounded by a linear function of g (depending on δ).*

By the remarks above, this yields:

Corollary 2. *Let U_ϵ be the collection of nonabelian simple groups with $\epsilon(L) < \epsilon$. If $\epsilon < (1/85)$, then $U_\epsilon \cap \mathcal{E}(g)$ is finite.*

Liebeck and Saxl [LS] have shown that if L is a simple Chevalley group defined over the field of q elements, then $\epsilon(L) \leq (4/3q)$, unless $L \cong L_2(q)$, in which case a similar bound holds or $L \cong L_4(2), PSp_4(3)$, or $P\Omega_4^-(3)$. Thus for a fixed g, there are only finitely many Chevalley groups defined over a field with more than 113 elements in $\mathcal{E}(g)$.

The nonsimple case that remains is handled in [GN]. The following is shown there:

Theorem 3. *There exists a function $N(g)$ such that if G acts primitively on a set Ω of order n with $g = g(G, \Omega)$, then one of the following holds:*
(i) $F^*(G)$ is simple,
(ii) $G'' = 1$, and $g = 0$,
(iii) $F^*(G) \cong A_l \times A_l$, with $n = l^2$,
(iv) $n \leq N(g)$.

It is worthing noting that the classification of simple groups can be avoided in the proof of Theorem 3. Moreover, $N(g)$ can taken to be quadratic in g. By using the [LS] result mentioned above, we see that (i) can be replaced by (i') $F^*(G)$ is an alternating group or Chevalley group defined over a small field. Also, using [LS], it can be shown that the genus grows linearly (as a function of the degree) unless (i) or (ii) holds or G has a component isomorphic to an alternating group.

Thus one basically needs to determine the genus of almost simple groups to classify large primitive groups of fixed genus g. As a consequence, the Guralnick-Thompson conjecture on composition factors reduces to the almost simple case. By [LS], we only need to verify this for classical groups over small fields. Shih [S2] has shown that there are only finitely many genus zero groups G with $L_n(q) \leq G \leq PGL_n(q)$.

In this article, we sketch a method which avoids the rather long proof of Theorem 3 but does reduce the Guralnick-Thompson conjecture to the almost simple case (but of course does not classify the groups).

One can regard this problem from a different point of view. Let $\mathcal{G}$ denote a collection of pairs (G, Ω), where G acts faithfully and transitively on Ω. Define

$$\mathcal{M}_g(\mathcal{G}) = \{X \in \mathcal{M}_g | \exists \phi : X \to P^1, Mon(\phi) \in \mathcal{G}\}.$$

The problem is to compute $dim \mathcal{M}_g(\mathcal{G})$. In particular, when is $\mathcal{M}_g(\mathcal{G})$ dense in $\mathcal{M}_g$? Zariski (see [FG],[Z]) showed that if G is solvable and $g > 6$, then $\mathcal{M}_g(G)$ is not dense. Since $\mathcal{M}_g = \mathcal{M}_g(S_4)$ for $g \leq 6$, this is best possible. However, it was left open as to whether $\mathcal{M}_g(Sol)$ is dense, where Sol is the set of finite solvable groups. Since $\mathcal{M}_g(G)$ is a quasiprojective (not necessarily irreducible) subvariety of $\mathcal{M}_g$ (see [F2]), the following result answers the question.

Theorem 4. $\mathcal{M}_g(Sol) = \mathcal{M}_g(\mathcal{G})$ for some finite collection $\mathcal{G}$ of solvable groups (depending on g). In particular for $g > 6, \mathcal{M}_g(Sol)$ is not dense in $\mathcal{M}_g$.

The proof depends on Zariski's result, the main theorem of [N], and some results in section 2. We also consider when $\mathcal{M}_g(G, \Omega)$ is dense for a single (G, Ω). By a result of Zariski (see [FG] for a more detailed discussion), the case of most interest is when G acts primitively on Ω. The following is proved in [GN]:

Theorem 5. If $g > 4$ and G acts primitively on Ω and $\mathcal{M}_g(G, \Omega)$ is dense in $\mathcal{M}_g$, then either $G \cong S_n, n \geq (g/2) + 1$, or $G \cong A_n, n \geq 2g$, and Ω is the set of k subsets of a set m for some $k < m/2$ (in particular, $n = \binom{m}{k}$).

We conjecture that in the above Ω has cardinality n. It is known (see [FG]) that S_n does satisfy the result for $n \geq (g + 2)/2$. It is still open whether the alternating group does (necessarily $n \geq 2g - 1$).

This article is organized as follows. In section 2, we consider coverings of surfaces of positive genus and show no difficulties arise. In particular, we prove Theorem 4. In section 3, we relate the genus of a group and a transitive normal subgroup. This is used in section 4 to study groups preserving a product structure. In section 5, we reduce the composition factor problem to the almost simple case. In the final section we discuss coverings of generic surfaces and prove a special case of Theorem 5.

2. Coverings of Surfaces of Positive Genus.

Let $\phi : X \to Y$ be a branched covering of degree n. Let g be the genus of X and h the genus of Y. Let $G = Mon(\phi)$. As we noted earlier, the Riemann Hurwitz formula yields:

Lemma 2.1. *If $h > 1$, then $n \leq [(g-1)/(h-1)] \leq (g-1)$.*

We now consider the situation when $h = 1$ and the cover is primitive. Let r be the number of branch points, and let $\{x_1, \ldots, x_r\}$ be the elements of G corresponding to those branch points. By equation (1), we see that $2(g-1) = \sum_{i=1}^{r} ind(x_i)$. Thus we need some bounds on possible indices.

Lemma 2.2. *Let G be a group acting primitively and faithfully on a set Ω of size n. Let x be a nontrivial element of G. Set $f(x) = |\Omega(x)|$.*
    (i) *If $F^*(G)$ is a p-group, then $f(x) \leq ((p-1)/p)n$.*
    (ii) *If case (B) or (C1) hold, then $f(x) \leq (1/10)n$.*
    (iii) *If case (C2) holds, then $f(x) \leq (4/15)n$.*
    (iv) *If $F^*(G)$ is not simple, then $ind(x) \geq \sqrt{n}$ or $F^*(G)$ is abelian and $n \leq 9$.*
    (v) *If G does not contain A_n, then $ind(x) \geq (1/4)\sqrt{n}$.*
    (vi) *If case (C3) holds and L is a component of G with $\epsilon(L) = \epsilon$, then $f(x) \leq \epsilon n$.*

Proof. In cases (A),(B), and (C1), then Q (see the introduction) is a regular normal subgroup of G. Thus G is a semidirect product of Q by H, where H is a point stabilizer. Since either $f(x) = 0$ or x is conjugate to an element of H, we can assume that x is in H. Then the permutation action of x on Ω is equivalent to that of conjugation on Q. Hence $f(x) = |C_Q(x)|$. Now (i) follows immediately. Also (ii) now follows from [A2,(2.1)]. See [A2,(5.7) and (5.8)] for (iii). Now (iv) and (vi) follow from [GT, Section 9]. Finally (v) follows from [B] or [L].

Proposition 2.3. *Let $\phi : X \to Y$ be a primitive branched cover of degree n of compact connected Riemann surfaces with X of genus g and Y of genus 1. Let $G = Mon(\phi)$. Then one of the following hold:*
    (i) *$G \cong C_p$ for some prime p and the cover is unramified.*
    (ii) *$G \cong A_n$ or S_n.*
    (iii) *$F^*(G) = A_d, d < n$ and $8(g-1) \geq \sqrt{n}$.*
    (iv) *$12(g-1) \geq n$.*

Proof. Note that the cover is unramified if and only if $g = 1$. Since the fundamental group of an elliptic curve is abelian, this implies G is abelian. Since it is primitive, (i) holds. So assume that there is at least one branch point. Then as we observed above, $2(g-1) = \sum ind(x_i) \geq ind(x_1)$. The result now follows from the previous lemma, noting that by [LS] if L is not an alternating group, then $\epsilon(L) \le (2/3)$ (note that $ind(x) \ge (1/2)(n - f(x))$).

Theorem 2.4. Let $\phi : X \to Y$ be a branched cover of degree n with Y of genus zero and X of genus $g > 1$. If $G = Mon(\phi)$ is solvable, then there exists a branched cover $\phi' : X \to Z$ with Z of genus zero, $Mon(\phi')$ solvable and $deg(\phi') \le 2^{10}(g - 1)$.

Proof. If the cover is primitive, then it follows by [N2, Theorem 1.6] that $deg(\phi)$ satisfies the inequality. So assume that the cover is not primitive. Thus $\phi = \alpha\beta$, where $\alpha : W \to Y$ and $\beta : X \to W$ for some surface W of genus h. Note that $Mon(\alpha)$ and $Mon(\beta)$ are solvable. We also can assume that β is primitive. If $h = 0$, then we are done as above. If $h = 1$, then by Lemma 2.2(i) and the argument of the previous result, it follows that $deg(\beta) \le 8(g - 1)$. Since there is a cover from W to a sphere of degree 2, there is a solvable cover from X to a sphere of degree at most $16(g - 1)$. If $h > 1$, then by Lemma 2.1, $deg(\beta) \le (g - 1)/(h - 1)$. By induction on g (note $(h - 1) \le (g - 1)/2$), it follows that there exists a solvable cover of W to a sphere of degree at most $2^{10}(h - 1)$. Thus, there exists a solvable cover of X to a sphere of degree at most $[(g - 1)/(h - 1)]2^{10}(h - 1) = 2^{10}(g - 1)$ as desired.

Note that Theorem 4 is an immediate consequence of the previous result and the previously mentioned result of Zariski. In particular, it follows that there are Riemann surfaces of any genus $g > 6$ defined over the rationals which do not have solvable branched coverings.

3. The Genus of a Normal Subgroup.

Let G be a finite group acting on a set Ω with $|\Omega| = n$. Recall $f(x) = |\Omega(x)|$ for $x \in G$. If x has order d and $c\,|\,d$, set

$$V(x, c) = \frac{1}{d}\sum \left\{ f(x^j) \mid 1 \le j \le d, c \nmid j \right\}.$$

Let $E \in G(r)$. Suppose $K \triangleleft G$ and K acts transitively on Ω. Let $\phi : X \to \mathbf{P}^1$ be a cover with $Mon(\phi) = (G, \Omega, E)$. Let $\hat{X}$ denote the Galois closure of the cover. Let Y denote the surface corresponding to K and Z the surface corresponding to $H \cap K$, where $H \le G$ corresponds to X (so H is a point stabilizer of a point in Ω and $G = KH$). We wish to relate $g(X), g(Y)$ and $g(Z)$. Let $\Gamma = G/H \cap K$ and $m = |G/K|$. Note $|\Gamma| = mn$.

Let $E = (x_1, \ldots, x_r)$. Set d_i equal to the order of x_i and $\bar{d}_i$ the order of $\bar{x}_i$ in $G/K = \bar{G}$. We can assume that $\bar{d}_i = 1$ if and only if $i > s$. The situation can be described pictorially by:

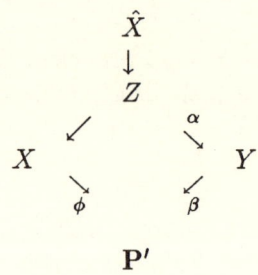

P'

Lemma 3.1. *Let $B = \{p_1, \ldots, p_r\}$ be the branch points of the cover ϕ where p_i corresponds to x_i. Let C be the set of branch points of the cover $\alpha : Z \to Y$. Then $C = \bigcup_{i=1}^{s} \beta^{-1}(p_i)$, $\left|\beta^{-1}(p_i)\right| = m/\bar{d}_i$, and all points of $\beta^{-1}(p_i)$ are conjugate under H.*

Proof. Clearly $C \subseteq \beta^{-1}(B)$. Suppose $\beta(q) = p_i$. Since the ramification of any point $w \in \hat{X}$ over p_i is d_i, we see that p_i ramifies to $\hat{X}$ if and only if $d_i = \bar{d}_i$. Since $\hat{X}$ is the Galois closure of $\alpha : Z \to Y$ (since $G = KH$, K acts faithfully on the cosets of $K \cap H$), we see that $C = \bigcup_{i=1}^{s} \beta^{-1}(p_i)$. Since Y is a Galois cover of $\mathbf{P}^1$ of degree m, $\left|\beta^{-1}(p_i)\right| = m/\bar{d}_i$.   Finally, it is clear that G acts transitively on $\{\beta^{-1}(p_i)\}$ and that K acts trivially on $\beta^{-1}(p_i)$. Thus as $G = HK$, the last statement holds.

Corollary 3.2. *If $g(Y) = 0$, then $g(Z) = g(K, \Omega, F)$ where $F = F_1 \dot\cup \ldots \dot\cup F_s$ and F_i consists of $m/\bar{d}_i$ elements which are all conjugate in G.*

Proof. This is just a restatement of the previous result, noting that $K = Mon(\alpha)$.

One can in fact show that the conjugacy class in G of F_i is just the class of $x_i^{\bar{d}_i}$. We can relate $g(Z)$ and $g(X)$.

Theorem 3.3. $g(Z) - 1 = m(g(X) - 1) + mV/2$, *where* $V = \sum_{i=1}^{r} V_i$ *and* $V_i = V(x_i, \bar{d}_i)$.

Proof. Note that $f(x, \Gamma) = 0$ if $x \notin k$ and $f(x, \Gamma) = mf(x, \Omega)$ for $x \in k$.

By the Riemann-Hurwitz formula,

$$2\left(mn + g\left(Z\right) - 1\right) = \sum_{i=1}^{r} ind\ \left(x_i, \Gamma\right)$$

$$= \sum \left(mn - \frac{1}{d_i} \sum_{j=1}^{d_i} f\left(x_i^j, \Gamma\right)\right)$$

$$= m\left(\sum n - \frac{1}{d_i} \sum_{j=1}^{d_i} f\left(x_i^j, \Omega\right)\right) + m\sum V_i$$

$$= m\left[2\left(n + g\left(X\right) - 1\right)\right] + mV.$$

4. Groups Preserving Product Structures.

Throughout this section, we assume that G acts faithfully and primitively on the set Ω of cardinality $n = \ell^t$ with $\ell \geq 5$ and $t \geq 2$. We also assume that $F(G) = 1$. Let H be a point stabilizer. We further assume that G preserves a product structure on Ω. So we can identify $\Omega = \Delta^{(t)}$ and $G \leq Sym(\Delta) \wr S_t = AS_t$ where S_t permutes the coordinates and A acts as permutations on each coordinate. Let $K = A \cap G \triangleleft G$. Note that $G = HK$. We further assume that G acts primitively on the t coordinates (i.e. G/K is a primitive subgroup of S_t) and that $F(G) = 1$. Assume $E = (x_1, \ldots, x_r) \in G(r)$ with x_i of order d_i with $1 < d_1 \leq \ldots \leq d_r$. Set $g = g(G, \Omega, E)$ and $h = g(\bar{G}, \bar{G}, \bar{E})$ where $\bar{G} = G/K$.

The following result is essentially [GT, Theorem 9.3]. The result here is slightly stronger but the proof is essentially identical.

Theorem 4.1. *Let $\bar{d}_i$ be the order of x_i (mod K). One of the following holds:*

(i) $h = 0$ and $\bar{G}$ is cyclic, dihedral, A_4, S_4, or A_5,
(ii) $h = 1$ and $\bar{G}'' = 1$,
(iii) $g - 1 \geq (1/2000)\,n$,
(iv) $g \geq 1, r = 3, (d_1, d_2, d_3) = (2, 4, 5), t = 5, \ell \leq 10$ and $\bar{G} = S_5$,
(v) $g \geq 1, r = 4, (d_1, d_2, d_3, d_4) = (2, 2, 2, 3), t = 4, \ell \leq 7$ and $\bar{G} = S_4$,
(vi) $g \geq 1, r = 4, (d_1, d_2, d_3, d_4) = (2, 2, 2, 4), t = 4, \ell \leq 6$ and $\bar{G} = S_4$.

In particular, it follows that if $h > 1$, then $g \geq 1$ and aside from a finite list of exceptions $g \geq 1$. We wish to obtain similar results for $h \leq 1$. Observe that if $x \in G - K$, then $f(x) \leq n/\ell$ and $ind\ x \geq ((\bar{d}-1)/\bar{d})\,(n - n/\ell)$ where $\bar{d}$ is the order of $\bar{x}$.

Theorem 4.2. *Let u be the genus of K acting on the first coordinate. Assume $u \geq 5$, $\ell \geq 20$, and $h \leq 1$. Then $g - 1 \geq (1/4)\sqrt{n}$.*

Proof. We use Theorem 3.3. First consider the case $h = 1$. Since the action on the coordinates is primitive, it follows that for $x \in G-K$, $f(x) \leq \ell^{(t+1)/2}$. Thus $V \leq 2\ell^{(t+1)/2}$. Since $(G/K)'' = 1$ and (G/K) is primitive, it follows from Theorem 0 that $m \leq 6t$. In the notation of Theorem 3.3, we know that the cover α is ramified (since $Mon(\alpha)$ is not abelian). By the assumption on ℓ and the fact that the action of K on a coordinate does not contain the alternating group, we know that $ind(x) \geq 10\ell^{(t-1)}$ and so $g(Z)-1 \geq 5\ell^{(t-1)}$. It follows from Theorem 3.3 that for $t > 4$,

$$(g - 1) \geq 5\ell^{(t-1)}/6t - \ell^{(t+1)/2} \geq \sqrt{n}.$$

A slightly more detailed analysis yields the same result for $t \leq 4$.

Now assume that $h = 0$. The cover α is the composition of t covers each of degree ℓ. The first cover of degree ℓ over Y has genus at least u. By the Riemann-Hurwitz formula and our assumption on u, we know that $g(Z) - 1 \geq 4\ell^{(t-1)}$. We give the argument for the case G/K is cyclic (of necessarily prime order t). It follows that for $x \in G - K$, $f(x) \leq \ell$. Thus $V \leq 2\ell$ and we can argue as in the previous case. A slight variation of this argument yields the same result when G/K is dihedral or one of the other three possible groups.

5. Reduction to the Almost Simple Case.

Let L be a nonabelian simple group. Define $g(L)$ to be the minimum genus of any group with $F^*(G) = L$ and $g'(L)$ to be the minimum genus of any group with L as a composition factor. Let $d(L)$ be the minimum degree of a faithful permutation representation of L.

Theorem 5.1. Let L be a finite nonabelian simple group with $g'(L) = g$. Let G be a group of minimal order subject to L is a composition factor of G and G has a permutation representation on a set Ω of genus g and degree n. Then G is primitive and one of the following holds:
 (i) $F^*(G) = L$ (and so $g(L) = g'(L)$),
 (ii) $F^*(G)$ is abelian, G is doubly transitive, and $g(L) = g + 1$.
(iii) $g(L) \leq 5$,
(iv) $d(L) \leq max\{20, 4(g - 1), 50(g - 1)^{1/2}\}$.

Proof. By [GT], we know that G is primitive. Suppose that N is a regular normal subgroup. Note this always exists in cases (A) and (B) (see the introduction). Thus $G = NH$ is a semidirect product where H is a point stabilizer. Let $E \in G(r)$ affording the genus g system. Say $E = (x_1, \ldots, x_r)$. Write $x_i = v_i h_i$ with $v_i \in N$ and $h_i \in H$. Then $F = (h_1, \ldots, h_r) \in H(r)$ (except that possibly some $h_i = 1$). Let Ω_j be the nontrivial orbits of H. Note that H acts faithfully on each Ω_j. Let g_j be the genus of (H, Ω, F).

Since $\text{ind}(h) \geq \text{ind}(hv)$ for $h \in H, v \in N$, $\sigma(g_j - 1) \leq g$. Suppose case (A) or (B) holds. Then L is also a composition factor of H and so $g_j > g$. This implies that H has only one nontrivial orbit and so G is doubly transitive. This eliminates case (B). In case (A), we can only conclude that $g(L) = g+1$ and so (ii) holds.

If case (C2) holds, then (iv) holds by [A].

Now assume case (C1) or (C3) hold. Thus G preserves a product structure on Ω unless (i) holds. We follow the notation of section 4. If $h > 1$, we apply Theorem 4.1 and note that $d \leq \sqrt{n}$. Thus (iv) holds. So assume $h \leq 1$. Then by Theorem 4.2, either (iii) or (iv) holds.

Thus if $g \geq 5$, it suffices to prove there are only finitely many almost simple groups (other than alternating or symmetric groups) of genus at most g to show that $\mathcal{E}(g)$ is finite. Of course, the finiteness of $\mathcal{E}(5)$ implies the finiteness of $\mathcal{E}(h)$ for $h < 5$.

6. Generic Groups.

Assume G acts transitively and faithfully on the set Ω of size n. Let $E \in G(r)$ and let g be the genus of (G, Ω, E). One can parametrize all covers of the sphere with the monodromy group corresponding to (G, Ω, E) (see [F2, section 1.2]) by an algebraic variety $\mathcal{H}$. Then there is an algebraic map $\Phi : \mathcal{H} \to \mathcal{M}(G, \Omega, E)$ which sends a cover $\phi : X \to \mathbf{P}^1$ to X. Φ is by definition surjective. We say G (or more precisely (G, Ω, E)) is generic of genus g if the image of Φ is dense in $\mathcal{M}_g$.

Assume that $g > 1$. The dimension of $\mathcal{H}$ is r. It is also easy to see that every fiber has dimension at least 3 (assuming $r \geq 3$). Thus:

Lemma 6.1. *dim* $\mathcal{M}(G, \Omega, E) \leq (r - 3)$.

It follows from a result of Zariski (see [Z] or [FG]) that one can essentially reduce to the case of primitive groups. So we now assume that G acts faithfully and primitively on Ω.

Theorem 6.2. *Let $d = \dim \mathcal{M}(G, \Omega, E)$. One of the following holds:*
(i) *$F^*(G)$ is a t-fold product of alternating groups of degree m and $n = \ell^t$, where $\ell = \binom{m}{k}$ for some $1 \leq k < (m/2)$.*
(ii) *$d \leq 9 + (2/3)(g - 1)$.*
(iii) *$G = S_4, n = 4$ and $d \leq 2g + 3$.*
(iv) *$n \leq 4$ and $d \leq 2g + 1$.*

Proof. It follows by [LS, Theorem 2] that if (i) does not hold, then for $x \neq 1$, $ind(x) \geq (1/6)n$. Thus

$$2(n + g - 1) \geq (1/6)rn. \tag{6.3}$$

First assume that $n \geq 18$. Then (6.3) implies that $r \leq 12 + (2/3)(g - 1)$ and (ii) holds.

Now assume that (i) does not hold and that $6 \leq n \leq 17$. It follows by inspection that the index of any nontrivial element is at least $(1/4)n$. Arguing as above, we see that $r \leq 8 + (2/3)(g - 1)$ and (ii) holds.

Finally, consider the case $n < 6$. If (i) or (iii) does not hold, then either $n = 5$ and the index of any nontrivial element is at least 2 or $n \leq 4$ and the index of any nontrivial element is at least $(1/3)n$. The result follows as above.

Corollary 6.3. *If $\mathcal{M}(G, \Omega, E)$ is dense in $\mathcal{M}_g$, then one of the following holds:*

(i) $g \leq 4$.

(ii) $g \leq 6$, $n = 4$, and $G = S_4$.

(iii) *Case (i) of Theorem 6.1 holds.*

In order to prove Theorem 5, one must analyze very closely case (C3). In fact, once this analysis is done, we find that the Liebeck-Saxl result is only needed for the case of almost simple groups. There should be a similar result for smaller g. Here one would expect a list of possible generic groups. Zariski conjectured that any primitive group satisfying the necessary condition that $r \geq 3g$ would be a generic group. This fails for solvable groups of genus 2 (see [FG]). It seems very difficult to decide when a group can be generic (see [F2] for more comments on this problem). There are some interesting problems even for genus one groups. Of course, in this case the moduli space is one dimensional.

References

[A] M. Aschbacher, "On Conjectures of Guralnick and Thompson", J. Algebra, to appear.

[AS] M. Aschbacher and L. Scott, "Maximal subgroups of finite groups", J. Algebra 92 (1985), 44-80.

[B] L. Babai, "On the order of uniprimitive permutation groups", Ann. Math. 113(1981), 553-568.

[FLS] W. Feit, R. Lyndon, and L. Scott, "A remark about permutations", J. Comb. Theory (A) 18(1975), 234-235.

[F1] M. Fried, "Galois groups and complex multiplication", Trans. Amer. Math. Soc. 237(1978), 141-162.

[F2] M. Fried, "Combinatorial computation of moduli dimension of Nielsen classes of covers", Contemporary Math 89(1989), 61-79

[FG] M. Fried and R. Guralnick, "The generic curve of genus > 6 is not uniformized by radicals", preprint.

[GN] R. Guralnick and M. Neubauer, "Monodromy groups of branched covering: composition factors and moduli dimension", preprint.

[GT] R. Guralnick and J. Thompson, "Finite groups of genus zero", J. Algebra 131(1990), 303-341.

[L] M. Liebeck, "On minimal degrees and base sizes of primitive permutation groups", Arch. Math. 43(1984), 11-15.

[LS] M. Liebeck and J. Saxl, "Minimal degrees of primitive permutation groups, with an application to monodromy groups of Riemann surfaces", Proc. London Math. Soc. 63(1991), 266-314.

[N1] M. Neubauer, "On solvable monodromy groups of fixed genus", USC Ph. D. Thesis, 1989.

[N2] M. Neubauer, "On monodromy groups of fixed genus", preprint.

[N3] M. Neubauer, "Primitive permutation groups of genus zero and one, I,II", preprint.

[S] L. Scott, "Matrices and cohomology", Ann. Math. 105(1977), 473-492.

[S1] T. Shih, "A note on groups of genus zero", preprint.

[S2] T. Shih, "Bounds of fixed point ratios of permutation representations of $GL_n(q)$ and groups of genus zero", Caltech Ph. D. Thesis, 1990.

[Z] O. Zariski, Collected Papers, vol III, MIT press, Cambridge, 1972

Primitive Permutation Characters

Robert M. Guralnick and Jan Saxl

Let G be a finite group. Consider subgroups H, K of G such that

$$1_H^G = 1_K^G. \tag{1}$$

It is well known that (1) is equivalent to:

$$|x^G \cap H| = |x^G \cap K| \text{ for all } x \in G \tag{2}$$

This property has been studied extensively and has applications in number theory and geometry. In this note, we consider the case that H is a maximal subgroup of G. It was asked by Wielandt in 1979 whether there are examples of (1) with H maximal in G and K not maximal. In [FK], it was shown that if there is such an example, then there will be an example with G almost simple. We first give one such example. This example was inspired by Borovik's discovery [B] of two nonconjugate subgroups which are locally conjugate (both isomorphic to A_6) in $E_8(\mathbf{C})$.

Example 1. Choose a prime p such that $p \equiv 1 \bmod 21$. Let J be the triple cover of M_{22}. Let Z denote the center of J. Then J has two irreducible representations of dimension 45 over the complex numbers, one of which is faithful and the other with kernel Z. Since p does not divide the order of J and all character values are defined over the field F of p elements, these representations are also defined over F. Denote the representations by $\phi_i, i = 1, 2$. Assume that ϕ_1 is faithful. Let H, K denote the images of the representations in $G = L_{45}(p)$. It follows by [AT] that if $x \in J$ and x has minimal order in the coset xZ, then $x_1 = \phi_1(x)$ and $x_2 = \phi_2(x)$ have the same trace and order. This observation also applies to every power of x. Thus the two representations are equivalent when restricted to $\langle x \rangle$. This implies that x_1 and x_2 are conjugate in G. By the equivalence of (1) and (2) above, this implies that (G, H, K) satisfies (1).

We claim that H is maximal and that K is not. The second part of the claim is easy to verify. The 45 dimensional representation of M_{22} extends to M_{24} (see [AT]). Suppose $H < M < G$. Since M_{22} has no nontrivial transitive permutation representation of degree dividing 45, it follows that J acts absolutely primitively on the 45 dimensional module. It follows from the character table of J in [AT] that J preserves no tensor product decomposition. It follows easily from these two observations that $S = F^*(M)$ is simple (for example, we can appeal to Aschbacher's classification of maximal subgroups of $L_n(p)$ which are not almost simple [A]). Clearly, 3 divides the order of the Schur multiplier of S and M_{22} is a subgroup of S. It follows by the classification of finite simple groups that if the Schur multiplier of S has order divisible by 3 and $|M_{22}|$ divides $|S|$, then S is among the following(cf [AT]):

(i) $S = McL, Suz, ON, Fi_{22}, Fi'_{24}$,
(ii) $S = L_n(q)$ with $3|(q-1,n)$,
(iii) $S = U_n(q)$ with $3|(q+1,n)$,
(iv) $S = E_6(q)$ with $3|(q-1)$, or
(v) $S = {}^2E_6(q)$ with $3|(q+1)$.

The last two possibilities are eliminated by [KL, Tables 5.3A, 5.4B] which show that those groups have no projective irreducible representations of dimension 45. Similarly no group in (i) has an irreducible 45 dimensional representation by [AT]. Now consider cases (ii) and (iii) with p not dividing q. Since M_{22} has no nontrivial 3-dimensional projective representation, $n \geq 6$. By [KL, Table 5.3A], in case (ii) the minimal dimension of a nontrivial projective representation is $q^{n-1} - 1 \geq 4^5 - 1 > 45$. A similar inequality holds in (iii) with the single exception of $U_6(2)$. However, $U_6(2)$ has no 45 dimensional projective irreducible representation in characteristic 0 or p [AT] (since p does not divide its order). So we need only consider cases (ii) and (iii) with $p|q$. Since $3|(p-1)$, case (iii) cannot occur. By [AT], the only possibility is $n = 21$. It is easy to see that $L_{21}(q)$ has no 45 dimensional irreducible representation (eg, see [KL, Proposition 5.4.11]). This proves the claim.

By taking subgroups of wreath products, one can construct infinitely many other examples (but this is not very exciting). If we replace the triple cover of M_{22} by the 6 fold cover of $L_3(4)$, we construct a similar example (by considering 36 dimensional representations). However, we can construct a different type of example with G alternating by using the first example. We make the following easy observation:

Lemma. *Suppose (G, H, K) satisfies (1). Set $n = [G : H] = [G : K]$. Let H_1 and K_1 be the images of G in S_n under the permutation representations of G on the cosets of H and K, respectively. Then (S_n, H_1, K_1) satisfies (1).*

Proof. Since (1) holds, each element x of G has the same number of fixed points on the cosets of H or the cosets of K. This implies that x has the same cycle type in each representation. This implies that (2) and so (1) hold for (S_n, H_1, K_1).

So consider our first example (G, H, K). First note that H is its own normalizer in Aut G (this follows from observing that the representation corresponding to an outer automorphism of G does not correspond to twisting the representation by an automorphism of H). We follow the notation in the lemma. The image of G acting on the cosets of H is obviously a primitive simple subgroup of A_n. By the remarks above, it is its own normalizer in S_n. It thus follows by [LPS, pp 132-6] that H_1 is a maximal primitive subgroup of A_n and H_1 is its own normalizer in S_n. Since K_1 is not primitive and is clearly not the stabilizer of the system of imprimitivity it stabilizes, it is not maximal.

By the lemma above, we know that (S_n, H_1, K_1) satisfies (1) and (2). We wish to replace S_n by A_n. The only difficulty is to check the elements of H_1 whose S_n conjugacy class is the union of two A_n classes. If $x \in A_n$ is such an element, it is trivial to observe that x has a cycle of length at least $2\sqrt{n}$. On the other hand, since $x \in L_{45}(p)$, the order of x is certainly less than $p^{45} < \sqrt{n}$.

We can iterate this process and get an infinite tower of examples

$$(G_i, H_i, K_i)$$

with $G_i \cong A_{n_i}$ and $H_i \cong K_i \cong A_{n_{i-1}}$ for $i > 1$. The only modification one needs to make is to observe that $n_i > 4^{n_{i-1}}$ and that no element of A_m has order more than 2^m. Thus one can descend from the symmetric group to the alternating group at each stage. This yields:

Theorem. *There exist infinitely many finite simple groups G with a maximal subgroup H and a nonmaximal subgroup K satisfying (1). In particular, G may be taken to be an (arbitrarily large) alternating group.*

We can translate this to number theory (cf [P]). Note that alternating groups are Galois groups over the $\mathbf{Q}$.

Corollary. *There exist pairs of number fields E, F of arbitrarily large degree with the same zeta function such that E is a minimal extension over $\mathbf{Q}$ and F is not.*

Let E, F be as in the corollary. It follows that $\mathrm{Br}(E|\mathbf{Q})$ and $\mathrm{Br}(F|\mathbf{Q})$ are commensurable, where $\mathrm{Br}(E|k)$ denotes the relative Brauer group of E/k (cf [FKS]). We do not know whether equality of the relative Brauer groups is possible in this situation. For this one would need to have more precise information about the ramified primes. Fein and Schacher have informed us that they have constructed examples of fields $F \subset E$ of degrees 3 and 6 respectively over $\mathbf{Q}$ with the same relative Brauer groups (but of course they do not have the same zeta functions).

The authors were reminded of this problem by the lecture of Thompson at the Durham conference. The key idea in the solution was inspired by the lecture of Borovik at the conference. The authors would like to thank the LMS for organizing such conferences and the SERC for its financial support.

The first author gratefully acknowledges the support of the National Science Foundation. He also wishes to thank the Institut für Experimentelle Mathematik in Essen for its generous hospitality.

References

[A] M. Aschbacher, "On the maximal subgroups of the finite classical groups", Invent. Math. 76(1984), 469-514.

[AT] J. H. Conway, R. T. Curtis, S. P. Norton, R. A. Parker, and R. A. Wilson, An ATLAS of finite groups, Oxford University Press, 1985.

[B] A. Borovik, "The structure of finite subgroups of simple algebraic groups", Algebra and Logic 28, No.3 (1989), 249-279 (in Russian).

[FKS] B. Fein, W. Kantor, and M. Schacher, "Relative Brauer groups II", J. Reine Angew. Math. 328(1981),39-57.

[FK] P. Förster and L. G. Kovács, "A Problem of Wielandt on finite permutation groups", J. London Math. Soc. 41(1990), 231-243.

[G] R. Guralnick, "Subgroups inducing the same permutation representation", J. Algebra 81(1983), 312-319.

[KL] P. Kleidman and M. Liebeck, The Subgroup Structure of the Finite Classical Groups, Cambridge University Press, Cambridge, 1990

[LPS] M. Liebeck, C. Praeger, and J. Saxl, "The maximal factorizations of the finite simple groups and their automorphism groups", Memoirs Amer. Math. Soc. 86(1990), no. 432.

[P] R. Perlis, "On the equation $\zeta_K(s) = \zeta_{K'}(s)$", J. Number Theory 9(1977), 342-360.

Closures of finite permutation groups and relation algebras. *

Cheryl E. Praeger

Abstract: There is a Galois connection between the lattice of subalgebras of the algebra $R(\Omega)$ of all binary relations on Ω and the lattice of subgroups of the symmetric group $\mathrm{Sym}(\Omega)$ on Ω. The Galois closed subgroups are those permutation groups on Ω which are 2-closed in the sense of Wielandt. This connection is discussed, and, in particular, the recent classification, by Liebeck and the author, of all maximal 2-closed finite permutation groups is described.

1. Introduction

A Galois connection between relation algebras on a set Ω and permutation groups on Ω was introduced by B. Jonsson in [10] as a means of investigating the lattice of relation algebras on Ω. In this paper the Galois connection is described, and recent results about the corresponding Galois closed subgroups for finite sets Ω are discussed.

Given a set Ω, the **converse** $\Delta^\star$ of a binary relation $\Delta \subseteq \Omega \times \Omega = \Omega^2$ is defined to be

$$\Delta^\star = \{(\beta, \alpha) \mid (\alpha, \beta) \in \Delta\}$$

and the **composition** $\Delta \circ \Gamma$ of two binary relations Δ and Γ on Ω is defined to be

$$\Delta \circ \Gamma = \{(\alpha, \gamma) \mid (\alpha, \beta) \in \Delta \text{ and } (\beta, \gamma) \in \Gamma \text{ for some } \beta \in \Omega\}.$$

The full relation algebra $R(\Omega)$ on Ω is defined as the algebra of all binary relations on Ω under the Boolean set-theoretic operations of meet, join and

---

* The research for this paper was supported by an Australian Research Council Grant number A68830358.

complementation, and under composition $\circ$ and conversion $\star$. A **relation algebra** R on Ω is a nonempty subset of $R(\Omega)$ which is closed under all of these operations. Let us examine a few examples of relation algebras.

Example 1.1 (a) For any set Ω the relation algebra generated by the empty relation $\emptyset$ is

$$\langle \emptyset \rangle = \{\emptyset, \Omega^2\}$$

and the relation algebra generated by the diagonal relation $\Delta_0 := \{(\alpha, \alpha) \mid \alpha \in \Omega\}$ is

$$\langle \Delta_0 \rangle = \{\emptyset, \Delta_0, \Omega^2 \setminus \Delta_0, \Omega^2\}.$$

These two relation algebras, $\langle \emptyset \rangle$ and $\langle \Delta_0 \rangle$, will be called **trivial relation algebras** and all others will be called **nontrivial**.

(b) For any nonempty subset Σ of $\Omega = \Sigma \cup \bar{\Sigma}$, the relation algebra $\langle \Sigma^2 \cup \bar{\Sigma}^2 \rangle$ generated by $\Sigma^2 \cup \bar{\Sigma}^2$ consists of $\emptyset$, $\Sigma^2 \cup \bar{\Sigma}^2$, $(\Sigma \times \bar{\Sigma}) \cup (\bar{\Sigma} \times \Sigma)$, and Ω^2.

It may seem somewhat arbitrary to call the relation algebra $\langle \Sigma^2 \cup \bar{\Sigma}^2 \rangle$ nontrivial, but the reasons for this will become clear when we introduce the Galois connection. The most natural class of relation algebras on Ω arises from permutation groups on Ω.

Example 1.1 (c) For each permutation group $G \leq \text{Sym}(\Omega)$ there is a natural action of G on $\Omega \times \Omega$ defined by

$$(\alpha, \beta)^g := (\alpha^g, \beta^g)$$

for $(\alpha, \beta) \in \Omega \times \Omega$ and $g \in G$. The relation algebra $S(G)$ generated by the set of G-orbits in $\Omega \times \Omega$ is just the set consisting of the empty relation, and all unions of the G-orbits in $\Omega \times \Omega$. In other words $S(G)$ consists of all subsets of $\Omega \times \Omega$ which are fixed setwise by G.

A generalization of this class of examples is obtained by replacing the set of G-orbits in $\Omega \times \Omega$ by the set of fundamental relations in an association scheme or coherent configuration or cellular ring (see [3, 8, 11, 24]).

In 1969, Wielandt [25] developed a theory of invariant relations for permutation groups. In [25] a subset of $\Omega \times \Omega$ fixed setwise by G was called a 2-**relation** of G and the set $S(G)$ was denoted 2-**rel** G. Wielandt studied the properties of the sets $S(G)$, and similar sets for higher order relations, with a view to obtaining information about (primitive) permutation groups. It is unfortunate that his work is not more widely known outside group theory.

We shall be interested in the lattice of subalgebras of $R(\Omega)$, and its relation to the lattice of subgroups of the symmetric group $\mathrm{Sym}(\Omega)$. First we observe that the group $\mathrm{Aut}\,R(\Omega)$ of algebra automorphisms of $R(\Omega)$ is isomorphic to $\mathrm{Sym}(\Omega)$.

Theorem 1.2 For any set Ω, $\mathrm{Aut}\,R(\Omega)$ is isomorphic to $\mathrm{Sym}(\Omega)$.

Proof Any permutation g of Ω induces a mapping $\hat{g} : R(\Omega) \to R(\Omega)$, given by

$$\hat{g} : \Delta \to \Delta^g$$

for $\Delta \subseteq \Omega \times \Omega$, where $\Delta^g := \{(\alpha^g, \beta^g) \mid (\alpha, \beta) \in \Delta\}$. This mapping $\hat{g}$ preserves all of the operations of $R(\Omega)$ so $\hat{g} \in \mathrm{Aut}\,R(\Omega)$. Conversely an automorphism $h \in \mathrm{Aut}\,R(\Omega)$ permutes amongst themselves relations of the form $\Delta_\alpha = \{(\alpha, \alpha)\}$, for $\alpha \in \Omega$, since these are the relations Δ such that $\Delta \circ \Gamma \subseteq \Gamma$ for all $\Gamma \in R(\Omega)$. Thus we can associate with h a permutation $h^\star \in \mathrm{Sym}(\Omega)$ defined by

$$\alpha^{h^\star} = \beta \text{ if and only if } \Delta_\alpha^h = \Delta_\beta.$$

Further, since h preserves composition, and since the relation $\Gamma = \{(\alpha, \beta)\}$ satisfies $\Gamma = \Delta_\alpha \circ \Gamma \circ \Delta_\beta$ it follows that $\Gamma^h = \{(aNpha^{h^\star}, \beta^{h^\star})\}$ for all α, β in Ω. It follows that the maps $g \to \hat{g}$ and $h \to h^\star$ are mutual inverses and consequently that $\mathrm{Aut}(\Omega)$ and $\mathrm{Sym}(\Omega)$ are isomorphic.

From now on we shall identify elements of $\mathrm{Aut}\,R(\Omega)$ with the corresponding permutations of Ω. The connection between subalgebras of $R(\Omega)$ and subgroups of $\mathrm{Sym}(\Omega)$ is defined below.

Definition 1.3 Mappings σ and ρ

$$\left\{ \begin{array}{c} \text{subsets } A \\ \text{of } \mathrm{Sym}(\Omega) \end{array} \right\} \begin{array}{c} \xrightarrow{\sigma} \\ \xleftarrow{\rho} \end{array} \left\{ \begin{array}{c} \text{subsets } S \\ \text{of } R(\Omega) \end{array} \right\}$$

are defined by

$$A^\sigma := \{\Delta \in R(\Omega) \mid \Delta^a = \Delta \text{ for all } a \in A\},$$
$$S^\rho := \{a \in \mathrm{Sym}(\Omega) \mid \Delta^a = \Delta \text{ for all } \Delta \in S\},$$

for $A \subseteq \mathrm{Sym}(\Omega)$ and $S \subseteq \mathrm{R}(\Omega)$.

It is not difficult to check that A^σ is a subalgebra of $\mathrm{R}(\Omega)$ for each $A \subseteq \mathrm{Sym}(\Omega)$, and S^ρ is a subgroup of $\mathrm{Sym}(\Omega)$ for all $S \subseteq \mathrm{R}(\Omega)$, and that

$$A^{\sigma\rho} \supseteq A, \ A^{\sigma\rho\sigma} = A^\sigma, \ S^{\rho\sigma} \supseteq S, \ S^{\rho\sigma\rho} = S^\rho.$$

Also the mappings σ and ρ are inclusion-reversing. Jonsson [10] called a subalgebra S and a subgroup A **closed** if $S^{\rho\sigma} = S$ or $A^{\sigma\rho} = A$ respectively. Thus the set of closed subalgebras or subgroups is the image of the map ρ or σ respectively. Wielandt [25, Definition 5.3] defined the **2-closure** $G^{(2)}$ of a permutation group G to be the group of all permutations which preserve every 2-relation of G, that is $G^{(2)} = G^{\sigma\rho}$, and he called a group G **2-closed** if $G = G^{(2)}$, that is if G is closed in Jonsson's sense. We shall use Wielandt's notation for 2-closures of groups. We observe that all closed subalgebras contain the trivial ones.

Lemma 1.4 If $|\Omega| \geq 2$ and $S \subseteq \mathrm{R}(\Omega)$, then S is trivial if and only if S^ρ is 2-transitive.

A permutation group is said to be **2-transitive** if it is transitive on ordered pairs of distinct points. Lemma 1.4 follows from the fact that $S \subseteq S^{\rho\sigma}$, that $\mathrm{Sym}(\Omega) = \langle \Delta_0 \rangle^\rho = \langle \emptyset \rangle^\rho$, and that, for a 2-transitive group G, $G^\sigma = \langle \Delta_0 \rangle$. The next lemma follows from the fact that σ is inclusion-reversing.

Lemma 1.5 For each $G \leq \mathrm{Sym}(\Omega)$, the subalgebra G^σ contains $\langle \Delta_0 \rangle$. In particular every closed subalgebra contains the two trivial subalgebras.

2. Extremal closed relation algebras.

The set of closed relation algebras on Ω, and, equivalently, the set of 2-closed permutation groups on Ω have not been classified, and the first step is to seek an understanding of the extremal elements of these sets. The maximal, proper, closed, relation algebras, and the corresponding minimal , nontrivial 2-closed, permutation groups are easily obtained. (A permutation group G on Ω is said to be **semi-regular** if the only element of G which fixes a point of Ω is the identity.)

Lemma 2.1 Let Ω be finite, $|\Omega| \geq 2$, and $G \leq \mathrm{Sym}(\Omega)$.

(a) ([25, 5.12] or [10] for cyclic groups.) If G has an orbit in Ω of length $|G|$ then G is 2-closed.

(b) ([10]) The group G is a minimal nontrivial 2-closed permutation group on Ω if and only if $|G|$ is prime.

Classifying the minimal, nontrivial, closed, relation algebras on Ω is more difficult. These correspond to the subgroups of $\text{Sym}(\Omega)$ which are maximal subject to being not 2-transitive on Ω.

Lemma 2.2 [10, 14] Let $|\Omega| \geq 2$, and let $G \leq \text{Sym}(\Omega)$, and $S \leq \text{R}(\Omega)$.

(a) If G is maximal subject to being not 2-transitive on Ω then G is 2-closed.

(b) The subalgebra S is a minimal nontrivial closed relation algebra on Ω if and only if S^ρ is maximal in $\text{Sym}(\Omega)$ subject to being not 2-transitive.

Proof.

(a) Since G is not 2-transitive, G^σ is nontrivial and hence $G^{\sigma\rho}$ is not 2-transitive, using Lemma 1.4. Then by the maximality of G, and as $G \subseteq G^{\sigma\rho}$, we have $G = G^{\sigma\rho}$.

(b) A subalgebra S is a minimal nontrivial closed subalgebra if and only if S^ρ is a 2-closed permutation group which is maximal subject to being not 2-transitive, using Lemmas 1.4 and 1.5. Part (a) shows that the maximality property implies 2-closure.

In 1986, Ivo Düntsch asked about such maximal, not 2-transitive, permutation groups, as Jonsson had suggested that he investigate some of the corresponding relation algebras. At that time the classification of maximal subgroups of $\text{Sym}(\Omega)$, by Liebeck, Saxl and the author, was about to be published. Exploiting the O'Nan-Scott Theorem [16, 23] for finite primitive permutation groups it is not difficult to show that a maximal subgroup G of $\text{Sym}(\Omega)$, where $|\Omega| = n$, belongs to one of the following six families.

(a) **intransitive**: $S_k \times S_{n-k}$, $1 \leq k \leq n/2$,

(b) **imprimitive**: $S_a \text{ wr } S_b$, $n = ab$, $a > 1$, $b > 1$,

(c) **affine**: $\text{AGL}_d(p)$, $n = p^d$, $d \geq 1$, p prime,

(d) **diagonal**: $T^k.(\text{Out } T \times S_k)$, $n = |T|^{k-1}$, $k > 1$, T a nonabelian simple group, (where the extension may be non-split),

(e) **product type**: $S_a \text{ wr } S_b$, $n = a^b$, $a \geq 5$, $b > 1$,

(f) **almost simple:** $T \leq G \leq \operatorname{Aut} T$, T a nonabelian simple group, and G primitive on Ω.

It was shown in [15] that all groups in (a) (if $k \neq n/2$) , (b), (c) (see [19] for $d > 1$), (d), and (e) (see also [9]) are maximal in $\operatorname{Sym}(\Omega)$. Also it was shown that, in case (f), the subgroup $G = \operatorname{N}_{\operatorname{Sym}(\Omega)}(T)$ is maximal in $\operatorname{Alt}(\Omega)G$ unless $G < H < \operatorname{Alt}(\Omega)G$ with (G, H, n) in an explicit list of exceptions. (These and subsequent results make use of the finite simple group classification for their proof.)

In order to get a satisfactory classification of the maximal, not 2-transitive permutation groups it is necessary to identify the 2-transitive groups in the discussion above and to examine their subgroups. It was shown already by Burnside [4] that a finite 2-transitive permutation group is either a subgroup of an affine group, or is almost simple. Moreover the affine and almost simple 2-transitive groups are known explicitly, using the finite simple group classification (see [5, 6, 7, 13]). It is therefore necessary, for each 2-transitive group H, to find all subgroups G of H which are maximal in H subject to being not 2-transitive, and to decide, for each such subgroup G, if it is indeed maximal in $\operatorname{Sym}(\Omega)$ subject to being not 2-transitive. The results in [21], about inclusion of one primitive gr oup in another, are sufficient to show that a primitive subgroup $G = \mathbf{Z}_p^d.G_0$ of $\operatorname{AGL}_d(p)$ which is maximal, not 2-transitive in $\operatorname{AGL}_d(p)$, is either maximal not 2-transitive in an almost simple 2-transitive group, or is a subgroup of $\operatorname{AGL}_r(p) \operatorname{wr} \operatorname{S}_k$ for some $k > 1$ with $d = rk$, and this group lies in $\operatorname{S}_{p^r} \operatorname{wr} \operatorname{S}_k$. Thus the classification of the groups G in the affine case reduces to the classification of irreducible, primitive subgroups G_0 of $\operatorname{GL}_d(p)$ which are maximal intransitive on the nonzero vectors of the underlying vector space. This latter classification was completed using Aschbacher's classification of "geometric" linear groups [1].

The strategy for dealing with maximal, not 2-transitive subgroups G of almost simple 2-transitive groups H is quite different. To find such groups G we choose H to be minimal by inclusion such that $G \subset H$ and H is 2-transitive (and almost simple). Then G is maximal in H. If G contains the socle of H then $G = \operatorname{L}_2(8) < H = \operatorname{Ree}(3)$. In all other cases we have a maximal factorization $H = GH_\alpha$, where $\alpha \in \Omega$, with G not containing the socle of H, and with $G_\alpha = G \cap H_\alpha$ maximal in G. All such factorizations are listed in [18], and for each group G in these lists we must check if

G is maximal, not 2-transitive in $\mathrm{Sym}(\Omega)$. An informal statement of the classification in [14] is given below.

Theorem 2.3 [14] Let $G \leq \mathrm{Sym}(\Omega)$ be maximal not 2-transitive on a set Ω of n points. Then one of the following holds.

(a) G is one of the groups in the families (a), (b), (d) or (e) above.

(b) G is almost simple, primitive but not 2-transitive, and maximal in $\mathrm{Alt}(\Omega)G$.

(c) $G = \mathbf{Z}_p^d.G_0 < \mathrm{AGL}_d(p)$, where $n = p^d$, p prime, and G_0 is maximal intransitive on nonzero vectors. Moreover one of the following holds.

   (i) G_0 preserves an extension field $\mathrm{GF}(p^d)$, or a nondegenerate quadratic form up to scalar multiplication, or a tensor product decomposition of the underlying vector space.

   (ii) G_0 normalizes an extraspecial r-group for some prime r dividing d.

   (iii) G_0 is one of a few exceptional groups with either $d = 2$, $p \leq 59$, or $(d, p) = (4, 3)$.

   (iv) G_0, modulo scalars, is almost simple and absolutely irreducible.

(d) G is a maximal subgroup of an almost simple 2-transitive group H where one of

   (i) $G = \mathrm{L}_2(8) < H = \mathrm{Ree}(3)$, $n = 28$,

   (ii) $H = \mathrm{Sp}_{2m}(2)$ acting on quadratic forms of type $\epsilon = \pm$ and G is $\mathrm{Sp}_{2a}(q^b).b$ (with $m = ab$), $O_{2m}^{-\epsilon}(2)$, S_{10} (with $m = 4, \epsilon = -$), $\mathrm{U}_3(3).2$ (with $m = 3$, $\epsilon = +$), or $\mathrm{L}_2(8)$ (with $m = 3$, $\epsilon = -$).

   (iii) $G = \mathrm{M}_{22} < H = \mathrm{HS}$, $n = 176$.

3. k-closed permutation groups.

As we mentioned above, finite 2-closed permutation groups have not been classified. In this section we consider how close a finite primitive permutation group is to its 2-closure. A conceptually helpful way of doing this is by looking at the sequence of k-closures of a permutation group, for $k \geq 2$. These were introduced by Wielandt [25].

For $k \geq 2$ a subset Δ of $\Omega^k = \Omega \times \ldots \times \Omega$ (k copies) is called a k-**relation** on Ω, and for $G \leq \mathrm{Sym}(\Omega)$, the set of G-invariant k-relations is denoted by k-**rel** G. Then Wielandt defined the k-**closure** $G^{(k)}$ of G as

$$G^{(k)} := \{a \in \mathrm{Sym}(\Omega) | \Delta^a = \Delta \text{ for all } \Delta \in k\text{-rel}G\}.$$

Of course the correspondence between groups and their sets of k-relations could be set up as a Galois correspondence between $\mathrm{Sym}(\Omega)$ and the algebra $R_k(\Omega)$ of all k-relations on Ω (under suitable operations). Wielandt observed that

$$G \leq \ldots \leq G^{(k)} \leq G^{(k-1)} \leq \ldots \leq G^{(2)}$$

for $k > 2$, and if $|\Omega| = n$ is finite then $G = G^{(n-1)}$. In fact Wielandt showed in [25, 5.12] that $G = G^{(k)}$ if there are points $\alpha_1, \ldots, \alpha_{k-1}$ in Ω such that $G_{\alpha_1, \ldots, \alpha_{k-1}} = 1$, that is if G has a base of length less than k. In [17, 22] it was shown that for finite primitive permutation groups the k-closure is close to G for quite small values of k.

Theorem 3.1 [17, 22] Let G be a finite primitive permutation group. Then, for any $k \geq 6$, the k-closure $G^{(k)}$ has the same socle as G.

Moreover if $G < H \leq G^{(k)}$, where $k \geq 2$ and G and H have different socles, then G and H are explicitly classified in [17, 22]. M.H. Klin asked if a stronger result might be true for primitive groups, namely if there was a constant c such that, for any finite primitive group G, and any $k \geq c$, the k-closure $G^{(k)}$ was equal to G. However Peter Neumann has pointed out that this is not the case: for any positive integers k and r, and for any $m \geq rk+2$, the groups $G := \mathrm{Alt}(m)$ and $H := \mathrm{Sym}(m)$, acting on the set of r-element subsets of a set of size m, have the same k-closure, $G^{(k)} = H^{(k)}$, so $G^{(k)} \neq G$. There are similar examples for affine and projective groups. Thus Theorem 3.1 is the best we might hope for, at least for primitive groups with nonabelian socles.

Finally in this section we mention Krasner algebras investigated by M. Krasner in 1938. Krasner [12] considered an algebra of relations on a set Ω which is the union $\bigcup_{k \geq 2} R_k(\Omega)$ under appropriate operations. A Galois correspondence for this algebra, see [20], can be defined in a manner similar to that discussed above. For this algebra the lattice of closed subalgebras corresponds precisely to the lattice of all subgroups of $\mathrm{Sym}(\Omega)$.

4. Reflexivity of operator algebras

The concept of a finite 2-closed permutation group is analogous to that of a reflexive algebra of linear operators on a finite dimensional Hilbert space. The analogy is like this. The finite set Ω corresponds to a finite-dimensional Hilbert space H, the group $\mathrm{Sym}(\Omega)$ corresponds to the algebra $\mathcal{L}(H)$ of all linear operators on H, the set 2-rel G of all invariant binary relations of a subgroup G of $\mathrm{Sym}(\Omega)$ corresponds to the lattice $\mathrm{lat}(\mathcal{A})$ of all invariant subspaces of H under a subalgebra $\mathcal{A}$ of $\mathcal{L}(H)$, and the 2-closure $G^{(2)}$ of G corresponds to the subalgebra $\mathrm{Alg}(\mathrm{lat}(\mathcal{A}))$ of all operators which preserve all the subspaces in $\mathrm{lat}(\mathcal{A})$. A subalgebra $\mathcal{A}$ of $\mathcal{L}(H)$ is called **reflexive** if $\mathcal{A} = \mathrm{Alg}(\mathrm{lat}(\mathcal{A}))$. Of course the difficult questions in operator theory do not always have analogues for relation algebras, but there is a rather curious result of Azoff [2] about k-reflexivity which provided new information about the centralizer algebra $\mathcal{A}'$ and second centralizer algebra $\mathcal{A}''$ of a subalgebra $\mathcal{A}$. The analogous problem for closures of permutation groups is quite interesting.

Given a subalgebra $\mathcal{A}$ of $\mathcal{L}(H)$, and an integer $m \geq 2$, there is a natural action of $\mathcal{A}$ as an algebra of linear operators on the direct sum H^m of m copies of H, namely for $A \in \mathcal{A}$ and $\mathbf{x} = (x_1, \ldots, x_m) \in H^m$, $\mathbf{x}A :=$ $(x_1 A, x_2 A, \ldots, x_m A)$. Let us denote the corresponding subalgebra of $\mathcal{L}(H^m)$ by $\mathcal{A}(m)$. (In [2] the notation $\mathcal{A}^{(m)}$ is used but this conflicts with Wielandt's notation for the m-closure of a permutation group.) Then $\mathcal{A}$ is said to be m-**reflexive** if $\mathcal{A}(m)$ is reflexive. Azoff showed that for Hilbert spaces H of dimension n, every algebra $\mathcal{A}$ is $(n-1)$-reflexive but there is an algebra which is not $(n-2)$-reflexive. He used the concept of m-reflexivity to find a commutative subalgebra $\mathcal{A}$ such that $\mathcal{A} \neq \mathcal{A}'' \cap \mathrm{Alg}(\mathrm{lat}(\mathcal{A}))$.

The analogue for permutation groups is as follows. Given $G \leq \mathrm{Sym}(\Omega)$, and an integer $m \geq 2$, let $G(m)$ denote the subgroup of $\mathrm{Sym}(\Omega^m)$ induced by G on Ω^m under the natural action:

$$\alpha^g := (\alpha_1^g, \ldots, \alpha_m^g), \text{ for } \alpha = (\alpha_1, \ldots, \alpha_m) \in \Omega^m, \ g \in G.$$

Then $G(m)$ corresponds to $\mathcal{A}(m)$ and the corresponding property of m-reflexivity is the property that $G(m)$ is 2-closed. So the basic question for permutation groups corresponding to Azoff's question for m-reflexivity is:

$$\text{For which } m \text{ is } G(m) = G(m)^{(2)}?$$

This admits a straightforward answer.

Proposition 5.1 Let $G \leq \mathrm{Sym}(\Omega)$ and let m and k be integers, $m \geq 2$, $k \geq 2$. Then

(a) For $\Delta \subseteq \Omega^{mk}$, $\Delta \in k\text{-rel}G(m)$ if and only if $\Delta \in mk\text{-rel}G$. Thus $G(m)^{(k)} = G^{(mk)}$. In particular $G(m)$ is 2-closed if and only if G is $2m$-closed.

(b) If G has a base in Ω of length less than r then $G(\lceil \frac{r}{2} \rceil)$ is 2-closed. In particular, if $|\Omega| = n$ then, for every group G, $G(\lceil \frac{n-1}{2} \rceil)$ is 2-closed, but, for $G = A_n$, $G(\lfloor \frac{n-2}{2} \rfloor)$ is not 2-closed.

Proof That a subset Δ of $\Omega^{mk} = (\Omega^m)^k$ is a $G(m)$-invariant k-relation if and only if it is a G-invariant mk-relation follows from the definitions. Consequently $G(m)^{(k)} = G^{(mk)}$. Part (b) follows from [25, 5.12].

The analogue of the question for centralizer algebras is the following and is not settled.

For an abelian subgroup G of $\mathrm{Sym}(\Omega)$ let C be the centralizer of G in $\mathrm{Sym}(\Omega)$, and C_2 the centralizer of C in $\mathrm{Sym}(\Omega)$. Then $G \leq C_2 \cap G^{(2)}$. When is $G = C_2 \cap G^{(2)}$? Of course if G is semi-regular on Ω then $G = G^{(2)} = G^{(2)} \cap C_2$. Similarly if G is the product of its transitive constituents then $G = G^{(2)} = G^{(2)} \cap C_2$. On the other hand if $G = \langle (12)(34), (34)(56) \rangle$ then $C = C_2 = G^{(2)} = \langle (12), (34), (56) \rangle \neq G$.

References

[1] Aschbacher, M., 'On the maximal subgroups of the finite classical groups', *Invent. Math.* **76** (1984), 469–514.

[2] Azoff, E.A., 'K-Reflexivity in finite dimensional spaces', *Duke Math. J.* **40** (1973), 821–830.

[3] Bannai, E. and Ito, T., *Algebraic Combinatorics I, Association Schemes*, (Benjamin/Cummings, Menlo Park, Calif., 1984).

[4] Burnside, W., *Theory of Groups of Finite Order*, (Cambridge Univ. Press, Cambridge, 1911).

[5] Cameron, P.J., 'Finite permutation groups and finite simple groups', *Bull. London Math. Soc.* **13** (1981), 1–22.

[6] Hering, C., 'Transitive linear groups and linear groups which contain irreducible subgroups of prime order', *Geom. Ded.* **2** (1974), 425–460.

[7] Hering, C., 'Transitive linear groups and linear groups which contain irreducible subgroups of prime order, II', *J. Algebra* **93** (1985), 151–164.

[8] Higman, D.G., 'Intersection matrices for finite permutation groups', *J. Algebra* **6** (1967), 22–42.

[9] Jones, G.A. and Soomro, K.D., 'The maximality of certain wreath products in alternating and symmetric groups', *Quart. J. Math. (Oxford)* **37** (1986), 419–435.

[10] Jonsson, B., 'Maximal algebras of binary relations', Contemp. Math. (eds. K. Appel et al) **33** (1984), 299–307.

[11] Klin, M.H., Muzichuk, M.E. and Faradzev, I.A., 'Cellular rings and groups of automorphisms of graphs', to appear in *Algebraic Theory of Combinatorial Objects, recent results and investigations*, Eds. I.A. Faradzev, A.A. Ivanov and M.H. Klin, (Mathematics and its Applications, Kluwer Acad. Pub., 1990/91).

[12] Krasner, M., 'Une généralisation de la notion de corps', *J. Math. Pures Appl.* **17** (1938), 367–385.

[13] Liebeck, M.W., 'The affine permutation groups of rank 3', *Proc. London Math. Soc. (3)* **54** (1987), 477–516.

[14] Liebeck, M.W. and Praeger, C.E., 'Relation algebras and finite permutation groups', *J. London Math. Soc.*(to appear).

[15] Liebeck, M.W., Praeger, C.E. and Saxl, J., 'A classification of the maximal subgroups of the finite alternating and symmetric groups', *J. Algebra* **111** (1987), 365–383.

[16] Liebeck, M.W., Praeger, C.E. and Saxl, J., 'On the O'Nan-Scott Theorem for finite primitive permutation groups', *J. Austral. Math. Soc. (A)* **44** (1988), 389–396.

[17] Liebeck, M.W., Praeger, C.E. and Saxl, J., 'On the 2-closures of primitive permutation groups', *J. London Math. Soc.* **37** (1988), 241–252.

[18] Liebeck, M.W., Praeger, C.E. and Saxl, J., 'The factorizations of the finite simple groups and their automorphism groups', *Memoirs, Amer. Math. Soc.* **86** (1990), 1-151.

19] Mortimer, B., 'Permutation groups containing affine groups of the same degree', *J. London Math. Soc.* **15** (1977), 445–455.

20] Pöschel, R. and Kaluznin, L.A., 'Funktionen und Relationenalgebren, ein Kapitel der Diskreten Mathematik', *Math. Monographien Herausgegeben* **15**, Eds. W. Gröbner and H. Reichardt, (VEB deutscherverlag der wissenschafter, Berlin, 1979).

21] Praeger, C.E., 'The inclusion problem for finite primitive permutation groups', *Proc. London Math. Soc. (3)* **60** (1990), 68–88.

22] Praeger, C.E. and Saxl, J., 'Closures of finite primitive permutation groups', (submitted) (1990).

23] Scott, L.L., 'Representations in characteristic p', *Proc. Symp. Pure Math.* **37** (1980), 318–331.

24] Weisfeiler, B. and Lehman, A. A., 'A reduction of a graph to a canonical form and an algebra arising during this reduction', (in Russian), *Nauchno-Technicheskaya Informatsia (Seriya 2)* **9** (1968), 12–16.

25] Wielandt, H.W., *Permutation groups through invariant relations and invariant functions*, (Lecture Notes, Ohio State Univ., 1969).

Symmetric presentations I: Introduction, with particular reference to the Mathieu groups M_{12} and M_{24}

R.T. Curtis

Abstract

In [7,8] we showed how the Mathieu group M_{12} is generated by five elements of order three whose set normalizer in M_{12} is isomorphic to the projective special linear group $PSL_2(5)$, and how the Mathieu group M_{24} is analogously generated by seven involutions whose set normalizer in M_{24} is isomorphic to $PSL_2(7)$. These two generating sets can be defined naturally as permutations of the twelve faces of the dodecahedron, and the twenty-four faces of the genus three Klein map [1,2,12] respectively. Thus we obtain directly analogous constructions of the Mathieu groups M_{12} and M_{24} from the two smallest non-abelian simple groups.

In this paper we make precise the concept of a group being 'symmetrically generated', obtain presentations for M_{12} and M_{24} implicitly in terms of these symmetric generating sets, and extend the ideas to other groups. It turns out that many of the sporadic simple groups are symmetrically generated in a particularly revealing manner.

0 Introduction

In the familiar notation of Coxeter [5] the symbol $\{p, q\}$ denotes a regular map whose faces are regular p-gons, q of which meet at each vertex. Consideration of the angle at a vertex rapidly shows us that the map is respectively spherical, Euclidean or hyperbolic according as the quantity

$$(p - 2)(q - 2)$$

is less than, equal to, or greater than 4. A systematic enumeration of the first two cases then follows; thus, for instance, the symbol $\{5, 3\}$ corresponds to the dodecahedron. Let us, in passing, recall that the group of rotational symmetries of $\{5, 3\}$ has order 60 and is isomorphic to the alternating group A_5.

The symbol $\{7, 3\}$ will, similarly, correspond to the regular tesselation of the hyperbolic plane into (infinitely many) heptagons meeting three at each vertex, and its group of rotational symmetries is the 'triangle group' (2,3,7) (see [5]).

Now a **Petrie polygon** of a regular map is defined to be a path along successive edges of the map in such a way that

(i) any two consecutive edges have a face in common, but

(ii) no three consecutive edges do.

If we take the tesselation $\{7,3\}$ and identify vertices which are distance eight apart along Petrie paths we obtain the finite **Klein map**, denoted $\{7,3\}_8$ by Coxeter, which has 24 heptagonal faces, 56 vertices and 84 edges. Klein [12] obtained this map on the genus three compact Riemann surface associated with the quartic

$$x^3 y + y^3 z + z^3 x = 0$$

whose group of automorphisms is isomorphic to the projective special linear group $PSL_2(7)$ and which thus attains the Hurwitz bound $84(g-1)$, where g is the genus. Geometrically the 24 faces correspond to the 24 points of inflexion of the curve, the 56 vertices to the 56 points of contact of its 28 bitangents, and the 84 edges to its 84 **sextactic points**, which is to say points where conics make six point contact with the curve. The linear transformations fixing the curve induce a faithful incidence-preserving action on the vertices of the map. The full group of symmetries of these two regular maps has, in each case, order twice that of the simple group given: $C_2 \times A_5$ for $\{5,3\}$, and $PGL_2(7)$ for $\{7,3\}_8$. Three dimensional models of the dodecahedron have been familiar and much-studied objects since the time of the Greeks. Moreover, while it is physically impossible for a three dimensional model to exhibit all the rotational symmetries of the Klein map, a fascinating model [see 13] of the dual map $\{3,7\}_8$ having 56 flat triangular faces, meeting seven at each vertex, and possessing full tetrahedral symmetry has been constructed.

Now it is well known that the 20 vertices of a dodecahedron can be coloured (in essentially two distinct ways), with five colours so that

(i) every face has a vertex of each colour, and

(ii) the four vertices of each colour determine a regular tetrahedron.

These five tetrahedra are permuted by the rotation group A_5, and the 'two distinct ways' are interchanged by reflections. This, in fact, sets up a correspondence between the faces of the dodecahedron and the 5-cycles in a conjugacy class of A_5, in such a way that opposite faces correspond to inverse permutations.

Analogously the 84 edges of the Klein map may be coloured with 7 colours (in essentially two distinct ways) so that

(i) every face has an edge of each colour, and

(ii) the edges of each colour form the blocks of imprimitivity under the action of the group of rotations.

For each of the five colours we define a permutation of the faces of the dodecahedron:

> *Each of the four red vertices has three faces incident with it; rotate these clockwise.*

This gives us five permutations of cycle-shape 3^4 which together generate M_{12}.

Similarly, for each of the seven colours, we define a permutation of the faces of the Klein map:

> *Each red edge is incident with two faces; interchange them.*

These seven involutions of cycle-shape 2^{12} together generate M_{24}.

A pleasing geometric aspect of the Klein quartic is well worth recording here: A straight line will have four points of contact with a quartic curve, and so tangents at points of inflexion must cut the curve once more. But the subgroup of $PSL_2(7)$ fixing one of the points of inflexion is a cyclic group of order seven which can easily be seen to fix a total of three points of inflexion (an *inflexional triangle*) and no further points [12]. Thus the tangent at a point of inflexion passes through another point of inflexion, and the edges of an inflexional triangle are tangents to the curve. The permutation of the points of inflexion which rotates each inflexional triangle with the sense thus defined corresponds to the permutation of the twenty-four 7-cycles in a conjugacy class of $PSL_2(7)$ induced by squaring. i.e. our element z in [7,8]. For further geometrical properties see [9,10].

It is our intention, in what follows, to produce a suitable generalization of the above which will give rise to other finite, and in particular sporadic, simple groups.

1   Symmetric generating sets

Accordingly, let G be a group, and $\mathcal{T} = \{t_1, t_2, t_3 \ldots, t_m\}$ be a subset of G. For $i = 1, 2, \ldots, m$ we then let $T_i = \langle t_i \rangle$, and $\overline{\mathcal{T}} = \{T_1, T_2, \ldots, T_m\}$. We define

$$N = N_G(\overline{\mathcal{T}})$$

i.e. N is the normalizer in G of the set of cyclic subgroups $\overline{T}$.

We refer to N as the **control subgroup**, and define T to be a **symmetric generating set** for G if, and only if, $G = \langle T \rangle$ and N permutes $\overline{T}$ doubly transitively by conjugation.

A presentation for G given explicitly, or implicitly, in terms of T is called a **symmetric presentation**. Similarly, if N permutes T *primitively*, we define T to be a **primitive generating set**, when a presentation given in terms of T would be a **primitive presentation**. The most significant difference between the above definition and previous usage of the phrase 'symmetric presentation' in the literature [3,11] is that we are requiring that all allowable permutations of the generators be realisable as inner automorphisms.

It is convenient to introduce the following notation:

$$N^i = C_N(t_i), \ N^{ij} = C_N(\langle t_i, t_j \rangle), \ \text{etc}\ldots.$$

$$N_i = N_N(T_i), \ N_{ij} = N_{N_i}(T_j), \ \text{etc}\ldots.$$

Clearly $N^i \leq N_i$, $N^{ij} \leq N_{ij}$, etc., and when N simply permutes the set of elements T without any normalizing action (e.g. when the t_i are involutions) the subscripts and and superscripts are interchangable. This is, in fact, the case for both M_{12} and M_{24}, but we prefer to have the more general notation available for future, subtler, investigations.

In the present paper we develop this approach and arrive at symmetric presentations for M_{12} and M_{24} which, although writable in terms of two generators, are based implicitly on the above generating sets. We give a number of other examples and lay the foundations for a systematic search for groups possessing such generating sets.

In practice our symmetric presentations will be given in terms of a set of generators for N(usually x and y will suffice), together with one of the symmetric generators t.

Our central problem will now be one of writing elements of the control subgroup N in terms of the symmetric generators. To do this we first note that any element of N which can be written in terms of t_i and t_j must commute with N^{ij}. We thus have the simple but powerful result:

Lemma 1
$$\langle t_i, t_j \rangle \bigcap N \leq C_N(N^{ij}).$$

In the case when the control subgroup is acting as straight-forward permutations of the symmetric generators, without normalizing action,

$$C_N(N^{ij})$$

is just the centralizer in the permutation group N of its two point stabilizer. Thus, starting with a given permutation group N we can immediately see which of its elements can be written in terms of two of the symmetric generators. Of course this result can readily be generalized to subgroups of G generated by more than two of our symmetric generators, and the general result is

Lemma 1′

$$\langle t_{i_1}, t_{i_2}, ..t_{i_m} \rangle \cap N \leq C_N(N^{i_1 i_2 ... i_m}).$$

In fact the condition is slightly stronger than this - as we shall see in the case of M_{24}.

The following are simply identities:

Lemma 2 *If G is a group and $x, t \in G$ then*

$$(xt)^m = 1 \quad \Leftrightarrow \quad x^{-m} = t^{x^{m-1}} t^{x^{m-2}} \ldots t^x t,$$
$$[t, x]^m = (t^{-1}.t^x)^m.$$

They are useful in observing how to write elements of the control subgroup N in terms of the symmetric generators $\mathcal{T}$.

2   The Mathieu group M_{12}

As our first example we take the Mathieu group M_{12}, whose behaviour was the motivation behind this investigation. As explained in the Introduction this group may be generated by five elements of order three which are normalized, as a subset, by a subgroup of M_{12} isomorphic to the alternating group A_5. Thus we are told by our Lemma 1 to look at $C_N(N^{01})$, the centralizer in A_5 of its two point stabilizer. Clearly

$$C_N(N^{01}) = \langle (2,3,4) \rangle \cong C_3,$$

and so the only elements in N which can be written in terms of two of the symmetric generators are the 3-cycles. Indeed we find that

$$(2,3,4) = (t_0^{-1}.t_1)^2.$$

We now take a presentation for A_5 and adjoin a further generator which we require to commute with a subgroup isomorphic to A_4. This generator will thus have just 5 (or 1 if the presentation collapses!) images under conjugation by the A_5. These will be our symmetric generators; they may be labelled so that the A_5 permutes their subscripts in the natural way. The above identity shows us how to write any 3-cycle of the A_5 in terms of the symmetric generators. Explicitly we are led to the following presentation:

$$\langle x, y, t; \ x^5 = y^3 = (xy)^2 = 1 = t^3 = [y, t] = [y, t^{x^3}] = (xt)^8, \ y = [x^3, t]^2 \rangle.$$

Here $x = (0, 1, 2, 3, 4)$, $y = (4, 2, 1)$ as permutations of the symmetric generators,

$$N = \langle x, y \rangle \cong A_5, \text{ and } C_N(t) = \langle y, y^{x^2} \rangle \cong A_4.$$

Thus t has only five images under conjugation by N viz. $t_i = t^{x^i}$. The relation

$$y = [x^3, t]^2 = (t_3^{-1} t_0)^2$$

tells us how to write a 3-cycle in terms of the symmetric generators, and the group defined by this much alone is isomorphic to $C_3 \times M_{12}$. Finally the relation $(xt)^8$ factors out the centre to give us M_{12}.

For the convenience of the reader we reproduce permutations in the usual version of M_{12} which correspond to the above generators. For more details, though, the reader is referred to [7,8].

	Action on 12 letters	Action on $\mathcal{T}$
$x =$	$(1, 3, 9, 5, 4)(2, 6, 7, X, 8)$	$(0, 1, 2, 3, 4)$
$y =$	$(\infty, 6, 7)(0, 3, 9)(1, 8, 5)(2, 4, X)$	$(4, 2, 1)$
$xy =$	$(\infty, 6)(0, 3)(1, 9)(2, 7)(4, 8)(5, X)$	$(0, 4)(2, 3)$
$t = t_0 =$	$(\infty, 8, X)(0, 3, 9)(1, 4, 7)(2, 6, 5)$	
$t^x = t_1 =$	$(\infty, 2, 8)(0, 9, 5)(3, 1, X)(6, 7, 4)$	
$t^{x^2} = t_2 =$	$(\infty, 6, 2)(0, 5, 4)(9, 3, 8)(7, X, 1)$	
$t^{x^3} = t_3 =$	$(\infty, 7, 6)(0, 4, 1)(5, 9, 2)(X, 8, 3)$	
$t^{x^4} = t_4 =$	$(\infty, X, 7)(0, 1, 3)(4, 5, 6)(8, 2, 9)$	

where the symbol X denotes 10.

Representatives of the two conjugacy classes of subgroups isomorphic to M_{11} are given by

$$\langle x, yt^{-1} \rangle \text{ and } \langle x^{yx^{-1}}, y^{-1}t \rangle.$$

As was described in [7,8] our control subgroup $N \cong A_5$ extends to a copy of S_5 in the automorphism group of M_{12}, Aut $M_{12} \cong M_{12}{:}2$, in which the odd elements permute and *invert* the symmetric generators. Using this information we are able to write down a presentation for $M_{12}{:}2$

$$\langle x, y, t; \ x^5 = y^2 = (xy)^4 = [x,y]^3 = 1 = t^3 = t^{x^{-1}y}.t^{x^3} = t^y.t = (xt)^8,$$

$$[x,y] = [x, t^x]^2 \rangle$$

where the action on the $\overline{T}$ is now given by

$$x = (0,1,2,3,4), \ y = (3,4), \ xy = (0,1,2,4) \text{ and } [x,y] = (0,3,4).$$

Subgroups isomorphic to M_{11} which are not conjugate in the simple group are

$$\langle x, t[y,x] \rangle \text{ and } \langle x, t[y, x^2] \rangle.$$

Perhaps one of the most pleasing things about the above presentation for M_{12} is the fact that it can be demonstrated, by hand, to give a group of order 95,040 as we assert. To see this we first let $K = \langle x, y, t_i^{-1} t_j \rangle$ and investigate the index of N in K. First observe that N, being a homomorphic image of A_5, has order 60 or 1, and that the latter case would lead to the whole group being trivial. Moreover if $K = N$ then $(t_i^{-1} t_j) \in N$ and, since $(t_0^{-1} t_1)^2 = (2,3,4)$, we have $(t_0^{-1} t_1) = (4,3,2)$, which commutes with t_0. Thus t_i commutes with t_j, and so $G = \langle x, y, t \rangle = \langle T \rangle$ is abelian. As G is visibly perfect this would imply that G is trivial. We may assume, therefore, that either G is trivial or $N \cong A_5$ and $N \neq K$.

To ease notation we shall let the symbol i stand for t_i, and $\overline{i}$ for t_i^{-1}. By a slight abuse of notation we shall allow an expression such as $\overline{i}j$ to stand not only for the *element* $t_i^{-1} t_j$, but also for the *coset* $N t_i^{-1} t_j$. However, to avoid any confusion, we shall denote equality of elements by the '=' sign, whereas equality of cosets will be signified by 'tilde'. Thus, since $(\overline{i}j)^2 \in N$, we have

$$\overline{i}j \sim \overline{j}i, \text{ although } \overline{i}j \neq \overline{j}i.$$

We, further, denote the identity coset N by $*$. Eleven, apparently distinct, cosets of N in K are now given by:

$$1 \ \rule{3cm}{0.4pt} \ 10$$

$$* \qquad\qquad \overline{i}j \sim \overline{j}i$$

We shall show that this is indeed all there are, when it will be apparent that either G is trivial, or K has order 660 and contains $N \cong A_5$ to index 11. From here it is straightforward to show that K is a homomorphic image of $PSL_2(11)$. To verify that the coset enumeration of K over N is complete we must show that right multiplication of any of the above by $\overline{0}1$ will return one of these 11 cosets. The symmetry of the situation shows that it will suffice to do this for $\bar{0}1$, $\bar{0}2$, $\bar{1}2$ and $\bar{2}3$. Now

$$\bar{0}1.\bar{0}1 \sim *, \quad \bar{0}2.\bar{0}1 \sim \bar{2}0.\bar{0}1 \sim \bar{2}1,$$

$$\bar{1}2.\bar{0}1 \sim \bar{2}1.\bar{0}1 \sim \bar{2}0.(\bar{0}1)^2 \sim \bar{2}0^{(2,3,4)} \sim \bar{3}0.$$

Finally
$$\bar{2}3.\bar{0}1 \sim \bar{3}2.\bar{0}1 \sim \bar{3}0.(\bar{0}2)^2.\bar{2}1 \sim \bar{3}0^{(1,4,3)}.\bar{2}1 \sim \bar{1}0.\bar{2}1$$

$$\sim \bar{0}1.\bar{2}1 \sim \bar{0}2.(\bar{2}1)^2 \sim \bar{0}2^{(0,4,3)} = \bar{4}2.$$

[As an interesting aside we exhibit $\bar{0}1 \in K$ as a permutation on the eleven symbols $\{*, \bar{\imath}j; i \neq j\}$:

$$\bar{0}1 : \quad (*, \bar{0}1)(\bar{0}2, \bar{2}1, \bar{3}0, \bar{3}1, \bar{4}0, \bar{4}1)(\bar{2}3, \bar{4}2, \bar{3}4),$$

from which we observe that $L_2(11)$ is generated by a set of elements of order 6 which correspond to the directed edges of a complete graph on five vertices. These elements are permuted by an A_5 which itself is generated by the squares of these 6-elements. Their 10 cubes, which must also generate $L_2(11)$, form a *primitive generating set* since they are permuted primitively, but not doubly transitively by $N \cong A_5$. Indeed, denoting the element $(\bar{\imath}j)^3$ above by t_{ij}, we see that

$$t_{01}t_{23}t_{01} = (0,3)(1,2) \text{ of } N.$$

From here we readily obtain

$$\langle x, y, t; x^5 = y^2 = (xy)^3 = 1 = t^2 = [t, xy] = [t^{[y,x^{-1}]}, y] = (yt)^3 \rangle$$

as a *primitive* presentation for $PSL_2(11)$, where $N \cong A_5$ with $x = (0,1,2,3,4)$, $y = (1,2)(3,4)$, $xy = (0,2,4)$ and $t = t_{13}$. The relator $(yt)^3$ is equivalent to $t_{13}t_{24}t_{13} = (1,2)(3,4)$].

To complete our demonstration that the original group has the order claimed we must perform a coset enumeration over K, where of course $t_i^{-1}t_j \in K$ for all i and j. Thus $Kt_i^{-1} = Kt_0^{-1}$ for all i, (which we shall denote by $\bar{\imath} \sim \bar{0}$). We shall leave the details of the enumeration to the reader but exhibit the Cayley diagram of G/K with respect to the symmetric generators.

Since $(t_0^{-1}t_1)^2$ commutes with t_0, it equals $(t_1t_0^{-1})^2$; it is now clear that the two copies of $L_2(11)$ contained in our M_{12} and containing N [see 7, figure 1] are $\langle N, \bar{\imath}j \rangle$ and $\langle N, i\bar{\jmath} \rangle$. They are interchanged by any *outer* automorphism of M_{12} which normalizes N e.g. $(1,2,4,3)(0)$ on $\mathcal{T}$ together with inversion.

3   The Mathieu group M_{24}

Before proceeding to M_{24} we need a symmetric presentation for $PSL_2(23)$. Here we take $N \cong S_4$, the symmetric group on four letters, and let $x =$

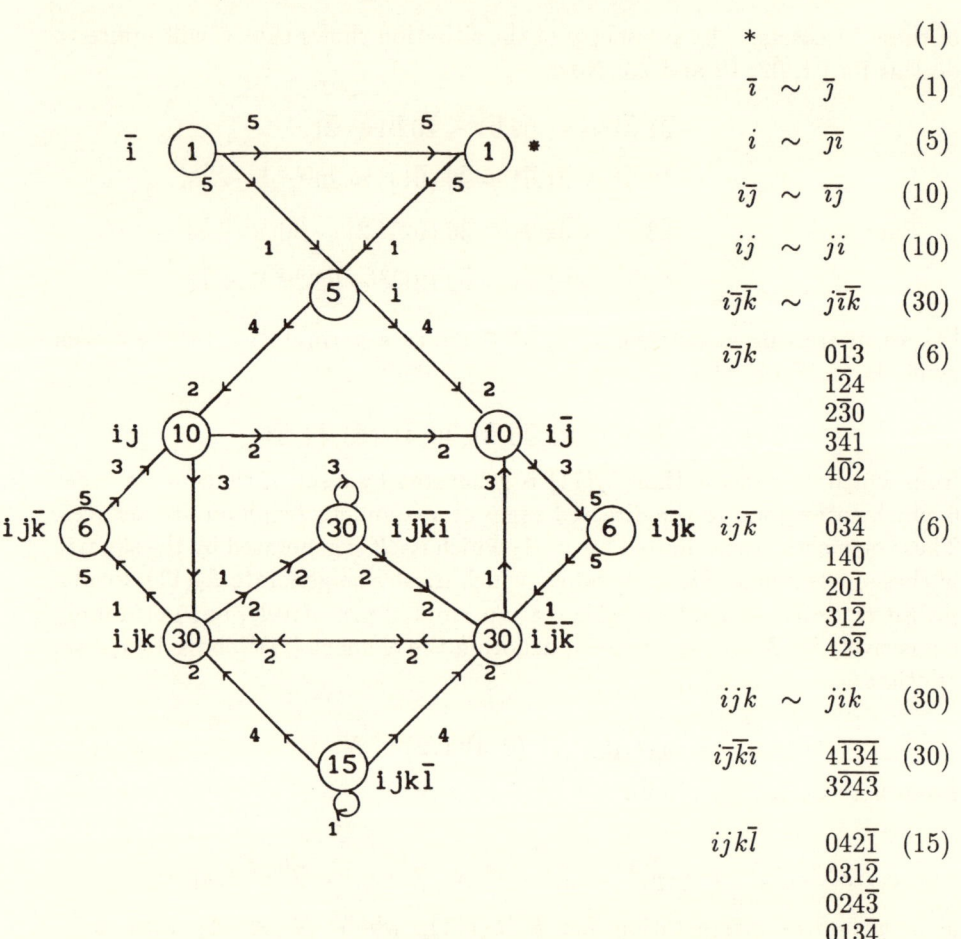

$$* \qquad (1)$$

$$\bar{\imath} \ \sim \ \bar{\jmath} \qquad (1)$$

$$i \ \sim \ \overline{\jmath\imath} \qquad (5)$$

$$i\bar{\jmath} \ \sim \ \overline{\imath\jmath} \qquad (10)$$

$$ij \ \sim \ ji \qquad (10)$$

$$i\bar{\jmath}\bar{k} \ \sim \ j\bar{\imath}\bar{k} \qquad (30)$$

$$i\bar{\jmath}k \qquad 0\bar{1}3 \qquad (6)$$
$$1\bar{2}4$$
$$2\bar{3}0$$
$$3\bar{4}1$$
$$4\bar{0}2$$

$$ij\bar{k} \qquad 03\bar{4} \qquad (6)$$
$$14\bar{0}$$
$$20\bar{1}$$
$$31\bar{2}$$
$$42\bar{3}$$

$$ijk \ \sim \ jik \qquad (30)$$

$$i\bar{\jmath}\bar{k}\bar{\imath} \qquad 4\overline{134} \qquad (30)$$
$$3\overline{243}$$

$$ijk\bar{l} \qquad 042\bar{1} \qquad (15)$$
$$031\bar{2}$$
$$024\bar{3}$$
$$013\bar{4}$$

Figure 1: The Cayley diagram of M_{12} over $L_2(11)$.

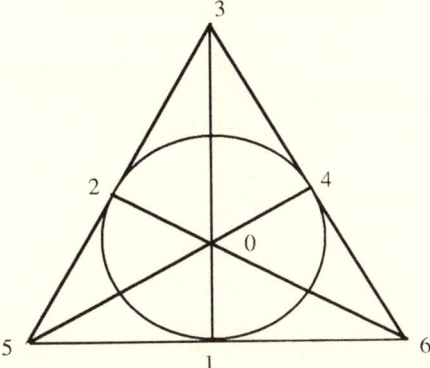

Figure 2: The 7-point projective plane

$(0,1,2,3)$, $y = (2,3)$, $xy = (0,1,3)$ be the action on the four involutory symmetric generators $\{t_0, t_1, t_2, t_3\}$. The usual observation shows us that

$$\langle t_0, t_1 \rangle \cap N \leq C_N(N^{01}) = C_N((2,3)) = \langle (0,1), (2,3) \rangle \cong V_4.$$

Now $\langle t_0, t_1 \rangle = H_{01}$, say, is a dihedral group, in this case of order six, and we have $y = (t_0 t_1)^3 = (t.t^x)^3 = [t,x]^3$. From here we can write down the surprisingly simple presentation:

$$\langle x, y, t; \; x^4 = y^2 = (xy)^3 = 1 = t^2 = (xyt)^{11}, \; y = [t,x]^3 \rangle \cong L_2(23),$$

where we have suppressed the redundant relations $[y,t] = [y,t^x] = 1$.

As we know M_{24} is generated by seven involutions whose set normalizer in M_{24} is isomorphic to $L_2(7)$. Furthermore the Lemma tells us that

$$\langle t_0, t_1 \rangle \cap N \leq C_N(N^{01}) = \langle (2,4)(5,6), (2,5)(4,6) \rangle \cong V_4,$$

where we are taking N to be the permutation group preserving the 7-point projective plane as illustrated in figure 2.

Since the t_i are involutions the subgroup $\langle t_0, t_1 \rangle$ is dihedral of order $2k$, say; it will thus have trivial centre unless k is even, in which case its centre will have order 2 and be generated by $(t_0 t_1)^{k/2} = (t_1 t_0)^{k/2}$. This element is visibly fixed by a permutation interchanging t_0 and t_1 and thus commutes with $(0,1)(4,6)$. But in the present case we have

$$C_N(N^{01}) \cap \langle t_0, t_1 \rangle = N^{01} \cap \langle t_0, t_1 \rangle \leq Z(\langle t_0, t_1 \rangle),$$

and so the only possible element of N writable in terms of t_0 and t_1 is $(2,5)(4,6)$. Note that this was deduced from $N \cong L_2(7)$ only and is independent of M_{24}. We check from [8] that

$$(2,5)(4,6) \ = \ (t_0 t_1)^3 \ = \ (t.t^x)^3 \ = \ [t,x]^3.$$

It is now natural to ask about the structure of the subgroup generated by three symmetric generators 'lying on a line', such as $\{t_0, t_1, t_3\}$. Clearly

$$\langle t_0, t_1, t_3 \rangle \ \leq \ C_{M_{24}}(\langle (2,4)(5,6), (2,5)(4,6) \rangle) \ \cong \ S_4 \times V_4.$$

In fact equality holds and the additional relation $[t_0 t_3 t_0, t_1]$ is sufficient to ensure this isomorphism. Our presentation for M_{24} then takes the form:

$$\langle x, y, t; \ x^7 = y^2 = (xy)^3 = [x,y]^4 = 1 = [y,t] = [y,t^x] = [y,t^{x^3}] = t^2$$

$$= (yxt)^{11} = [t^{x^3 t}, t^x], \ y = (t^x t^{x^3})^3 \rangle \ \cong \ M_{24},$$

where $x = (0,1,2,3,4,5,6)$, $y = (2,6)(4,5)$, $xy = (0,1,6)(2,3,5)$ in $N \cong L_2(7)$, and

$$t_i \ = \ t^{x^i}.$$

As for M_{12} we give permutations in the usual version of M_{24} which satisfy these relations. The reader who refers to [7] will notice that we have taken

a^{-1} as our x, c as y, and $\pi_1^{a^3}$ as t_0.

	Action on 24 letters	Action on $\mathcal{T}$
$x =$	$(\infty)(14,3,15,20,0,8,18)$	$(0,1,2,3,4,5,6)$
	$(17)(7,4,13,2,11,16,10)$	
	$(22)(21,5,1,9,6,19,12)$	
$y =$	$(\infty,7)(17,14)(22,21)(18,3)(10,4)(12,5)$	$(2,6)(4,5)$
	$(8,11)(16,6)(19,0)(20,1)(2,15)(9,13)$	
$t_0 =$	$(\infty,20)(1,7)(3,18)(11,8)(9,14)(13,7)$	
	$(5,0)(19,12)(4,16)(10,6)(2,15)(21,22)$	
$t_1 =$	$(\infty,0)(9,4)(15,14)(16,18)(6,3)(2,17)$	
	$(1,8)(12,21)(13,10)(7,19)(11,20)(5,22)$	
$t_2 =$	$(\infty,8)(6,13)(20,3)(10,14)(19,15)(11,17)$	
	$(9,18)(21,5)(2,7)(4,12)(16,0)(1,22)$	
$t_3 =$	$(\infty,18)(19,2)(0,15)(7,3)(12,20)(16,17)$	
	$(6,14)(5,1)(11,4)(13,21)(10,8)(9,22)$	
$t_4 =$	$(\infty,14)(12,11)(8,20)(4,15)(21,0)(10,17)$	
	$(19,3)(1,9)(16,13)(2,5)(7,18)(6,22)$	
$t_5 =$	$(\infty,3)(21,16)(18,0)(13,20)(5,8)(7,17)$	
	$(12,15)(9,6)(10,2)(11,1)(4,14)(19,22)$	
$t_6 =$	$(\infty,15)(5,10)(14,8)(2,0)(1,18)(4,17)$	
	$(21,20)(6,19)(7,11)(16,9)(13,3)(12,22)$	

Coset enumeration is slightly more difficult as we lack an identifiable sub-group of low index. However the presentation at the beginning of this section shows that

$$\langle t_2, t_4, t_5, t_6 \rangle \cong L_2(23)$$

and a Todd-Coxeter enumeration using the CAYLEY package [4] verifies that this subgroup has index 40,320 as required. It is worth noting that, since $\langle t_2, t_4 \rangle \ni (t_2 t_4)^3 = (0,3)(5,6)$,

$$\langle t_2, t_4, t_5 \rangle = \langle t_2, t_4, t_5, t_6 \rangle,$$

and, by the same argument, any four symmetric generators three of which lie on a line generate the whole group M_{24}. Finally

$$\langle x, yxtxy \rangle \cong M_{23},$$

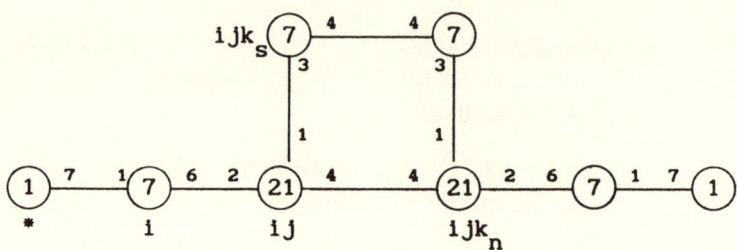

Figure 3: The Cayley diagram of $L_3(2)_2$.

thus enabling us to obtain the familiar permutation representation on 24 points.

Before leaving this example we recall that, if G is *any* group symmetrically generated by 7 involutions permuted within G by a subgroup $N \cong L_2(7)$, then the only element of N writable in terms of t_0 and t_1 is $(2,5)(4,6)$. Moreover $\langle t_0, t_1 \rangle$ is dihedral and $(2,5)(4,6)$, being in its centre, must have form $(t_0 t_1)^k$. We denote the well-defined group, which is independent of the particular presentation of $L_2(7)$ used and which has only this one additional relator, by $L_3(2)_k$.[We choose the form $L_3(2)$ rather than $L_2(7)$ to convey the permutation action on 7 letters]. Thus M_{24} is a homomorphic image of the group $L_3(2)_3$ obtained by factoring out the normal subgroup generated by the relators $[t_3^{t_0}, t_1]$ and $(xyt)^{11}$. It is natural to attempt to identify the group $L_3(2)_2$. The enumeration over N is again readily performed by hand to give Figure 3.

We have $(ij)^2 \in N \Rightarrow ij \sim ji$. In the diagram ijk_s stands for a *special* triple i.e. a line in the projective plane, and ijk_n for a *non-special* triple. G plainly has order $168 \times 36 \times 2 = 12,096$ and it is easily seen that:

$$L_3(2)_2 \cong G_2(2) \cong U_3(3) : 2,$$

where the outer automorphism corresponds to the obvious symmetry of the Cayley graph.

4   The Symmetric and Alternating groups

If we take our control group N isomorphic to the full symmetric group S_n permuting n symmetric generators without normalizing action, then our Lemma 1 tells us that

$$\langle t_i, t_j \rangle \cap N \leq C_{S_n}(S_{n-2}) = \langle (i,j) \rangle \cong C_2 \text{ for all } n \geq 5.$$

If the symmetric generators are involutions then, as above, $\langle t_i, t_t \rangle$ is dihedral but now (i,j) *interchanges* the generators and so has the form

$$(i,j) = t_i t_j \ldots t_i$$

of *odd* length. We thus obtain a family of well-defined groups S_n^k, k odd, in which transpositions are given in terms of the symmetric generators by such an expression of odd length k. The first non-trivial case is $k = 3$:

Theorem 1

$$S_n^3 \cong S_{n+1}.$$

Proof. Since $t_i t_j t_i \in N$ we have $ij \sim i$, and so under right multiplication t_j interchanges the cosets N and Nt_j and fixes all the others, i.e. t_j induces a transposition on the $n + 1$ cosets $\{*, i\}$.

Note that this proof implies the following:

Corollary 1 *If G is a group symmetrically generated by a set of n involutions $T = \{t_0, t_1, .., t_{n-1}\}$ such that $t_i t_j t_i \in N$, the control subgroup of G which faithfully permutes T doubly transitively , then $G \cong S_{n+1}$.*

Proof. We have $t_i t_j t_i : (* i)(* j)(* i) = (i\ j)$, and so $N \cong S_n$ and we are in the situation of Lemma 1.

The family of groups S_n^k is being investigated by A. Hammas and we shall make available details of his findings at a later date. For the moment, though, we record that

$$S_2^5 \cong D_{10}, \; S_3^5 \cong C_2 \times A_5, \; S_4^5 \cong 2^{\cdot}(A_5 \times A_5):2$$

$$\text{and } S_5^5((xt)^{17}; \; x = (0,1,2,3,4)) \cong Sp_4(4),$$

where, by the above, we mean S_5^5 factored by the additional relator indicated. Thus we see that the groups S_n^5 are closely related to the orthogonal groups $O_n(4)$. In the 3-dimensional case the three symmetric generators may be seen as reflective symmetries of the dodecahedron:

Let A be a vertex of the regular dodecahedron, and let F_1, F_2, F_3 be the three faces meeting at A. Then t_i is the reflection in the edge of F_i opposite A.

The factors of the group S_3^7 include many linear fractional groups.

Now let $G \cong A_{n+2}$ be the alternating group acting on $\{1, 2, \ldots, n+2\}$, and let $t_i = (i, n+1, n+2) \in G$. Clearly $T = \{t_i; i = 1, 2, \ldots, n\}$ is a symmetric generating set for G with control subgroup $N \cong S_n$, the symmetric group on n letters, permuting $\overline{T}$ naturally. As in the case of $\text{Aut} M_{12}$, odd elements

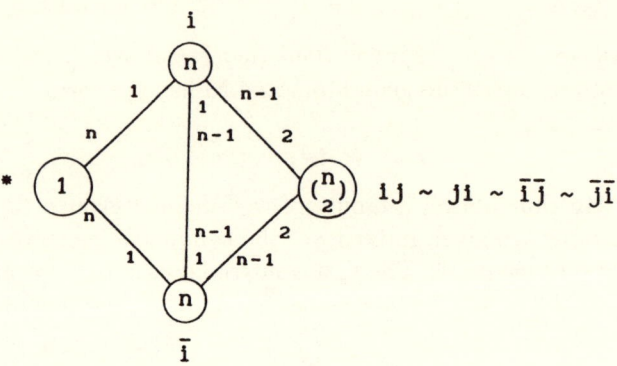

Figure 4: The Cayley diagram of A_{n+2} over S_n.

of N, that is those involving the transposition $(n+1, n+2)$, permute and invert the elements of $\mathcal{T}$. Our presentations will depend slightly on the parity of n.

Case I. n odd.

We let $x = (1, 2, \ldots, n)$, $y = (n-1, n)(n+1, n+2)$ when we see that $x^{-2}yx = (n, n-1, \ldots, 2)(n+1, n+2)$, and so

$$\langle x, y \rangle = N \cong S_n, \langle x^{-2}yx, y \rangle \cong S_{n-1}.$$

Moreover $(t_i t_j)^2 = 1$, $t_i^{-1} t_j t_i^{-1} = (i, j)(n+1, n+2)$, $t_k^{t_i t_j} = (k, j, i)$. In coset notation this says

$$* \sim ijij \sim \overline{i}j\overline{i} \sim \overline{j}\overline{i}kij,$$

from which we may readily deduce the Cayley diagram given in Figure 4.

Thus, given a presentation of S_n, we obtain a presentation of A_{n+2}:

$$\langle x, y, t;\ \langle x, y\rangle \cong S_n,\ t^3 = 1,\ t^y = t^{x^{-2}yx} = t^2,$$

$$y^x = t^{-1}[x^{-1}, t^{-1}],\ [y, x] = t^{x^2(tx^{-1})^2}\rangle.$$

Case II. n even

We now let $x = (1, 2, \ldots, n)(n + 1, n + 2)$, y and t be as above, and so $x^{-2}yx = (n, n - 1, \ldots, 2)$. The presentation is then modified to read:

$$\langle x, y, t;\ \langle x, y\rangle \cong S_n,\ t^3 = 1 = [t, x^{-2}yx],\ t^y = t^2,$$

$$y^x = t[x^{-1}, t],\ [y, x] = t^{x^2 tx^{-1}t^{-1}x^{-1}}\rangle.$$

Conclusion

The huge improvement in computing facilities and the availability of packages, such as CAYLEY and the MEATAXE, which contain coset enumeration programs has led to large numbers of new presentations of groups. These vary greatly both in form and purpose, many being found with a particular objective in mind. Efficient presentations, while perhaps telling us little about the structure of the group, often amaze us that so much intricate information can be conveyed with such economy. At the other extreme are the Coxeter presentations represented by weighted graphs in which each vertex represents a generating involution, and the weight of an edge gives the order of the product of the involutions incident with it (supplemented, in general, by further relations). By 'covering up' parts of the Coxeter diagram one can recognise various subgroups (see the Y-diagrams in [6]) and brevity is sacrificed for this additional information.

In a sense these symmetric presentations attempt to combine reasonable efficiency with considerable stuctural information. The examples given above all have three generators x, y and t, but in each case y is defined in terms of x and t. Thus they are two generator presentations. However, underlying the presentation given, and readily recoverable, are the symmetric generators $\mathcal{T}$, and it is natural to ask what subgroups are generated by subsets of $\mathcal{T}$. For instance in M_{12} we know that $(2, 3, 4) \in \langle t_0, t_1\rangle$, and so

$$\langle t_0, t_1\rangle \leq C_{M_{12}}(\langle(2, 3, 4)\rangle).$$

In fact

$$\langle t_0, t_1\rangle = C_{M_{12}}(\langle(2, 3, 4)\rangle) \cong C_3 \times A_4.$$

The same fact shows that in this case any three symmetric generators generate the whole of M_{12}.

The main problem with these presentations is that, in the absence of a known subgroup of low index, mechanized coset enumeration may prove difficult, although this can often be overcome by exploiting the symmetry.

This paper has been devoted to establishing notation and concepts, and investigating their role in a number of interesting cases. In future we shall start by specifying the control group N and then seek possible G containing it. Following this procedure we shall, in a sequel to this paper, construct the smaller Janko groups J_1 and J_2 complete with graphs and symmetric presentations.

References.

1. H.F. Baker. Note introductory to the study of Klein's group of order 168.

2. W. Burnside. *Theory of groups of finite order* (Cambridge University Press, 1911).

3. C.M. Campbell. Symmetric Presentations and linear groups. *Contemp. Math.* (1985), **45** 33-9.

4. J. Cannon. An introduction to the group theory language, Cayley. In *Computational Group Theory* (Academic Press, 1984), pp.145-83

5. H.S.M. Coxeter and W.O.J. Moser. *Generators and Relations for Discrete Groups* 4th Ed.(Springer-Verlag, 1984).

6. J.H. Conway, R.T. Curtis, S.P. Norton, R.A. Parker, R.A. Wilson. *ATLAS of Finite Groups* (Oxford University Press, 1985).

7. R.T. Curtis. Natural constructions of the Mathieu groups. *Math. Proc. Cam. Phil. Soc.*(1989), **106** 423-9.

8. R.T. Curtis .Geometric interpretations of the 'natural' generators of the Mathieu groups. *Math. Proc. Cam. Phil. Soc.*(1990), **107** 19-26.

9. W.L. Edge. Acta Mathematica **79** (1947) 153-223

10. W.L. Edge. Octadic surfaces and plane quartic curves. *Proc. L.M.S.(2)* **34** (1932) 492-523.

11. D.L. Johnson. Presentations of groups. *London Math. Soc. Lecture Notes 22 (Cambridge University Press,* 1976).

12. F. Klein. Uber die Transformation siebenter Ordnung der elliptischen Functionen. *Gesammelte Math. Abhandlungen III* 90-135.

13. E. Schulte and J.M. Wills. A polyhedral realization of Felix Klein's map $\{3,7\}_8$ on a Riemann surface of genus 3. *J. London Math. Soc.*(1985) (2) **32** 539-547.

Finite and locally finite groups containing a small subgroup with small centralizer

B. HARTLEY

1. Elements with bounded centralizer order. Let G be a finite group containing a subgroup A of order $\leq n$ with centralizer $C_G(A)$ of order $\leq k$. What can be said about G in terms of these parameters, possibly also involving the structure of A? In this paper we describe, without proofs, some recent results on this question. Some of them, not surprisingly, have applications to locally finite groups. A slightly different question, with a similar flavour, is discussed in §2.

The most famous theorem of this type is the classical one of Brauer and Fowler [4], which asserts that if G is a finite simple group containing an involution i, then $|G|$ is bounded in terms of $|C_G(i)|$. More generally, similar arguments show that the same holds if i is an involutory automorphism of G and $C_G(i)$ is the fixed point group of i, provided of course that G is non-abelian. The proofs of these facts are elegant and essentially elementary. We give below the generalization to automorphisms of arbitrary order. This uses the classification of finite simple groups, which we have no compunction in using where convenient in this paper.

Theorems that bound some invariant of finite groups in terms of others often have consequences in locally finite group theory. For example, it is not too difficult to deduce from the automorphism form of the Brauer-Fowler Theorem that a locally finite group containing an involution with finite centralizer has a locally soluble subgroup of finite index, and much stronger results are true (see below). The kind of results we look for are selected with such applications in mind.

THEOREM 1.1. *[13] Let G be a finite non-abelian simple group admitting an automorphism α of order n with k fixed points. Then $|G|$ is n, k-bounded.*

This means that $|G|$ is bounded above by some function of n and k. We shall use such terminology frequently. The case when n is prime was dealt with by Brauer and Fong [3], and inner automorphisms of arbitrary order were handled by Kuzucuoğlu and the author [11].

The following two results are quite easily deduced (see [13]).

THEOREM 1.2. *Let G be an arbitrary finite group admitting an automorphism α of order n with at most k fixed points. Then G contains a soluble normal α-invariant subgroup S of n, k-bounded index.*

We could take S to be the soluble radical of G, of course, but the above phrasing is necessary for the further discussion below.

THEOREM 1.3. *If G is a locally finite group containing an element with finite centralizer, then G has a locally soluble normal subgroup of finite index.*

It should be possible in Theorem 1.2 to choose S to satisfy further restrictions, along the lines of the conjecture below. For example, if $n = 2$, we can arrange that S is nilpotent of class at most 2 and $|S'|$ is k-bounded [7]. As is well known, a finite group is abelian if it admits a fixed point free involutory automorphism α. In general, if α is an arbitrary involutory automorphism of G, we can say that G departs from being abelian by amounts at the top and bottom that are bounded in terms of the number of fixed points of α.

CONJECTURE 1. *Let $c = c(n)$ denote the number of primes, counting multiplicities, in the prime factorization of n. Let S be a finite soluble group admitting an automorphism α of order n with at most k fixed points. Then $|S : F_{h(c)}(S)|$ is n, k-bounded, where $h(c)$ is some function of c.*

Here, $\{F_i(X)\}$ is the Fitting series of the finite group X. More optimistically, we may hope that $h(c) = c$.

We proceed to review some of the evidence ond related results. When n is a prime power p^m, we have that S contains a subgroup T of p^m, k-bounded index and such that $T = F_m(T) = O_{p'p}(T)$ [10]. If $m = 1$, this tells us that $F = F(G)$ has p^m, k- bounded index. Work of Khukhro [15], [16] then shows that G has a nilpotent normal subgroup of p-bounded class and p, k-bounded index.

As further evidence, consider the fixed point free case, in which $k = 1$. Then $< \alpha >$ is a Carter subgroup of $G < \alpha >$, and work of Shamash and Shult [18] shows that the Fitting height of G is $c(n)$-bounded. When n is the product of two primes, $G = F_2(G)$ (Cheng [5]).

It is also natural to consider the influence of the condition $(n, |G|) = 1$. Here, Turull [20] shows that $|S : F_{c(n)}(S)|$ is n, k-bounded.

It seems to be unknown whether the conjecture holds when n is the product of three primes, not all equal, $k = 1$ and $(n, |G|) \neq 1$; also when n is the product of two primes, $k > 1$, and $(n, |G|) \neq 1$.

2. Bounds on the Lie rank. Let x be an element of order n in a finite symmetric group $S = \text{Sym}(m)$. Then it is well known that

$$C_S(x) \cong \text{Dr}_{r|n} C_r \wr \text{Sym}(m_r)$$

is a direct product of wreath products $C_r \wr \text{Sym}(m_r)$, where C_r is a cyclic group of order r and the wreath product is a permutational one in which the symmetric group acts in its natural permutation representation. We have $m_r \geq 0$, $\sum r m_r = m$, and the convention is that if $m_r = 0$, the corresponding factor is omitted. We clearly have $n \sum m_r \geq m$, and so a bound on $\sum m_r$ gives us one on m (in terms of n).

A similar result is known for classical groups in almost all cases, and is probably true without exception. It can be deduced from [11] Theorem C and results on centralizers of unipotent elements in Springer-Steinberg [19], Chapter IV. See [8] for a fuller discussion.

PROPOSITION 2.1. *There exists a function $f(n, r)$ with the following properties. Let G be a finite classical group of rank $l(G)$ over a field of characteristic p, and let x be an element of order $n = p^a m$ of G, where $(p, m) = 1$. If n is even, assume that G is not an even-dimensional orthogonal group in characteristic 2. If $l(G) \geq f(n, r)$, then $C_G(x)$ contains a central product $S_1 \ldots S_t$ of quasisimple quotients of classical groups, such that $t \leq n + p^a \left[\frac{4}{m}\right]$ and the sum of the ranks of the S_i is at least r.*

The structure of the centralizers of unipotent elements in even dimensional orthogonal groups in characteristic 2 does not seem to have been worked out in the literature, but the climate of opinion seems to be that the above result should remain true in that case also. With the notation of the proposition, one can see that there is a function $g(n, r)$ such that, irrespectively of the type of G, x centralizes an elementary abelian p-subgroup of rank $\geq r$ if $l(G) \geq g(n, r)$.

The following can be deduced from Proposition 2.1 with some work.

THEOREM 2.2. *[8]. Let G be a simple locally finite group containing an element of odd order with Černikov centralizer. Then G is finite.*

Recall that a group E is Černikov, if it contains a normal subgroup D of finite index such that

$$D \cong C_{p_1^\infty} \times \cdots \times C_{p_t^\infty}$$

is the direct product of a finite number of Prüfer p-groups C_{p^∞}. The restriction to elements of odd order is to avoid the exceptional situation of Proposition 2.1 and so is presumably unnecessary.

A much stronger result than Theorem B was conjectured in [11], namely that if G is any non-linear simple locally finite group, then the centralizer in G of every element involves a non-linear simple group. Theorem B would follow because of [11], Corollary C1. For direct limits of finite simple groups, and slightly more generally, a version of this conjecture was proved as Theorem B of [11]. This shows that many centralizers, rather than all, have the desired structure. Taking account of a slightly stronger version of Proposition 2.1 that comes from Springer-Steinberg [19], Chapter IV, and the methods of [11], one can prove that if G is a direct limit of finite simple groups, and G is not linear, then the centralizer of every element of odd order in G involves a non-linear simple group [8]. Again, the restriction to elements of odd order is presumably unnecessary.

A major difficulty in proving results like Theorem 2.2 lies in taking account of the "Kegel kernels" (we follow R.E. Phillips in using this terminology). If G is simple, locally finite and countable, then G is the union of a "Kegel sequence" $G_1 < G_2 < \ldots$ of finite subgroups. Each G_i contains a maximal normal subgroup M_i such that $G_i \cap M_{i+1} = 1$, so that the natural map embeds G_i in the simple group G_{i+1}/M_{i+1}. The M_i are the "Kegel kernels". The classification of finite simple groups gives strong information about centralizers of elements of G_i/M_i, but unless the M_i can be pinned down in some way, it is difficult to translate this into information about centralizers in the G_i themselves.

Presumably the following should be true.

CONJECTURE 2. *If G is a locally finite group containing an element with Černikov centralizer, then G has a locally soluble subgroup of finite index.*

This is known for elements of prime power order [9]. For involutions it was proved by Asar [1] without using the classification of finite simple groups. The same result was announced independently by Pavlyuk [17]. If our element has order p^m, then we get immediately that G satisfies the minimal condition on p-subgroups, from which much structural information flows. The absence of this seems to be a serious obstacle for elements whose order is not a prime power.

3. Groups of automorphisms. It is well known that if $q \equiv 1 \bmod 4$, then $PSL(2, q)$ contains a self-centralizing elementary abelian subgroup of order 4. Thus, generalizations of Theorem 1.1, in which α is replaced by a non-cyclic group of automorphisms, will be hard to come by.

A version of Proposition 2.1 holds for arbitrary subgroups of classical groups. Roughly speaking, it says that in a finite classical group, a small (relative to the rank) subgroup centralizes a large simple section.

THEOREM 3.1. *[12] There exists a function $f(n, r)$ such that if H is any finite group of order n, G is a finite simple classical group of rank $\geq f(n, r)$, and $H \leq G$, then H centralizes a section $U/V \cong Alt(r)$ of G. Further, if G is defined over a field of characteristic p, then V is a p-group of class at most 2 and exponent dividing p^2.*

By saying that H centralizes U/V, we mean that $[H, U] \leq V$. The obvious version for G alternating holds with $V = 1$. There is no reason to expect the centralizer of H in G to cover U/V in general, and the next result shows that it may fail quite dramatically to do so.

THEOREM 3.2. *[12] Let p be an odd prime. Then there exists a finite group H of order $2p^3$ that can be embedded in infinitely many groups $PSL(n, p)$ as a subgroup with trivial centralizer.*

This H has a normal elementary abelian subgroup of order p^3. Thus, Proposition 2.1 fails for non-cyclic subgroups. Nevertheless, the following appears to be unknown.

QUESTION. *Let G be a simple locally finite group containing a finite subgroup with finite centralizer. Does it follow that G is linear?*

In Theorem 3.2, the group H actually embeds in $SL(n, p)$. One might try to get a negative answer to the above question by making the embeddings of H commute with the maps in some direct system of groups $SL(n, p)$. However this does not seem easy. We do have that if G is simple locally finite and non-linear, then each finite subgroup of G centralizes an infinite section that is a direct product of finite alternating groups of unbounded ranks [12].

Theorem 3.2 can be looked at from a different point of view, yielding a consequence in representation theory. As we have noted, the embeddings of H come from embeddings in $GL(n, p)$. The centralizer of the image is just the non-zero scalars. Thus, we have an action of H on an n-dimensional vector space V_n over the field of p elements, such that $\operatorname{Hom}_H(V_n, V_n) = F_p$. This gives us a collection of finite dimensional indecomposable F_pH-modules of unbounded dimension. Of course, there is nothing new about that. But in all previous examples, as far as I am aware, the endomorphism rings have large radicals, and indeed this is inevitable for p-groups. Here, the endomorphism rings are as small as they can be.

The appropriate form of Conjecture 1 for groups of automorphisms is probably obtained by replacing α by a group A of automorphisms and $c(n)$ by $c(A)$, the length of the longest chain of subgroups of A. If A is soluble, this is the number of primes (counting multiplicities) in the prime factorization of $|A|$. In this generality it is false. Bell and the author [2] have shown that any finite non-nilpotent group can act fixed point freely on finite soluble groups of arbitrarily large Fitting height (though with large fixed point groups on quotients). For nilpotent automorphism groups the situation is unclear except in the fixed point free case. Here, Dade's results on Carter subgroups [6] can be applied along th lines of the cyclic case. The assumption $(|A|, |G|) = 1$ improves matters. Here, Isaacs and the author [14] have shown that $|G : F_{2c+1}(G)|$ is $|A|, k$-bounded if A is soluble, and Turull [20] has shown that $|G : F_c(G)|$ is $|A|, k$-bounded if A acts "with regular orbits". This hypothesis holds for instance if A is nilpotent and has no section isomorphic to $C_p \wr C_p$ for any prime p.

REFERENCES

1. A.O. Asar, *The solution of a problem of Kegel and Wehrfritz*, Proc. London Math. Soc. (3)**45** (1982), 337-364.

2. S.D. Bell and B. Hartley, *A note on fixed point free actions of finite groups*, Quarterly J. Math. (Oxford) **41** (1990), 127-130.

3. R. Brauer and P. Fong, *On the centralizers of p-elements in finite groups*, Bull. London Math. Soc. **6** (1974), 319-324.

4. R. Brauer and K.A. Fowler, *On groups of even order*, Ann. of Math. **(2) 62** (1955), 563-583.

5. K.N. Cheng, *Finite groups admitting automorphisms of order pq*, Groups- St. Andrews, 1985, Proc.
Edinburgh Math. Soc. **(2) 30** (1987), 51-56.

6. E.C. Dade, *Carter subgroups and Fitting heights*, Illinois J. Math. **13** (1969), 449-514.

7. B.Hartley and T. Meixner, *Periodic groups in which the centralizer of an involution has bounded order*, J. Algebra **64** (1980), 285-291.

8. B. Hartley, *Simple locally finite groups containing an element of odd order with Černikov centralizer*, (to appear).

9. B. Hartley, *Fixed points of automorphisms of certain locally finite groups and Chevalley groups*, J. London Math. Soc. **(2) 37** (1988), 421-436.

10. Brian Hartley and Volker Turau, *Finite groups admitting an automorphism of prime power order with few fixed points*, Math. Proc. Cambridge Philos. Soc. **102** (1987), 431-441.

11. B. Hartley and M. Kuzucuoğlu, *Centralizers of elements in locally finite simple groups*, Proc. London Math. Soc. (to appear).

12. B. Hartley, *Centralizing properties in simple locally finite groups and large finite classical groups*, J. Austral. Math. Soc. (to appear).

13. Brian Hartley, *A general Brauer-Fowler Theorem, and centralizers in locally finite groups*, (to appear).

14. B. Hartley and I.M. Isaacs, *On characters and fixed points of coprime operator groups*, J. Algebra **131** (1990), 342-358.

15. E.I. Khukhro, *Finite p-groups admitting an automorphism of order p with a small number of fixed points*, Mat. Zametki (Russian) **38** (1985), 652-657.

16. E.I. Khukhro, Proc International Group Theory Conference, Bressanone, 1989 (Springer Lecture Notes) (to appear).

17. I.I. Pavlyuk, *On a problem of Kegel and Wehrfritz*, Seventh All-Union Symposium on Group Theory, Krasnoyarsk, Abstracts of Talks (1980), 83.

18. J. Shamash and E. Shult, *On groups with cyclic Carter subgroups*, J. Algebra **11** (1969), 564-597.

19. T.A. Springer and R. Steinberg, *Conjugacy classes*, in "Seminar on algebraic groups and related finite groups, Springer Lecture Notes 131," Springer, Berlin, 1970 reprinted 1986.

20. Alexandre Turull, *Groups of automorphisms and centralizers*, Math. Proc. Cambridge Philos. Soc. **107** (1990), 227-238.

Some Topics in Asymptotic Group Theory

WILLIAM M. KANTOR*

University of Oregon

1. ENUMERATION

There is nothing unusual about asymptotics in finite group theory: there are a number of known (or even well-known) asymptotic results. While these are not really the subject of this paper, it seems appropriate to begin with some especially intriguing examples (the first and last of which will be need later).

1.1. If p is prime then the number of isomorphism classes of groups of order p^k is at least $p^{\frac{2}{27}k^3 - 6k}$ (Higman [Hi]), and asymptotically $= p^{\frac{2}{27}k^3 + O(k^{8/3})}$ (Sims [Si]).

1.2 (Neumann [Ne, MN]).
The number of isomorphism classes of groups of order n is less than $n^{\frac{1}{2}(\log n)^2}$.

Here, and throughout this paper, logarithms will be to the base 2. The preceding result, as well as the next three, depend on the classification of finite simple groups.

1.3 (Holt [Ho]).
$$\frac{\text{\# isomorphism classes of perfect groups of order} \leq n}{\text{\# isomorphism classes of groups of order} \leq n} \rightarrow 0 \text{ as } n \rightarrow \infty.$$

1.4 (Cameron-Neumann-Teague [CaNT]).
For almost all n (in the sense of density), the only primitive permutation groups of degree n are S_n and A_n.

1.5 (Cameron [Ca]). If G is a primitive subgroup of S_n, then either
    (i) $n = \binom{m}{k}^{\ell}$ and G is a subgroup of S_m *wreath* S_ℓ with S_m acting on the k-sets of an m-set and the wreathed product having the product action, or
    (ii) $|G| \leq n^{C \log n}$ for some constant C.

---

* Research supported in part by NSF grant DMS 87-01794 and NSA grant MDA 904-88-H-2040.

Here **1.1**-**1.4** are, in some sense, in a classical enumerative vein: estimate the number of groups of a certain sort. The remainder of this paper is concerned with rather different types of asymptotic questions, involving lengths of presentations or of words given in terms of generators, or the proportion of pairs of elements of a simple group that generate the group. These questions (or at least those in §§2,4) are motivated, to some extent, by questions that arose in Theoretical Computer Science. Sketches of some proofs will be given, especially in the cases of some results not in print. For a survey of related results see [BHKLS].

2.   SHORT PRESENTATIONS

The *length* of a presentation $\langle X \mid R \rangle$ is the sum of $|X|$ together with the sum of the lengths of all of the members of R as words in $X \cup X^{-1}$. This is motivated, in part, by thinking of inputting $\langle X \mid R \rangle$ into a computer.

Stupid-looking Example: The usual presentation for a cyclic group of order n has length $1 + n$. A shorter presentation is $\langle X \mid R \rangle$ with $X = \{x_0,...,x_m\}$ and $R = \{ x_{i+1}x_i^{-2}, x_0^{a_0}x_1^{a_1}\cdots x_m^{a_m} \mid i = 0,...,m \}$, where $m = [1 + \log_2 n]$ and $n = \Sigma a_i 2^i$ in base 2. Its length is at most $(m + 1) + 3m + (m + 1) \le 5\log n + 7$. In fact, however stupid-looking this may seem to be as a way to represent a cyclic group, this method has, indeed, been used in practise.

2.1 Conjecture: *Every finite group G has a presentation of length* $O((\log|G|)^3)$.

Note that the constant 3 is best possible here. For, using Higman's bound **1.1**, for all $\epsilon > 0$ and all $C > 0$ it is easy to check that

$$\frac{\# \text{ p-groups of order } p^k}{\# \text{ presentations of length } \le C(\log p^k)^{3-\epsilon}} \longrightarrow \infty \text{ as } k \longrightarrow \infty$$

since the denominator is straightforward to calculate.

2.2 Theorem (Babai-Kantor-Luks-Pálfy [BKLP]).   *The Conjecture is true except, perhaps, if some composition factor of G is isomorphic to* $^2A_2(q)$, $^2B_2(q)$ *or* $^2G_2(q)$.

The remainder of this section is devoted to an indication of the ideas involved in the proof of this theorem, along with comments on the difficulties encountered with the groups $^2A_2(q)$, $^2B_2(q)$ and $^2G_2(q)$. The proof falls into two main steps: I. Simple Groups, and II. Glueing.

STEP I.   Simple Groups.
2.3 Proposition.   *Every simple group has a presentation of length* $O((\log|G|)^2)$, *except perhaps for the groups* $^2A_2(q)$, $^2B_2(q)$ *and* $^2G_2(q)$.

Sporadic groups can be ignored here. The usual presentation for A_n has length $< n^2$. Therefore, it remains only to consider a group G of Lie type over $\mathbb{F}_q$, of characteristic p and rank ℓ, say, in which case the order of magnitude of $\log|G|$ is $\ell^2\log q$. We presuppose various parts of [Car] here and in later portions of this paper.

Groups of Rank $\ell \geq 2$. Here the obvious approach is to try to use the Curtis-Steinberg-Tits presentation [Cu], but this is much too long (its length involves q instead of $\log q$). Nevertheless, it is not at all surprising that this presentation can be modified so as to behave as desired. The details are as follows.

Assume, for the moment, that G is untwisted. Then the Curtis-Steinberg-Tits presentation for some perfect central extension $\tilde{G}$ of G uses generators $x_\alpha(t)$ for certain roots α, where $t \in \mathbb{F}_q$. (Specifically, α belongs to the union of the rank 2 subsystems determined by pairs of fundamental roots, so the number of these roots α has order of magnitude ℓ^2.) The relations are

$$x_\alpha(t)x_\alpha(u) = x_\alpha(t + u)$$

$$[x_\alpha(t), x_\beta(u)] = \prod_{i,j>0} x_{i\alpha+j\beta}(C_{ij\alpha\beta}t^i u^j)$$

for all relevant α and β with $\beta \neq \pm\alpha$, and all t, $u \in \mathbb{F}_q$, where i, j and $C_{ij\alpha\beta}$ are integers (and $|C_{\alpha\beta ij}| \leq 3$). In order to shorten this presentation, let $\theta_1,...,\theta_e$ be a basis of $\mathbb{F}_q$ over $\mathbb{F}_p$. Then use only the generators $x_\alpha(\theta_k)$, together with the relations of the form

$$x_\alpha(\theta_k)^p = 1, \qquad [x_\alpha(\theta_k), x_\alpha(\theta_m)] = 1$$

$$[x_\alpha(\theta_k), x_\beta(\theta_m)] = \prod_{i,j>0} x_{i\alpha+j\beta}(C_{ij\alpha\beta}\theta_k^i\theta_m^j),$$

which are interpreted as follows. For any $s_k \in \mathbb{F}_p$ and any root γ, expand $x_\gamma(\Sigma_k s_k\theta_k)$ as $\Pi_k x_\gamma(\theta_k)^{s_k}$, where expressions such as $x_\alpha(\theta_k)^p$ and $x_\gamma(\theta_k)^{s_k}$ are themselves expanded as in the Stupid-looking Example by adjoining up to $7\log p$ additional generators and relations for each such term. The length of this presentation is dominated by that of the commutator relations, which is $O(\ell^2\ell^2 \cdot ee \cdot e\log p)$. Thus, this is a presentation for $\tilde{G}$ of length $O((\log|G|)^3)$.

In order to shorten this presentation somewhat, and at the same time kill the center of $\tilde{G}$, choose each θ_k as θ^k for a generator θ of $\mathbb{F}_q^*$. In addition to the generators $x_\alpha(\theta^k)$, introduce generators $w_\alpha(1)$, $w_\alpha(\theta)$ and h_α, together with the following relations for all α and $\beta \neq \pm\alpha$ restricted as above:

$$w_\alpha(1) = x_\alpha(1)x_{-\alpha}(1)^{-1}x_\alpha(1) \qquad\qquad w_\alpha(\theta) = x_\alpha(\theta)x_{-\alpha}(-\theta^{-1})x_\alpha(\theta)$$

$$h_\alpha = w_\alpha(\theta)w_\alpha(1)^{-1} \qquad\qquad\qquad [h_\alpha, h_\beta] = 1$$

$$x_\alpha(\theta^k)^{h_\alpha} = x_\alpha(\theta^2\theta^k) \qquad\qquad\qquad x_\alpha(\theta^k)^{h_\beta} = x_\alpha(\theta^{2(\alpha,\beta)/(\beta,\beta)}\theta^k)$$

(where $x_{-\alpha}(-\theta^{-1})$, $x_\alpha(\theta^2\theta^k)$ and $x_\alpha(\theta^{2(\alpha,\beta)/(\beta,\beta)}\theta^k)$ are computed by expansion as above, and (α, β) denotes the usual inner product of roots). Then our earlier relations can be shortened to

$$x_\alpha(\theta^k)^p = 1, \qquad [x_\alpha(\theta^k), x_\alpha(\theta^m)] = 1$$

$$[x_\alpha(\theta^k), x_\beta(\theta^m)] = \prod_{i, j > 0} x_{i\alpha+j\beta}(C_{ij\alpha\beta}\theta^{ik+jm})),$$

with $0 \le k, m \le 1$. For, these and conjugation by the various elements h_α imply the remaining ones given earlier. Now there are $O((\ell e + \ell + \ell)\log p)$ generators, and the relations have total length $O(\ell + \ell \cdot e\log p + \ell + e\ell + \ell\log p + \ell e + \ell^2 \cdot e\log p) = O(\log|G|)$. The center of $\tilde{G}$ is killed using products of the elements h_α, where the required relations are explicitly written down on a case by case basis.

All of this involves an unfortunate loss of the beauty of the original Curtis-Steinberg-Tits presentation in order to achieve efficiency.

Twisted Groups of Rank $\ell \ge 2$. While there are straightforward modifications of the above presentations valid for twisted groups, an annoying snag does occur. Namely, we used the fact that $x_\alpha(t)^{h_\alpha} = x_\alpha(\theta^2 t)$ in order to see that $\langle h_\alpha \rangle$ had very few orbits on X_α. Such a situation does not occur for odd-dimensional unitary groups, in which one type of root group is nonabelian. Nevertheless, in this case a presentation of length $O((\log|G|)^2)$ can still be obtained.

Very briefly, $G = PSU(2\ell + 1, q)$, $\ell \ge 2$, is best viewed as having a "root system" of type BC_ℓ, namely, the union of a B_ℓ and a C_ℓ system. However, [Gr] provides a presentation suitable for our purposes using a C_ℓ system Φ. His generators are $x_\alpha(t)$ with α short, $t \in \mathbb{F}_{q^2}$, as well as $x_\alpha(t,u)$ with α long, $t, u \in \mathbb{F}_{q^2}$ and $u + \bar{u} = \varepsilon_\alpha t\bar{t}$, where $\varepsilon_\alpha = \pm 1$ and the involutory field automorphism of $\mathbb{F}_{q^2}$ is $t \to \bar{t}$. The obvious sorts of commutator relations then suffice for a presentation (cf. [Gr]). These can be shortened by using generators $x_\alpha(\theta^k)$ for short roots α, as well as suitable generators $x_\alpha(0, \theta^m)$ and $x_\alpha(\theta^k, \theta^m)$ for long roots α, where $q^2 = p^e$ and θ is a generator of $\mathbb{F}_{q^2}^*$; and then introducing further generators h_α for all α as well as additional generators required in order to take various powers of generators. Relatively little care is needed to obtain a presentation of length $O((\log|G|)^2)$. It is presently not known how to obtain a presentation for G of length $O(\log|G|)$. This is the only "bad" twisted case of rank $\ell > 2$: in all other cases all root groups are abelian, and the usual group H has at most 3 orbits on each root group. Similar considerations reappear as we examine rank 1 groups, but there they produce a much more serious obstacle.

Groups of Rank $\ell = 1$. The standard presentations in this case involve all of the elements of a field $\mathbb{F}_q$, but this time *it is not at all clear how to cut the presentations down to the desired size* -- assuming, of course, that these groups do, indeed, have presentations of the desired lengths! The easiest way to explain the problem is to give a suitably short presentation for the group PSL(2, q) (similar to [To]).

A Presentation for PSL(2, q).
Let $q = p^e$ with p odd, and $\kappa := (q - 1)/2$.
Generators:
h, r, x_i, $i = 0,...,e - 1$.
Relations:
$h^\kappa = 1$, $r^2 = 1$, $h^r = h^{-1}$,

$[x_0, x_i] = [x_1, x_i] = 1$, $x_0^p = x_1^p = 1$, $r = x_0 x_0^r x_0$, $h^{-1}r = x_1 x_1^{h^{-1}r} x_1$,

$x_{i+2} = x_i^h$ for $0 \le i \le e - 2$, $x_{e-2}^h = \Pi_k x_k^{a_k}$, $x_{e-1}^h = \Pi_k x_k^{b_k}$,

with the a_k, $b_k \in \mathbb{F}_p$ dictated by a single irreducible polynomial over $\mathbb{F}_p$ used to define $\mathbb{F}_q$ (i. e., h corresponds to multiplication by the square of a generator θ of $\mathbb{F}_q^*$ and $\theta^e = \Sigma_k a_k \theta^k$, $\theta^{e+1} = \Sigma_k b_k \theta^k$), where powers $x_k^{a_k}$, $x_k^{b_k}$ are viewed as being expanded as in the Stupid-looking Example (i.e., by adjoining $O(\log p)$ additional generators and relators for each such term). The length of the resulting presentation for PSL(2,q) is $O(\log q)$.

In order to understand more clearly the preceding presentation, consider any rank 1 group $G = $ PSL(2, q), $^2A_2(q)$, $^2B_2(q)$ and $^2G_2(q)$. There is a standard presentation for G. Namely, let $B = UH$ be a Borel subgroup, where $H = \langle h \rangle$ is (isomorphic to) a subgroup of index 1, 2 or 3 in $\mathbb{F}_q^*$ ($\mathbb{F}_{q^2}^*$ in the unitary case) and hence $\kappa = |H|$ is explicitly known. Moreover, $N = H\langle r \rangle$ is dihedral except in the unitary case, where it has the presentation $\langle h, r \mid h^\kappa = 1, r^2 = 1, h^r = h^q \rangle$. Then a presentation for G is obtained by starting with ones for U and N, by giving the action of h on U, and finally by giving all the relations of the form $w = uvu'$ with $w \in \langle h \rangle r$, $u, u' \in U$, and $v \in U^r$ [St2].

When $G = $ PSL(2,q) we greatly decreased the number of generators by building in conjugation by h: there are at most 2 orbits of $\langle h \rangle$ on $U - \{1\}$, and we gave the action of h on a representative of each such orbit. Then all q of the relations $w = uvu'$ could be deduced from at most two of them, simply by conjugating by h. For each of the remaining rank 1 groups $\langle h \rangle$ has at least q orbits on $U - \{1\}$. The problem in those cases is to find some way to *deduce* most of these relations from a *bounded* number of them. Until some way is found to deal with this problem (or somehow circumvent it by using a different type of presentation for G), 2.1 will remain open: *the existence of short presentations of these rank 1 groups is the **only** obstacle to 2.1.*

STEP II.   Glueing.

Now consider any finite group G, and let N be a maximal normal subgroup of G. We may assume that $N \neq 1$. Then, by induction, there are presentations

$$G/N = \langle\, X \mid R \,\rangle \text{ and } N = \langle\, Y \mid S \,\rangle$$

each of which is suitably short (i.e., of respective lengths $O((\log|G/N|)^2)$ and $O((\log|N|)^3)$. The problem is to glue these together to form a new presentation that is itself sufficiently short. Glueing together the two presentations is standard, so once again it is necessary to find a way to proceed efficiently. This is less obvious and more interesting than Step I.

By abuse of language, view X as a subset of G and Y as a subset of N. Then R consists of elements of N, so each $r \in R$ is a word in $Y \cup Y^{-1}$. However, the presentations $\langle\, X \mid R \,\rangle$ and $\langle\, Y \mid S \,\rangle$ have nothing to do with one another, so there is no reason to expect that r will be a "nice" word in $Y \cup Y^{-1}$. In particular, it may have very large length as a word in $Y \cup Y^{-1}$, which would be unacceptable for our purposes. Fortunately, there is a way around this difficulty using the following surprising result:

2.4 Proposition [BS]. *Let N be a finite group and Y a set of generators of N. Let $r \in N$. Then there is a sequence $w_1,...,w_k = r$ of elements of N such that*
>   *each w_i is either in Y*
>>   *or is the product of two previous w_j's*
>>   *or is the inverse of a previous w_j,*
>   *and $k < 2(\log|N| + 1)^2$.*

Proof (based on remarks by E. M. Luks). We may assume that $N \neq 1$. We will construct a sequence A of elements of N and a subsequence $B \subseteq A$ such that the following all hold: each term in A is either in Y or is the product or inverse of terms occurring earlier in the sequence, $|A| < 2(\log |N|)^2$, $|B| \leq \log|N|$, and $N = \Pi(B)^{-1}\Pi(B)$.   Here, for any sequence $B = (b_1,...,b_k)$ of elements of N we write

$$\Pi(B) := \{\, b_1^{\varepsilon(1)}\cdots b_k^{\varepsilon(k)} \mid \text{each } \varepsilon(i) = 0 \text{ or } 1 \,\}.$$

The construction of the sequences A and B will be accomplished by successive increasing approximations.

Start with $A = B$ consisting of one element $\neq 1$ of Y (so initially $|\Pi(B)| = 2$). If, after several increases, we still have $N \neq \Pi(B)^{-1}\Pi(B)$, then $\Pi(B)^{-1}\Pi(B)Y \neq \Pi(B)^{-1}\Pi(B)$, so there exist $u,v \in \Pi(B)$ and $y \in Y$ such that $z := u^{-1}vy \notin \Pi(B)^{-1}\Pi(B)$. Then extend A and B to the following sequences by appending the indicated terms or sequences:

A': A, $\boxed{\text{via B}}$, u, $\boxed{\text{via B}}$, v, u^{-1}, $u^{-1}v$, z

B': B, z.

Here $\boxed{\text{via B}}$ refers to the fact that a product such as $b_1 \cdots b_k$ (with $b_1,...,b_k$ in B, in order) can be embedded in a sequence $b_1b_2, b_1b_2 \cdot b_3,...,b_1 \cdots b_{k-1} \cdot b_k$ of $k - 1$ terms, each of which is a product of terms either in B or occurring earlier in this appended sequence.

Now observe that $|\Pi(B')| = 2|\Pi(B)|$ since $\Pi(B) \cap \Pi(B)z = \emptyset$, so that at most $\log|N| - 1$ increases can take place: $|B| \leq \log|N|$. Also, $|A'| \leq |A| + 2(|B| - 1) + 3$, where A is increased at most $\log|N| - 1$ times, so at the end of all the increases we have $|A| \leq 1 + (\log|N| - 1)(2\log|N| + 1)$. This completes the construction of the desired sequences A and B.

Finally, each element of B (in fact, of A) occurs in a sequence of the sort required in **2.4**, of length $\leq |A|$; and we saw that each element of $\Pi(B)$ occurs in such a sequence of length $\leq |A| + (|B| - 1)$. For the same reason, each element of $N = \Pi(B)^{-1}\Pi(B)$ occurs in such a sequence of length $\leq |A| + |B| + (|B| - 1) \leq 2(\log|N|)^2 + \log|N| - 1$. $\square$

Note that this proof is short and ingenious while not looking at all like standard group theory. For somewhat sharper bounds and an effective version of **2.4**, see [BCFS].

Returning to the situation preceding the Proposition, adjoin a sequence using **2.4** in order to obtain additional generators for *each* $r \in R$, together with the relations implicit in the sequence. Similarly, adjoin further generators and relations in order to express the fact that $N \lhd G$. This readily produces a presentation of length $O((\log|G|)^5)$. Much more careful bookkeeping turns the exponent 5 into a 3 [BKLP].

3. THE PROBABILITY OF GENERATING

If G is a finite group generated by 2 elements, what proportion of the pairs of elements of G generate G? In other words, what is the probability that two randomly chosen elements of G generate G? This section will consider this question in the case of nearly simple groups. The most lovely result along these lines is due to Dixon:

3.1 Theorem [Di]. $\Pr(x, y \in S_n$ generate $S_n) \to \frac{3}{4}$ *as* $n \to \infty$,

$$\Pr(x, y \in S_n \text{ generate } A_n) \to \frac{1}{4} \text{ } as \text{ } n \to \infty.$$

In other words, x and y "almost always" generate A_n or S_n, depending upon the parity of x and y. In order to show that

$$\text{Pr}(\ x,\ y \in S_n \text{ do } \textit{not} \text{ generate } A_n \text{ or } S_n\ ) \to 0 \text{ as } n \to \infty,$$

Dixon used two ingredients:

1. *Number Theory* ([Di], based on [ET]):
$$\text{Pr}\left(\begin{array}{l} x \in S_n \text{ has a cycle of length a prime} \le n - 3 \text{ while} \\ \text{all other cycles have length relatively prime to this one} \end{array}\right) \to 1 \text{ as } n \to \infty.$$

2. 1873 *Group Theory* [Jo]:
If $G \le S_n$ is a primitive permutation group containing a p-cycle for some prime $p \le n - 3$, then G is A_n or S_n.

(*Historical comment*: The preceding result appears to be the first published application of Sylow's Theorem, which had been published only a year earlier. It is, of course, only the conjugacy part required here -- in the case of Sylow subgroups of prime order.)

In Dixon's situation,

Pr(x, y do not generate A_n or S_n)
$$\le \text{Pr}(\ x,\ y \text{ generate a primitive group} \ne A_n,\ S_n\ ) + \Sigma |L|^2/|G|^2$$

summed over all subgroups L of S_n maximal with respect to being intransitive or imprimitive. (This is a very crude estimate: equality would require that the various subgroups be pairwise disjoint sets!) By **1**, if x and y are randomly chosen in S_n then each of them probably has a power that is a p-cycle for some prime $p \le n - 3$, and then **2** implies that Pr(x, y generate a primitive group $\ne A_n, S_n$) is negligible. The terms in $\Sigma |L|^2/|G|^2$ involve the orders of obvious subgroups $S_k \times S_{n-k}$ and S_k *wreath* S_ℓ of S_n. Estimating this sum is made slightly simpler by noting that an upper bound is $\Sigma(|L|^2/|G|^2) \cdot (|G|/|N_G(L)|) \le \Sigma(|L|^2/|G|^2) \cdot (|G|/|L|) = \Sigma|L|/|G|$ where L ranges over one representative $S_k \times S_{n-k}$ or S_k *wreath* S_ℓ from each conjugacy class of such subgroups. Thus, it was only necessary for Dixon to check that this latter sum $\to 0$ as $n \to \infty$.

Almost 20 years after Dixon's paper, Babai [Ba] showed that

$$\text{Pr}(\ x,\ y \text{ do not generate } A_n \text{ or } S_n\ ) = 1/n + O(1/n^2),$$

where the leading term 1/n corresponds to the fact that 2 elements not generating A_n or S_n "probably" have a common fixed point! However, in this case the proof no longer used **1** and **2** above: Babai used the classification of finite simple groups. Dixon had shown that $\Sigma|L|/|G|$, summed over one representative $S_k \times S_{n-k}$ or S_k *wreath* S_ℓ from each conjugacy class, is $1/n + O(1/n^2)$. Consequently, it was only necessary to obtain a bound on Pr(x, y generate a primitive group $\neq A_n, S_n$) significantly better than one obtained in [Di]. (Better bounds had been known [Bo; BoW], obtained using number theory and generating functions; but they are not quite good enough to produce the desired result.)

Babai's argument runs as follows. It is only necessary to estimate $\Sigma|L|/|G|$ summed over one representative of each subgroup $L \neq A_n, S_n$ of S_n maximal with respect to being primitive. The possibility **1.5**(i) occurs with miniscule probability and hence can be ignored. Then $|L| \leq m := n^{C\log n}$ by **1.5**. Let K be a minimal normal subgroup of L, so that $L = N_G(K)$. By the classification, there are at most m characteristically simple groups of order m. Any such group K has at most $m^{\log m}$ subgroups (since any group of order at most m is generated by at most log m elements, by Lagrange's Theorem), and hence has at most $m^{\log m}$ transitive permutation representations. Consequently, there are at most $m \cdot m^{\log m}$ transitive characteristically simple subgroups K of S_n, and hence the desired sum $\Sigma|L|/|G|$ is at most $m \cdot m \cdot m^{\log m}/n! = O(1/n^2)$ (the upper bound $n^{\sqrt{n}}/n!$ is obtained in [Ba]).

At the same time that Babai was making Dixon's theorem more precise, a result for classical groups corresponding to Dixon's was being proved:

3.2 Theorem [KaLu]. *Let* G_0 *denote a finite simple classical group, and let* $G_0 \leq G \leq \text{Aut}(G_0)$. *If* P(G) *is the probability that two randomly chosen elements of* G *do **not** generate a group containing* G_0, *then* $P(G) \rightarrow 0$ *as* $|G| \rightarrow \infty$.

The methods used in the proof were similar to those of Dixon and Babai. A theorem of Aschbacher [As] asserts that each maximal subgroup L of G falls into one of nine families of subgroups of G. Eight of the families are defined very explicitly in terms of the vector space V over $\mathbb{F}_q$ used to define G_0 (the stabilizer of a subspace; the stabilizer of a direct sum or a tensor decomposition; the stabilizer of a field extension or the centralizer of a field automorphism; a classical group embedded as usual; the normalizer of a symplectic-type r-group for a prime r other than the characteristic p of $\mathbb{F}_q$). In the ninth family, $L = N_G(S)$ with S a nonabelian simple subgroup of PSL(V) such that $S \leq L \leq \text{Aut}(S)$, and the projective representation of S on V is absolutely irreducible and is defined over no proper subfield of $\mathbb{F}_q$.

The number of conjugacy classes within each of the eight explicit families is discussed in [As] (and in greater detail in [KlLi]), which makes it easy to obtain a suitable upper bound on $\Sigma|L|/|G|$ restricted to each such family. By [Li], $|L| \leq q^{3n}$

for L in the ninth family, so that it is only necessary to show that there are not too many summands $|L|/|G|$ arising from this family -- exactly the same sort of question we saw Babai had to deal with. For this purpose, once again we will see that ridiculously crude estimates suffice.

Namely, as above, there are $\leq q^{3n}$ simple groups of order $\leq q^{3n}$. Fix such a simple group S. The number of (equivalence classes of) absolutely irreducible projective representations of S in characteristic p is at most the order of the universal cover of S. For each such representation, maximality forces L to be the normalizer of (the image of) S; and L is isomorphic to a subgroup of Aut(S) containing S, so that $|L| \leq |S||\log|S|$. These crude estimates are enough to yield a proof of **3.2** when $n \geq 21$. Slightly more care is needed for the remaining small values of n.

An examination of the argument in [KaLu] gives slightly more information than in **3.2**. For purposes of the next result we assume, temporarily, that $PSp(2\ell, 2^e)$ with $\ell \geq 2$, $P\Omega(3, q)$, $P\Omega^+(6, q)$ and $P\Omega^-(6, q)$ are replaced by the respective isomorphic groups $P\Omega(2\ell + 1, 2^e)$, $PSL(2, q)$, $PSL(4, q)$ and $PSU(4, q)$. As above let V be the underlying vector space.

3.3 Theorem. *In the situation of* **3.2**, $P(G) = \Sigma|G_0{:}M|^{-1} + O(q^{-7(n-1)/6})$, *where M ranges over a representative of each G_0-conjugacy class of maximal subgroups of G_0 of each of the following types*:

    (i) *The stabilizer of a point or hyperplane of* V;

    (ii)*The image of a group* (i) *under a triality automorphism of* $G_0 = P\Omega^+(8, q)$;

    (iii) *The stabilizer of a totally isotropic 2-space when* $G_0 = PSp(4, q)$ *or* $PSU(5, q)$, *or of a totally singular 3-space when* $G_0 = P\Omega(7, q)$.

For example, if $G_0 = PSL(n, q)$ then, in the unlikely event that $\langle g, h \rangle$ does not contain G_0, $\langle g, h \rangle$ "probably" fixes a point or hyperplane. It should be noted that the constant 7/6 best possible in **3.3**, as is seen when $G_0 = P\Omega(7, q)$ and M is either the stabilizer of a totally singular line or $G_2(q)$ (but for no infinite collection of pairs G_0, M disjoint from this one).

There is no doubt that, for any simple group G_0 and any group G such that $G_0 \leq G \leq Aut(G_0)$, the probability that two randomly chosen elements of G do not generate a group containing G_0 approaches 0 as $|G| \to \infty$. There is sufficient published information to prove this conjecture for various choices of G $(^2B_2(q)$, $^2G_2(q)$, $G_2(q)$, $^3D_4(q)$ or $E_6(q))$. Recent work [LS] reported in Seitz's lectures at this Symposium probably handles all the exceptional groups G_0 for characteristic p not too small (namely, $p > 113$).

4.   WORD LENGTH.

While the preceding results say something about how *often* two elements generate a given group, they say nothing about *how* this generation takes place. In order to explain the difference, consider the following standard

Example. $S_n = \langle (1,...,n), (1,2) \rangle$.   Write $S = \{ (1,...,n), (1,2) \}$.   Then every element S_n has length $\leq n^2$ in $S \cup S^{-1}$; but some elements have length $\geq n^2/6$ (e.g., the involution $z \rightarrow n + 1 - z$).   (N.B. -- The length of each element of S_n does not seem to be known -- which is rather surprising in view of the standard nature of this pair S of generators.)

This leads to the consideration of the *diameter* of a group G with respect to a set S of generators of G.   Temporarily write $T = S \cup S^{-1}$, and enumerate the elements of G as follows:

T	$\lvert T \rvert$ elements
TT	$\leq \lvert T \rvert^2$ elements (actually, $\leq \lvert T \rvert (\lvert T \rvert - 1)$ elements $\neq 1$)
$\vdots$	
TT$\cdots$T	$\leq \lvert T \rvert^d$ elements.

The diameter is the smallest d such that these sets cover G, and in that case $\lvert G \rvert \leq 1 + \Sigma_1^d \lvert T \rvert^i$ (or, more precisely, $\lvert G \rvert \leq 1 + \Sigma_1^d \lvert T \rvert (\lvert T \rvert - 1)^i))$, so that $d \geq \frac{\log \lvert G \rvert - 1}{\log 2 \lvert S \rvert}$. Note that d is the same as the diameter of the (undirected) Cayley graph determined by the pair G, T; and the preceding inequality is essentially the "Moore bound" for this graph.

For example, when $G = S_n$ and $\lvert S \rvert = 2$ we have $d \geq \frac{\log n! - 1}{2}$, which suggests that one might be able to do better than in the above Example.   That this is, indeed, the case, is seen both in the next result and in **4.4**.

4.1 Theorem [BKL].   *If G is a nonabelian finite simple group then there is a set S of at most 7 generators of G such that the corresponding diameter is $O(\log \lvert G \rvert)$* (better: the diameter is $\leq 10^{10} \log \lvert G \rvert$).

Note that this result is false for cyclic groups.   Namely, if G is cyclic and $G = \langle S \rangle$ then the diameter is easily seen to be greater than $\frac{1}{2}(n^{1/\lvert S \rvert} - 1)$ -- in fact, this holds for any abelian group.   Thus, this is a particularly useless way to distinguish between nonabelian and abelian simple groups.

The theorem is constructive: a set S is more or less explicitly constructed; and, implicit in the proof, there is an *algorithm* which, given $g \in G$, will compute an expression for g as a word in S using $O(\log \lvert G \rvert)$ group operations.   However, this is not the same question as determining an expression for g of shortest length in the generators (the S-*length* of g) -- nor even the exact length of g as a word in the generators.

Two generators. Of course, in **4.1** one naturally expects that there is a set S of 2 generators producing diameter O(log|G|). This has been verified in "most" cases (a few of these are discussed are in [Ka2]):

4.2 Theorem. (i) *If* G *is an alternating group, or a group of Lie type and rank* > 1, *then there is a set* S *of 2 generators of* G *such that the corresponding diameter is* O(log|G|).

(ii) *If* G *is an alternating group, or a group of Lie type and rank* ≥ 20, *then there is a set* S *of 2 generators of* G, *one having order 2, such that the corresponding diameter is* O(log|G|).

In (ii), the corresponding undirected graph is *trivalent.* The rank assumption is unfortunate, and in many instances the arguments sketched below can be modified so as to work in somewhat lower ranks (much lower when the characteristic is 2); see [Ka2] for examples of this. However, despite the more tractable appearance of the smaller rank cases, the general version of (ii) remains open and seems to require a less naive approach than will be presented below. In both (i) and (ii) there is an associated algorithm in the sense indicated previously.

Question 1: Clearly (i) is aimed at extending Steinberg's result [St1] that groups of Lie type have 2-element generating sets. Do Steinberg's 2 generators produce diameter O(log|G|)? The proof in [St1] uses roughly $\ell = \text{rank } G$ commutations, therefore producing words of length $> 2^\ell$, which is too large. Note, however, that if ℓ is *bounded* and > 1, then Steinberg's proof shows that his generators do, indeed, produce diameter O(log|G|).

Question 2: Is **4.2**(ii) true for all $\ell \geq 2$? Presumably it is in all such cases, and also when $\ell = 1$. However, the latter is open even for **4.2**(i) even in the most familiar rank 1 instance:

Question 3: If $G = PSL(2, q)$ with q *not* a prime, find a set S of 2 generators producing diameter O(log q). (For a set S of 3 generators producing this diameter see **4.3**.)

Question 4: Give a *constructive* proof that, when p is prime, PSL(2, p) has diameter O(log p) with respect to $\left\{ \begin{pmatrix} 1 & 1 \\ 0 & 1 \end{pmatrix}, \begin{pmatrix} 0 & -1 \\ 1 & 0 \end{pmatrix} \right\}$. The fact that the diameter is O(log p) -- in fact, ≤ 500log p -- is due to Lubotzky and Sarnak (see

[BKL, 8.1]). However, their proof is nonconstructive, using [We]. In order to see the difficulty inherent in this question, consider the much more restricted -- but also open -- question:

$$\text{Write} \begin{pmatrix} 1 & \frac{1}{2}(p-1) \\ 0 & 1 \end{pmatrix} \text{as a word of length } O(\log p) \text{ in the above generators.}$$

Question 5: Prove that "most" S produce small diameter. For example, prove that $\Pr\begin{pmatrix} x,y,z\in S_n, \ \langle x,y\rangle = S_n \\ z \text{ has length } O(\log n!) \text{ in } \{x,y,x^{-1},y^{-1}\} \end{pmatrix} \to 1 \text{ as } n \to \infty.$

On the other hand, all S ought to come close to working. For example, there is the following conjecture: if $S_n = \langle x,y\rangle$, then the corresponding diameter is $O(n^2)$.

Sketch of the parts of the proof of 4.2.
See 4.4 for the case of alternating groups. When G is classical we will replace it by the corresponding linear group, which will then also be called G. We will generally assume that q is odd, the even case being similar but simpler.

Example I. $G = SL(2, q)$, q *odd*.
Write $x(t):= \begin{pmatrix} 1 & t \\ 0 & 1 \end{pmatrix}$ for $t\in \mathbb{F}_q$, $h(b):= \begin{pmatrix} b^{-1} & 0 \\ 0 & b \end{pmatrix}$ for $b\in\mathbb{F}_q^*$, and $r:= \begin{pmatrix} 0 & -1 \\ 1 & 0 \end{pmatrix}$. Then
$$x(t + u) = x(t)x(u) \text{ and } x(t)^{h(b)} = x(tb^2) \text{ for all } b \neq 0, t, u\in\mathbb{F}_q.$$

4.3 Proposition: (i) *If* q *is an odd prime then* G *has diameter* $O(\log |G|)$ *with respect to* $S:= \{x(1), r'\}$, *where* $r':= h(\frac{1}{2})r$.
(ii) *If* q *is odd, and if* θ *generates* $\mathbb{F}_q^*$, *then* G *has diameter* $O(\log |G|)$ *with respect to* $S:= \{x(1), r', h(\theta)\}$.

Proof. If ad - bc = 1 then a straightforward calculation yields that, for $c \neq 0$,
$$g = \begin{pmatrix} a & b \\ c & d \end{pmatrix} = x(-c^{-1} + ac^{-1})x(-c)^r x(-c^{-1} + dc^{-1}).$$
In case c = 0 use rg instead of g. This reduces the proof to showing that the S-length of each x(a), $a\in\mathbb{F}_q$, is $O(\log q)$ with respect to the given set S.

If q = p is an odd prime write $\theta = 2$; if q > p let θ be as in (ii). In either case, $\mathbb{F}_q = \mathbb{F}_p(\theta^2)$. Every element $t\in\mathbb{F}_q$ can be written in the form
$$t = \Sigma_0^m a_i\theta^{2i} = (\cdots(a_m\theta^2 + a_{m-1})\theta^2 + \cdots)\theta^2) + a_0$$
where either
$$q = p, \ m + 1 \leq \tfrac{1}{2}\log q, \text{ and } a_i\in\{0,1,2,3\} \text{ (base 4 representation of t), or}$$
$$q > p, \ m + 1 \leq \log_p q, \text{ and } a_i\in\mathbb{F}_p.$$

Suppose that $q = p$. Each $x(t)$ is a word $x(t) = (\cdots(x(a_m)^{h(2)}x(a_{m-1}))^{h(2)}\cdots)^{h(2)}x(a_0)$ in $m + 1$ elements $x(a_i)$ and $2m$ elements $h(\theta)^{\pm 1}$. Here, each $x(a_i) = x(1)^{a_i}$ has length ≤ 3, while (by matrix multiplication) $h(2)^{-1} = x(1)^{-2}(x(1)^2)^{r'}x(1)(x(1)^{-4})^{r'}$ has length ≤ 13. Thus, $x(t)$ has length $O(\log p)$, as required.

Suppose that $q > p$. As above, $x(t) = (\cdots(x(a_m)^{h(\theta)}x(a_{m-1}))^{h(\theta)}\cdots)^{h(\theta)}x(a_0)$, where we just saw that each $x(a)$, $a \in \mathbb{F}_p$, has S-length $O(\log p)$. Thus, each $x(t)$ has length $m \cdot O(\log p) = O(\log q)$. $\square$

Remark. By crudely counting the lengths in the above arguments, it is easy to check that the diameters are $\leq 45 \log|G|$ in (i) and $\leq 135 \log|G|$ in (ii). Namita Sarawagi has observed that $h(2) = x(1)r'x(1)^4r'x(1)r'^{-1}$ has length ≤ 9, thereby improving these estimates.

Example **IIa**. SL(n, q), q *odd.*

Let $s := r_{n-1}\cdots r_1$, so sH is an n-cycle within W. Let d_1 denote an involutory diagonal automorphism of G centralizing a hyperplane, normalizing H and L_{α_1}, and inverting X_{α_1}; write $d_{i+1} := d_1^{s^i}$.

If $g := r_1 d_1 \cdot h_{\alpha_3}(2) r_3 d_3 \cdot h_{\alpha_5}(2\theta) r_5 d_5 \cdot d_7 \cdot x_{\alpha_9}(1) d_9$ *then* $S := \{s, g\}$ *behaves as required* (cf. [Ka2]). The point here, and in the other examples of **4.2**(ii) sketched below, is that g is chosen so that its eigenspaces and those of suitable shifts (conjugates by powers of s) will have very well-behaved overlaps: if $g' := gg^{s^2}$ then $[g'^{4s^{-1}}, g'^4]^{s^{-7}}g' = x_{\alpha_2}(a)$ with $a \in \mathbb{F}_p^*$. Then $x_{\alpha_3}(a)$ has length $O(1)$, while $x_{\alpha_3}(a)^{g'} = x_{\alpha_3}(4a)$. As in **4.3** we can use conjugation by g' in order to see first that all elements of $x_{\alpha_3}(\mathbb{F}_p)$ have length $O(\log p)$ and then that all elements of X_{α_5} have length $O(\log q)$; then so do all elements of X_{α_1} and $X_{-\alpha_1} = (X_{\alpha_1})^g$. As in **4.3** it follows that all elements of L_{α_1} have length $O(\log q)$, and then so do $z := sr_1$ and all elements of H_{α_1}. Note that $U \subset YY^s \cdots Y^{s^{n-1}}$ where $Y := X_{\alpha_1}X_{\alpha_1}^z \cdots X_{\alpha_1}^{z^{n-2}}$, and there are cancellations occurring in these products since $s^k(s^{k+1})^{-1} = s^{-1}$ and $z^k(z^{k+1})^{-1} = z^{-1}$. It follows that each element of Y has length $O(n \cdot \log q)$, so that each element of U has length $O(n \cdot n\log q)$. Each element of $H = H_{\alpha_1}H_{\alpha_1}^s \cdots H_{\alpha_1}^{s^{n-2}}$ also has length $O(n\log q)$. On the other hand, each element of $W = N/H$ has $\{r_1, s\}$-length $O(n^2)$. Then each element of N has S-length $O(n^2\log q) = O(\log|G|)$, and hence so does each element of $G = UNU$.

The proof is always easier when q is even because root elements have order $p = 2$ and hence are more readily accessible:

Example **IIb**. SL(n, q), q *even*:

This time let $g := r_1 \cdot h_{\alpha_4}(\theta) r_4 \cdot x_{\alpha_7}(1)$ and $S := \{s, g\}$; again write $g' := gg^s$. Then

$(g'^6)^{s^{-6}}g = x_{\alpha_2}(1)$, so that $gx_{\alpha_7}(1) = r_1 \cdot h_{\alpha_4}(\theta)r_4$ has length $O(1)$, as does $u :=$ $gx_{\alpha_7}(1)(gx_{\alpha_7}(1))^{s^3} = r_1 \cdot h_{\alpha_4}(\theta) \cdot h_{\alpha_7}(\theta)r_7$. Since $x_{\alpha_4}(b)^u = x_{\alpha_4}(b\theta^2)$ for all b, as above we find that all elements of X_{α_4} and $X_{-\alpha_4} = (X_{\alpha_4})^g$ have length $O(\log q)$. Now proceed as before.

Example IIc. Sp(2ℓ, q), q *odd*.

This time $s := r_\ell \cdots r_1$ induces a 2ℓ-cycle. The support V_{α_1} of L_{α_1} is a nonsingular 4-space of V. Let d_{α_1} denote an involution in G that normalizes L_{α_1}, induces the identity on $V_{\alpha_1}^\perp$ and inverts X_{α_1}; write $d_{i+1} := d_1^{s^i}$. If $\sigma := \alpha_{\ell-1}^{r_\ell}$ then $\alpha_\ell = (\sigma + \alpha_{\ell-1})^s$. Recall that ℓ ≥ 16 and write $\alpha := \sigma^{s^{14}}$ and $\beta := \alpha^{s^2}$; define d_α and $d_\beta := d_\alpha^{s^2}$ in the obvious manner. Note that $V_\alpha = V_{\alpha_{13}}$.

If $g := r_1 d_1 \cdot h_{\alpha_3}(2) r_3 d_3 \cdot h_{\alpha_5}(2\theta) r_5 d_5 \cdot d_7 \cdot x_{\alpha_9}(1) d_9 \cdot r_\alpha d_\alpha$ *then* $S := \{s, g\}$ *behaves as required*. Once again $g' := gg^{s^2}$ satisfies $[g'^{4s^{-1}}, g'^4]^{s^{-7}g'} = x_{\alpha_2}(a)$ with $a \in \mathbb{F}_p^*$. As before we can use conjugation by s and g' in order to see that all elements of X_{α_5} and $X_{-\alpha_5}$ have length $O(\log q)$; and then (as in 4.3) so do z := sr_1 and all elements of H_{α_5}. Moreover, so do all elements of $(X_{\alpha_{12}})^{gr_{12}s^{-12}} = X_{-\alpha_\ell}$ and $(X_{\alpha_{14}})^{gr_{14}s^{-14}} = X_{\alpha_\ell}$; and then so do r_ℓ and all elements of H_{α_ℓ}. After suitably ordering the positive roots we find that $U \subset YY^s \cdots Y^{s^{2\ell-1}} X_{\alpha_\ell} X_{\alpha_\ell}^s \cdots X_{\alpha_\ell}^{s^{2\ell-1}}$ with $Y := X_{\alpha_1} X_{\alpha_1}^z \cdots X_{\alpha_1}^{z^{2\ell-3}}$. Also, H and N/H are easily handled exactly as in the previous Examples.

Example IId. Ω⁻(2ℓ + 2, q), q *odd*.

Define s as in Example IIc, as well as the support V_γ for every root γ of the B_ℓ root system for G. Let V_0 be the anisotropic 2-space $\langle V_\gamma \mid \gamma$ is long$\rangle^\perp$, and let j denote an involution in G that interchanges V_0 with a subspace of $V_{\alpha_{\ell-3}}$ while inducing the identity on $\langle V_0, V_0^j \rangle^\perp$; note that $|jj^{s^2}| = 3$. Let d_{α_1} be an involution in G that normalizes L_{α_1}, induces the identity on $V_{\alpha_1}^\perp$ and inverts X_{α_1}. Define σ, α, β, d_i, d_α and d_β as in Example IIc; once again $V_\alpha = V_{\alpha_{13}}$. Since ℓ ≥ 20, α and β are perpendicular to $\alpha_{\ell-3}$ and $\alpha_{\ell-1}$.

If $g := r_1 d_1 \cdot h_{\alpha_3}(2) r_3 d_3 \cdot h_{\alpha_5}(\theta) r_5 d_5 \cdot d_7 \cdot x_{\alpha_9}(1) d_9 \cdot r_\alpha d_\alpha \cdot j$ *then* $S := \{s, g\}$ *behaves as required*. This time $g' := gg^{s^2}$ satisfies $[g'^{6s^{-1}}, g'^6] = x_{\alpha_8+\alpha_9}(36)$, so if p ≠ 3 then we can proceed as in 4.3 in order to see that all elements of X_{α_5} and $X_{-\alpha_5}$ have length $O(\log q)$. Then so do all elements of $(X_{\alpha_{12}})^g$ and $(X_{\alpha_{14}})^g$. Conjugating by s^{-1} we find that, if $Y \cong \Omega^+(8, q)$ denotes the orthogonal group on $V_{\alpha_1} \perp V_{\alpha_3}$, then all the long root groups X_γ lying in Y have length $O(\log q)$. Using the usual method we see that all elements of Y have length $O(\log q)$; the same is then true for the orthogonal group $Y^{s^{-4}}$ on $V_{\alpha_{\ell-3}} \perp V_{\alpha_{\ell-1}}$. However, $Y^{s^{-4}g} = Y^{s^{-4}j}$ contains L_{α_ℓ}! Then r_ℓ and all elements of H_{α_ℓ} and X_{α_ℓ} have length $O(\log q)$, and hence we can proceed exactly as before (cf. [BKL]).

If $p = 3 < q$ write $u := g'^2 s^4$ and $v := [g'^2, u] = x_{\alpha_9}(-\theta^2 + 1)$, note that $v^u = x_{\alpha_9}((-\theta^2 + 1)\theta^2)$, and obtain all of X_{α_9} as in **4.3**. If $q = 3$ then $[g'^4, g'^{3s^{-2}}] = x_{\alpha_9}(1)$. Now proceed as before. $\square$

We conclude with a purely combinatorial argument.

4.4 Proposition. *There are trivalent Cayley graphs for A_n and S_n having diameter $O(n \log n)$.*

The following proof is motivated by an idea due to Quisquater ([Qu]; cf. [BHKLS]). My original approach was slightly more complicated, very similar in spirit to the partitioning method of [BKL] but using [Ka2]. The two generators constructed below have the added property that their orders are bounded -- 2 and 15 -- whereas one of those obtained as in [BKL] has order roughly $C \log n$.

Proof. Let $m \geq 4$, and consider the m-set $X = \{0,1,2,...,m-1\}$; expressions such as $x, 2x+1$, etc., are always assumed to refer to elements of X. Write

$$b_0 := \prod_{\substack{2^j \leq x < 2^{j+1} \\ j \text{ even}}} (x,2x,2x+1) \quad \text{and} \quad b_1 := \prod_{\substack{2^j \leq x < 2^{j+1} \\ j \text{ odd}}} (x,2x,2x+1) \quad \text{if } m \text{ is even,}$$

$$b_1 := \prod_{\substack{2^j \leq x < 2^{j+1} \\ j > 0 \text{ even}}} (x-1,2x-1,2x) \quad \text{and} \quad b_0 := \prod_{\substack{2^j \leq x < 2^{j+1} \\ j \text{ odd}}} (x-1,2x-1,2x) \quad \text{if } m \text{ is odd.}$$

Note that each product consists of pairwise commuting 3-cycles. In each case, $\langle b_0, b_1 \rangle$ fixes 0 and b_1 fixes 1. If $x \in X$ and $x > 1$ then $b_i^{\pm 1}$ moves x to a smaller member of X (in fact, to a member $\leq \frac{1}{2}x$) for some i. Thus, $1 = x^w$ for a word w in $\{b_0, b_1\}$ of length $\leq \log m$. It follows that $\{(0,1), b_0, b_1\}$ generates S_m with diameter $O(m \log m)$ [Qu]. Namely, each transposition $(0,x) = (0,1)^{w^{-1}}$ has length $\leq 2 \log m + 1$; and it is easy to see that each element of S_m has length $\leq m$ in these $m - 1$ transpositions.

Now consider an n-set, $n \geq 11$, which we may assume has the form $\{\infty, \infty', p\} \cup X \cup X'$ where $X' = \{x' \mid x \in X\}$, ∞ and ∞' are new symbols, and so is p if $n = 2m + 3$ is odd while $p = 0'$ if $n = 2m + 2$ is even. Let

$$t := \prod_x (x,x') \quad \text{or} \quad (\infty,\infty') \prod_x (x,x') \quad \text{depending on the parity desired, and}$$

$$g := (\infty',p,\infty,0,1') b_0 b_1',$$

where, for example, $b_1' = b_1^t$ denotes the permutation of X' behaving as b_1 does on X. (In particular, b_0 fixes 0 while b_1' fixes 0' and 1'.) We will show that $S := \{t, g\}$

generates S_n *with diameter* $O(n\log n)$.

Clearly $g^3 = (\infty,\infty',0,p,1')$, $g^{-5} = b_0 b_1'$ and $(g^{-5})^t = b_0' b_1$. As seen above, for each $x \in X-\{0\}$ there is a word w of length $\leq \log m$ in $\{g^2, (g^2)^t\}$ fixing $\infty, \infty', 0, p$ and sending $1'$ to x'. Thus, $(\infty,\infty',0,p,x')$ has S-length $O(\log n)$. Since $(\infty,\infty',0,p,1')^{-2}(\infty,\infty',0,p,x')(\infty,\infty',0,p,1') = (\infty,\infty',x')$ for $x > 1$, it follows that (∞,∞',u) has length $O(\log n)$ for each $u \in \{0,p\} \cup (X'-\{0'\})$. Now conjugate by t in order to see that (∞,∞',u) also has length $O(\log n)$ for each $u \in \{0'\} \cup (X-\{0\})$. Each element of A_n has length $\leq 2n$ in these 3-cycles, while parity can be adjusted if needed by using t. $\square$

Postscript (January 31, 1991): There are certainly many further directions one can go in asymptotic group theory. The following very recent result, concerning the number $k(G)$ of conjugacy classes of a group G, uses the classification of finite simple groups in order to greatly improve estimates (essentially $k(G)>C\log\log |G|$) obtained by Landau and Brauer [Br] using only the class equation of G:

(Pyber [Py]) $k(G) > c\log |G|/(\log\log |G|)^8$ for some constant c.

References

[As] M. Aschbacher, On the maximal subgroups of the finite classical groups. Invent. Math. 76 (1984) 469-514.

[Ba] L. Babai, The probability of generating the symmetric group. J. Comb. Theory(A) 52 (1989) 148-153.

[BCFS] L. Babai, G. Cooperman, L. Finkelstein and Á. Seress, Nearly linear time algorithms for permutation groups with a small base, pp. 200-209 in *Proc. 1991 Int. Symp. Symbolic and Algebraic Computation*.

[BHKLS] L. Babai, G. Hetyei, W. M. Kantor, A. Lubotzky and Á. Seress, On the diameter of finite groups, pp. 857-865 in *Proc. 31st IEEE Symposium on Foundations of Computer Science* (1990).

[BKL] L. Babai, W. M. Kantor and A. Lubotzky, Small diameter Cayley graphs for finite simple groups. European J. Combinatorics 10 (1989) 507-522.

[BKLP] L. Babai, W. M. Kantor, E. M. Luks and P. P. Pálfy, Short presentations for simple groups (in preparation).

[BS] L. Babai and E. Szemerédi, On the complexity of matrix group problems, I, pp. 229-240 in *Proc. 25th IEEE Symposium on Foundations of Computer Science* (1984).

[Bo] J. D. Bovey, The probability that some power of a permutation has small degree. BLMS 12 (1980) 47-51.

[BoW]   J. D. Bovey and A. Williamson, The probability of generating the sym-
         metric group. BLMS 10 (1978) 91-96.

[Br]   R. Brauer, Representation theory of finite groups, pp. 133-175 in
         Lectures on Modern Mathematics (ed. T. L. Saaty). Wiley, New
         York 1963.

[Ca]   P. J. Cameron, Finite permutation groups and finite simple groups.
         BLMS 13 (1981) 1-22.

[CaNT]   P. J. Cameron, P. M. Neumann and D. N. Teague, On the degrees of
         primitive permutation groups. Math. Z. 180 (1982) 141-149.

[Car]   R. Carter, *Simple groups of Lie type.* Wiley, London-New York-
         Sydney-Toronto 1972.

[Cu]   C. W. Curtis, Central extensions of groups of Lie type. J. reine angew.
         Math. 220 (1965) 174-185.

[Di]   J. D. Dixon, The probability of generating the symmetric group. Math.
         Z. 110 (1969) 199-205.

[ET]   P. Erdös and P. Turán, On some problems of a statistical group theory
         II. Acta Math. Acad. Sci. Hung. 18 (1967) 151-163.

[Gr]   R. L. Griess, Schur multipliers of finite simple groups of Lie type.
         TAMS 183 (1973) 355-421.

[Hi]   G. Higman, Enumerating p-groups, I: Inequalities. PLMS 10 (1960)
         24-30.

[Ho]   D. F. Holt, Enumerating perfect groups. JLMS 39 (1989) 67-78.

[Jo]   C. Jordan, Sur la limite du degré des groupes primitifs non alternées.
         Bull. Soc. Math. France 1 (1873) 40-71.

[Ka1]   W. M. Kantor, Some Cayley graphs for simple groups. *Proc. Conf.
         Combinatorics and Complexity*, Chicago 1987 = Discrete Applied
         Math. 254 (1989) 99-104.

[Ka2]   W. M. Kantor, Some large trivalent graphs having small diameters (to
         appear).

[KaLu]   W. M. Kantor and A. Lubotzky, The probability of generating a finite
         classical group. Geom. Ded. 36 (1990) 67-87.

[KlLi]   P. B. Kleidman and M. W. Liebeck, *The subgroup structure of the finite
         classical groups*. LMS Lecture Note Series 129, Cambridge
         University Press 1990.

[Li]   M. W. Liebeck, On the orders of maximal subgroups of the finite classi-
         cal groups. PLMS 50 (1985) 426-446.

[LS]   M. W. Liebeck and G. M. Seitz, Maximal subgroups of exceptional
         groups of Lie type, finite and algebraic. Geom. Ded. 35 (1990) 353-
         387.

[MN]   A. McIver and P. M. Neumann, Enumerating finite groups. Quart. J.
         Math. 38 (1987) 473-488.

[Ne]   P. M. Neumann, An enumeration theorem for finite groups. Quart. J. Math. 20 (1969) 395-401.

[Py]   L. Pyber, Every finite group has many conjugacy classes, preprint, Math. Inst. Hung. Acad. Sci. (1990).

[Qu]   J-J. Quisquater, Structures d'interconnexion: Constructions et applications, Ph. D. Thesis, Université de Paris - Sud (Orsay), July 1987.

[Si]   C. C. Sims, Enumerating p-groups. PLMS 15 (1965) 151-166.

[St1]   R. Steinberg, Generators for simple groups. Can. J. Math. 14 (1962) 277-283.

[St2]   R. Steinberg, Generators, relations and coverings of algebraic groups, II. J. Algebra 71 (1981) 527-543.

[To]   J. A. Todd, A second note on the linear fractional group. JLMS 11 (1936) 103-107.

[We]   A. Weil, Sur les courbes algébriques et les variétés que s'en déduisent. Act. Sci. Ind. 1041 (1948).

The 3-modular characters of the McLaughlin Group McL and its Automorphism Group McL.2

Ibrahim A. Suleiman and Robert A. Wilson

Introduction

McLaughlin's sporadic simple group McL was originally constructed (see [9]) as a permutation group on 275 letters. It is a simple group of order $898128000 = 2^7.3^6.5^3.7.11$. It is now known to be the pointwise stabilizer of a 2-dimensional sublattice ([2], [6]) in the Leech lattice. Its maximal subgroups were found by Finkelstein (see [7]). The modular character tables for the relevant primes $p = 2, 7$ and 11 were found by Thackray [16]. The 5-modular character tables were found by Hiss, Lux and Parker, up to a few ambiguities (see [8], [10]). These ambiguities together with others in the values of the 5-modular characters 560 and 3038 of the automorphism group of McL, denoted by McL.2, were resolved by Suleiman (see [13]). The main purpose of this paper is to complete the 3-modular character table of McL and to find the 3-modular character table of McL.2.

I   The 3-modular character table of McL

In this section we are going to complete what has been done by R. Parker [10] on the 3-modular characters of McL. To do so we have to work out again most of the 3-modular characters using the techniques of the 'Meat-Axe' which is the main tool in our work. We then use the method of 'condensation' (see [12], [15], [17]) to complete the 3-modular character table of McL.

The central characters modulo 3 give the block distribution of the ordinary irreducible characters. There are three blocks of defect zero. These blocks are $B_1 = \{5103\}, B_2 = \{8019_a\}$ and $B_3 = \{8019_b\}$. Hence, $5103, 8019_a$ and 8019_b are three 3-modular irreducible characters in McL. The remaining 21 ordinary irreducible characters are in the principal block B_0 of defect 6. Since there are thirteen 3-regular classes, it follows that there are ten 3-modular irreducible representations in the principal block.

To start with, we find from the Leech lattice two 21×21 matrices over GF_3 which generate McL (see [13]). Then using the Meat-Axe, we find that:

$$21 \otimes 21 = 21 + 2(1) + 104_a + 104_b + 210$$

$$21 \otimes 104_a = 1 + 21 + 104_a + 560 + 1498$$

$$21 \otimes 104_b = 1 + 21 + 104_b + 560 + 1498$$

Hence we find the following 3-modular irreducible characters:

$$1, 21, 104_a, 104_b, 210, 560 \text{ and } 1498.$$

Calculating the character values

We find representatives for all the 3-regular classes of McL as words in our generators. We then work out the character values on these classes using the program 'EV' of the Meat-Axe which works out the eigenvalues of a matrix (see [13]). This program is designed specifically to calculate Brauer character values. If a is an element of order n, first factorize the polynomial $(x^n - 1)$ as a product of irreducible polynomials $f_i(x) = x - \alpha_i$ over the required field GF_q. Then the nullity of $f_i(a)$ will give the number of times (say n_i) that a root of unity α_i is an eigenvalue of a. If β_i is the complex root of unity corresponding to α_i, then the character value of the required representation on that element will be $\sum_i \beta_i . n_i$. The character values of the above six representations on all the 3-regular classes are given in rows 1-6 and 9 of table II.5.

The primitive permutation representations of McL

The primitive permutation representations on 275, 2025, 7128, 15400 and 22275 points were found by using the program 'VP' of the Meat-Axe (see [13]). This program reads two matrices and a vector and finds the orbits of the vector under the matrices. Most of the time, we take a vector fixed by a particular subgroup. The program also writes the action of the group generators as permutations of the vectors in this orbit. The permutation characters are found by inducing up the trivial character of the relevant maximal subgroups. These permutation characters are shown in Table I.1.

Using the ordinary character table of McL (see [4]), we can see that these characters decompose as the sum of ordinary irreducible characters of McL

SOME PERMUTATION CHARACTERS OF McL AND McL.2

1A	2A	3A	3B	4A	5A	5B	6A	6B	7A	7B	8A	9A	9B
275	35	5	14	7	0	5	5	2	2	2	1	2	2
2025	105	0	27	9	0	5	0	3	2	2	1	0	0
7128	168	0	27	12	3	3	0	3	2	2	2	0	0
15400	280	10	37	20	0	5	10	1	0	0	2	1	1
22275	435	0	54	15	0	5	0	6	1	1	1	0	0

1A	10A	11A	11B	12A	14A	14B	15A	15B	30A	30B
275	0	0	0	1	0	0	0	0	0	0
2025	0	1	1	0	0	0	0	0	0	0
7128	3	0	0	0	0	0	0	0	0	0
15400	0	0	0	2	0	0	0	0	0	0
22275	0	0	0	0	1	1	0	0	0	0

1A	2B	4B	6C	8B	8C	10B	12B	12C	20A	20B	22A	22B	24A	24B
275	11	15	2	1	5	1	3	0	0	0	0	0	1	1
2025	–	–	–	–	–	–	–	–	–	–	–	–	–	–
7128	66	6	3	12	0	1	0	3	1	1	0	0	0	0
15400	66	70	3	0	4	1	4	1	0	0	0	0	0	0
22275	121	0	4	3	1	0	0	0	0	0	0	0	0	0

Table I.1

as follows:

$$
\begin{array}{lllll}
P1 & = & 275 & = & 1 + 22 + 252 \\
P2 & = & 2025 & = & 1 + 22 + 252 + 1750 \\
P3 & = & 7128 & = & 1 + 22 + 252 + 1750 + 5103 \\
& & 15400_a & = & 1 + 252 + 4500 + 5103 + 5544 \\
P4 & = & 15400_b & = & 1 + 22 + 252 + 252 + 1750 + 3520_a + 4500 + 5103 \\
P5 & = & 22275_a & = & 1 + 22 + 2(252) + 2(1750) + 3520_a + 5103 + 9625 \\
& & 22275_b & = & 1 + 252 + 1750 + 5103 + 5544 + 9625 \\
& & 22275_c & = & 1 + 22 + 2(252) + 2(1750) + 3520_a + 5103 + 9625
\end{array}
$$

The first two permutation representations $P1$ and $P2$ are chopped directly using the Meat-Axe to 3-modular irreducibles in McL. We find that:

$$
275 = 4(1) + 3(21) + 104_a + 104_b
$$
$$
2025 = 6(1) + 5(21) + 2(104_a) + 2(104_b) + 1498
$$

All the above irreducibles are known, therefore to find more 3-modular irreducibles we use condensation of permutation modules.

Condensation of permutation modules

As our implementation of the Meat-Axe has an effective limit of around 3000 dimensions over GF_3, we can proceed no further with a direct attack. Therefore, as we still have to find three more 3-modular irreducible characters for McL, we use another method, namely 'condensation' of the permutation representations of McL on 275, 2025, 7128, 15400 and 22275 points. First we summarise the theoretical basis for the calculations (see [12], [14], [17]).

Let G be a group, and let V be a kG-module, where k is assumed to be a finite field of characteristic p. Let H be a subgroup whose order is not divisible by p. Define the idempotent $e = \frac{1}{|H|} \sum_{h \in H} h$ of kG. Then $e.kG.e$ is a sub-algebra of kG known as a Hecke algebra.

From any kG-module V, we obtain an $e.kG.e$-module Ve. We say that Ve is condensed from V since Ve consists of the fixed points of the action of H on V. Now $\dim Ve \approx \frac{\dim V}{|H|}$, so it should be much easier to apply the 'Meat-Axe' to Ve rather than to V. Moreover, any information about Ve which we obtain with the 'Meat-Axe' give rise to information about V, via the following proposition:

Proposition : *Let $\chi_1, \chi_2, \ldots, \chi_r$ be the irreducible constituents of the module V; then the irreducible constituents of Ve are the non-zero members of the set $\{\chi_1 e, \chi_2 e, \ldots, \chi_r e\}$.(see [12], [17]).*

The permutation modules $P1, P2, P3, P4$ and $P5$ can be condensed using a suitable condensation subgroup H of order not divisible by 3. We chose a condensation subgroup $H \cong C_{14}$ generated by a particular word in our generators for McL.

(i)   The permutation representation on 275 points

The condensation of the permutation module $P1$ is to identify the condensed irreducibles corresponding to the 3-modular irreducibles. We condense $P1$ over the subgroup H of order 14 to get a condensed module $M1$ of dimension 23. $M1$ is chopped up to irreducibles as follows:

$$23 = 4(1_a) + 3(1_b) + 8_a + 8_b$$

and therefore, we have the following correspondence:

The permutation $P1$ on 275 points	1	21	104_a	104_b
The 23-dimensional condensed module $M1$	1_a	1_b	8_a	8_b

Table I.2

(ii)   The permutation representation on 2025 points

$P2$ is condensed as well, using the same subgroup H of order 14 to get a condensed module $M2$ of dimension 153. Then $M2$ is chopped up using the Meat-Axe to:

$$153 = 6(1_a) + 5(1_b) + 2(8_a) + 2(8_b) + 110$$

Therefore we get the following correspondence (Table I.3) between the irreducibles of the condensed module and the 3-modular irreducibles of McL:

The permutation $P2$ on 2025 points	1	21	104_a	104_b	1498
The 153-dimensional condensed module $M2$	1_a	1_b	8_a	8_b	110

Table I.3

(iii)   The permutation representation on 7128 points

The permutation module $P3$ will not give us new 3-modular irreducibles since 5103 is in the block B_1 of defect zero, but we condense it to identify the corresponding irreducible for 5103 in the condensed module. We get a condensed module $M3$ of dimension 552, which is chopped up using the Meat-Axe to the following irreducibles:

$$522 = 6(1_a) + 5(1_b) + 2(8_a) + 2(8_b) + 110 + 369$$

Therefore we get the following table of correspondences:

The permutation $P3$ on 7128 points	1	21	104_a	104_b	1498	5103
The 552-dimensional condensed module $M3$	1_a	1_b	8_a	8_b	110	369

Table I.4

(iv) The permutation representation on 15400 points

The permutation character $P4$ decomposes as a sum of ordinary characters as follows:

$$P4 = 15400_b = 1 + 22 + 252 + 252 + 1750 + 3520_a + 4500 + 5103$$

Using the same cyclic subgroup H we condense $P4$ to get a condensed module $M4$ of dimension 1120. The condensed module $M4$ is chopped up to the following irreducibles:

$$1120 = 15(1_a) + 12(1_b) + 7(8_a) + 7(8_b) + 3(16) + 4(38)+$$
$$+41_a + 41_b + 3(110) + 369$$

We therefore suspect there are new irreducibles corresponding to the condensed constituents 41_a and 41_b.

Now, we know that a condensed module Ve is a vector subspace of the original module V, so any submodule of the condensed module Ve can be embedded in the original module V. Therefore, we can construct the corresponding invariant subspace of the original module V by 'spinning up' the subspace containing the condensed module, under the group generators. This can be done using a program 'UK' of the Meat-Axe (see [13]). In this case, by spinning up the subspace 41_a, we constructed the corresponding representation of dimension 605. This representation is called 605_a. Similarly, 41_b corresponds to 605_b in the original module $P4$. Therefore we have the following correspondence (Table I.5) between the irreducibles in the condensed module $M4$ and the 3-modular irreducibles in the original module $P4$:

perm $P4$ on 15400 points	1	21	104_a	104_b	210	560	605_a	605_b	1498	5103
1220-dim condensed module $M4$	1_a	1_b	8_a	8_b	16	38	41_a	41_b	110	369

Table I.5

Using the program 'EV' of the Meat-Axe as described above, we calculate the character values of 605_a and 605_b on the 3-regular classes of McL (see Table II.5).

(v)　The permutation representation on 22275 points

The last 3-modular irreducible representation of McL can be shown to be a constituent of the permutation module $P5$ on 22275 points. The permutation character $P5$ decomposes as a sum of ordinary characters as follows:

$$P5 = 22275_a = 1 + 22 + 2(252) + 2(1750) + 3520_a + 5103 + 9625$$

Using the cyclic group H of order 14, we condense the permutation module $P5$ to get a condensed module $M5$ of dimension 1623. Then $M5$ is chopped up to irreducibles as follows:

$$1623 = 18(1_a) + 18(1_b) + 9(8_a) + 9(8_b) + 6(16) + 6(38) + 5(110) + 369 + 200$$

This gives an indication that there is a 3-modular irreducible of degree at least 2794. By spinning up a small subspace containing 200 as described above, we proved that the corresponding 3-modular irreducible representation is of degree 2794. Therefore we have Table I.6, which gives the correspondences between the irreducibles of the condensed module $M5$ and those of the original module $P5$:

perm $P5$ on 22275 points	1	21	104_a	104_b	210	560	605_a	605_b	1498	2794	5103
1623-dim condensed module $M5$	1_a	1_b	8_a	8_b	16	38	41_a	41_b	110	200	369

Table I.6

Therefore the ordinary irreducibles in this permutation module can be written as sums of 3-modular irreducibles as indicated below:

$$1 = 1$$
$$22 = 1 + 21$$
$$252 = 2(1) + 2(21) + 104_a + 104_b$$
$$1750 = 2(1) + 2(21) + 104_a + 104_b + 1498$$
$$3520_a = 3(1) + 2(21) + 2(104_a)2(104_b) + 2(210) + 2(560) + 1498$$
$$5103 = 5103$$

It follows that

$$9625 = 5(1) + 6(21) + 3(104_a) + 3(104_b) + 4(210) + 4(560) + 2(1498) + 2794$$

Using this decomposition of 9625 into 3-modular irreducibles, and the characters we already know, we can work out the character values of 2794 on all the 3-regular classes.

Theorem 1 *The complete 3-modular character table of McL is as in the first part of Table II.5.*

Corollary : *The decomposition matrix of the principal block B_0 is as in Table I.7.*

II   The 3-modular character table of McL.2

In this section, we are going to find the 3-modular character table of McL.2. To do so, we first find generators for McL.2 over GF_3(see [13]). Then, using the Meat-Axe directly as before, we find the following 3-modular characters: $1, 21, 104_a, 104_b, 210, 560$ and 1498. We find representatives for all the 3-regular classes of McL.2 \ McL as words in our generators. We then work out the character values on these classes using the program 'EV' of the Meat-Axe. These are given in Table II.5.

The primitive permutation representations of McL.2

The primitive permutation representations on 275, 7128, 15400 and 22275 points were found by using 'VP'. The permutation characters are displayed in Table I.1. Using this table together with the ordinary character table of McL.2 (see [4]), we can write down each of the above permutation characters as sums of ordinary characters. The decomposition of these characters on the simple group McL is the same as shown in section I. We write χ^+ to indicate the character of $G.2$ whose values on $G.2 \backslash G$ are printed in the 'ATLAS' and χ^- to indicate the character whose values on $G.2 \backslash G$ are the negatives of these. These permutation characters can be written as sums of ordinary characters as follows:

$$275^+ = 1^+ + 22^+ + 252^+$$
$$7128^+ = 1^+ + 22^- + 252^+ + 1750^+ + 5103^+$$
$$15400^+ = 1^+ + 22^+ + 252^+ + 252^+ + 1750^- + 3520_a^+ + 4500^+ + 5103^+$$
$$22275^+ = 1^+ + 22^+ + 2(252^+) + 1750^+ + 1750^- + 3520_a^- + 5103^+ + 9625^+$$

Using the ordinary character table of McL.2 (see [4]) and the characters we already know, we can write some of the above ordinary irreducibles as sums of 3-modular irreducibles as follows:

$$
\begin{aligned}
1^+ &= 1^+ \\
22^+ &= 1^+ + 21^- \\
252^+ &= 2(1^+) + 21^+ + 21^- + 104_a^+ + 104_b^+ \\
1750^+ &= 2(1^+) + 21^+ + 21^- + 104_a^+ + 104_b^+ + 1498^+ \\
3520_a^+ &= 2(1^+) + 1^- + 21^+ + 21^+ + 104_a^+ + 104_b^+ + 104_a^- + 104_b^- + \\
&\quad + 210^+ + 210^- + 560^+ + 560^- + 1498^-
\end{aligned}
$$

In addition to the 3-modular irreducibles $1, 21, 104_a, 104_b, 210, 560$ and 1498 found explicitly, we know that the ordinary character 5103 is irreducible modulo 3, since it is in a block of defect zero. Moreover, 8019_a and its complex conjugate 8019_b, which were irreducible in the simple group, fuse to give a new character of McL.2 with character values zero on all the 3-regular classes of McL.2\ McL. Similarly, the representations 605_a and 605_b have b_7-irrationalities, so they fuse in McL.2 to give an irreducible character of degree 1210. Finally, the last 3-modular irreducible representation, namely 2794, of the simple group McL was found by condensation of permutation modules. Therefore, we condense the corresponding permutation modules to find this representation for McL.2.

Condensation of permutation modules

This time we condense the permutation modules on 275, 7128, 15400 and 22275 points using a cyclic subgroup K of order 20.

(i)   The permutation representation on 275 points

On the simple group McL, the permutation character can be written as sum of 3-modular irreducible characters as follows:

$$275 = 4(1) + 3(21) + 104_a + 104_b$$

Also this permutation character can be written as a sum of ordinary irreducible characters of McL.2 as follows:

$$275^+ = 1^+ + 22^+ + 252^+$$

where these break into 3-modular irreducibles as follows:

$$
\begin{aligned}
1^+ &= 1^+ \\
22^+ &= 1^+ + 21^- \\
252^+ &= 2(1^+) + 21^+ + 21^- + 104_a^+ + 104_b^+
\end{aligned}
$$

Therefore 275^+ can be written as 3-modular irreducibles of McL.2 as:

$$275^+ = 3(1^+) + 1^- + 2(21^+) + 21^- + 104_a + 104_b$$

The decomposition matrix of the principal block Bo of McL

The 3-modular irreducible characters of McL

	1	21	104_a	104_b	210	560	605_a	605_b	1498	2794
22	1	1	0	0	0	0	0	0	0	0
231	0	1	0	0	1	0	0	0	0	0
252	2	2	1	1	0	0	0	0	0	0
770_a	0	0	0	0	1	1	0	0	0	0
770_b	0	0	0	0	1	1	0	0	0	0
896_a	1	1	1	0	1	1	0	0	0	0
896_b	1	1	0	1	1	1	0	0	0	0
1750	2	2	1	1	0	0	0	0	1	0
3520_a	3	3	2	2	2	0	0	0	1	0
3520_b	2	0	1	1	2	3	1	1	0	0
4500	4	2	2	2	1	2	1	1	1	0
4752	4	2	2	2	4	4	1	1	1	0
5544	2	0	1	1	1	2	1	1	0	1
8250_a	4	3	2	2	3	4	1	0	1	1
8250_b	4	3	2	2	3	4	0	1	1	1
9625	5	6	3	3	4	4	0	0	1	2
9856_a	6	4	3	3	4	5	1	1	1	1
9856_b	6	4	3	3	4	5	1	1	1	1
10395_a	6	3	3	3	4	6	1	1	1	1
10395_b	6	3	3	3	4	6	1	1	1	1

The ordinary characters of McL

Table I.7

Now, we condense the permutation module on 275 points to identify the above constituents in the condensed module. Using the subgroup K of order 20, we obtain a condensed module $M1$ of dimension 17, which is chopped up using the Meat-Axe to:

$$17 = 3(1_a) + 2(1_b) + 6_a + 6_b$$

Therefore, we have the following correspondences between the constituents of $M1$ and the 3-modular irreducibles in the original module:

The permutation $P1$ on 275 points	1^+	21^+	104_a^+	104_b^+
The 17-dimensional condensed module $M1$	1_a	1_b	6_a	6_b

Table II.1

Here the constituents of 1^- and 21^- disappear on condensation, because the elements of order 20 are fixed point free on 1^- and 21^-.

(ii)   The permutation representation on 7128 points

On the simple group McL, the permutation character 7128 can be written as a sum of 3-modular irreducibles as follows:

$$7128 = 6(1) + 5(21) + 2(104_a) + 2(104_b) + 1498 + 5103$$

On McL.2, the permutation character can be written as a sum of ordinary characters as follows:

$$7128^+ = 1^+ + 22^- + 252^+ + 1750^+ + 5103^+$$

where these break up into 3-modular irreducibles as follows:

$$\begin{aligned}
1^+ &= 1^+ \\
22^- &= 1^- + 21^+ \\
252^+ &= 2(1^+) + 21^+ + 21^- + 104_a^+ + 104_b^+ \\
1750^+ &= 2(1^+) + 21^+ + 21^- + 104_b^+ + 104_a^+ + 1498^+ \\
5103^+ &= 5103^+
\end{aligned}$$

Therefore, we can write 7128^+ as a sum of 3-modular irreducibles of McL.2 as follows:

$$7128^+ = 6(1^+) + 2(21^+) + 3(21^-) + 2(104_a^+) + 2(104_b^+) + 1498^+ + 5103^+$$

Using condensation of this module over the same subgroup K of order 20, we get a condensed module $M2$ of dimension 367, which is chopped up to irreducibles as:

$$367 = 6(1_a) + 2(1_b) + 2(6_a) + 2(6_b) + 75 + 260$$

Therefore, the correspondence between the irreducibles of the condensed module and those of the original module of degree 7128 can be expressed as in Table II.2:

The permutation P1 on 7128 points	1^+	21^+	104_a^+	104_b^+	1498^+	5103^+
The 552-dimensional condensed module $M2$	1_a	1_b	6_a	6_b	75	260

Table II.2

(iii)   The permutation representation on 15400 points

On the 3-regular classes of McL, the permutation character 15400 can be written as sum of 3-modular irreducibles as follows:

$$15400 = 5(1) + 12(21) + 7(104_a) + 7(104_b) + 3(210)$$
$$+4(560) + 605_a + 605_b + 3(1498) + 5103$$

Now, we condense the permutation module of degree 15400 for McL.2 over the same cyclic subgroup K of order 20 to get a condensed module $M3$ of dimension 790. Then $M3$ was chopped up to:

$$790 = 9(1_a) + 6(1_b) + 4(6_a) + 4(6_b) + 3(6_c) + 3(6_d) + 2(12_a)$$
$$+14 + 2(24_a) + 2(24_b) + 61 + 3(79) + 260$$

By spinning up the subspace containing the 61 in the condensed module $M3$, we constructed the corresponding representation of degree 1210 in the original module, which restricts to McL as $605_a + 605_b$. Therefore, we can write 4500^+ as:

$$4500^+ = 2(1^+) + 2(1^-) + 21^+ + 21^- + 104_a^+ + 104_b^- + 104_a^- + 104_b^+$$
$$+210^+ + 560^+ + 560^- + 1498^-$$

Hence, we can write the correspondence between the irreducibles in $M3$ and the those of the permutation module on 15400 points, as in Table II.3.

permutation on 15400	1^+	21^+	104_a^+	104_b^+	104_a^-			
790-dimensional module	1_a	1_b	6_a	6_b	6_c			
permutation on 15400	104_b^-	210^+	560^+	210^-	560^-	1498^-	1210	5103^+
790-dimensional module	6_d	14	24_a	12	24_b	79	61	260

Table II.3

(iv)   The permutation representation on 22275 points

This permutation module on 22275 points contains the last 3-modular irreducible, of degree 2794, for the simple group McL. This is a constituent of the ordinary irreducible 9625.

The character of this permutation module can be written as a sum of ordinary irreducibles in McL.2 as follows:

$$22275^+ = 1^+ + 22^+ + 2(252^+) + 1750^+ + 1750^- + 3520_a^- + 5103^+ + 9625^+$$

Therefore, by determining the signs of the constituents of 9625, we can work out the character values of 2794 on the 3-regular classes of McL.2\McL.

The condensation of the above permutation module over the same subgroup K of order 20 gives rise to a condensed module $M4$ of dimension 1140. $M4$ is chopped up using the Meat-Axe to the following irreducibles:

$$1140 = 12(1_a) + 10(1_b) + 6(6_a) + 6(6_b) + 3(6_c) + 3(6_d) + 4(12)$$
$$+2(14) + 3(24_a) + 3(24_b) + 2(75) + 3(79) + 143 + 260$$

The correspondences between the irreducibles of $M4$ and those of the original module on 22275 points are given in Table II.4:

permutation on 15400	1^+	21^+	104_a^+	104_b^+	104_a^-	104_b^-	210^+
1140-dimensional module $M5$	1_a	1_b	6_a	6_b	6_c	6_d	14
permutation on 15400	560^+	210^-	560^-	1498^-	2794	5103^+	
1140-dimensional module $M5$	24_a	12	24_b	79	143	260	

Table II.4

In McL.2, we have seen that the above permutation module on 22275 points can be written as:

$$22275^+ = 1^+ + 22^+ + 2(252^+) + 1750^+ + 1750^- + 3520_a^- + 5103^+ + 9625^+$$

The 3-modular character table of McL and McL.2

Ind	deg	1A	2A	4A	5A	5B	7A	7B B**	8A	10A	11A	11B B**	14A	14B B**	fus ind	2B	4B	8B	8C	10B	20A	22A	22B B**
		898128000 40320	96	750	25	14	14	30	11	11	14	14		7920 720	48	16	5	10	10	11	11	
+	1	1	1	1	1	1	1	1	1	1	1	1	1	+	1	1	1	1	1	1	1	1
+	21	5	1	1	-4			-1	-1	-1	-1			++	1	1	5	-1	1	1	-1	-1
+	104	8	0	4	-1		-1	-1	0	-1	-1	-1	-1	oo	4	4	0	0	0	-1	-1+15	-1-15
o	104	8	0	4	-1		-1	-1	-2 b11	0	-1	-1	b11	1	oo	4	4	0	0	0	-1	-1-15	-1+15-2-b11
o	210	2	-2	10	0			••	-2 b11	2	1	1	b11	1	++	4	4	0	0	0	-1	**-2-b11	••
-	560	-16	0	-15	0	0	0	0	0	-1	-1	-2	-2	oo	10 -10	0	0	-4	0	0	15	-15	111 -111
o	605	-19	5	5	0	-b7-1		-1	-1	1	-1	-1	-2	+	0	0	0	0	0	0	0	0
o	605	-19	5	5	0	-b7-1	-1	-1	1	0	-1	-1	0b7-1	••	0	0	0	0	0	0	0	0
+	1498	42	-2	1	5	0	-1	0	2	2	0	0	**b7-1	++	0 -20	2	-2	0	0	0	0	0
+	2794	-6	-2	19	-1	-2	1	0	-1	2	1	1		++	24 -4	-2	2	-1	1	2	2
+	5103	63	3	3	-2	0	1	3	-1	0	0	1	1	++	45 9	3	-1	0	-1	-1	1	1
o	8019	-45	3	-6	-1	0	0	-1	0	0	0	0		o	0	0	0	0	0	0	0	0
•	8019	-45	3	-6	-1	-b7	-1	0	0	-b7	-1	0		•	0	0	0	0	•• -b7	0	0	0
o	8019	-45	3	-6	-1		-1	0	0	-b7					0	0	0	0	-b7	0	0	0

Table II.5

where the ordinary characters break up as 3-modular characters as follows:

$$
\begin{aligned}
1^+ &= 1^+ \\
22^- &= 1^- + 21^+ \\
252^+ &= 2(1^+) + 21^+ + 21^- + 104_a^+ + 104_b^+ \\
1750^+ &= 2(1^+) + 21^+ + 21^- + 104_a^+ + 104_b^+ + 1498^+ \\
1750^- &= 2(1^-) + 21^- + 21^+ + 104_a^- + 104_b^- + 1498^- \\
3520_a^+ &= 2(1^+) + 1^- + 21^+ + 21^+ + 104_a^+ + 104_b^+ + 104_a^- + 104_b^- + \\
&= +210^+ + 210^- + 560^+ + 560^- + 1498^- \\
5103^+ &= 5103^+
\end{aligned}
$$

Therefore,

$$
9625^+ = 3(1^+) + 2(1^-) + 4(21^+) + 2(21^-) + 2(104_a^+) + 2(104_b^+) + 104_a^-
$$

$$
+104_b^- + 3(210^+) + 210^- + 560^+ + 560^- + 1498^+ + 1498^- + 2794
$$

Thus we can obtain the character values of 2794 on all the 3-regular classes of McL.2\ McL.

Theorem 2 *Table II.5 represents the complete 3-modular character table of McL and McL.2, in 'ATLAS' notation.*

References

[1] J. ALPERIN, 'Local representation theory', Cambridge studies in advanced mathematics 11, Cambridge University Press, 1984.

[2] J.H. CONWAY, A group of order 8 315 553 613 086 720 000, Bull. London Math. Soc. 1 (1969) 79-88.

[3] J.H. CONWAY, Three lectures on exceptional groups, in 'Finite simple groups', edited by Powell and Higman, pp. 215-147, Academic Press (1971).

[4] J.H. CONWAY, R.T. CURTIS, S.P. NORTON, R.A. PARKER and R.A. WILSON, 'An ATLAS of finite groups', Clarendon Press, Oxford, 1985.

[5] C.W. CURTIS and I. REINER, 'Representation theory of finite groups and associative algebras', Interscience publishers, New York, 1962.

[6] R.T. CURTIS, On subgroups of .0, *I* - Lattice Stabilizers, J. Algebra 27 (1973) 549 -573.

[7]   L. FINKELSTEIN, The maximal subgroups of Conway's Group C_3 and McLaughlin's Group, J. Algebra 25 (1973) 58-89 .

[8]   G. HISS, K. LUX and R. PARKER, The 5-modular characters of the McLaughlin group, (unpublished).

[9]   J.E. McLAUGHLIN, A simple group of order 898 128 000, in 'Theory of Finite Groups', Brauer and Sah, eds., pp. 109-111, Benjamin, 1969.

[10]   R.A. PARKER, Modular character tables (unpublished).

[11]   R.A. PARKER and R.A. WILSON, The computer construction of matrix representations of finite groups over finite fields, J. of Symb. Comp. 9 (1990) 583-590.

[12]   A.J.E. RYBA, Computer condensation of modular representations, J. of Symbolic Comp. 9 (1990) 591-600.

[13]   I.A. SULEIMAN, Modular representations of finite simple groups, Ph.D. thesis, Birmingham University, England, 1990.

[14]   I.A. SULEIMAN and R.A. WILSON, Computer construction of matrix representations of the covering group of the Higman-Sims group, J. Algebra, to appear.

[15]   I.A. SULEIMAN and R.A. WILSON, The 3- and 5-modular characters of the covering and the automorphism groups of the Higman-Sims group, J. Algebra, to appear.

[16]   J.G. THACKRAY, Modular representations of some finite groups, Ph.D. Thesis, University of Cambridge, England, 1981.

[17]   R.A. WILSON, The 2- and 3-modular characters of J_3, its covering group and automorphism group, J. of Symbolic Computation, to appear.

The Orbifold Notation for Surface Groups

J.H. Conway

"Even quite ungainly objects, like chairs and tables, will become almost spherical if you wrap them in enough newspaper."

The symmetries of any finite object, such as a chair or a table, all fix a point, say the centre of gravity of the object, and so act on the surface of a sphere, for example any sphere centred on the centre of gravity.

The symmetries of a repeating pattern on a carpet or tiled floor, or on a wall, supposed continued to infinity, will probably constitute one of the 17 plane crystallographic groups.

Among the works of the Dutch draughtsman Maurits C. Escher, one can find examples of all these 17 groups, and also some even more interesting designs such as *Circle Limit I, II, ...*, whose symmetries are various discrete groups of isometries of the hyperbolic plane.

In this paper a **surface group** will be a discrete group of isometries of one of the following three surfaces:

1. the sphere

2. the Euclidean plane

3. the hyperbolic plane.

These are all the simply-connected surfaces of constant Gaussian curvature.

We shall present a simple and uniform notation that describes all three types of group. Since this notation is based on the concept of *orbifold* introduced by Bill Thurston, we shall call it **the orbifold notation.**

Roughly speaking, an orbifold is the quotient of a manifold by a discrete group acting on it. It therefore has one point for each orbit of the group on the manifold (*Orbifold = Orbit-manifold*).

Mirrors and mirror-boundaries

An orbifold may have boundary curves even though our three original surfaces do not. The boundary points arise from points lying on mirrors.

A **mirror** or **mirror-line** for a group is the line fixed by some reflection in the group. A point that lies on a mirror is called a **mirror-point**, an

ordinary mirror-point if it lies on just one mirror, and an **m-fold mirror-point** if it lies on exactly m mirrors.

The points of an orbifold that correspond to mirror-points are boundary points. An **ordinary** boundary point is the image of an ordinary mirror-point, and a **type** m **corner-point** is the image of an m-fold mirror-point. At a type m corner, the boundary has an angle $\frac{\pi}{m}$.

We shall say that a boundary-curve has type $*ab\ldots c$ to mean that its corners have types $a, b, \ldots, c$, reading around the curve in some consistent direction.

Tables and chairs

A plain rectangular table has two planes of symmetry, which divide the sphere into four segments. We can consider any one of these segments as the orbifold– its boundary consists of two semicircles that intersect each other at the zenith and at the nadir, at angle $\frac{\pi}{2}$. So this boundary curve has type $*22$, and indeed $*22$ is the orbifold notation for the symmetry group of the table.

A plain square table has two further (diagonal) planes of symmetry, and the four symmetry planes divide the sphere into eight segments, the typical segment having two corners at angle $\frac{\pi}{4}$. This time the symmetry group is $*44$.

A chair has a single plane of symmetry, which cuts the sphere in a great circle, that is to say, a boundary curve without corners, type $*$. We might also write $1*$, so as to give the star something to hang on to– digits 1 have no significance in this notation, except as place-fillers.

Gyration-points and cone points

An orbifold may have some special points that do not lie on boundary curves. A **gyration** is a rotation in the group whose centre does *not* lie on any mirror. A point of the surface is called an m-**fold gyration point** if it is the centre of some gyration of order m, but not of any gyration of higher order. The image in the orbifold of an m-fold gyration point is called a **cone-point** of order m– it is a point around which the angle is $\frac{2\pi}{m}$ rather than 2π.

Some brick walls

The simple brick wall consisting of bricks laid directly above and directly to the side of each other in a rectangular array, has four types of mirror-line, namely
vertical in brick,　horizontal in brick
vertical in cement,　horizontal in cement.

The orbifold can be identified with one quarter-brick (with a bit of cement adhering to it), whose boundary is a rectangle, type $*2222$, and this is the orbifold notation for the symmetry group of the wall.

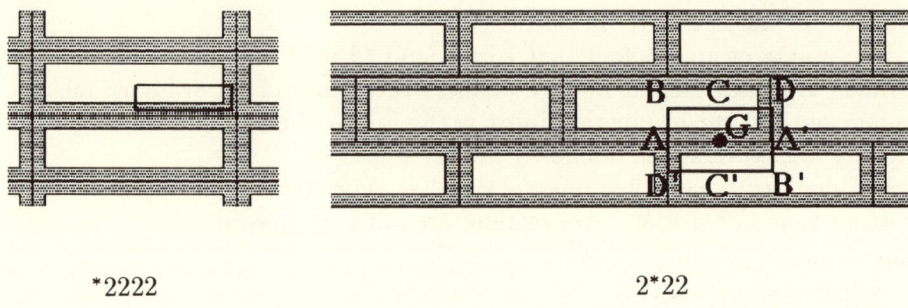

*2222   2*22

Figure 1

A more common, because stronger, kind of wall, has the bricks staggered by half a length in adjacent layers.

This time we cannot directly identify the orbifold with a portion of the wall. In Figure 1, we have outlined a rectangle $ABCDA'B'C'D'$ bounded by mirrors, but the centre G of this rectangle does not lie on any mirror, and is a 2-fold gyration point, since the appearance of the wall would be undisturbed if it were rotated through half a revolution about G. The orbifold is obtained from this rectangle by identifying each point of the rectangle (including the interior points) with its image under the gyration about G. It is realisable as a conical surface with a cone point of order 2 at G, and a boundary curve that has just *two* right-angled corners, at $B = B'$ and $D = D'$. The orbifold notation for this group is 2*22, the initial 2 representing the cone-point, and the remainder of the symbol the boundary curve with its two corners.

The orbifold notation in general

The orbifold notation for any surface group consists of a number of digits $A, B, \ldots, C,\ a, b, \ldots, c,\quad \alpha, \beta,, \ldots, \gamma, \ldots$, together with some circles, stars and crosses:

$$\circ \circ \ldots \circ \quad AB \ldots C \quad *ab \ldots c \quad *\alpha\beta \ldots \gamma \quad \ldots \quad \times \times \ldots \times$$

Each initial circle represents a **handle**, and each final circle a **crosscap**. The digits $A, B, \ldots, C$ *not* preceded by stars are the orders of the distinct cone-points on the orbifold, and each star together with all the digits that immediately follow it indicates the type of a boundary curve.

We remind the reader that the classification theorem for 2-manifolds assures us that any connected compact 2-manifold can be obtained by adding handles and crosscaps to a sphere, and then punching a hole in it for each boundary curve. If we also take into account the various possibilities for local collapse, we can see that the above notation covers all possibilities for the orbifold of a surface group.

It can also be shown that every such notation does correspond to the orbifold of some group, with just the exceptions mn and $*mn$ $(m > n \geq 1)$. (If $n = 1$, these exceptions appear as m and $*m$.)

The Euler characteristic of an orbifold

The Euler characteristic of an orbifold generalises the Euler characteristic $(V - E + F)$ of a manifold in the natural way. It is a rational number which coincides with the usual integer-valued characteristic for an ordinary manifold without cone-points, and whose value gets divided by $|G|$ when we take the quotient by a discretely acting group G, of finite order $|G|$. From a 'map' it can be computed by the usual formula $V - E + F$, provided we make proper allowances for the divided nature of the vertices, edges, and faces, namely:-

a 'vertex' at a cone-point of order m counts as $\frac{1}{m}$ of a vertex, while

a 'vertex' at an ordinary boundary point is $\frac{1}{2}$ of a vertex, and

a 'vertex' at a type m corner-point is $\frac{1}{2m}$ of a vertex. Also

an 'edge' running along a boundary curve is $\frac{1}{2}$ of an edge, and finally

a 'face' with just one internal cone-point, of order m, is $\frac{1}{m}$ of a face.

For example, for the orbifold from our second brick wall, we shall take the 'map' to consist of

$\frac{1}{4} + \frac{1}{4}$ vertices, namely the corner-points $B = B'$ and $D = D'$

$\frac{1}{2} + \frac{1}{2}$ edges, the images of $D'B$ and BD, and

$\frac{1}{2}$ of a face, the image of the rectangle $ABCDA'B'C'D'$.

The Euler characteristic is therefore

$$(\frac{1}{4} + \frac{1}{4}) - (\frac{1}{2} + \frac{1}{2}) + \frac{1}{2} = \frac{1}{2} - 1 + \frac{1}{2} = 0.$$

There are other ways to compute the characteristic, namely from the Gauss-Bonnet formula as the integrated Gaussian curvature over the entire orbifold (in this example the curvature was everywhere 0), or directly from the orbifold notation, in the manner we shall now describe.

A visit to SymmetryLand

We suppose that every day of our holiday, we start with $2 in our pocket (because the Euler characteristic of the sphere is 2), and go on a visit to SymmetryLand.

SYMMETRYLAND TICKET CHARGES

Ticket type	Symbol	Cost of ticket for	
		Adult:	Child:
2-trip	2	$\frac{1}{2}$	$\frac{1}{4}$
3-trip	3	$\frac{2}{3}$	$\frac{1}{3}$
4-trip	4	$\frac{3}{4}$	$\frac{3}{8}$
5-trip	5	$\frac{4}{5}$	$\frac{2}{5}$
6-trip	6	$\frac{5}{6}$	$\frac{5}{12}$
n-trip	n	$\frac{(n-1)}{n}$	$\frac{(n-1)}{2n}$
TOP ticket	o or ×	2	1
Chaperone's	*	–	1

SYMMETRYLAND RULES:

- Children not in possession of TOP tickets must be chaperoned.

- A chaperone's ticket entitles the bearer to enter SymmetryLand alone, or in charge of any number of children, in which case the chaperone is responsible for keeping their behaviour within acceptable boundaries.

- SymmetryLand extends credit to regular visitors.

Every orbifold notation can be regarded as specifying a possible day's tickets at SymmetryLand, and its Euler characteristic is the amount of change we shall have in our pocket at the end of the day.

For example, 2*22 corresponds to a day when we purchase one adult's 2-trip ticket, one chaperone's ticket, and two child's 2-trip ones.

The total cost in dollars is $\frac{1}{2} + 1 + \frac{1}{4} + \frac{1}{4} = 2$ so that the change from our $2 is 0, agreeing with our earlier answer.

(It is fairly easy to justify this rule. It is well-known that the Euler characteristic of the sphere is 2, and that handles and crosscaps cause reductions by 2 and 1, respectively. Now punching a hole can be viewed as removing 1 face, so causing a decrease of 1, while changing an ordinary point to a conical point of order n can be viewed as replacing 1 whole vertex by $\frac{1}{n}$ of a vertex, causing a decrease of $\frac{(n-1)}{n}$ etc. The map can be chosen so that any point of particular interest is a vertex of it.)

The 17 plane crystallographic groups

It turns out that there are just 17 SymmetryLand outings from which we return flat broke, and these correspond to the 17 discrete groups acting on the flat(Euclidean) plane so that the orbifold has finite area, namely:-

Notation	English name	
*632	hexascope group	(hexatropic kaleidoscope group)
632	hexatrope group	(cheiral hexatropic group)
*442	tetrascope group	(tetratropic kaleidoscope group)
4*2	tetragyro group	(tetratropic gyrational group)
442	tetratrope group	(cheiral tetratropic group)
*333	triscope group	(tritropic kaleidoscope group)
3*3	trigyro group	(tritropic gyrational group)
333	tritrope group	(cheiral tritropic group)
*2222	discope group	(ditropic kaleidoscope group)
2*22	dirhomb group	(ditropic rhomboidal group)
22*	digyro group	(ditropic gyrational group)
22×	diglide group	(ditropic gliding group)
2222	ditrope group	(cheiral ditropic group)
** or 1**	monoscope group	(monotropic kaleidoscope group)
× or 1×	monorhomb group	(monotropic rhomboidal group)
×× or 1××	monoglide group	(monotropic gliding group)
° or °1	monotrope group	(cheiral monotropic group)

To verify the completeness of the enumeration, we consider applying the following modifications to a given group of characteristic 0:

1. Replace a group $AB \ldots C$ by $*AB \ldots C$ – this halves the characteristic

2. Replace an adult's TOP ticket (°) by two child's ones

3. Replace a child's TOP ticket (×) by a chaperone's ticket (*)

4. Since a chaperone (*) is now present, replace an adult's n-trip ticket by two child's ones.

After these modifications we have at least one star (and at most two, since they cost $1), and it is easy to see the group is one of

$$*632, *442, *333, *2222, **.$$

So the other groups are obtainable from these by reversing the above moves, namely:

replacing any two equal digits after the star by one before it;

replacing any final star by a final circle;

replacing two final crosses by one initial circle, or

deleting an initial star that was followed only by digits.

In the table, the groups obtained in these ways from one of the five starting groups above are listed immediately after that group.

The "English names" we give are new, but we recommend them since some thought has gone into their selection. The **tropicity** of a group is the maximal order of any rotation in it. A group is **cheiral** if it contains no reflections or glide-reflections, and **gyrational** if it contains *both* gyrations and reflections.

The finite spherical groups

A finite spherical group corresponds to a day in SymmetryLand from which we return with a positive amount of change. The cases are:-

Geometric name	Coxeter	Orbifold	Algebraic	Algebraic name
icosahedral reflection	$[3,5]$	$*532$	$\pm I$	diplo-icosahedral
icosahedral rotation	$[3,5]^+$	532	I	icosahedral
octahedral reflection	$[3,4]$	$*432$	$\pm O$	diplo-octahedral
octahedral rotation	$[3,4]^+$	432	O	octahedral
tetrahedral reflection	$[3,3]$	$*332$	TO	tetra-octahedral
gyro-octahedral	$[3^+,4]$	$3*2$	$\pm T$	diplo-tetrahedral
tetrahedral rotation	$[3,3]^+$	332	T	tetrahedral
prismatic	$[2,n]$	$*22n$	$\pm D_{2n}$	diplo-dihedral
antiprismatic	$[2^+,2n]$	$2*n$	DD_{4n}	dihedro-dihedral
dihedral	$[2,n]^+$	$22n$	D_{2n}	dihedral
pyramidal	$[n]$	$*nn$	CD_{2n}	cyclo-dihedral
gyro-prismatic	$[2,n^+]$	$n*$	$\pm C_n$	diplo-cyclic
gyro-antiprismatic	$[2^+,2n^+]$	$n\times$	CC_{2n}	cyclo-cyclic
cyclic	$[n]^+$	nn	C_n	cyclic

Here we have given no fewer than two systems of English names, and three notations. The reason is that the groups are classified differently according as we take the basic type of reflecting operation to be a **plain reflection**, or the **central inversion** (represented by the matrix -1). The first choice is usually appropriate for more geometrical purposes, and corresponds well to our orbifold notation, and to Coxeter's notation which regards the groups as subgroups of groups generated by reflections. The "geometric" names express the groups in terms of automorphisms of certain polyhedra.

However, for algebraic purposes, it is often useful to take the second approach, whereby the groups are obtained by modifying the pure rotation groups G by multiplying some of their elements by -1. We use the standard adjectives "cyclic", "dihedral", ...,"icosahedral" in the names of these groups.

If we straightforwardly adjoin -1 to G, we obtain the group $\pm G$, called "diplo-G", meaning "the double of G". Instead, we can multiply all those elements of G that lie outside H by -1, where H is a subgroup of index 2 in G, to obtain the group called HG. The "algebraic name" for this group is "adverbo-G", where the adverb describes the subgroup H.

As abstract groups, $\pm G$ is isomorphic to $C_2 \times G$, while HG is isomorphic to G, with rotation subgroup H. So, in the "adverbo–adjectival" name for HG, the adjective refers to the whole group G, and the adverb to its rotation subgroup H. (The rotation subgroup is the subgroup of elements of determinant $+1$.) The subscripts on our groups are **orders**.

So for example, CD_{24} is abstractly a dihedral group of order 24, whose rotation subgroup is a cyclic group of order 12, and we call it the cyclo-dihedral group of order 24.

The braces indicate that the correspondence between the geometric and algebraic systems is not always the same. One should interchange the two lines to the right of the brace whenever the parameter n is odd. Thus the symmetry group of a regular hexagonal prism, is the prismatic group $[2,6] = *226$, which is the *diplo*-dihedral group $\pm D_{12}$. However the symmetry group of the regular pentagonal prism, namely $[2,5] = *225$ is the *dihedro*-dihedral group DD_{20}. In general, we get the diplo-types from polyhedra with central symmetry, and otherwise the more subtle types.

The hyperbolic groups

We shall only briefly mention these here. They correspond to days at SymmetryLand when we overspend our \$2 allowance, and so return home in debt to the SymmetryLand owners. One such pattern is Escher's *Circle Limit IV*, which consists of alternating black devils and white angels. If one crawls along a mirror of this pattern until one hits another mirror and then turns right along this mirror and continues in the same way, one's path is a quadrilateral with four corners of $\frac{\pi}{3}$, whose centre appears as a 4-fold gyration point. This makes the orbifold notation 4*3, since all four corners of the quadrilateral are identified by the gyration.

However, a more careful examination of the figure shows that every fourth figure– either angel or devil– is facing away from us rather than towards us, so that the indicated gyration is not actually a symmetry at this level of detail, and the group drops to *3333. Let us compute characteristics:

For 4*3 we get $2 - \frac{3}{4} - 1 - \frac{1}{3} = -\frac{1}{12}$.

For *3333 we get $2 - 1 - \frac{1}{3} - \frac{1}{3} - \frac{1}{3} - \frac{1}{3} = -\frac{1}{3}$.

The ratio between these numbers is indeed 4, indicating correctly that the second group *3333 has index 4 in 4*3. The latter group has index 2 in the reflection group ∗642, which could be obtained by adjoining the reflexive symmetries of one of our quadrilaterals (which would interchange angels and devils).

The work *Circle Limit III* is equally intriguing. It shows a pattern of fishes in four colours. There are some white lines that are misleading, since one's first guess is that they are hyperbolic straight lines, whereas they are in fact hypercycles ("circles of super-infinite radius"). These outline some "squares" and "triangles" which alternate rather like the faces of a cuboctahedron, except that at each vertex there are three triangles and three squares. The group is 433, there being gyration points of orders 4 and 3 at the centres of the squares and triangles, and another type of gyration point of order 3 at the vertices of the tessellation.

That, at least, is what we see if we are colourblind. If we take account of the colours, the group drops to 222222, since the square regions are now of six distinct types– each has two fish, each of two colours, and each of the six pairs of colours happens.

The characteristic of 433 is $2 - \frac{3}{4} - \frac{2}{3} - \frac{2}{3} = -\frac{1}{12}$, while that of 222222 is $2 - \frac{1}{2} - \frac{1}{2} - \frac{1}{2} - \frac{1}{2} - \frac{1}{2} - \frac{1}{2} = -1$. The index is 12, corresponding to the fact that every *even* permutation of the 4 colours is induced by some symmetry of the tessellation.

Stepping around the world, or across the plane

Please regard ⌐ as a left footprint, and ⌐ as a right one. Now hop around the equator, taking n hops to do so. Your footprints:

⌐ ⌐ ⌐ ⌐ ⌐ ⌐

have the symmetry group nn. If you had walked, taking $2n$ full paces, the group would have been $n\times$. The table shows that the 7 infinite families of **axial** groups (the finite groups that fix an axis) correspond to the 7 ways to proceed around the world:

⌐⌐⌐⌐⌐⌐	hop	nn	⌐⌐⌐⌐ spinning hop	$22n$
⌐⌐⌐⌐⌐⌐	step	$n\times$		
⌐⌐⌐⌐⌐⌐	jump	$n\ast$	⌐⌐⌐⌐ spinning jump	$\ast 22n$
⌐⌐⌐⌐⌐⌐	sidle	$\ast nn$	⌐⌐⌐⌐ spinning sidle	$2\ast n$

But these figures could be used with more justice also for the trail left by a (doubly infinite) progression across the Euclidean plane. The resulting groups, which have been called the **frieze groups**, or the **2-dimensional line groups**, are naturally symbolised by replacing the digit n by ∞. They are

$$\ast 22\infty, \quad 2\ast\infty, \quad 22\infty, \quad \ast\infty\infty, \quad \infty\ast, \quad \infty\times, \quad \infty\infty$$

The only other discrete groups acting on the Euclidean plane are the **2-dimensional point groups**. I call them (n) and $(*n)$. Here the parentheses may be considered as standing for the open-ended nature of the resulting orbifold, or as apologising for the slight illegitimacy of the notation. These groups can be thought of as obtained from nn and $*nn$ by letting the radius of the sphere tend to infinity.

The group (n) is generated by a rotation of order n, and the group $(*n)$ by two reflections in lines at angle $\frac{\pi}{n}$. So (4) is the group of the swastika, and $(*5)$ the group of the regular pentagon.

Note

The philosophy that geometrical groups should be studied through their orbifolds is Bill Thurston's. I claim originality only for the simple and elegant notation introduced here. David Singerman tells me that Murray MacBeath has long described the orbifolds of surface groups in a less compact but essentially equivalent way.

A remark on two diophantine equations

of Peter Cameron

by

Christoph Hering

During the 12th British Combinatorial Conference in Norwich, 1989, Peter J. Cameron suggested to investigate two diophantine equations, namely

$$\binom{k}{2} - 1 = \frac{q^n - 1}{q - 1}$$

and

$$\binom{k}{2} - 1 = q^n + 1 \,,$$

where q, n, k are positive integers and $q > 1$, $n > 3$ (see [1]). For every particular value of q, these equations lead to Thue- Mahler equations of degree 2. Such second degree Thue-Mahler equations can be solved in general by an algorithm developed in [2]. Clearly, it is possible to apply this algorithm here to obtain at least partial results. For the first equation we can decide the question concerning the possibility of a solution for every given value of q. In the second case it even is sufficient to specify the set of prime divisors of q. Note, that by results of Siegel, for every given q there are only finitely many solutions. In this paper we solve the first equation for prime powers $q < 47$ and the second equation for powers of a prime p, where $p < 47$. It is not at all difficult to treat further cases.

While solving Peter Cameron's equations we also solve a collection of Thue-Mahler equations of the form

$$x^2 - m = c_0 c_1^y,$$

where m, c_0, c_1 are rational integers. These equations might be of independent interest. A list of the cases which we treated here can be found at the very end of the paper.

The second equation

Let $X, q \geq 2$ and $Y \geq 2$ be natural numbers and assume that

$$\binom{X}{2} - 1 = q^Y + 1$$

Then

$$X^2 - X - 4 = 2q^Y.$$

We denote $x = 2X - 1$ and obtain

$$x^2 - 17 = 8q^Y.$$

Let p be a prime divisor of q different from 2 and 17. We have $x^2 \equiv 17 \pmod{p}$ so that 17 is a square modulo p. By the Gauss reciprocity law, p likewise is a square modulo 17, so that $p \equiv 1, 2, 4, 8, 9, 13, 15$ or $16 \pmod{17}$. If $17|q$, then $17|x^2, 17|x, 17^2|x^2$ and $17^2 \nmid q^Y$ so that $Y = 1$. Therefore we may assume that **every prime divisor of q is**

$$\equiv 1, 2, 4, 8, 9, 13, 15 \text{ or } 16 \pmod{17}. \tag{1}$$

We now consider the special case $3|Y$. Here $x^2 - 17 = z^3$, where $z = 2q^{Y/3}$. By a result of Nagell [3], the integral solutions of the diophantine equation $x^2 - 17 = z^3$ are

$$
\begin{array}{rccccccccc}
x = & 3, & 4, & 5, & 9, & 23, & 282, & 375, & 378661 \\
z = & -2, & -1, & 2, & 4, & 8, & 43, & 52, & 5234
\end{array} \cdot
$$

For our situation Nagell's result implies: **If Y is divisible by 3, then we have one of the cases**

$$
\begin{array}{rccccc}
X = & 5, & 12, & 12, & 188, & 189331 \\
q = & 2, & 2, & 4, & 26, & 2617 \\
Y = & 3, & 6, & 3, & 3, & 3
\end{array} \cdot
$$

Assume that Y is even or q is a square. The diophantine equation $x^2 - 17 = 8z^2$ has infinitely many solutions. The initial solutions are $(5,1)$ and $(7,2)$. By [2, Theorem 1], $(x, q^{Y/2}) = (5,1)M^i$ or $(x, q^{Y/2}) = (7,2)M^i$ for $i \geq 0$, where

$$
M = \begin{bmatrix} 3 & 1 \\ 8 & 3 \end{bmatrix}.
$$

Of course, the first elements of the series of values possible for $q^{Y/2}$, which increases rapidly, can be easily computed.

Finally, we assume that q is a power of a prime p. This leads to the diophantine equation

$$
x^2 - 17 = 8p^z
$$

which is a Thue-Mahler equation of degree 2. For each particular prime p it has only finitely many solutions, and these can be determined by the algorithm described in [2]. We investigate all primes less than 47. Of these only $2, 13, 19$, or 43 can occur, because of (1). Our choice of the various parameters and the most important intermediate results are listed in Table 2. The last three columns of this table contain solutions. Note that each case branches into the two subcases $\mathfrak{X} = \emptyset$ and $\mathfrak{X} = \{1\}$. Each subcase in general again branches according to the various initial solutions (x_0, y_0). If, for a pair of initial solutions belonging to the same fundamental solution we have in a given column the same entry, then this entry sometimes is listed only once. Always $m = 17$. Also, we choose $d = 1$ unless $p = 2, \mathfrak{X} = \{1\}$, where we choose $d = 4$.

As an example we describe the case $p = 2$. This leads to the Thue-Mahler equation $x^2 - 17 = 8 \cdot 2^z$. We must distinguish the cases $z \equiv 0 (mod\ 2)$ and $z \equiv 1 (mod\ 2)$. Assume at first that z is even. Then we are looking for solutions (x, y) of the diophantine equation

$$x^2 - 17 = 8y^2 \tag{2}$$

such that y is a power of 2. Using the notation of [2], we have $r = 1, c_0 = 8, c_1 = 2, m = 17, \mathfrak{X} = \emptyset$ and $D(\mathfrak{X}) = 8$. The largest square dividing 8 is 4 so that for d we have the choice 1 or 2. We choose $d = 1$ so that $D = 8$. The smallest positive solution of $x^2 - 1 = 8y^2$ is $(\xi, \eta) = (3, 1)$. Thus

$$M = \begin{bmatrix} \xi & \eta \\ D\eta & \xi \end{bmatrix} = \begin{bmatrix} 3 & 1 \\ 8 & 3 \end{bmatrix}.$$

As the fundamental boundary $s = \sqrt{(\xi - sign(m))|m|/2D} < 2$, the only fundamental solution is $(5,1)$, and the initial solutions are $(5,1)$ and $(5, -1)$ $M = (7, 2)$. By [2, Theorem 1] every non-negative integral solution of (2) is of the form $(5, 1)\ M^i$ or $(7, 2)M^i$ for some $i \geq 0$. We choose $T = 16$, and our choice for the modulus is 257. Computing modulo $n = 257$ we see that the conditions $16|y$ and y is a power of 2 can't be satisfied at the same time. So $q \leq 8$. If $z \equiv 1 (mod\ 2)$, then $\mathfrak{X} = \{1\}$ and $D(\mathfrak{X}) = 16$. We choose $d = 4$, and hence $D = 1$. The only non-negative solution of $x^2 - 17 = y^2$ is $(x, y) = (9, 8)$ (see [2, Theorem 3]). We also have proved

Theorem 1. The diophantine equation

$$x^2 - 17 = 2^y$$

has solutions only for $y = 3, 5, 6$ and 9.

We summarize our results.

Theorem 2. Let x, q and n be natural numbers such that

$$\binom{x}{2} - 1 = q^n + 1,$$

where $q \geq 2$ and $n \geq 3$. If q is a power of a prime not larger than 46 or if $3|n$, then (x, q, n) is one of

$$
\begin{array}{rccccc}
x = & 5, & 12, & 12, & 188, & 189331 \\
q = & 2, & 2, & 4, & 26, & 2617 \\
n = & 3, & 6, & 3, & 3, & 3
\end{array}
\quad .
$$

Also, every prime divisor of q is $\equiv 1, 2, 4, 8, 9, 13, 15$ or $16 (mod\ 17)$.

The first equation

Let $X, q \geq 2$ and $Y \geq 2$ be natural numbers and assume that

$$\binom{X}{2} - 1 = \frac{q^Y - 1}{q - 1}.$$

Then

$$(q - 1)X^2 - (q - 1)X - 2q + 4 = 2q^Y. \qquad (3)$$

Let $q - 1 = a_1^2 a_2$, where a_2 is square free. Denote $t = (2a_1, q - 1)$, $x = \frac{2(q-1)}{t} X - \frac{q-1}{t}$ and $m = \frac{9q^2 - 26q + 17}{t^2}$.

This leads to

$$x^2 - m = \frac{8(q - 1)}{t^2} q^Y.$$

Note that $t = a_1$ or $t = 2a_1$. Also $t^2 | 4a_1^2 | 4(q-1)$ and $t^2 | (q-1)^2 + 4(q-1)(2q - 4) = 9q^2 - 26q + 17$. Let p be a prime divisor of q different from 2 and 17. We have $(tx)^2 - (9q^2 - 26q + 17) = 8(q - 1)q^Y$, so that $(tx)^2 \equiv 17(mod\,p)$. Hence 17 is a square modulo p, and as above we obtain: **Every prime divisor different from 17 of q is**

$$\equiv 1, 2, 4, 8, 9, 13, 15 \text{ or } 16(mod\,17).$$

Let $q = 17A$, where $A \in \mathbb{N}$. Then $(tx)^2 \equiv 17(mod\,17)$ and $17^2 | (tx)^2$. Hence $0 \equiv 9q^2 - 26q + 17 = 9 \cdot 17^2 A^2 - 26 \cdot 17A + 17 \equiv 17(-26A + 1)(mod\,17^2)$, $9A \equiv 1(mod\,17)$ and $A \equiv 18A \equiv 2(mod\,17)$. Thus we proved: **If $17|q$, then $q/17 \equiv 2(mod\,17)$. In particular, q is not a power of 17.**

We now consider the case $Y = 3$, i.e.

$$\binom{X}{2} - 1 = \frac{q^3 - 1}{q - 1} = q^2 + q + 1.$$

Denoting $x = 2X - 1$ and $Q = 2q + 1$, we find

$$4X^2 - 4X - 8 = 8(q^2 + q + 1)$$

and

$$x^2 - 15 = 2Q^2 \tag{4}$$

The smallest positive solution of the equation $x^2 - 1 = 2y^2$ is $(3,2)$. So the fundamental boundary is $\sqrt{15/2} < 3$. Therefore the diophantine equation (4) has no solution, which implies that $Y \neq 3$.

For every particular q, the solutions of (3) can be determined by the algorithm described in [2]. For prime powers less than 47 this has been documented in Table 2. (Actually, for $q = 4$ and $q = 16$ we slightly modified the algorithm in [2]. If one of the $c_i, i \geq 1$ in [2,§1] is a square, then it suffices to consider the subsets $\mathfrak{X} \subseteq \{1, \cdots, r\} \backslash \{i\}$ if furtheron all solutions of the diophantine equations $x^2 - m = (D(\mathfrak{X})/d^2) y^2$ are determined, for which $y = d c_1^{z_1} \cdots c_{i-1}^{z_{i-1}} \bar{c}_i^{z_i} c_{i+1}^{z_{i+1}} \cdots c_r^{z_r}$ for suitable non-negative integers $z_1, \cdots, z_r$, where $\bar{c}_i^2 = c_i$. Thus, in our case we can do without the $\mathfrak{X} = \{1\}$-column for $q = 4$ or 16.) In two cases $D(\mathfrak{X})$ is a square: $q = 2$ and $\mathfrak{X} = \{1\}$ leads to $x^2 - 1 = y^2$. The only non-negative integral solution is $(x, y) = (1, 0)$. Also, $q = 19$ and $\mathfrak{X} = \emptyset$ leads to $x^2 - 77 = y^2$ and $(x, y) = (9, 2)$ or $(39, 38)$ by Theorem 3. We obtain the solution $(X, Y) = (7, 2)$.

We summarize our results:

Theorem 3. Let x, q and n be natural numbers such that

$$\binom{x}{2} - 1 = \frac{q^n - 1}{q - 1},$$

where $q \geq 2$ and $n \geq 3$. Then actually $n \geq 4$. There is no solution

if q is a prime power less than 47. **Every prime divisor of q is**
$\equiv 0, 1, 2, 4, 8, 9, 13, 15$ **or** $16 (mod\ 17)$. **If** $17|q$, **then** $q/17 \equiv 2(mod\ 17)$.

Remark. The tables 1 and 2 actually contain the solution of various Thue-Mahler equations of the form

$$x^2 - m = c_0 c_1^y.$$

The table below contains the various triples (m, c_0, c_1) which we considered, and for each case the values possible for y.

m	17	17	17	17	1	57	385	1905	8401	75	77	3885
c_0	8	8	8	8	8	24	56	120	248	6	4	84
c_1	2	13	19	43	2	4	8	16	32	13	19	43
y	$0, 2,$	$0, 1, 2$	$0, 1$	$0, 1$	0	$0, 2$	$0, 2$	0	0	$0, 2$	$0, 2$	$0, 2$
	$3, 6$											

p	$\mathfrak{X}$	$D(\mathfrak{X})$	D	ξ	η	s	(x_0, y_0) (x'_0, y'_0)	T	$l(T)$	I	n
2	$\emptyset$	8	8	3	1	<2	5 1	16	16	9	257
							7 2			6	
	$\{1\}$	16	1								
13	$\emptyset$	8	8	3	1	<2	5 1	13^2	14	$\emptyset$	
							7 2				
	$\{1\}$	104	104	51	5	<3	11 1	13	26	9,22	$8857 =$
							41 4			3,16	$17\cdot 521$
19	$\emptyset$	8	8	3	1	<2	5 1	19	20	7,17	2
							7 2			2,12	
	$\{1\}$	152	152	37	3	<2	13 1	19	38	1,20	$565 = 5\cdot 113$
							25 2			17,36	3
43	$\emptyset$	8	8	3	1	<2	5 1	43	44	8,30	$1068 =$
							7 2			13,35	$2^2\cdot 3\cdot 89$
	$\{1\}$	344	344	10405	561	<17	19 1	43	86	17,60	$4369 =$
							4711 254			25,68	$17\cdot 257$

Table 1, part 1. $m = 17$.

p	$l(n)$	I'	S	X	q	Y
2	64	$9,25,41,57$	$75,109,148,182$	12	2	6
				12	4	3
				12	8	2
		$6,22,38,54$		4	2	2
				5	2	3
13				19	13	2
	52	$9,22,35,48$ $3,16,29,42$	$3059,3557,5300,5798$			
19	2	1 0	0			
	114	$1,20,39,58,77,96$	$76,268,531,376,523,486$			
	1	0	2			
43	44	I	$431,637$			
	258	$17,60,103,146,189,232$ $25,68,111,154,197,240$	$800,4166,1310,3826,460,3316$ $1053,3909,543,3059,203,3569$			

Table 1, part 2

Table 2, part 1

	q = 2	q = 4	q = 8	q = 16
t	1	1	1	1
m	1	57	385	1905

	q=2	q=2	q=4	q=8	q=16	q=16
$\mathfrak{X}$	∅	{1}	∅	∅	{1}	∅
$D(\mathfrak{X})$	8	16	24	56	448	120
d	2	4	2	2	8	2
D	2	1	6	14	7	30
ξ	3		5	15	8	11
η	2		2	4	3	2
s	< 1		< 5	< 14	< 14	< 18

	q=2	q=4	q=4	q=8	q=8	q=8	q=8	q=16	q=16	q=16
x_0	1	9	21	21	203	49	63	∄	45	375
y_0	0	2	8	2	54	12	16		2	68
T	4	4	16	16	16	16	32		8	8
$l(T)$	2	2	8	4	4	4	8		4	4
I	0	1	4	∅	∅	3	4		1	2
n	3	15	7			7	431		33	33
$l(n)$	4	4	8			1	24		4	4
I'	0, 2	1, 3	I			0	4, 12, 20		I	I
S	0	13, 7	6			5	207, 415, 240		13	20
X		4				5				
Y		2				2				

Table 2, part 1. The last two rows contain solutions. ∄ means no solution.

Table 2, part 2

	q = 32	q = 13	q = 19
t	1	4	6
m	8401	75	77

	q=32	q=32	q=13	q=19	q=19	q=19
$\mathfrak{X}$	∅	{1}	∅	{1}	∅	{1}
$D(\mathfrak{X})$	248	7936	6	78	4	76
d	2	16	1	1	2	2
D	62	31	6	78	1	19
ξ	63	1520	5	53		170
η	8	273	2	6		39
s	< 65	< 454	5	5		< 19

	q=32	q=32	q=32	q=32	q=13	q=19	q=19	q=19	q=19
x_0	93	4867	124	61535	9	33	15	45	∄
y_0	2	618	15	11052	1	13	5	5	
T	4	4	8	8	13	169	1	1	
$l(T)$	2	2	4	4	7	91	1	1	
I	∅	∅	∅	∅	6	49	0	0	
n					451	181	5	5	
$l(n)$					42	91	1	1	
I'					6, 13, 20, 27, 34, 41	49	I	I	
S					364, 254, 218, 210, 320, 438	167	0	0	
X						6		7	
Y						2		2	

Table 2, part 2

q	43				

t	2	
m	3885	

$\mathfrak{x}$	$\emptyset$			$\{1\}$
$D(\mathfrak{x})$	84			3612
d	1			2
D	84			903
ξ	55			601
η	6			20
s	< 36			< 36

x_0	63	2961	273	399	$\S$
y_0	1	323	29	43	
T	43	43	43	43^2	
$l(T)$	14	14	14	602	
I	$\emptyset$	$\emptyset$	6, 13	140, 441	
n			3	5409	
$l(n)$			1	602	
I'			0	I	
S			2	5218, 1393	
X				10	
Y				2	

Table 2, part 3

References

[1] Cameron, P.J. (editor): Problems from the 12th British Combinatorial Conference Norwich 1989.

[2] Hering, C.: On the diophantine equations $ax^2 + bx + c = c_0 c_1^{y_1} \cdots c_r^{y_r}$. To appear

[3] Nagell, T.: Einige Gleichungen von der Form $ay^2 + by + c = dx^3$. Vid. Akad. Skri Oslo Nr. 7 (1930).

Testing for isomorphism between finitely presented groups.

D.F.Holt and Sarah Rees

1 Introduction

The purpose of this paper is to describe a computer program that attempts to decide whether two given finitely presented groups are isomorphic or not. This problem has of course been proved to be undecidable in general, and so any such program is bound to fail on some inputs, but we might nevertheless hope to achieve success in many cases. The basic idea is that we try to prove isomorphism and nonisomorphism alternately, for increasing periods of time, hoping that eventually we will succeed in one of these two aims.

We attempt to prove isomorphism, by running the Knuth-Bendix procedure on the group presentations, in order to generate a word reduction algorithm for words in the generators. In cases of success, this will enable us to verify that a particular map from one group to the other is an isomorphism. In order to find this map, we have to use an exhaustive search, which often becomes rapidly impractical as the word lengths of the images of the generators under the map increase. We can offset this a little by using various additional tests, such as checking that the induced map between the abelian quotients is an isomorphism, or capable of being extended to an isomorphism when only some of the generator images are known. In addition, if both groups can be mapped onto a suitable finite permutation group, then we can check that our map or partial map can induce an automorphism of the permutation group. In principle, if the two groups are isomorphic, then this algorithm would eventually find an explicit isomorphism, but in practice the required time would eventually be prohibitive in many cases.

We attempt to prove nonisomorphism by examining the finite quotients of the two groups, and checking to see if they correspond. In contradistinction to the isomorphism test, this approach cannot be guaranteed to work in general, even in principle, since there are certainly nonisomorphic groups whose finite quotients are the same; for example, finitely presented infinite simple groups. Furthermore, for reasons of space, it is only possible to look for finite quotients of reasonably small order (up to about 50,000), and in the case of finite nonabelian simple images, we have to look for epimorphisms onto each simple group individually. However, for various classes of groups that arise naturally from geometry and topology, such as fundamental groups of manifolds, this approach seems to be quite successful. Indeed, this is one of the applications that we had in mind when planning this program.

In Sections 2 and 3, respectively, we shall describe the isomorphism and nonisomorphism parts of the procedure, and in Section 4 we shall discuss some examples and results. As a general point, it is worth mentioning that, in order to ensure that the program produces the result as quickly as possible (where this is possible at all), it is essential to restrict the time that we spend on each individual calculation. All of the particular programs involved have optional time parameters, which force them to abort after a certain amount of time has passed (or, in some cases, a certain amount of data has been generated). If we abort a program for this reason, then we may have to rerun it later with a larger allowable time. The correct tuning of these time parameters is very important for efficiency. It is also important, when we are attempting to prove that two groups G and H are isomorphic, to look alternately for isomorphisms from G to H and from H to G since, in some examples, one direction is considerably easier than the other.

2   Testing for isomorphism

In this section the program **findisoms**, used to search for an explicit isomorphism between two groups, is described.

2.1   The basic algorithm.

Suppose that $G = \langle g_1, g_2, \ldots, g_m | r_1, r_2, \ldots, r_s \rangle$ and $H = \langle h_1, h_2, \ldots, h_n | s_1, s_2, \ldots, s_t \rangle$ are two finitely presented groups. In this section, whenever we refer to a word in the group generators g_i (with $1 \le i \le m$), we shall mean a (possibly empty) string in the symbols g_i and g_i^{-1}. For two such words, w and v, we write $w = v$ only if they are equal as strings, but we shall say that they are equal in the group G if they map onto the same element of G.

Fundamental to the procedure that searches for an isomorphism from G to H is the fact that for each of the two groups we are able to construct, using the Knuth-Bendix procedure (see Section 2.5), a reduction machine M. To each such machine, there is an associated language L_M, which consists of a set of words in the group generators with the property that, for each group element g, there is at least one word in L_M that maps onto g. For efficiency of the algorithm, it is preferable that as few words as possible in L_M should map onto g. The machine M, given any word w in the group generators, identifies a word v in L_M, equal to w as an element of the group, certainly no longer than w as a word, and known as the *reduction* of w by M. The words in L_M itself are called *reduced* and are left unchanged by M. Given such a machine, it is straightforward (using a depth-first search) to generate a list of all reduced words (i.e. all words in L_M) up to a given length.

Ideally L_M should be a set of shortest representatives of the elements of the group, with a unique representative per group element. This of course may not be possible (and will certainly not be for a group without soluble word problem), but in fact is not strictly necessary. This will be discussed further. For now, suppose that machines M_G and M_H exist.

Now, given the machines M_G and M_H for G and H respectively, we proceed as follows.

1. Construct a homomorphism θ from G to H.

   Using the machine M_H we systematically generate 'all' sequences of reduced words $w_1, w_2, \ldots, w_m$ in the generators of H. We order such sequences according to the sum of the lengths of the words. We can then define a map θ from the generators of G into H by $g_i\theta = w_i$, and try and extend it by multiplication to give a homomorphism from G to H. Images of the relators r_i of G can be calculated as products of the w_i's, and thus as words in the h_i. If they reduce using M_H to the trivial word, then θ has been shown to be a homomorphism, and we proceed.

2. The 'onto' test.

   Given that θ is a homomorphism, we generate 'all' reduced words (up to some fixed length) in the g_i given by the machine M_G and map them, using θ, into H, and then reduce them using M_H. If in this way we find every generator h_i of H as an image of some word u_i, then θ has been shown to be surjective.

3. Construct an inverse homomorphism $\hat{\theta}$.

   When the 'onto' test shows θ to be surjective, we can define an inverse map $\hat{\theta}$ from the generators of H into G by $h_i\hat{\theta} = u_i$. Images in G of the relators s_i of H can be calculated as products of the u_i, and reduced using M_G. If these are all trivial, then $\hat{\theta}$ is a homomorphism from H to G and $\hat{\theta}\theta$ is the identity map.

4. The 'one to one' test.
   For each i, we apply $\hat{\theta}$ to the word w_i and reduce using M_G. Provided we get g_i every time, we know that $\theta\hat{\theta}$ is the identity, so that θ is one to one. In this case, θ is an isomorphism.

Of course, if any of the above processes fails for a particular sequence $w_1, \ldots, w_m$, then we stop and proceed immediately to the next sequence. This basic procedure is run repeatedly, for longer and longer time periods, and with the various controlling parameters increased each time. An increasing amount of time is given each time for the construction of the machines M_G and M_H (unless they are already known to be optimal), so that with time they generate words with less and less repetition of group elements, and reduce words more effectively. The maximum length of words given by M_G allowed for the 'onto' test also increases each time. Because of the fact that, at any given time, the machines M_G and M_H may not reduce a given word to its shortest possible form, it is perfectly possible that an isomorphism may not be recognised as such the first time that it is encountered. The idea is that it must be recognised after some finite amount of time.

The basic procedure is clearly intrinsically slow, and much needs to be done to speed it up.

We try to get greater efficiency by building the maps θ up as partial maps, defined initially on only the first few generators, and rejecting them as maps which could extend to isomorphisms at as early a stage as possible. We currently use three tests to reject a partial map θ, defined on the first l generators, $g_1, g_2, \ldots, g_l$, all of which are described in detail below.

1. We compute the relators involving just the generators $g_1, g_2, \ldots, g_l$, and check to see if they reduce to the trivial word in H. The generators have already been ordered using the program **ordrels** to ensure that this test is as efficient as possible.

2. The abelian quotient test.
   We check to see whether the partial map induced by θ on the torsion free parts of the abelian quotients of G and H could extend to an isomorphism between the two abelian groups.

3. The permutation group quotient test.
   For some finite permutation group P that arises as a finite quotient

of G and H, we check to see whether the partial map induced by θ between such finite quotients could extend to an isomorphism between the two permutation groups.

Once a partial map is rejected we backtrack and move onto the next possibility.

2.2   Ordering the generators of G with ordrels.

ordrels eliminates any clearly redundant generators of G, and then reorders the remaining generators and relators of G such that, in the resulting presentation $G = \langle g_1, g_2, \ldots, g_m | r_1, r_2, \ldots, r_s \rangle$, there are nonnegative integers $0 \le a_1 \le a_2 \le \ldots \le a_m = s$, with the property that the relators $r_1, \ldots, r_{a_i}$ involve only the generators $g_1, \ldots, g_i$, for $i = 1, \ldots, m$, and the sequence $(a_1, a_2, \ldots, a_m)$ is lexicographically as large as possible subject to this property. (In fact, if m is larger than about 6, it becomes too time-consuming to consider all possible orderings of the g_i, and so we use a more heuristic procedure.)

2.3   Explanation of the abelian quotient test.

Any isomorphism between two groups should induce an isomorphism between their free abelian quotients. (By the free abelian quotient of a group G we mean the quotient G/G_T where G_T/G' is the torsion subgroup of the abelian quotient G/G'. This is computed initially for the groups G and H using the standard integer matrix diagonalization algorithm.) Thus, when the abelian quotient test is being used, a partial map between G and H is rejected as a candidate for isomorphism if the partial map it induces between the free abelian quotients of G and H cannot be extended to an isomorphism. The test proceeds as follows.

Let $\overline{g_1}, \overline{g_2}, \overline{g_3}, \ldots, \overline{g_m}$ and $\overline{h_1}, \overline{h_2}, \overline{h_3}, \ldots, \overline{h_n}$ be the images of the g_i's and h_i's in the free abelian quotients of G and H respectively. Then, if w_1, w_2, w_3, $\ldots$, w_l are the images in H of the generators $g_1, g_2, g_3, \ldots, g_l$ under a partial mapping θ from G to H, the images $\overline{w_1}, \overline{w_2}, \overline{w_3}, \ldots, \overline{w_l}$ of $\overline{g_1}, \overline{g_2}, \overline{g_3}, \ldots, \overline{g_l}$ are simply the abelianized versions of these words (i.e. if $w_i = h_{r_1}^{s_1} h_{r_2}^{s_2} h_{r_3}^{s_3} \ldots h_{r_k}^{s_k}$, $\overline{w_i} = s_1 \overline{h_{r_1}} + s_2 \overline{h_{r_2}} + \ldots s_k \overline{h_{r_k}}$). The $\overline{w_i}$'s are then possible isomorphic images of the $\overline{g_i}$'s provided that, for some choice of generators for the two free abelian groups, the matrices of coordinates of $\overline{w_1}, \overline{w_2}, \overline{w_3}, \ldots, \overline{w_l}$ and $\overline{g_1}, \overline{g_2}, \overline{g_3}, \ldots, \overline{g_l}$ with respect to these generating sets are the same. The following lemma specifies a lower triangular canonical form for the matrix of coefficients of a set of words in an abelian group in terms of a minimal generating set.

Lemma 2.1 *For any set of words $a_1, a_2, a_3, \ldots, a_s$ in a free abelian group A, there is a unique set $f_1, f_2, f_3, \ldots, f_t$ of elements of A with the property that it is part of a set of free generators for A and, for $1 \leq k \leq s$,*
(1) if $rk\langle a_1, a_2, \ldots, a_k \rangle = i_k$, then $\langle a_1, a_2, a_3 \ldots, a_k \rangle \subseteq \langle f_1, f_2, \ldots, f_{i_k} \rangle$,
(2) if $i_k = rk\langle a_1, a_2, \ldots, a_k \rangle > rk\langle a_1, a_2, \ldots, a_{k-1} \rangle$, then $a_k = x_{k,1} f_1 + x_{k,2} f_2 + x_{k,3} f_3 + \ldots + x_{k,i_k} f_{i_k}$, where $x_{k,i_k} > 0$ and, for all $j < i_k, 0 \leq x_{k,j} < x_{k,i_k}$.

PROOF: If $a \in A$, then there is a unique positive integer n and a unique element f of A such that $a = nf$, and f is part of a free generating set for A. We'll call n the *index* of a, and f the *root* of a. Now, where A_k is used to denote the subgroup $\langle a_1, a_2, a_3, \ldots, a_k \rangle$ of rank i_k, and F_{i_k} the subgroup $\langle f_1, \ldots, f_{i_k} \rangle$ of A, we simply define f_1 to be the root of a_1, and more generally, where $i_k > i_{k-1}$, $F_{i_{k-1}} + f_k$ to be the root of $F_{i_{k-1}} + a_k$ in $A/F_{i_{k-1}}$. Then x_{k,i_k} is the index of $F_{i_{k-1}} + a_k$, and f_k is uniquely specified within the coset $F_{i_{k-1}} + f_k$ by condition (2).   □

Thus we simply have to calculate and compare the canonical forms for the matrices of coefficients of the sets $\overline{g_1}, \overline{g_2}, \ldots, \overline{g_l}$ and $\overline{w_1}, \overline{w_2}, \ldots, \overline{w_l}$. If these do not match, then θ is rejected.

2.4   The permutation group quotient test

For the permutation group quotient test we use information we have collected during the non-isomorphism part of the program (see Section 3). We identify a finite permutation group P, which can be found as a finite quotient of both G and H, chosen to be as large as possible subject to the constraint that the number k of inequivalent epimorphisms from each of G and H onto P should not be too big. (The epimorphisms are calculated using the program **permim**, described in Section 3. Two epimorphisms are said to be equivalent if one is equal to the other followed by an inner automorphism of P.) Let $\phi_1, \phi_2, \ldots, \phi_k$, and $\psi_1, \psi_2, \ldots, \psi_k$ be a complete set of inequivalent epimorphisms from G and H, repectively, onto P. If θ is an isomorphism from G to H then, for some i, ϕ_i must be equivalent to $\theta\psi_1$. Therefore, if θ is a partial map mapping $g_1, g_2, g_3, \ldots, g_l$ to $w_1, w_2, w_3, \ldots, w_l$ that extends to an isomorphism from G to H then, for some i with $1 \leq i \leq k$, the cycle-types of $(g_j)\phi_i$ and $w_j\psi_1$ must be the same for all $1 \leq j \leq l$. Thus we calculate the cycle types of the images of the generators $g_1, g_2, \ldots, g_l$ under $\phi_1, \phi_2, \ldots, \phi_k$, and also of the words $w_1, w_2, \ldots, w_l$ under ψ_1. If there is not at least one ϕ_i so that the cycle types of $g_1\phi_i, g_2\phi_i, \ldots, g_l\phi_i$ are the same as those of $w_1\psi_1, w_2\psi_1, \ldots, w_l\psi_1$, then θ is rejected.

2.5   The reduction machines.

Let $G = \langle g_1, g_2, \ldots, g_m | r_1, r_2, \ldots, r_s \rangle$ be a finitely presented group. We define a total order $\succ$ on the words in the generators $g_1, g_2, \ldots, g_m$ and their inverses as follows. If $w = g_{i_1}^{\alpha_1} g_{i_2}^{\alpha_2} \ldots g_{i_k}^{\alpha_k}$ and $v = g_{j_1}^{\beta_1} g_{j_2}^{\beta_2} \ldots g_{j_l}^{\beta_l}$, with $\alpha_i, \beta_j \in \{\pm 1\}$, then $w \succ v$ if either $k > l$ or $k = l$ and, for some h, $i_1 = j_1$ and $\alpha_1 = \beta_1, i_2 = j_2$ and $\alpha_2 = \beta_2, \ldots i_{h-1} = j_{h-1}$ and $\alpha_{h-1} = \beta_{h-1}$, and either $i_h = j_h$ and $\alpha_h = -1, \beta_h = 1$ or $i_h > j_h$. This is the common 'shortest, lex-least' ordering.

Suppose that $R = \{(u_1, v_1), (u_2, v_2), \ldots, (u_k, v_k)\}$ is a set of pairs of words in the generators of G such that, for each i, $u_i \succ v_i$ and u_i and v_i map onto the same element of G. Then R is called a set of *rewrite rules* for G, and the pairs (u_i, v_i) are known as *rules*.

We define a *reduction machine* $M(G, R)$ for G with respect to R to be a finite, directed, edge-labelled, graph, with the following properties.

1. One vertex of the graph is identified as the *start* vertex, and some of the remaining vertices are identified as *rewrite* vertices. There is a given one-one correspondence between the rewrite vertices and the rules in R.

2. From each vertex which is *not* a rewrite vertex, there is one outgoing edge labelled with each generator g_i and one with each g_i^{-1}.

3. A word in the g_i can be completely traced out from the start vertex to a non-rewrite vertex if and only if it contains none of the u_i's as a subword. Such a word is called *reduced*. (The set of reduced words is also called the *language* L_M of the machine $M = M(G, R)$.)

4. If a word w in the g_i is not reduced, then there is a rule (u_i, v_i) in R and words x and y such that $w = x u_i y$, every proper prefix of $x u_i$ is reduced, and the path along w from the start vertex terminates after the prefix $x u_i$ at the rewrite vertex corresponding to (u_i, v_i).

Given any word w in the g_i that is not reduced already, the machine $M(G, R)$ can be used to find a reduced word v that is equal in G to w, with $w \succ v$. The algorithm is quite simple. The word w is traced out as far as is possible. If $w = zy$, where z leads to the rewrite vertex with rule (u_i, v_i), then $z = x u_i$, and w is equal in G to $x v_i y$. Clearly $w \succ x v_i y$. The procedure is repeated using $x v_i y$ instead of w, until finally v is obtained which traces out to a non-rewrite vertex. This procedure must terminate, because there are only finitely many words v with $v \prec w$.

It is also straightforward to generate, in order, the reduced words of the machine, as those words that can be traced out without hitting a rewrite vertex.

The accuracy of the machine $M(G, R)$ depends on the set R. An initial set R_0 can be obtained from the relators of G. Any relator r_i is equivalent to an equation $u_i = v_i$, where $r_i = u_i^{-1} v_i$ and $u_i \succ v_i$. R_0 is defined to be the set of such pairs (u_i, v_i), corresponding to the relators of G, together with all pairs $(g_i g_i^{-1}, \varepsilon)$ and $(g_i^{-1} g_i, \varepsilon)$, where ε is the empty word. A sequence of sets of rewrite rules $R_0, R_1, R_2, \ldots$ is then obtained using the Knuth-Bendix procedure. The basic idea of this is that R_{n+1} is obtained from R_n by looking, in R_n, for two rules (u_i, v_i) and (u_j, v_j), where $u_i = xy$, and $u_j = yz$ with y nonempty. Then xyz is equal in G to both $v_i z$ and $x v_j$, and so $v_i z$ and $x v_j$ are equal in G. Both $v_i z$ and $x v_j$ are then reduced as far as possible using the existing rules in R_n, and by cancelling common prefixes and suffixes. This results ultimately in two reduced words a and b that are equal in G. If either a or b is non-trivial, then (a, b) (if $a \succ b$) or (b, a) (if $b \succ a$) is a new rule, which is adjoined to R_n to give R_{n+1} (and existing rules are modified, if necessary, using the new rule). If no such pair of rules (u_i, v_i) and (u_j, v_j) results in a new rule, then the process terminates with the finite set R_n. The procedure is described in more detail in [Gil, 1979].

The following property of the Knuth-Bendix procedure can be proved. Let u be any word, and let v be the the minimal word with respect to $\succ$ that is equal to u in G. Then there is some integer n so that the reduction from u to v may be deduced only from the rewrite rules in R_n. Thus the machines built from successive R_n's give increasingly accurate reduction. If the Knuth-Bendix procedure actually terminates in a finite set R_n, then the corresponding reduction machine gives accurate reduction to the minimal representatives with respect to $\succ$. This will happen eventually for a finite group G, and it occasionally happens for infinite groups, but we cannot rely on it in general.

It should be observed that, where reduction is needed in **findisoms**, it is always used to verify whether or not a word reduces to either the trivial word or a single generator. The rules necessary simply to check reductions of this type for all words up to a given length are often relatively quickly obtained.

Examples of reduction machines are given in Fig.1 and Fig.2, for the free abelian group on two generators, with presentation $\langle a, b | [a, b] \rangle$. (We have written A in place of a^{-1} and B in place of b^{-1}. The machine shown in Fig. 1, built from the initial set of reduction rules $R_0 = \{(ba, ab), (aA, \varepsilon), (Aa, \varepsilon),$ $(bB, \varepsilon), (Bb, \varepsilon)\}$ does not give complete reduction with respect to $\succ$ (notice that the word bA does not reduce). However the Knuth Bendix procedure very rapidly terminates to give a finite set of 8 rules. The machine in Fig. 2, built from this complete set of rules, therefore gives complete reduction.

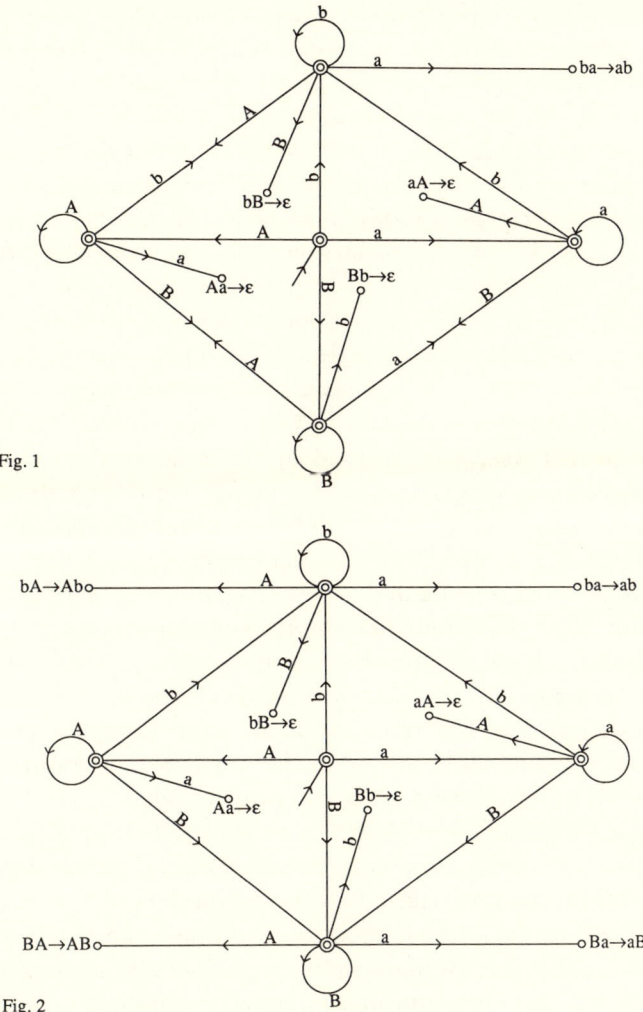

Fig. 1

Fig. 2

3   Testing for nonisomorphism

We attempt to prove that two finitely presented groups are not isomorphic, by systematically looking for their finite quotients of small finite order (up to a few thousand). If these are not the same for the two groups, then they cannot be isomorphic. This approach to the problem was also used by Havas and Kovács in [HaK, 1984] to distinguish between various knot groups.

Let G be a fixed finitely presented group. The first program that we use, **permim**, finds the epimorphisms of G onto a fixed finite permutation group P. More precisely, it finds a representative of each equivalence class of epimorphisms, where two are said to be equivalent if one is equal to the other followed by an inner automorphism of P. This, and the other programs mentioned here, will be described in greater detail below. Let K be the kernel of an epimorphism of G onto P. When P is not too big, we can compute the abelian quotient invariants of K/K', using standard algorithms. Furthermore, our program for this purpose **subabquot**, can work modulo a given prime p, and compute the largest elementary abelian p-factor group $K/K^{(p)}$ of K, where $K^{(p)} = \langle K'K^p \rangle$. If $G/K^{(p)}$ is not too large (of order up to about 50,000, as a guideline), then we can compute $G/K^{(p)}$ explicitly as a regular permutation representation, and then repeat the process with $G/K^{(p)}$ in place of $G/K = P$. (A regular permutation representation of a group is a transitive representation in which no non-identity element fixes any point.) In this way, we can build up finite quotient groups of G that are extensions of soluble groups by P. Finally, we have a simple program **orders**, which computes the numbers of elements of each possible order in a regular permutation group. This enables us to distinguish between quotient groups (such as the dihedral and quaternion groups of order 8), which would otherwise look alike. In our current implementation, we try **permim** on all finite nonabelian simple groups of order up to about 10,000 (although this could easily be increased) together with some almost simple groups like automorphism groups and covering groups of simple groups, and a few small examples like S_4, which have proved useful in particular examples. We only attempt the extension process using **subabquot** when P is either trivial or one of a few small examples, such as S_4, A_5, S_5, $PSL(2,7)$, $PGL(2,7)$ or A_6. In these cases, we have to work with the regular permutation representation of P. In all other cases, we simply count the number of equivalence classes of epimorphisms from G and H to P, and compare the two answers.

In order to assist the comparison between the finite quotients of the two given groups, we store the information about extensions of soluble groups by P in a graph-like structure. The top vertex of this graph is labelled $G/K(= P)$, and a general vertex is labelled G/L, where L is a characteristic subgroup of K. Together with a vertex, we may store the numbers of elements of each order of G/L, and the abelian quotient invariants of L/L'. For a given prime p, we may define a directed edge from the vertex with label G/L to that with label $G/L^{(p)}$. This edge is labelled with the order p^n of $L/L^{(p)}$ with $n \geq 0$. Any one of the above items of information may not be present, either because we have not tried to calculate it yet or because we have attempted to calculate it and failed as a result of integer overflow or lack of space. In the former case, we may attempt to compute it later. If G has several inequivalent epimorphisms onto P, then we get one such graph for each of these, and so, when we are comparing two groups for nonisomorphism, we have to check if there is a pairwise correspondence between their graphs.

We turn now to a more detailed description of the individual programs involved. **permim** is also described in Section 7.1 of [HoP, 1989]. Roughly speaking, we use a backtrack search to test possible images of the generators of G in P as candidates for epimorphisms from G to P. We need representatives c_i of the conjugacy classes Cl_i of P, together with their centralizers $C_i = C_P(c_i)$ in P. In our implementation, P must be one of a fixed list of groups (including suitable permutation representations of all nonabelian simple groups of order up to a million, together with some almost simple groups and some small groups such as S_4), for which we have computed and stored this information in advance (together with a base and strong generating set for P, to facilitate computation within P - see, for example, [Sims, 1971]). The first step is to run the preliminary filter **ordrels** on the presentation of G, which is also used prior to running the **findisoms** program described in Section 2. **ordrels** eliminates any clearly redundant generators of G, and then reorders the remaining generators and relators of G such that, in the resulting presentation $G = \langle g_1, g_2, ..., g_m | r_1, r_2, ..., r_s \rangle$, there are nonnegative integers $0 \leq a_1 \leq a_2 \leq ... \leq a_m = s$, with the property that the relators $r_1, ..., r_{a_i}$ involve only the generators $g_1, ..., g_i$, for $i = 1, ..., m$, and the sequence $(a_1, a_2, ..., a_m)$ is lexicographically as large as possible subject to this property. The idea of this is that, when we are testing a partial set of images $\phi(g_1), \phi(g_2), ..., \phi(g_i)$ in P as a candidate for completion to a homomorphism from G to P, we can attempt to rule out this partial set immediately by checking to see if the relators $r_1, ..., r_{a_i}$ are satisfied by these images. This helps to prune down the backtrack search for epimorphisms. Since we are looking for representatives of equivalence classes of epimorphisms, it is clear that we can restrict our candidates for $\phi(g_1)$ to the representatives c_i of the conjugacy classes in P. For a given $\phi(g_1) = c_i$, we compute the orbits of the centralizer C_i of c_i acting by conjugation on P. Then we can restrict our candidates for $\phi(g_2)$ to representatives of these orbits. If $m > 2$, then we compute the $C_P(\langle \phi(g_1), \phi(g_2) \rangle)$ and the orbits of this group acting by conjugation on P. Representatives of these orbits are our candidates for $\phi(g_3)$, and so on. Unfortunately, these centralizers are likely to rapidly become very small or trivial, in which case we shall essentially be searching through the whole group for images of the generators. This restricts the size of P to which the algorithm is applicable, and the maximum feasible size diminishes rapidly as the number n of generators increases, unless the numbers a_i are favourably large. However, searching through all elements of P is a very efficient process using a base and strong generating set, and testing whether relators hold is also fast, since we normally only need to look at the images of a few points in the set being permuted by P.

We turn now to the program **subabquot**. Let P be a permutation group on the finite set $\Omega = \{1, 2, \ldots, t\}$, and let $\phi : G \longrightarrow P$ be an epimorphism, where $G = \langle g_1, g_2, \ldots, g_m | r_1, r_2, \ldots, r_s \rangle$ is as above. We shall regard the generators g_i as acting on Ω via ϕ. Since only the orbit of 1 under P is involved, we may as well assume that P is transitive on Ω. Let $Q = G_1$ be the stabilizer of the point 1 in P, and let $H = \phi^{-1}(Q)$. We shall give a brief description of the standard Reidemeister-Schreier algorithm to obtain a presentation of H. An original implementation of this by Havas is described in [Hav, 1974]. See also Section 4 of [Neu, 1982] for a discussion of related algortihms and implementations. For each i in $\Omega - \{1\}$, we can regard one particular equation $j^{g_k} = i$ as being the definition of i, where j is already defined, and 1 is regarded as being defined initially. This associates an element $g(i)$ of G to each point i in Ω, with the property that $1^{g(i)} = i$, where $g(1) = 1$ and $g(i) = g(j)g_k$ with $j^{g_k} = i$ a definition. To each equation $j^{g_k} = i$ that is not a definition, we can define an element $h(j, k)$ of H by $g(j)g_k = h(j, k)g(i)$. The $h(j, k)$ then generate H (they form the set of Schreier generators). We obtain the relators for H as follows. For each relator r_k of G and each point i in Ω, we can use the equations $g(j)g_k = h(j, k)g(i)$ to obtain an equation $g(i)r_k = w(i, k)g(i)$, where $w(i, k)$ is a word in the $h(j, k)$. Then the $w(i, k)$ form a set of defining relators for H with respect to the Schreier generators. Having obtained a presentation of H in this way, we can compute the abelian quotient invariants of H by the standard integer diagonalization algorithm, although it is important to check for integer overflow, since this becomes a serious danger when the index of H in G grows beyond a few hundred. (In [HaS, 1979], an implementation is described that attempts to overcome this problem.) Alternatively, we can work modulo a fixed prime p to obtain the largest elementary abelian p-factor group $H/H^{(p)}$ of H. The problem of integer overflow no longer occurs, and so this is applicable to subgroups of much larger index.

In the applications to nonisorphism testing, we only make use of the case in which P acts regularly on Ω; i.e. $Q = 1$, so let us assume that this is the case. Then H is equal to the kernel of ϕ. Let ψ be the epimorphism from H to $H/H^{(p)}$. Then the elements of $G/H^{(p)}$ have the form $\overline{h}g(j)$, for $\overline{h}$ in $H/H^{(p)}$ and $j = 1, 2, \ldots, t$, where $\overline{h}g(j)g_k = \overline{h}\psi(h(j, k))g(i)$, and this enables us to construct the epimorphism from G to the regular permutation representation of $G/H^{(p)}$.

Finally, the program **orders**, which is only written to work on regular permutation groups, is straightforward. For each i in Ω, it constructs the unique permutation in P that takes 1 to i, and computes the length of the orbit of 1 in this permutation, to give its order.

4   Results

Our programs can be used in various ways. The simplest approach is simply to provide two group presentations, and then run the general **testisom** program that attempts to decide if the groups are isomorphic or not. Either this program will run forever (or run out of space), or it will finish by providing an explicit isomorphism (with inverse maps) between the two groups, or show that the groups are not isomorphic by describing a way in which their finite quotient structures differ. This approach is particularly suitable if one has a large collection of pairs of groups for testing, since it can then be left to run for a long period, but it should probably be prevented from running for too long on any single example. For example, in [Kel, 1990], a collection of about 30 pairs of link groups is considered, each pair of which came from identical links with different orientations. All of these are groups with four generators and up to four relations. This is not the easiest situation for the isomorphism searching program, since all of the relations involve all four generators, which means that the relations can only be checked when all four images are known. However, in all cases but two, the program was successful very rapidly. Of the two remaining cases, one pair turned out to be isomorphic and the other not. The isomorphic pair was

$$G = \langle\, a,b,c,d \mid\ (cAbA)^2(cDcA)^2, (cAbA)^2 bAcDcAcDcB,$$
$$(cAbA)^2(cAcD)^2, (cAbA)^2 dAcDcAcDcD \,\rangle$$

and

$$H = \langle\, a,b,c,d \mid\ cA(cAbAcD)^2 cA, cAcAbAbAcDcAbAcDcB,$$
$$(cA)^3 bAcDcAbAcD, cAcAdAbAcDcAbAcDcD \,\rangle.$$

(Here, and elsewhere in this section, we have used A for a^{-1}, etc.) **testisom** eventually found the isomorphism

$$a \to a, b \to cDc, c \to b, d \to bCb,$$

after running for about three hours on a Sun 3/60. The nonisomorphic pair was

$$G = \langle\, a,b,c,d \,|\quad c^2a^2c^2aC^2A^2C^2A, c^2a^2c^2A^6ba^6C^2A^2C^2B,$$
$$a^2c^2a^2cA^2C^2A^2C, c^2a^2c^2a^2C^8dc^8A^2C^2A^2C^2D \,\rangle$$

and

$$H = \langle\, a,b,c,d \,|\quad c^2A^2c^2aC^2a^2C^2A, c^2A^2c^2A^2ba^2C^2a^2C^2B,$$
$$A^2c^2A^2ca^2C^2a^2C, c^2A^2c^2A^2da^2C^2a^2C^2D \,\rangle.$$

These two groups turned out to have the same soluble quotients of elementary abelian length 2, and the general tree construction for building soluble quotients described in Section 3 could not get to length 3, since these quotients were too large. However, we eventually found that G and H had respectively 5616 and 6048 equivalence classes of epimorphisms onto the symmetric group S_4, which of course proves their nonisomorphism. Another suitable sequence of examples were certain fundamental groups of one-cusp manifolds provided by Jeff Weeks. These have the presentations

$$\langle\, t,x \,|\, t^l x T^l x t^l X T^{l+m} X \,\rangle$$

for various values of l and m. Certain pairs of these, such as $(l,m) = (r,s)$ and $(-r-[s/2], s)$ where $4|s$ and $(r,s) = 1$ could not be distinguished easily by geometrical invariants. All examples that we tried were readily shown to be nonisomorphic, since their soluble quotients of elementary abelian length 2 were distinct.

It is also possible to instruct the **testisom** program to attempt to prove either isomorphism only, or nonisomorphism only. This generally makes everything run at least twice as fast, if one's suspicion turns out to be right. For example, for the first example described above, when searching for isomorphisms only, **testisom** took just over an hour. (This was using the abelian quotient test, described in Section 2.3. Without this test, it took nearly 12 hours.)

When attempting to prove isomorphism between two groups G and H, it is important to alternate the attempts to construct the isomorphism from G to H and from H to G, since it may turn out to be much easier in one direction than the other. A spectacular example of this phenomenon is provided by the two presentations of Listing's Knot Group given by $G = \langle a,b,c,d,e|cEaB, DaE, aDcB, EcD\rangle$ and $H = \langle u,v|u^3vUV^2Uv\rangle$. The program took just under three hours on a Sun 3/60 to find the isomorphism

$$a \to uv^2u, b \to vu^2vu^2v, c \to vu^2v, d \to uv, e \to vu.$$

from G to H, but only a few seconds to find the isomorphism

$$u \to aC, v \to AbA$$

from H to G.

This example is also useful for measuring the effect of the abelian quotient test and the permutation group quotient test described in Sections 2.3 and 2.4. When searching for isomorphisms only, the time taken to find the isomorphism from G to H using both of these tests, with the permutation group S_5 was about 10,400 seconds. With the abelian quotient test alone it was about 15,800 seconds, and with neither it had still not completed after more than 60 hours! (The largest abelian factor group of G and H is infinite cyclic.) In many examples, the effect of the abelian quotient test seems to be dramatic, whereas the permutation group test is helpful, but perhaps less so than we had hoped.

For difficult cases of nonisomorphism, the experienced user can abandon the general program, and investigate the finite quotients in greater detail, using the individual programs. This enables the user to concentrate on the particular quotients that seem to be relevant, which will not necessarily be covered by the general program. For example, J.J. Seidel asked if the group

$$G = \langle a, b, c, d, e, f \mid \quad a^3, b^3, c^3, d^3, e^3, f^3,$$
$$(ac)^2, (ad)^2, (ae)^2, (bd)^2, (be)^2, (bf)^2, (ce)^2, (cf)^2,$$
$$(df)^2, (aB)^2, (aF)^2, (bC)^2, (cD)^2, (dE)^2, (eF)^2 \rangle$$

was isomorphic to a certain space group H defined as follows. Let W be the derived group of the Weyl group $W(D_5)$. (Then W is an extension of an elementary abelian group of order 2^4 by A_5). Let H be the semi-direct product of two copies of the 5-dimensional integral root lattice for W by W. We found that both of these groups had epimorphisms ϕ onto A_5, and that, in both cases, the kernel of ϕ had an elementary abelian quotient of order 2^{12}. The general program could not go beyond this. However, by introducing the new target group P, isomorphic to W, we found that both G and H have 14 equivalence classes of epimorphisms ψ onto P, but the elementary abelian quotients of the kernels of the maps ψ do not correspond. G cannot therefore be isomorphic to H.

For another difficult example, let V be the natural 3-dimensional module over the field of 2 elements for the group $PSL(3,2)$, and let W be its dual. Then define G and H to be the semidirect products of $V \oplus V$ and $V \oplus W$ by $PSL(3,2)$, respectively. Then G and H are nonisomorphic groups (of order 10,752), but none of our current techniques serve to distinguish between them, since they are both extensions of elementary abelian groups of order

2^6 by $PSL(3,2)$, and they contain the same numbers of elements of each order. As in the previous example, we can distinguish between these groups by finding a suitable new target group P. (The best is the semidirect product of V with $PSL(3,2)$.) In order to be able to cope with this kind of phenomenon more systematically in future, we plan to extend our programs so that they can investigate the submodule structure of reasonably small modules extended by groups.

As a final example, let G be the Fibonacci group

$$F(2,7) = \langle\, a,b,c,d,e,f,g \mid ab = c, bc = d, cd = e, de = f,$$
$$ef = g, fg = a, ga = b \,\rangle$$

and H a cyclic group of order 29. This is a well-known difficult example. For a recent survey of results on Fibonacci groups, see [Tho, 1991]. Although we did not really undertake this project with pathological examples like this in mind, we tried **testisom** on this, and it took about ten hours to prove that the groups were isomorphic. This is an example in which the Knuth-Bendix process needs to be run for a long time in order to generate the correct reduction machine M_G, but of course **testisom** does not know this.

5   References

[HaK, 1984] G. Havas and L.G. Kovács, 'Distinguishing eleven crossing knots', in: M. Atkinson (ed.), *Computational Group Theory*, Academic Press, London (1984), 367-373.

[Hav, 1974] G. Havas, 'A Reidemeister-Schreier program', in: M.F. Newman (ed.) *Proceedings of the Second International Conference on the Theory of Groups, Canberra 1973*, Lecture Notes in Mathematics, vol. 372, Springer, Berlin (1974), 347-356.

[HaS, 1979] G. Havas and L. Sterling, 'Integer matrices and abelian groups', in: E.W. Ng (ed.), *Symbolic and Algebraic Computation*, Lecture Notes in Computer Science, vol. 72, Springer, Berlin (1979), 431-451.

[HoP, 1989] D. Holt and W. Plesken, "Perfect Groups", Oxford University Press (1989).

[Gil, 1979] R.H. Gilman, 'Presentations of groups and monoids', J. Alg. 57 (1979), 544-554.

[Kel, 1990] A.J. Kelly, 'Groups from link diagrams', Ph.D. Thesis, University of Warwick (1990).

[Neu, 1982] J. Neubüser, 'An elementary introduction to coset table methods in computational group theory', in: C. Campbell and E. Robertson (eds.) *Groups - St Andrews 1981*, LMS Lecture Note Series 71, CUP (1982), 1-45.

[Sims, 1971] C.C. Sims, 'Computing with permutation groups', in: S.R. Petrick (ed.), *SYMSAM '71 Proc. 2nd Symposium on Symbolic and Algebraic Manipulation, Los Angeles, 1971*, ACM, New York (1971), 23-28.

[Tho, 1991] R.M. Thomas, 'The Fibonacci groups revisited', in: C. Campbell and E. Robertson (eds.) *Groups St Andrews 1989 Volume 2*, LMS Lecture Note Series 160, CUP (1991), 445-454.

Discrete Groups and Galois Theory

J.G. Thompson

1. Introduction.

Suppose $r \geq 3$. To each r–element subset S of the Riemann sphere $\mathbf{C} \cup \{\infty\}$ is associated by Riemann's Existence Theorem [S] a Fuchsian group Γ and a covering map $\lambda : h \to \mathbf{C} \cup \{\infty\} \backslash S$ such that

1. $\Gamma = < g_1, \ldots, g_r \mid g_1 \cdots g_r = 1, \; \mathrm{tr} g_i = -2, \; i = 1, \ldots, r > \subseteq SL(2,\mathbf{R})$.

2. The fibres of λ are Γ-orbits.

3. $\quad \lim_{\tau \to q_i} \lambda(\tau) = p_i,$

   where q_i is the fixed point of g_i on $\mathbf{R} \cup \{\infty\}$, and $S = \{p_1, \ldots, p_r\}$.

4. Setting
$$Cp(\Gamma) = \cup \Gamma(q_i),$$

   λ induces an analytic bijection between $\Gamma \backslash h \cup Cp(\Gamma)$ and $\mathbf{C} \cup \{\infty\}$.

5. $\mathbf{C}(\lambda)$ is the field of meromorphic functions on $\Gamma \backslash h \cup Cp(\Gamma)$.

This theorem reduces the problem of determining field extensions E of $\mathbf{C}(\lambda)$ with $[E : \mathbf{C}(\lambda)] < \infty$ which are unramified outside S to the study of the subgroups of Γ of finite index. Since Fuchsian groups Γ can be constructed easily, it is attractive to attack the inverse problem of Galois theory by studying these Fuchsian groups. There are at least two serious obstacles to this approach. First, given Γ, we do not know how to determine S constructively, and *a fortiori*, we don't know the algebraic properties of S as a function of Γ. Second, it is at present unknown to what extent the trace spectrum of Γ determines Γ to within conjugation in $SL(2,\mathbf{R})$. Both of these difficulties are formidable, but the state of knowledge about the inverse problem of Galois theory is fragmentary, especially as concerns the realizability of non soluble finite groups as Galois groups over $\mathbf{Q}$, and so it seems worthwhile to try to determine as much as possible about $\lambda(Cp(\Gamma))$ and the trace spectrum of Γ. In case Γ is determined to within conjugacy by its trace spectrum, the orbit of $\lambda(Cp(\Gamma))$ under $SL(2, \mathbf{C})$ is determined by the trace spectrum, too, and one might hope to construct explicit functions of the trace spectrum whose values are $\lambda(Cp(\Gamma))$. It is pointless, however, to try to imagine the nature of such functions until the precise relationship between Γ and its trace spectrum is determined.

2. Fricke rings.

Let F_n be the free group on $\{u_1, \ldots, u_n\}$. For each non empty subset I of $\{1, \ldots, n\}$, let $t(I)$ be an indeterminate, and set $R_n^* = \mathbf{Z}[t(I) \mid \emptyset \subset I \subseteq \{1, \ldots, n\}]$. For each $\rho \in hom(F_n, SL(2, \mathbf{C}))$, define $\rho^* : R_n^* \to \mathbf{C}$ by sending $t(I)$ to $\mathrm{tr}\rho(u_{i_1} \cdots u_{i_k})$ if $I = \{i_1, \ldots, i_k\}$, $i_1 < i_2 < \ldots < i_k$, where $\mathrm{tr} : SL(2, \mathbf{C}) \to \mathbf{C}$ is the usual trace. Let

$$I_n^* = \cap \ker \rho^*$$

and set $R_n = R_n^*/I_n^*$; R_n is called the nth Fricke ring [M]. Denote by $t_{i_1 \ldots i_k}$ the image of $t(I)$ in R_n, and denote by $R_n^{(d)}$ the subring of R_n generated by all the $t_{i_1 \ldots i_k}$ with $k \le d$.

    Lemma.   If $2 \le d \le n - 1$, then $R_n^{(d+1)}$ is integral over $R_n^{(d)}$.

    Proof.   From [M], $t_{123}^2 - \alpha t_{123} + \beta = 0$, where $\alpha, \beta \in R_3^{(2)}$. Fix I with $\mid I \mid = d + 1$, and set

$$v_1 = u_{i_1} \cdots u_{i_{d-1}}, \; v_2 = u_{i_d}, \; v_3 = u_{i_{d+1}}.$$

Then $< v_1, v_2, v_3 >$ is free, and so we have, with $t = t_{i_1 \ldots i_{d+1}}$, $t^2 - \alpha' t + \beta' = 0$, where $\alpha', \beta' \in R_n^{(d)}$. The proof is complete.

The subring $R_n^{(2)}$ of R_n is quite complicated. The ring $R_4^{(2)}$ is defined by a single relation in the ten variables $t(I)$ [M].

The appropriate quotient ring of R_n for the study of the groups Γ of the Introduction is R_n/I_n, where I_n is generated by $t_i + 2$, $1 \le i \le n$, and by $t_{i_1 \ldots i_n} + 2$. It would be helpful to have necessary and sufficient conditions on $\phi \in hom(R_n^*, \mathbf{C})$ which guarantee that $\phi = \rho^*$ for some $\rho \in hom(F_n, SL(2, \mathbf{C}))$ such that $\rho(F_n) = \Gamma$ is one of the Fuchsian groups of the Introduction. Such conditions are easy to find when $n = 3$, that is, when $r = 4$, but for $n \geq 4$, they appear to be more complicated.

3. An exact sequence associated to (Γ, λ).

Let k be the field generated by the elementary symmetric functions of $\lambda(Cp(\Gamma))$, where Γ, λ are as in the Introduction. For each group G, let $\widehat{G}$ denote its profinite completion and for each field F, let $G_F = Gal(\overline{F}/F)$, where $\overline{F}$ is a separable algebraic closure of F. Let $F = F(\Gamma, \lambda) = \{f \mid$

    1. f is meromorphic on $h \cup Cp(\Gamma)$.

    2. f is algebraically dependent on $\overline{k}(\lambda)\}$.

Then F is a Galois extension of $k(\lambda)$, and if $\mathcal{G} = GalF/k(\lambda)$, there is an exact sequence

$$1 \longrightarrow \widehat{\Gamma} \longrightarrow \mathcal{G} \longrightarrow G_k \longrightarrow 1.$$

This extension splits. The proofs of these assertions are not difficult, depending essentially on choosing particular local variables at each element of $h \cup Cp(\Gamma)$.

If G is a finite group, the centre of G is 1, and $\phi \colon \widehat{\Gamma} \longrightarrow G$ is a surjective homomorphism, then from (1), we conclude that G is the Galois group of a Galois extension field E of $k_1(\lambda)$, where $E \frown \overline{k} = k_1$ and where G_{k_1} is the largest subgroup of G_k which induces inner automorphisms of $\widehat{\Gamma}/ker\ \phi$.

So we have a gadget for constructing Galois groups, and the given description exhibits the relevance of $\lambda(Cp(\Gamma))$.

4. The trace spectrum of Γ.

An element γ of Γ is said to be primitive if $< \gamma > = C_\Gamma(\gamma)$. The trace spectrum of Γ is the multiset $\{| \text{ tr } \gamma |\}$, where γ ranges over a set of representatives for the conjugacy classes of primitive elements of Γ.

Suppose G is a finite group, and H_1, H_2 are non conjugate subgroups of G which induce the same permutation character. Let $\phi : \Gamma \longrightarrow G$ be a surjective homomorphism and let Γ_i

be the inverse image in Γ of H_i. It follows that Γ_1 and Γ_2 have the same trace spectrum. Such examples were given by Sunada [Su] , and it is not difficult to find G such that Γ_1, Γ_2 are of genus 0. A result of Margulis [Z] implies that we may choose Γ so that the commensurability group of Γ in $GL(2, R)$ is $\Gamma.Z(GL(2, R))$, and in such a situation, Γ_1, Γ_2 are not conjugate in $GL(2, R)$. One may conjecture that if Γ_1, Γ_2 are Fuchsian groups of genus zero with the same trace spectrum, then $g^{-1}\Gamma_1 g$ and Γ_2 are commensurable for some $g \in GL(2, R)$. If such a commensurability result holds for Fuchsian groups of genus zero, it would perhaps be exploitable.

References

[M]   Magnus, W.,   Rings of Fricke Characters and Automorphism Groups of Free Groups, Math. Zeit., 170, (1980) 91-103.

[S]   Siegel, C.L.,   Topics in Complex Function Theory, Vol. I, Wiley and Sons, (1969).

[Su]   Sunada, T.,   Riemannian coverings and isospectral manifolds, Ann. of Math. (2) 121, No.1, (1985) 169-186.

[Z]   Zimmer, Robert J.,   Ergodic Theory and Semisimple Groups, Birkhäuser, (1984).

Smooth Coverings of Regular Maps *

Steve Wilson

Abstract

This paper describes a method, algorithmic in part, for determining those regular maps N which are smooth covers of a given regular map M. In the case where the base map M is reflexible, the method is able to distinguish between the chiral and the reflexible covers. The method is illustrated with an important example, $M = K_2(O)$, a 9-fold covering of which is one of the two smallest chiral maps with triangular faces.

Preliminaries

The paper [W2] contains a fuller account of the following definitions and preliminary results: a *map* M is an embedding of a (very general) graph into a surface. We consider the map to be barycentrically subdivided into triangular regions called flags, and choose one flag to be special; this is the *root* flag, I. Each flag f has three neighbors, denoted fr_0, fr_1, fr_2 as in Figure 1:

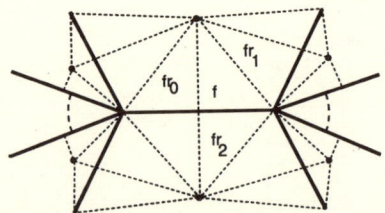

Figure 1: The flags adjacent to f

Then r_0, r_1, r_2 are permutations on Ω, the collection of flags. These three involutions generate a group C, the *connection group*. A *symmetry* of M is a permutation of Ω which commutes with C, and $G(M)$ is the group of all symmetries. We want to discuss two kinds of regularity: (1) M is *rotary* if $G(M)$ has symmetries R and S which send the root flag I to Ir_1r_0 and Ir_1r_2, respectively. $G^+(M)$ is the subgroup of $G(M)$ generated by R and S. This is a $(2, p, q)$ group, where p and q are the orders of R and S respectively. (2) If M is rotary, then it is *reflexible* provided that $G(M)$ also contains a symmetry X which sends I to Ir_1, and otherwise it is *chiral*. Each symmetry corresponds to the flag to which it sends I, and we often refer to the flag as if it were a group element and vice versa.

A *projection* from a map N to a map M is a continuous function Θ which preserves flags, and which carries the root flag in N to that in M. In this case, we say N is a *covering* of M. Then Θ, considered as a function from the flags of N onto the flags of M is a homomorphism $\Theta : G(N) \to G(M)$. The kernel K of this homomorphism is called the *group of covering transformations*.

A previous paper [W1] discussed the case where Θ is *totally ramified* , i.e., where Θ is one-to-one at each vertex. In that case, K is cyclic, and we can determine, in an absolutely constructive way, all the maps N which so cover M, and which ones are reflexible, which chiral. We can also determine those in which the projection is smooth at the face centers rather than branched.

In this paper, we wish to look at the *smooth* case, that is, the case in which Θ is a local homeomorphism at every point of N. We wish, in this case, to achieve the same goals: given a rotary map M, we want to find all rotary maps N which are smooth coverings of M, and determine which are reflexible, which chiral. We also want the process to be as constructive as possible.

Our plan is that of classical Riemann surfaces: we cut up the base until it is simply-connected, and then glue copies of this (the "sheets" of the covering) together. As we do so, we look for necessary conditions for the result to be regular.

Step I *Sheets*

Let F_1 be (the interior of) the face containing I, and let $S_1 = F_1$. Recursively, let F_i be a face adjacent to S_{i-1}, let e_i be (the relative interior of) any edge joining S_{i-1} and F_i and let $S_i = S_{i-1} \cup e_i \cup F_i$.

Let $Z = S_F$, where F is the number of faces. This is the *under-sheet*. It is open, simply-connected, contains all of the faces, but none of the vertices. $M \backslash Z$ is a union of edges, and is called the *hem*.

The *sheets* are the connected components of $\Theta^{-1}(Z)$. The elements of K permute the sheets (i.e. if Z_a is one of the sheets and $k \in K$, then $Z_a k$ is also a sheet), and the corresponding flags lie one on each sheet. We adopt a system of labels, giving to each sheet as a label the element of K which lies on that sheet. If W is any face of N let $Z(W)$ be the label of the sheet on which W lies. A consequence of this is that if $k \in K$ then $Z(Wk) = Z(W)k$.

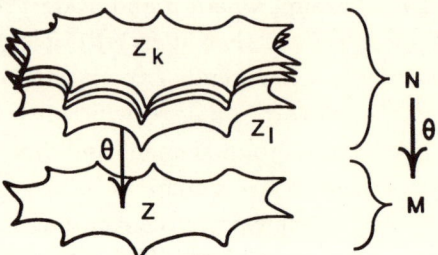

Figure 2: The covering of the under-sheet by the sheets

Step II   *Edge Variables*

Suppose that e is an edge of M which is in the hem, and that U and V are the faces on either side of it. Then U has a pre-image U_I in N which lies on sheet I, and U_I has an edge e' which is a pre-image of e. The edge e' separates U_I and V_x where V_x is a pre-image of V lying on some sheet x.

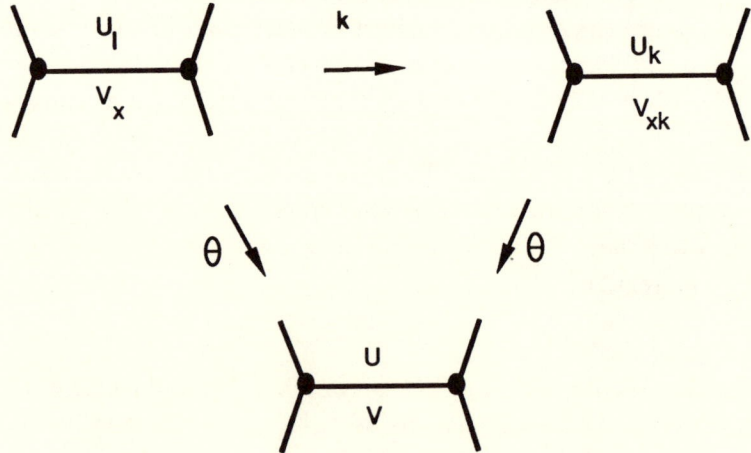

Figure 3: Two pre-images in N of an edge of M

If we now apply any symmetry $k \in K$, then $e'' = e'k$ is another pre-image of e and it separates pre-images of U and V which lie on sheets
$$Z(U_I k) = Z(U_I)k = Ik = k$$
and
$$Z(V_x k) = Z(V_x)k = xk,$$
respectively. Thus, every pre-image of e separates sheets whose labels differ

by x. We indicate this by labelling e in M with an arrow marked "x" pointing across e from U to V (or "x^{-1}" pointing from V to U). These labels are variables which stand for group elements. With these variables assigned, we can trace a path in N through many sheets by following the projection of the path in M.

Let us introduce an example. Figure 4 is a picture of the map $M = K_2(O)$; this map is a two-fold cover of the octahedron, branched at the vertices only. M is of type $\{3,8\}$ and so any smooth cover will be of the same type. The group $G^+(M)$ is the (2,3,8) group of order 48 given by the presentation $\langle R, S - I = (RS)^{-1} = R^3 = S^8 = (RS^3)^2 \rangle$.

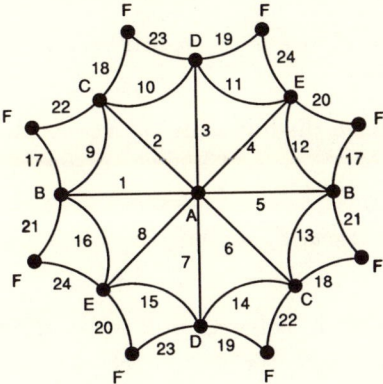

Figure 4: The map $K_2(O)$

One way to assign the under-sheet is to let the hem consist of all the edges around the outside in Figure 4 together with edge 5. Assigning edge variables, we have Figure 5, below.

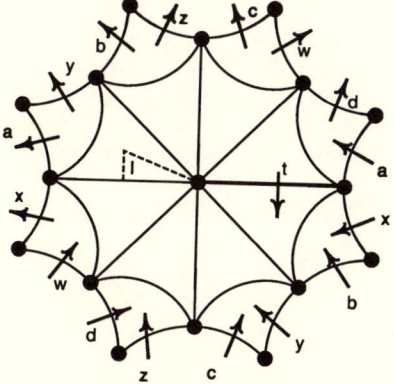

Figure 5: $K_2(O)$ with edge-labels

Step III   *Vertex equations*

Because the covering is to be *smooth*, a path which goes once around a vertex must lead, in N, from any one sheet back to the same sheet. Thus the path product must evaluate to I:

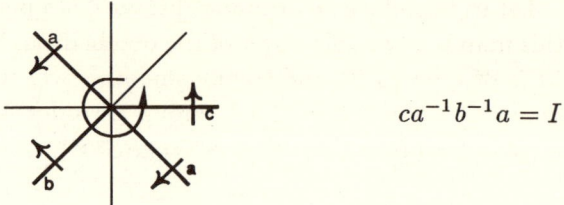

$$ca^{-1}b^{-1}a = I$$

Figure 6: A typical vertex

In $M = K_2(O)$, we have the following: at vertex A, $t = I$. At vertices B, C, D, E, we have $I = ax^{-1} = by^{-1} = cz^{-1} = dw^{-1}$; i.e., $a = x, b = y, c = z, d = w$, as in Figure 7.

Looking at the last vertex, F, we have

$$ab^{-1}cd^{-1}a^{-1}bc^{-1}d = I. \tag{$*$}$$

In general, we will be able to solve all but one of the vertex equations, and the substitution of these into the remaining equation will yield a surface symbol for the surface of M.

Because N is connected, K is generated by a, b, c, d. And conversely, if we have any group generated by four elements satisfying $(*)$, we can form a map N which is a smooth covering of M by gluing copies of the under-sheet according to the scheme of the edge-labels. The question we need to ask is: under what conditions on a, b, c, d is the result rotary? When is it reflexible?

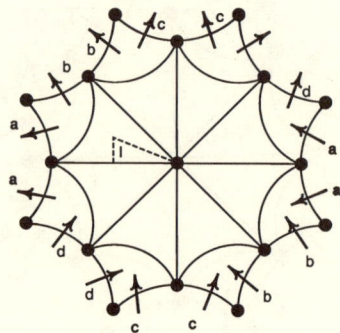

Figure 7: After solving for vertex equations

Step IV *Symmetry-connection functions*

Recall that R and S are symmetries which act as rotations one step around the face and vertex, respectively, which are incident with the root flag I and assume for the moment that, as in our example, neither the edge adjacent to I nor the one incident with Ir_1 are in the hem; then the flags R, S and $\gamma = RS^{-1}$ are in sheet I. For x in K, define the functions ρ, σ, and g as follows:

$$\rho(x) = Z(xR)$$
$$\sigma(x) = Z(xS)$$
$$g(x) = Z(x\gamma)$$

The values of these functions can be determined, at least for the generators, by chasing paths through M. For example, Figure 8 shows a path from the flag I to the flag a.

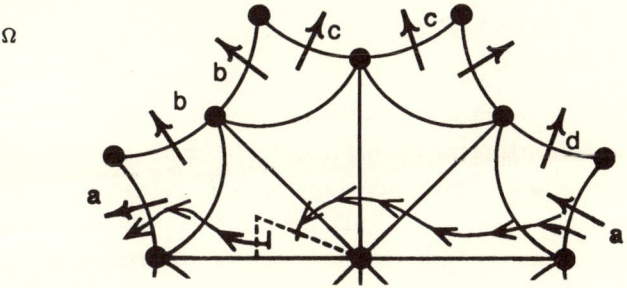

Figure 8: A path to a.

If we apply the symmetry R to this path, the result is shown in Figure 9, below.

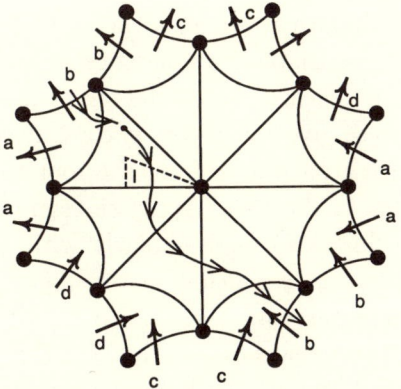

Figure 9: The path to a rotated by R

This path leaves sheet I through the edge labelled "b", and against the arrow. This gives us: $\rho(a) = b^{-1}$.

Doing the same with the other generators and functions, we get:

x	$\rho(x)$	$\sigma(x)$	$g(x)$
a	b^{-1}	b	a^{-1}
b	$b^{-1}a$	c	$a^{-1}d^{-1}$
c	$c^{-1}da$	d	$b^{-1}cd^{-1}$
d	$c^{-1}b$	a^{-1}	$b^{-1}a$

It is, perhaps, surprising that these functions, essentially geometric in nature, have an algebraic aspect: they are homomorphisms, in fact automorphisms of K.

For example,

$$\rho(x) = Z(xR) = Z(RR^{-1}xR)$$
$$= Z(R)R^{-1}xR \quad [\text{because } R^{-1}xR \in K]$$
$$= IR^{-1}xR$$
$$= R^{-1}xR$$

So in order for N to be rotary, K must be a group generated by a, b, c, d, satisfying $(*)$ and admitting automorphisms ρ, σ, and g whose effects on the generators is as above. Conversely, any such K can serve as the group of covering transformations for a rotary cover of M. In fact, any two of the three automorphisms is sufficient to make N rotary. If we drop the assumption at the beginning of this section about which edges are in the hem, we can still define $\rho(x) = R^{-1}xR$, etc, and then determine the effect of these homomorphisms by tracing paths through the map as before.

Summary

To find all the rotary smooth covers of a rotary map M, first choose an undersheet Z. Assign edge-variables to each edge in the hem. The product in the correct order around each vertex of the directed variables of incident edges is the identity. These give generators and relations for the group K. For each variable x, trace a path in Z from the root to the edge x, across it and on through Z back to the root. Apply the symmetry R to the path and trace the result to find $\rho(x)$, and similarly with S and/or γ. These determine automorphisms ρ, σ, g which K must allow.

There, of course, is the small rub. There is no algorithm for determining which groups K might satisfy such a description. But some special cases often yield coverings. We might ask for K to be abelian, or the cyclic group C_N, or dihedral, or S_n, and determine which groups in each class could serve as coverings.

For instance, in our continuing example, $M = K_2(O)$, what maps N do we get if we assume that K is abelian? In this case, $(*)$ is null, and, writing the group additively, we have

x	$\rho(x)$	$\sigma(x)$
a	$-b$	b
b	$a - b$	c
c	$a + d - c$	d
d	$b - c$	$-a$

(a) The reader might like to examine the case in which K is assumed to be cyclic. In C_N, an automorphism is multiplication by a unit mod N. Thus there are units r and s such that $\rho(x) = rx$ and $\sigma(x) = sx$. From this, it is not hard to show that $a = b = c = d = 0$; i.e., that no cyclic covering exists (except, of course, the trivial covering of M by itself).

(b) Assume $K = C_N \times C_N$, $a = (1,0)$, $b = (0,1)$, $c = (r,s)$. Then $\rho(x) = xP$, $\sigma(x) = xQ$, where P and Q are the matrices

$$P = \begin{bmatrix} 0 & -1 \\ 1 & -1 \end{bmatrix} \quad Q = \begin{bmatrix} 0 & 1 \\ r & s \end{bmatrix}$$

Then $d = cQ = (rs, r + s^2)$ and from $-a = dQ$, we get:

$$r^2 + s^2 r = -1 \tag{1}$$

$$s^3 + 2rs = 0 \tag{2}$$

From $cP = a + d - c$, we get $rs + 1 - r = s$, which implies:

$$rs = r + s - 1 \tag{3}$$

and $r + s^2 - s = -r - s$, which implies:

$$s^2 = -2r \tag{4}$$

Then
$$\begin{aligned}
-1 &= r^2 + s^2 r \quad \text{by (1)} \\
&= r^2 + s(sr) \\
&= r^2 + s(s + r - 1) \quad \text{by (3)} \\
&= r^2 + s^2 + sr - s \\
&= r^2 + s^2 + s + r - 1 - s \quad \text{by (3)} \\
&= r^2 + s^2 + r - 1 \\
&= r^2 - 2r + r - 1 \quad \text{by (4)} \\
&= r^2 - r - 1
\end{aligned}$$

We conclude that

$$r^2 = r. \tag{5}$$

One the one hand, $s^2 r = (-2r)r = -2r^2 = -2r$ by (5) while on the other,

$$s^2 r = s(sr) = s(r + s - 1)$$
$$= sr + s^2 - s$$
$$= r + s - 1 - 2r - s$$
$$= -r - 1.$$

Comparing our two hands, we see that $r = 1$, and it follows that $s^2 = -2$, $a = (1, 0)$, $b = (0, 1)$, $c = (1, s)$, $d = (s, -1)$.

Now, the equation $s^2 = -2$ has a solution mod$N$ iff N is a product of primes $\equiv 1$ or 3 (mod 8) or twice such a number. The first few of these solutions are:

N:	2	3	6	9	11	17	18	19	22	...
s:	0	± 1	± 2	± 4	± 3	± 7	± 4	± 6	± 8	...

So, our example map $M = K_2(O)$ has at least a 4-fold smooth covering, a 9-fold, a 36-fold, etc. The rotation group of each of these coverings is a $(2, 3, 8)$ group of order $48N^2$. Query: Is it possible, in this example, for K to be not cyclic, generated by a and b, and yet *not* be $C_N \times C_N$?

Step V *Reflexibility*

Before continuing with our example, let us consider the question of reflexibility: A rotary cover N of a rotary map M might be reflexible or chiral regardless of whether the base map M is reflexible or chiral. Given a map M and a smooth cover N produced as above, how can we tell whether N is reflexible or chiral?

There are two cases: M is either reflexible or chiral.

In the case where M is reflexible, we have an easy answer. To require that the symmetry X, a reflection in the hypotenuse of the root flag, which exists in M, extends to an X on N, we only need require the function $\chi(x) = Z(xX)$, whose effects can be traced in M, to be an automorphism of K, just as we did with ρ, σ, and g. In our example:

x	$\chi(x)$
a	b
b	a
c	d^{-1}
d	c^{-1}

Looking at the subcase $K = C_N \times C_N$, if χ switches a and b, then $\chi(x, y) = (y, x)$. Then $\chi(c) = -d$ implies $(s, 1) = (-s, 1)$. Then $2s = 0$ and it follows that N must be 2. So of all the coverings of M listed above, only the 4-fold covering is reflexible. The 9-fold covering is then interesting as a candidate for the smallest chiral map with triangular faces.

In general, one can use any two of the three rotary automorphisms ρ, σ, or g, and similarly, one need use only one of the reflectional symmetries $\alpha = RX$, $\beta = SX$ or X, whichever is most convenient, to specify an automorphism that K must admit in order for the resulting map to be reflexible.

However, in the second case, where the base map M is chiral, no clear method suggests itself for determining which of its rotary covers are reflexible. Since no X symmetry exists in M, one does not know how to trace the effect of an X in N by looking at M.

M does have at least one reflexible cover, namely the parallel product of M and its mirror-image [W2]. Perhaps the only possibility is to perform the algorithm on this covering map as well, and compare the results.

To continue with our example, abelian covers of $K_2(O)$: if we assume $K = C_N^3$, we derive a contradiction, while if we assume that $K = C_N^4$, then we get one map for each N, and each such map is reflexible.

Are there other abelian possibilities? These would correspond to subgroups of C_N^4 which are invariant under ρ and σ. Are there any?

Like the algorithm in [W1], the process of solving the vertex equations and finding the values of the automorphisms on the generators can be easily automated; though the remaining problem of determining which K satisfy such a description remains hard.

A number of examples have been worked on and partially solved, and the author welcomes any query from readers.

Acknowledgements

This research was done while the author enjoyed the hospitality of the Faculty of Mathematical Studies at the University of Southampton, Southampton, England.

References

[W1] Wilson, S. E., *Riemann surfaces over regular map*, Canadian J. Math., Vol. XXX, No.4, 763-782.

[W2] Wilson, S. E. , *Parallel products in groups and maps*, To appear.